T0339936

Practical Reservoir Engineering and Characterization

Practical Reservoir Engineering and Characterization

Richard O. Baker
Independent Consultant

Harvey W. Yarranton
Department of Chemical and Petroleum Engineering, University of Calgary, Canada

Jerry L. Jensen
Department of Chemical and Petroleum Engineering, University of Calgary, Canada

AMSTERDAM • BOSTON • HEIDELBERG • LONDON
NEW YORK • OXFORD • PARIS • SAN DIEGO
SAN FRANCISCO • SINGAPORE • SYDNEY • TOKYO
Gulf Professional Publishing is an imprint of Elsevier

Gulf Professional Publishing is an imprint of Elsevier
225 Wyman Street, Waltham, MA 02451, USA
The Boulevard, Langford Lane, Kidlington, Oxford, OX5 1GB, UK

ISBN: 978-0-12-801811-8

Library of Congress Cataloging-in-Publication Data
A catalogue record for this book is available from the Library of Congress

British Library Cataloguing-in-Publication Data
A catalogue record for this book is available from the British Library

For information on all Gulf Professional Publishing publications visit our website at http://store.elsevier.com/

Dedications

To my family, my friends, and to God ... R.B.

To my family and the many colleagues who inspired the work ... H.W.Y.

To the glory of God and to my parents, Jean, Jim, and Mary ... J. L. J.

Contents

Preface

The most valuable reservoir engineers are those who see the clearest and the most and who know what they are looking for.

Dake (1992)

Reservoir characterization sounds simple: determine the size, shape, and property distribution of a reservoir. And yet, an engineer's first encounter with reservoir characterization can be a shock. He cannot see or touch the reservoir. Like the proverbial blind man feeling an elephant, he must construct a mental picture of the reservoir from indirect information. This information must be interpreted from logs and cores, pressure measurements, fluid properties, and production data. These data are often sparse, incomplete, noisy, and sometimes nonexistent. From this murky picture, he must answer questions that impact the value of the company, such as the following:

- What is the original oil in place?
- What is the remaining recoverable oil in place?
- Where is the remaining oil located and under what conditions (pressure and saturation)?
- How can the remaining oil be recovered?
- What is the drive mechanism of the reservoir?
- What is needed to optimize recovery?
- Can oil rates and reserves be economically increased?

Clearly, there can be considerable subjective judgment in reservoir characterization and reservoir engineering. The goal of this book is not only to teach the ideas and methods of reservoir characterization, but also to provide a guide for some of the subjective judgments. The book is divided into three parts. The first part reviews the engineering fundamentals needed for reservoir characterization. The second part addresses the sources and analysis of reservoir engineering data including methods to estimate unknown properties. The third part presents reservoir characterization methods and demonstrates how to integrate results from different methods into a self-consistent reservoir characterization.

We emphasize that reservoir characterization is an integrated, iterative process that must contend with uncertainty. The focus of this text is on understanding and using commonly available data to contribute to this process. It is necessary to make some assumptions to even begin a reservoir characterization. For example, it may not be possible to determine the strength of an aquifer or the connectivity in the reservoir from initial static data sources (such as logs and cores). These characteristics are assumed and later refined based on dynamic data sources (such as pressure and production). It is extremely important that the engineer or geoscientist should not be afraid to make an assumption and see how that assumption and the corresponding calculations fit the data. It is also important to periodically check the underlying assumptions and the data interpretation. This constant active feedback loop continuously improves the reservoir concept as new data are collected and economics change. Initial estimates to the previous list of questions will, at best, be in the plus or minus 40% range but, with more wells and dynamic data, our answers should converge to be in the plus or minus 10% range at least for field scale parameters. Unfortunately, for local regions within the reservoir and at individual wells, the errors increase again. Dealing with uncertainty is one of the main challenges in reservoir characterization.

There are many excellent books on reservoir engineering, most focusing on engineering principles. This book is different because reservoir engineering and geological principles are demonstrated on many examples of real field data with all its inherent gaps and inconsistencies. It is important to see and use real field data because one of the challenges subsurface scientists face is interpreting noisy and incomplete data and transforming it to knowledge of fluid flows. There are large gaps in our reservoir knowledge because we sample only approximately one ten-billionth of the reservoir with core and logs and pressure and fluid property data are often incomplete. Therefore, methods to estimate properties when data are missing are presented. We emphasize the integration and cross-checking of data and methods. It is our strong opinion that both static data (such as facies and permeability) and dynamic data (such as pressure and production rates) must be analyzed and interpreted together to reduce uncertainty and cross-validate reservoir and fluid parameters.

One note of caution: we have used many field examples and, in many cases, provided an interpretation of the data. We cannot guarantee that the interpretation is correct or that the methods we propose will provide the best characterization of a given reservoir. Old interpretations can always be overturned by new data. Each reservoir is unique and each engineer must fashion a characterization from the data and methods available as best as he or she can. Solving the puzzle of reservoir characterization is a creative act and one of the most satisfying in our engineering experience. For the novice, we hope this book can help guide you in this experience. For the veteran, we hope you find this a useful reference with some new insights.

The authors would like to express gratitude to the many people who have contributed to this text. Richard would like to thank Shelin Chugh, Rod Batycky, Edwin Jong, Kerry Sandhu, Cameron McBurney, Nathan Meehan, Robert Jobling, and Gord Moore who have contributed greatly to the thought process. Harvey is in addition grateful to Susan Biolowas, Rupam Bora, Enrico DeLauretis, Dennis Beliveau, Bette Harding, Sonja Malik, Bob McKishnie, Greg Osiowy, Vladimir Vikalo, and Claudio Virues for their suggestions and assistance and to Mehran Pooladi-Darvish and Steve Ewan from Fekete & Associates for their help with the PTA figures and discussion. Jerry would like to thank Patrick Corbett, Larry Lake, Chris Clarkson, Rudi Meyer, Steve Hubbard, Fed Krause, Per Pedersen and his students for lively, thought-provoking discussions. We are especially indebted to our wives and our families, Karen Baker, Stewart and Dorothy Baker, Maureen Hurly, and Jane Jensen for their patience and encouragement.

Introduction

1

Chapter Outline

Petroleum is a hydrocarbon mixture derived from organic material. It can exist as a solid (coal), a liquid (oil), or a gas (natural gas). This book is primarily concerned with oil, although, as we shall see, gas and water are always associated with oil. Before considering oil reservoir engineering, let us review where oil is found and how oil is produced.

A common misconception is that oil and natural gas are found in underground caverns. In fact, oil and gas are found within the microscopic pores of rocks, Figure 1.0.1. A rock formation that contains petroleum is termed a petroleum-bearing reservoir. Not all petroleum reservoirs are productive. Petroleum must be able to flow through the pore spaces of the formation. Hence, the pores must form a connected network. The term permeability is defined as a measure of the flow capacity of this pore network. Petroleum can only be economically produced from a reservoir with sufficient permeability. The permeable rock formation must also be overlain by impermeable rock, forming a trap that prevents the petroleum from migrating out of the reservoir. Figure 1.0.2 shows a schematic of a trapped hydrocarbon deposit.

To produce petroleum, wells are drilled into the reservoir. The pressure in the wellbore is lower than in the reservoir, and reservoir fluid flows into the wellbore and up to surface. As shown in Figure 1.0.3, there are several types of wells,

Practical Reservoir Engineering and Characterization. http://dx.doi.org/10.1016/B978-0-12-801811-8.00001-8

Figure 1.0.1 Photograph of a core cut from a reservoir and a micrograph of a thin section from a core. The black regions in the micrograph are the pore space, while the dark and light grey areas are the solid rock.
Images from: http://rockhou.se/page/3/ and http://ior.senergyltd.com/issue8/pnp/herriot_watt/, January 7, 2012.

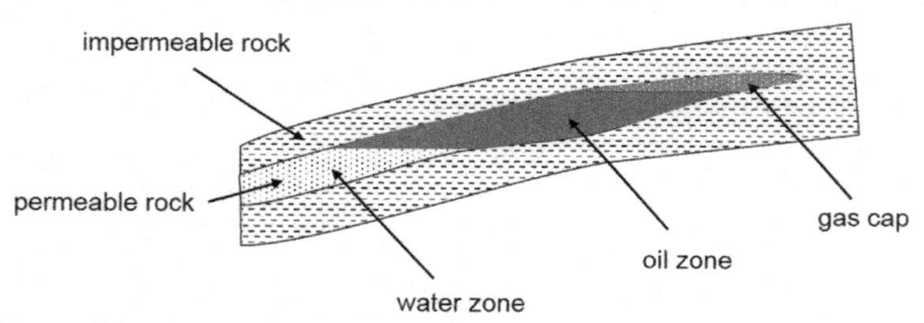

Figure 1.0.2 Hydrocarbon trap containing oil and gas.

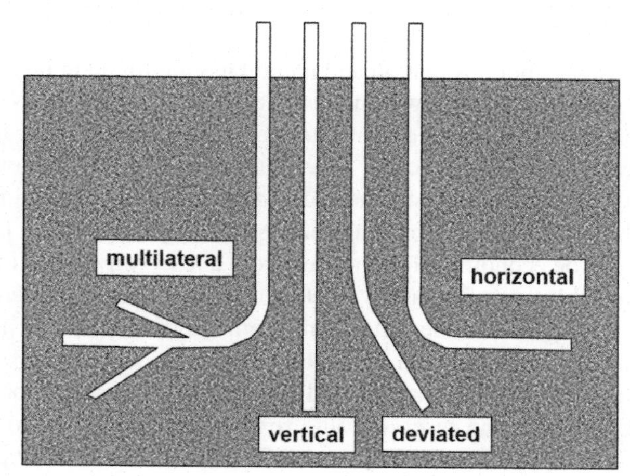

Figure 1.0.3 Types of wells.

including vertical, deviated, horizontal, and multilateral. Historically, most wells are vertical wells. Vertical wells contact the full height of the reservoir, but through a single hole that is usually less than a foot in diameter. Deviated wells are vertical wells drilled at an angle up to about 65°. Deviated wells are used when it is necessary to drill underneath a surface obstacle, such as a lake, or when many wells are drilled from a single drilling platform. Horizontal wells are a relatively recent technical advance. They can contact a large reservoir area, but may not contact the full height of a reservoir. Multilaterals are horizontal wells with extensions added to the main bore hole.

There are also different approaches to making the wellbore ready to produce fluids, that is, completing the well. Some wells are open hole at the formation of interest. Most wells are cased; that is, steel pipe is cemented in the drilled hole to prevent hole collapse and fluid migration from one formation to another. The casing is then perforated; holes are made through the casing into the formation so that reservoir fluid can reach the wellbore. Schematics of some different completion types are provided in Figure 1.0.4.

In some cases, the formation around the well is stimulated, typically through acid injection or hydraulic fracturing. Acid injection can dissolve material near the wellbore that may be restricting production. Hydraulic fracturing involves injecting fluid at high pressure to crack open the formation. Proppants (solid particles such as sand or ceramic beads) are injected into the open fractures to hold the fracture open after the pressure is reduced. The propped fractures create two planar conduits for fluid flow. Once the well is drilled and completed, production tubing is placed in the

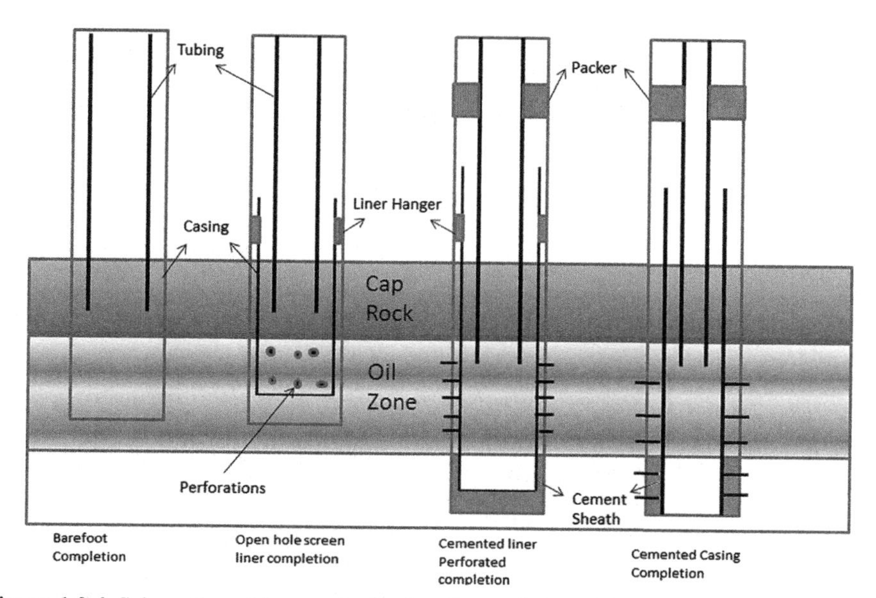

Figure 1.0.4 Schematics of four methods of well completion. Other configurations are also possible.

well, and the reservoir fluids are produced. A pump may be added to reduce the pressure in the wellbore and increase production rates. A schematic of a producing well is provided in Figure 1.0.5.

Once reservoir fluids reach the surface, they are separated into gas, oil, and water streams. An oilfield surface facility is shown in Figure 1.0.6. Gas, liquid, and water flow rates are measured for each well or group of wells so that the produced volumes can be allocated to the owners of the wells. Gas is compressed and sent by pipeline to a gas plant for further processing. Sometimes in remote locations or due to lack of market for gas, the gas is flared. Oil is sent to an oil pipeline and eventually to a refinery. Water is usually re-injected into a suitable formation.

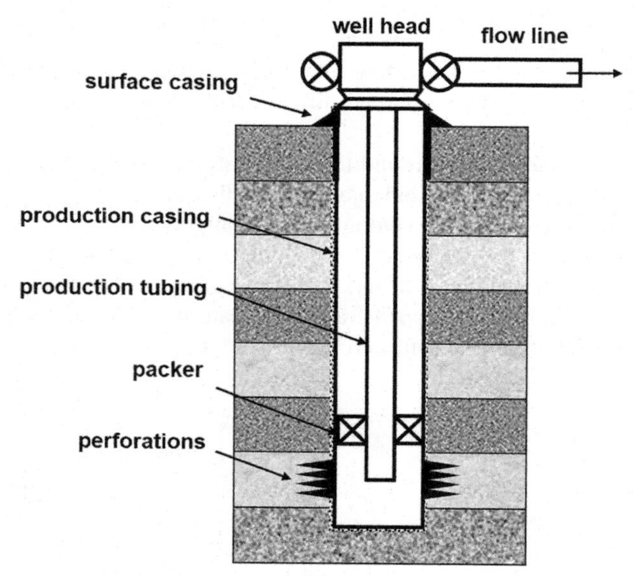

Figure 1.0.5 Schematic of a producing oil well.

Figure 1.0.6 Photograph of a large oil battery in Kuwait. http://www.en-fabinc.com/en/gathering.shtml, 2012.

1.1 Overview of Reservoir Engineering

Petroleum reservoirs and their associated wells and facilities make up the assets of petroleum-producing companies. The main objective of a petroleum-producing company is to increase the value of its petroleum reservoirs for its stakeholders. The value of a petroleum reservoir depends on several factors: the amount of petroleum in the reservoir; the amount that can be produced; how rapidly the petroleum can be produced; the capital and operating costs involved in recovering the petroleum; royalties and taxes; and the price paid for the petroleum. Petroleum assets can also be discovered, acquired, or sold to increase the value of the company. A petroleum producer usually attempts to maximize its value in all of these areas and to do so calls on various disciplines, including land management, geophysics, geology, engineering, economics, marketing, and accounting.

Roughly speaking, geologists, geophysicists, and petrophysicists describe rock properties and reservoir structure. Production engineers manage wells and surface facilities. Reservoir engineers manage the reservoir. The objective of a reservoir engineer is to produce as much of the petroleum in the reservoir as possible, as quickly as possible, and at the lowest cost, but at the same time maximize economic value. In other words, the role of the reservoir engineer is to determine the maximum amount of petroleum that can be recovered economically, the optimum production rate under existing operations, and the applicability of waterflooding, gas flooding, or enhanced oil recovery (EOR) for the reservoir.

1.1.1 Estimation of Volumes in Place, Reserves, and Rates of Recovery

There is an important distinction to be made between the hydrocarbon in the reservoir (original oil in place and original gas in place, or OOIP and OGIP) and the recoverable hydrocarbon (oil and gas reserves). OOIP and OGIP are what nature provides, while the reserves are what we can economically extract from the reservoir. The OOIP and OGIP partially dictate what recovery schemes can be used and how much of those reserves can be recovered. The ratio of reserves to hydrocarbon in place is defined as the recovery factor.

Two general methods can be used to predict oil in place volumes:

- Volumetrics,
- Material balance.

Five general forecasting methods can be used to determine recoverable volumes (reserves) and production rates:

- Analogy,
- Decline analysis,
- Data mining,
- Analytical,
- Simulation.

Hydrocarbons-in-place

The first step is to determine how much petroleum is in the reservoir. The reservoir can be thought of as a packed bed that is full of fluid at high pressure. To determine the amount of fluid in the bed, it is necessary to determine the volume of the bed and the porosity of the bed, that is, the fractional pore space. In the case of reservoir rock, there is a film of water on the rock. This initial water saturation (the fractional volume of the water) must be accounted for. Also, as oil is withdrawn from the bed and produced to the surface, its pressure and temperature decreases to approximately atmospheric pressure and surface temperature. Hence, the relationship among the fluid volume, pressure, and temperature (PVT) must be determined. Any phase changes that may occur as the pressure changes must also be accounted for. For example, does the oil enter a two-phase region and evolve gas? The data required to determine the volume of petroleum in the reservoir include:

- Reservoir volume,
- Reservoir porosity,
- Reservoir water saturation,
- Fluid properties (PVT relationships).

The volume of petroleum determined using this approach is the volumetric OOIP or OGIP. The method of estimating volumetric reserves is essentially the same for all reservoirs. However, reservoir properties are not uniformly distributed. Therefore, both empirical methods and simulation methods may break down. The understanding of reservoir heterogeneity is one the largest challenges reservoir engineers and geoscientists face.

A second method to estimate in-place volumes is to perform a material balance on the reservoir. As fluids are withdrawn, the reservoir pressure declines. The volume of oil, gas, and water initially present in the reservoir can be determined from the extent of pressure decrease for a given fluid withdrawal. The required data are:

- Fluid properties (density, viscosity, PVT relationships),
- Production and injection rates over time,
- Reservoir pressure over time.

The form of the material balance depends on the drive mechanism of reservoir (discussed later). The volume of petroleum determined with this method is the material balance OOIP or OGIP. As a rule of thumb, approximately 10% depletion of the original reservoir pressure is required to use this method with accuracy. In general, the material balance method works well in reservoirs with high permeability or transmissibility, because the reservoir pressure over a reservoir scale tends to equilibrate. Material balances usually work best in higher permeability gas fields (>1 mD). A second constraint of using material balance is the availability and quality of measured pressure data.

Often reservoirs are not closed systems, and an aquifer model is required to complete the material balance. These models relate aquifer influx to aquifer size, pressure, and rock and fluid properties. Aquifer models are often used in water-drive reservoirs in which water influx may provide a significant portion of the energy of the system (>10% of the total drive energy in the system). Generally, there is little information on aquifer parameters because companies usually do not target wells in an aquifer.

A recovery factor is required to convert oil-in-place or gas-in-place values (volumetric and material balance reserves) to recoverable reserves for a particular reservoir. This brings us to the second step, forecasting production and determining the recovery factor.

Recovery factors

The analogy method is used most often when there is little production history for the pool of interest. An analogous reservoir is a reservoir with similar rock and fluid properties, geological characteristics, and drive mechanism, but with a more extensive production history. A representative well and/or pool forecast is determined for an analogous reservoir, and a recovery factor is calculated. The forecast and recovery factor are then applied to the reservoir of interest.

The most commonly used method to estimate recoverable reserves is decline analysis. For example, in North America, more than 95% of the small pools have their reserves calculated using decline methods. Decline analysis is simply the extrapolation of existing trends in the production data. In some cases, there is a stronger theoretical basis for the decline trend, such as homogeneous gas reservoirs or oil reservoirs in single-phase flow (Petroleum Society of the Canadian Institute of Mining, Metallurgy and Petroleum, 1994; Araya and Ozkan, 2002; Arps, 1945; Hale, 1986; Camacho-Velazquez and Raghavan, 1989; Masoner, 1998; van Poollen, 1966). Generally, for oil reservoirs that have multiple phases flowing, decline analysis is empirical in nature, but it is still useful as a first estimate of recovery. Often multiphase flow as well as multilayering and heterogeneity can cause complexity and departure from linear trends from decline models.

Another method is to use data-driven models or data mining techniques that correlate injection and production rates by using neutral networks, statistics, or artificial intelligence methods (da Silva et al., 2007; Khazaeni and Mohaghegh, 2011). These techniques can be especially useful in an experienced geoscientist's or engineer's hands. Some production history is required to establish the correlations, and therefore these methods are best suited to mid to late development of a reservoir.

Analytical methods assume certain idealized flow conditions in the reservoir. Analytical models are often useful when reservoir flow is either one- or two—dimensional (2D) in nature. Analytical models can be further classified as:

• Radial/linear inflow equations	Relate production/injection rate to pressure gradient, permeability, and fluid properties for horizontal flow field
• Coning models	Relate production/injection rate to pressure gradient, permeability, and fluid properties for vertical flow field
• Displacement models	Prediction of water flood or gas flood performance from rock and fluid properties
• Inflow performance models	Relates bottom-hole flow potential to flow rate

Radial inflow equations are generally applicable to vertical wells, whereas linear inflow equations are generally applicable to horizontal wells. Inflow equations can be applied to any producing or injecting well to determine the flow rate at a given reservoir pressure or saturation. A separate prediction of the pressure and saturations is required to predict production rates over time. For example, in gas reservoirs, the inflow equations can be coupled with a material balance to create a "flowing material balance" production forecast.

Coning-type models are used for reservoirs in which the oil is immediately above a water zone or below a gas cap. In this case, the pressure drawdown around a producing well distorts the fluid level near the wellbore so that water or gas cones (for a vertical well) or crests (for a horizontal well) up or down to the perforations. Most coning models are based on steady state flow which may take long periods of time to evolve.

One- or two-dimensional displacement models represent the microscopic displacement of oil with water or immiscible gas. These models often can mimic fluid flow in reservoirs and can help predict production rates for developing reservoirs (Cronquist, 2001; Dake, 1994). The use of analytical models is strongly recommended (Dake, 1994), because analytical methods require far fewer assumptions than the alternative technique, reservoir simulation modeling. Also, the variables are clearly identified in analytical models. Therefore, one- or two-dimensional analytical models are often used as a guide for reservoir simulation. The weakness of the analytical methods is that they usually assume homogenous flow units.

The final method is to use reservoir flow simulation. Reservoir simulation is often considered to be the most sophisticated method of analysis. In brief, a representation of the reservoir is constructed based on the best estimates of the reservoir properties, including the porosity and permeability distributions, reservoir structure, fluid properties, and rock–fluid properties. Model wells represent the real wells completed in the reservoir, and the pressure response and flow rates into, within, and out of the reservoir are simulated.

Simulation is a powerful technique when there are enough data to construct the model and some historical dynamic data with which to calibrate the model. However, the reservoir engineer must be aware of the limitations of the data. For example, we can often subdivide the reservoir into many regions, but we do not know the exact parameters for each subregion. This lack of resolution leads to a non-uniqueness in the model. All reservoirs are heterogeneous, and sorting out this heterogeneity is a key challenge in reservoir characterization.

Reservoir simulation can provide an illusion of accuracy, because the reservoir is represented in such apparent detail. However, simulation is as prone to error as any other method. Errors in models are typically caused by three factors (Poeter and Hill, 1997):

1. Data error caused by inaccuracy and imprecision by the measuring device or human error;
2. Error as a result of the model being an over-simplification of the physical system (simplification of heterogeneity);

3. Error due to neglecting physical phenomena (omitting physical phenomena, for example, near wellbore damage particle movement or waxes).

Usually in both the geological (static) model and in the flow simulation (dynamic) model the second type of error occurs. Sometimes, especially in near wellbore phenomena, the third type error occurs (waxes, fractures, stress sensitive permeability, etc.)

Often complicating factors, such as fracture/fault flows and near wellbore phenomena such as waxes or particulate plugging, are not built into the flow simulation models. In fracture flow, fluid flows much more rapidly through the fracture system than the reservoir matrix. When plugging occurs, a wellbore or near wellbore phenomena may control flow rates rather than the overall flow through the reservoir that the simulation represents. In either case, a reservoir flow simulation may not be able to match the real world process. It is extremely common, unfortunately, for engineers and geoscientists to assume that all of the physics are represented in flow simulation models and to neglect the limitations of these models.

Nonetheless, simulation models have great potential as a finishing tool because they can integrate the views of both geoscientists and engineers. Even if there are not sufficient data to calibrate the models, simulation models remain a useful tool if a "what if" range of answers is needed. The drawback of simulation is that complex models are sometimes hard to update, and sometimes business decisions require faster answers or quicker adaption to new data.

1.1.2 Determining the Field Development Plan

It is extremely important to note the "whys" of what we are doing. Usually we are not doing reservoir characterization work and building models for the sake of reservoir characterization work and building models. We are engaging in reservoir characterization and model-building to: (1) determine future recovery, and (2) optimize the reservoir exploitation.

After determining the volume in place and a forecast of natural drive production rate and pressure, it is necessary to find the best method to extract the petroleum from the reservoir, that is, to design a development strategy. Well type is an important initial decision that needs to be made. From an analytical point of view, wells are simply points at which fluid can be withdrawn (produced) or added (injected) to the reservoir. Hence, the design of a development strategy concerns the number of wells, the location of each well, well spacing, the choice of injection fluid (if any), and the production and injection rate at each well.

The first decision in designing a development strategy is whether to implement an injection scheme. In the absence of a robust natural water drive or a large gas cap, the idea is to inject a low-cost fluid to maintain pressure in the reservoir while displacing and producing the petroleum. If injection is chosen, then the appropriate injection fluid must be identified. The choice of injection fluid divides production strategies into five categories:

• Primary production	No injection
• Waterflooding	Water injection
• Immiscible gas flooding	Injection of a gas that does not mix with the oil
• Miscible flooding	Injection of a liquid or gaseous solvent that mixes with the oil to form a single phase
• Chemical EOR	Injection of chemicals
• Thermal methods	Injection of steam
	Injection of air or oxygen that is ignited underground (fireflood)

Waterflooding and gas flooding are sometimes referred to as secondary recovery. Water is the most widely used injected fluid; more than 60% of the world's oil production comes from reservoirs that have some water injection.

A miscible flood implemented after a waterflood is sometimes referred to as tertiary recovery. The term enhanced oil recovery is used to describe any method apart from primary and secondary recovery. There are also a number of variations on the short list of methods given above, for example, water after gas floods, surfactant floods, and polymer floods. This book deals only with primary and secondary recovery methods.

Once the injection fluid is selected, the next choices are: the type of well (vertical, horizontal, or multilateral); the number and location of the wells; and the production and injection rate for each well. The available production strategies include:

- Adding production or injection wells (infill or step-out drilling);
- Converting production to injection wells;
- Changing injection fluid;
- Changing production and/or injection rates;
- Changing location of perforations and/or completion method.

All these strategies involve a trade-off between the cost to implement the strategy and the anticipated benefit of increased production rates and petroleum recovery. The best choice of development strategy depends partly on the nature of the reservoir, partly on local economic and regulatory conditions, and partly on the amount of data taken.

1.2 Reservoir Classifications

A key component of the reservoir engineer's work and the focus of this book is reservoir characterization. Reservoir characterization is the construction of a mental or mathematical model representation of the reservoir based on reservoir data. This representation is used to capture the features that determine the general fluid flow in the reservoir and affect production profiles and hydrocarbon recovery. To get an initial understanding of the reservoir or mental model, there are four considerations:

- Fluid type (bitumen, heavy oil, conventional oil, volatile oil, retrograde condensate, gas);
- Reservoir architecture (size and structure; porosity, permeability, and saturation distributions);
- Drive mechanism (expansion, solution gas, gas cap, water, combination); and
- Flow characterization.

Once these four factors are generally known, methods for analyzing the reservoir, determining reserves, and forecasting production can be determined.

Each factor is discussed in more detail below, but first consider the number of possible reservoir types. Figure 1.2.1 shows the possible classification of oil reservoirs according to fluid type, drive mechanism, reservoir architecture, and flow patterns. There are 6000 possible combinations just for oil reservoirs:

(4 oil types) × (5 drive mechanisms) × (4 types of structure) × (5 permeability types) × (3 bulk reservoir flow patterns) × (5 near wellbore flow patterns)

This is a daunting prospect! Fortunately, in practice, there are common behaviors and a much more limited set of methods that need to be applied.

Why then do we recommend such a classification system? The main reason is that following the classification scheme leads to an appropriate subset of methods/tools that have been developed for that reservoir type. The steps to optimizing the reservoir production are often well known, and there are usually excellent analogous reservoirs and examples in the literature that can be used for guidance. For example, for a compressible drive reservoir having an undersaturated oil, waterflood or pressure maintenance must be considered early in the reservoir life. For a tight oil reservoir, formation permeability will be a key factor in determining well spacing and recovery.

The second reason is that within a drive type/fluid type/architecture type, some common parameters often control the process. For example, in a solution gas drive, the critical gas saturation and amount of solution gas in the oil will control the production profile. For edge water drive systems, the strength (compressible energy) and permeability of the aquifer largely controls the recovery profiles. In reservoir and field performance, there are usually a large number of potential parameters that may control

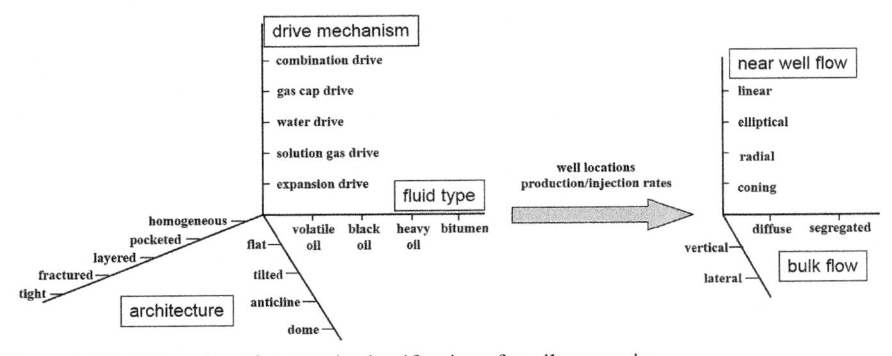

Figure 1.2.1 Categories of reservoir classifications for oil reservoirs.

the process. In our experience, only 1–3 parameters usually control the process. If we know those critical parameters, we can focus on those parameters first. To know what the critical parameters are, you need to know drive type/fluid type/architecture type.

The third reason for a broad classification system is that it reminds us that each reservoir typically has some unique aspects that can lead to surprises. The system helps us to recognize that we do not have complete knowledge about the reservoir. Thus, any reservoir characterization method must have some flexibility to accommodate these atypical features.

1.2.1 Fluid Type

The type of fluid is defined based on the phase behavior and properties of the fluid. There is an obvious distinction between liquid (oil) and gas reservoirs, but fluid type also dictates the expected phase behavior of the reservoir as the pressure declines. Different phase behavior data and models are required for different fluid types. Fluid properties are a part of most reservoir engineering calculations and are closely related to the reservoir drive mechanism.

There are six petroleum fluid classifications: dry gas, wet gas, condensate, volatile oil, black oil, and heavy oil. Brief definitions are:

- Dry gas: always in the gas phase at and above ambient temperatures and pressures;
- Wet gas: forms a two-phase mixture in the well and production facilities as temperature decreases;
- Retrograde condensate gas: forms a two-phase mixture as pressure decreases, resulting in liquid drop out in the reservoir;
- Volatile oil: initially in liquid (oil) phase, the volume and density of the oil phase vary significantly with changes in pressure;
- Black oil: initially in oil phase, oil phase has nearly constant compressibility and viscosity;
- Heavy oil: initially in oil phase, oil phase has nearly constant compressibility and high viscosity, which increases significantly as solution gas evolves;
- Bitumen: initially in oil phase, oil phase has nearly constant compressibility and very high viscosity >100,000 mPa·s, such that oil is initially immobile.

Volatile oils require a detailed fluid model to predict their phase behavior and often require compositional simulation to accurately forecast reservoir performance. Black oils have been the most commonly developed oil reservoir fluid type. Most of the methods presented in this book were developed for black oils. Heavy oil reservoirs are growing in importance as conventional oil supplies are depleted. Heavy oils can often be treated as black oils with high viscosity; however, the increase in viscosity as solution gas evolves must be taken into account. Also, the viscosity of heavy oil and bitumen can be so high that some conventional non-EOR development strategies are not applicable, and other methods such as steam injection are used to reduce the viscosity and improve recovery. The following discussion of reservoir classification focuses on black oil reservoirs.

1.2.2 Drive Mechanism

The pressure in the reservoir decreases when fluids are withdrawn from the reservoir. As pressure drops, the reservoir rock and the remaining fluids expand. The expansion of the rock and fluids provide energy to drive the movement of fluid through the reservoir. Depending on the situation, different expansion terms will dominate, leading to different analytical approaches for the reservoir model. Therefore, it is useful to a reservoir engineer to classify reservoirs according to the dominant component of expansion, that is, the drive mechanism.

Understanding the drive mechanism is critical because it controls the pressure versus withdrawal profile and thus dictates whether primary production is sufficient or whether injection and/or EOR processes are needed. The drive mechanism also dictates the timing of certain decisions. For example, if a pool pressure is well above bubble point pressure, then because the pressure decline is expected to be rapid, a decision must be made early whether or not to introduce water or gas injection. Note because well production rates depend on the pool pressure, understanding the drive mechanism is also critical for interpreting well rate versus time data.

The four drive mechanisms are: undersaturated oil expansion, solution gas drive, gas cap expansion, and water drive. It is also possible to have combined gas gap and water drive.

Undersaturated oil expansion: An undersaturated oil reservoir initially contains only an oil phase. The drive mechanism is a combination of formation compaction and connate water and oil expansion. As the reservoir pressure decreases, the oil reaches its bubble point, and a gas phase is formed. At this point, the drive mechanism changes to solution gas drive.

Solution gas drive: When an oil reservoir reaches its bubble point, solution gas begins to evolve from the oil in the reservoir. The liberated solution gas, the formation rock, connate water, and oil all contribute to expansion. Because gas is much more compressible than the oil, water, or rock, the solution gas expansion dominates. Hence, the drive mechanism below the bubble point is termed solution gas drive.

Gas cap expansion: When the initial pressure of the oil in a reservoir is equal to its bubble point pressure, the oil is saturated with gas and is usually part of a two-phase system; that is, there is a gas zone associated with the oil zone. Note that if an oil reservoir has a gas cap, then the initial pressure must be the bubble point pressure at the gas-oil contact (GOC). For thick reservoirs, a common feature is the presence of compositional gradients. These gradients result because, while the initial pressure must be the bubble point pressure at the GOC, as we down structure away from the GOC, the oil can be undersaturated. Because gas density is lower than oil density, the gas resides at the top of the reservoir structure and is termed a gas cap. A gas cap reservoir has the same expansion terms as a solution gas drive reservoir and an additional term accounting for the expansion of the gas cap. Again, the gas expansion terms dominate.

Water drive: A water drive reservoir consists of undersaturated oil overlying a water zone or aquifer. The drive mechanism for this type of reservoir depends on the strength of the aquifer. If the aquifer is strong and maintains nearly constant reservoir pressure, then the drive mechanism is the water influx from the aquifer. If the aquifer is weak, then the other expansion terms and, in particular, solution gas drive will dominate.

Combined gas cap and water drive: These oil reservoirs contain a gas cap and underlying water. The drive mechanisms are the same as described above. However, many of the simplifying assumptions used to analyze gas cap or water drive reservoirs cannot be made. Hence, combination drives are particularly challenging to analyze.

1.2.3 Reservoir Architecture and Bulk Fluid Flow

Reservoir architecture includes a number of geological characteristics, including the structure and permeability distribution of a reservoir. Reservoir structure is the shape of the reservoir and includes flat, tilted, anticlinal, and domed structures. The permeability distribution relates to the level and variability of permeability and how the permeability changes throughout the reservoir. Types of permeability distribution include homogeneous, layered, pocketed, cyclical, fractured, and tight. We will examine reservoir architecture in more detail, but we must first take into consideration the relationship between architecture and bulk fluid flow.

Bulk fluid flow refers to the primary direction of the fluid flow (lateral or vertical) and how the gas flows relative to the oil (diffuse or segregated). The direction of fluid flow is a critical factor for reservoirs containing more than one fluid phase. Reservoir architecture dictates the direction and magnitude of fluid flow in the reservoir, and therefore both must be considered together as the three examples given below demonstrate.

Example 1: Consider an oil reservoir with an underlying water zone, Figure 1.2.2. If a well is partially completed in the oil zone, oil will tend to flow horizontally and water will tend to flow vertically toward the well perforations through a water cone. This is the near wellbore flow.

When vertical and gravity segregation flow is significant, coning models are used to describe the flow field. In the reservoir developed with vertical wells in which the horizontal flow dominates, radial flow models are usually used to describe the flow field. In some cases, however, for hydraulically fractured wells, horizontal wells, and naturally fractured reservoirs, linear flow may dominate the reservoir bulk flow or more complex flow models may need to be used. The reservoir performance and the production and development strategy are different for each type of flow field.

Example 2: Consider a saturated oil reservoir that is at its bubble point, Figure 1.2.3. The evolved solution gas may flow with the oil; that is, both the oil

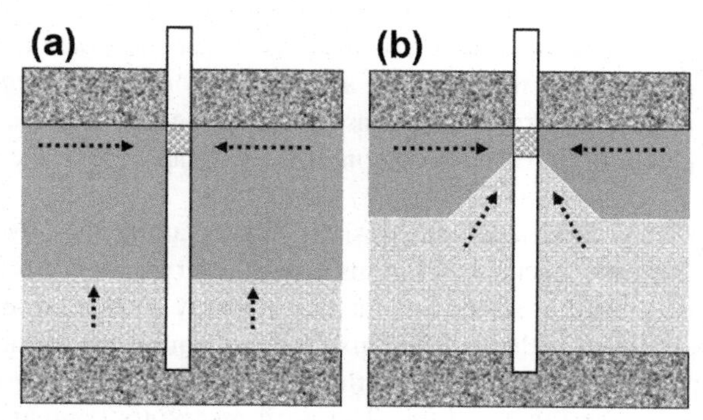

Figure 1.2.2 Flow direction in oil zone with underlying water: (a) thick oil zone, horizontal radial inflow models apply; (b) thin oil zone, coning models apply.

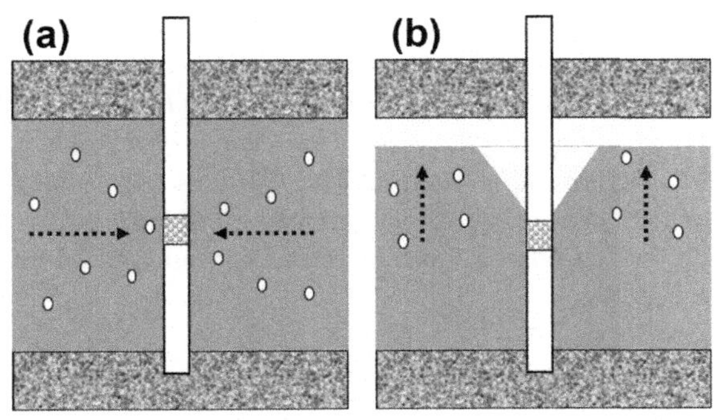

Figure 1.2.3 Flow of oil and gas in an oil reservoir below its bubble point; (a) gas flows horizontally with the oil and is produced at well (dispersed flow); (b) gas rises and forms a secondary gas cap (segregated flow) where gas may cone down to perforations from the gas cap.

and gas may flow radially to the wellbore. On the other hand, the solution gas may rise due to its lower density and form a secondary gas cap at the top of the reservoir. Here we have a combination of far field flow (gas segregating) and near well flow (gas coning) that both affect performance. Whether the gas segregates in the reservoir or not will result in different reservoir performance, require different forecasting techniques, and will have different optimum completion strategies.

Example 3: Consider a water drive reservoir containing a conventional oil and in which the water is located at one end of the reservoir, Figure 1.2.4. If the reservoir has high, uniform permeability, the water will efficiently sweep the oil, and high recovery factors will be realized. If the permeability varies considerably from layer to layer, the water will rapidly sweep the highest permeability layers and only gradually sweep the lower permeability layers. In this case, there will usually be earlier water production and lower ultimate recovery.

Reservoir architecture interacts with fluid properties and drive mechanism to determine how individual wells and regions will behave. Permeability architecture describes the continuity and permeability of the flow units; the combined structural model and permeability model dictate the geometry of reservoir flow. A critical

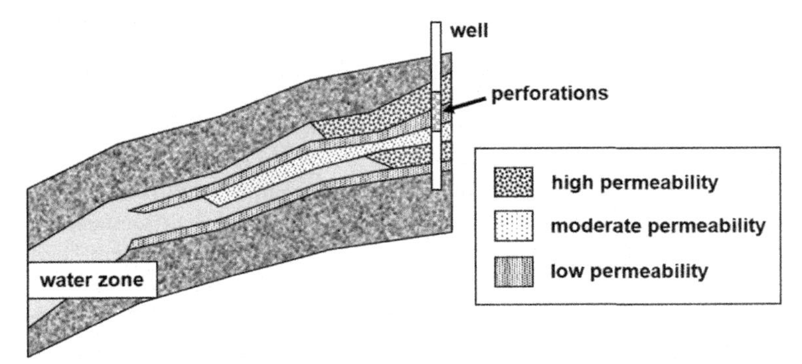

Figure 1.2.4 Uneven water sweep through a tilted, layered reservoir when the layers have different permeabilities.

component of reservoir engineering is the reconciliation of the dynamic data (pressures, rates) to the geological model. *Reservoir engineering and reservoir characterization can then be thought of as a geometry problem. In other words, what is the shape or size of the drainages volume and/or swept regions, and what and where are the internal flow baffles and conduits?* The volumes and shapes of the drainage or swept volumes are critical to well spacing and waterflood/gas flood recovery.

With this in mind, we combine reservoir architecture and bulk fluid flow to classify reservoirs by three criteria:

1. Direction of fluid flow
  a. Horizontal displacement
  b. Vertical displacement
2. Uniformity of flow (permeability variation and distribution)
  a. Uniform
  b. Layer
  c. Pocket (jigsaw or labyrinth or other words patchy distribution of permeable reservoir)
  d. Fracture
  e. Low permeability
3. Flow of solution gas/expansion drive or injected fluid
  a. Nonsegregated
  b. Segregated
  c. Gravity drainage

1.2.3.1 Direction of Fluid Flow

Consider a vertical well perforated in an oil zone above a water zone (Figure 1.2.2). Initially, oil flows radially toward the wellbore in the horizontal plane, while the water does not flow. In this case, the dominant direction of flow is horizontal. Eventually, as oil is withdrawn from the reservoir, the water zone expands, and the water contact rises toward the well perforations. Due to hydrostatic forces, water is drawn toward the perforations in a cone around the wellbore. When the cone reaches the wellbore, vertical flow of water occurs. While the oil still flows radially, this flow pattern is termed vertical displacement.

The flow situation can be more complex, however. For example, for a horizontal well in a perfectly homogeneous reservoir, a uniform crest of water would form along the well; however, both measurements and models suggest that the crest is not usually uniform. The displacement is therefore more complex and in situ measurements are required to interpret the flow behavior. In contrast, consider an undersaturated oil reservoir undergoing a waterflood with vertical wells. Oil flows radially to the producing well. Water is injected in patterns around the oil producers and sweeps horizontally through the reservoir. This flow pattern is termed horizontal displacement.

The dominant direction of fluid flow is determined primarily by the reservoir structure. The many possible reservoir structures can be classified into four main categories, Figure 1.2.5.

Flat or table-top reservoir: A table-top reservoir has little change in elevation across the reservoir. Therefore, the fluid contacts extend across most of the reservoir.

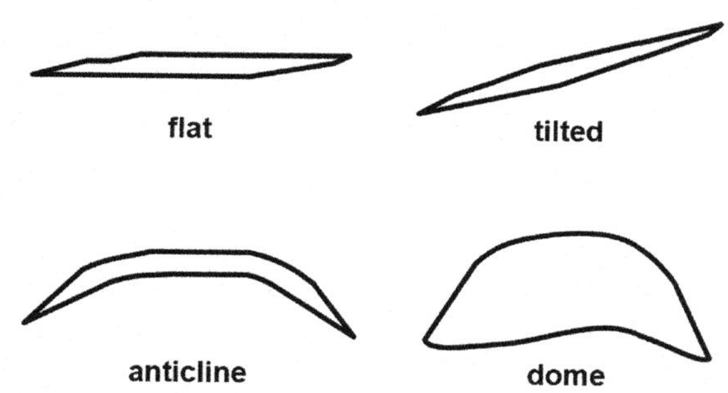

Figure 1.2.5 Main types of reservoir structure.

If the oil zone is thick, the initial flow field may be horizontal. Eventually, as the oil is withdrawn, the fluid contacts will move, the oil zone will thin, and vertical flow will become significant.

Tilted reservoir: A tilted reservoir has a significant dip; that is, the elevation of the reservoir changes along a given axis. Fluid contacts in a tilted reservoir are limited in area. Hence, angled flow (a combination of horizontal and vertical flow along the layers or bed boundaries) is dominant over most of the reservoir.

Anticlinal: Anticlines formed when tectonic forces fold rock formations. The water–oil contact is typically limited in area. The extent of the GOC is sensitive to the size of the gas cap. The dominant flow field typically depends on the extent of the GOC and the thickness of the oil zone.

Dome: Many carbonate reservoirs have a dome or pinnacle structure. The water–oil contact is usually extensive in area. As with the anticlinal reservoir, the extent of the GOC is sensitive to the size of the gas cap. Again, the dominant flow field typically depends on the extent of the GOC and the thickness of the oil zone.

The direction of flow is also strongly influenced by permeability. For example, there may be limited permeability in the vertical direction, perhaps due to shale layers. In this case, coning may not occur even in a structure that favors vertical flow, and the reservoir flow may be better classified as horizontal. Indeed, sometimes wells are perforated only above these shale layers to help reduce the risk of coning.

Another example is when there are faults in the reservoir. Faults can take a flat reservoir and make it tilted or create disconnected regions (compartments) where the pressures and fluids are different. In faulted reservoirs, the amount of sand-on-sand, or permeable interval-to-permeable interval, contact between fault blocks controls pressure communication and volumetric sweep.

1.2.3.2 Uniformity of Flow

Fluid flows through the path of least resistance in the reservoir, that is, through the highest permeability regions. Therefore, the distribution of permeability in the reservoir dictates how fluid flows through the reservoir. The major styles of permeability distributions are shown in Figure 1.2.6 and are defined below. Keep in mind that many reservoirs will have combinations of distribution types.

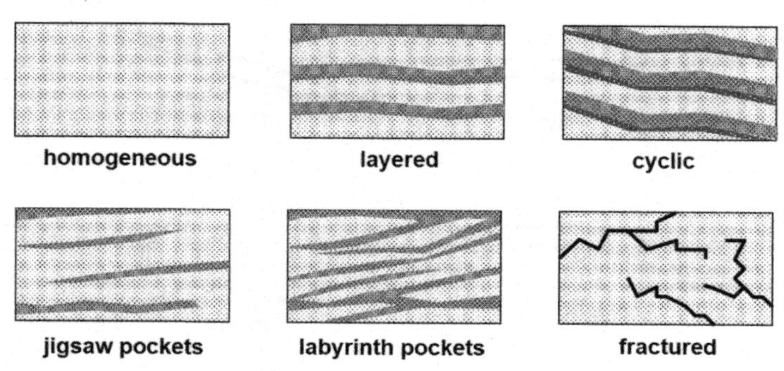

Figure 1.2.6 Types of permeability distribution.

Homogeneous permeability: When the permeability is nearly uniform across the reservoir (homogeneous), there are no preferential pathways for flow. Most flow models used in reservoir engineering are designed based on homogeneous permeability. In reality, no reservoir is completely homogeneous. In some cases, the variations in permeability are small, and the reservoir behaves as if homogeneous. Also, while the permeability may vary considerably on the scale of centimeters or meters, the average permeability on the scale of tens of meters may be nearly uniform. In this case, a homogeneous assumption usually gives reasonable results.

Layered permeability: Many sandstone reservoirs are deposited in layers (Weber and van Geuns, 1990). In some cases, the permeability changes significantly (perhaps by a factor of 100 or 1000) with each layer. Fluid will then flow preferentially along the highest permeability layers. These layers will experience high oil recovery and will have lower pressure than the low permeability layers. Fluid in the low permeability layers may flow vertically toward the high permeability layer. Also, if a layered oil reservoir is below the bubble point, solution gas may flow rapidly from the low permeability areas, resulting in rapid pressure decline and low oil recovery. The standard analytical models must be modified for layered permeability.

For single layer or layer-cake reservoirs with nearly constant and high permeability, high recovery factors are to be expected even with large well spacing, as long as there is sufficient drive energy in the system. If there is not sufficient drive energy, then usually secondary recovery waterflooding or gas flooding will be successful.

Permeability pockets: In heterogeneous reservoirs, there may be low permeability areas that act as total or partial barriers to flow, and there may be areas of highly contrasting permeability. Such permeability distributions can be caused by geological features such as shale drapes or faults and are common in carbonate and interbedded sandstone reservoirs. The contrast in permeability leads to differential flow in different parts of the reservoir. Differential flow can create pockets (volumes) of high pressure and low oil recovery where the permeability is low and pockets of low pressure and high oil recovery where the permeability is high. The analysis of this type of reservoir is complicated because different flow patterns may be encountered in different parts of the reservoir, the reservoir pressure is not uniform, and the fluid movement between pockets of high permeability is difficult to model.

There are two main types of permeability pockets or patches in sandstone reservoirs: jigsaw and labyrinth (Weber and van Geuns, 1990).

Jigsaw: A jigsaw permeability distribution arises when different sand bodies are touching, with only a few major gaps. There are partial permeability barriers and some discontinuous zones, but there are also frequent areas of contact and communication. Jigsaw reservoirs can often be modeled analytically with minor modifications to standard approaches.

Labyrinth: A labyrinth permeability distribution arises when there is a more complex arrangement of sand bodies or in a carbonate reservoir that has undergone extensive diagenesis. A labyrinth permeability distribution includes extensive permeability barriers and reservoir discontinuities and relatively few three-dimensional interconnections between the permeable pockets. Labyrinth reservoirs cannot easily be modeled with analytical methods. In some cases in which communication between permeable zones is limited, the permeability pockets can be modeled individually.

Multilayered reservoirs with substantial permeability contrast between the layers and some layers being discontinuous (jigsaw reservoirs) will have lower recovery factors than homogeneous reservoirs. To increase recovery from jigsaw or labyrinth type reservoirs, smaller spacing and profile control maybe needed to maximize recovery.

Cyclic permeability: In sandstone reservoirs, heterogeneous permeability distributions can appear repetitively because of cyclic conditions during their deposition. These conditions can be caused by short-term events, such as tides, or longer-term seasonal (e.g., springtime floods) or multiyear climate changes driven by sunspots and the Earth's orbit around the sun. Whatever the cause, a complicated-looking "random" permeability distribution may, in fact, be a collection of repetitive changes in permeability. Identifying these cycles makes for a more deterministic reservoir model in which the flow patterns and recoveries are linked to the effects of each cycle.

Naturally fractured reservoir: Some reservoirs fracture under tectonic stresses. The fractures form a network of highly permeable conduits through the reservoir. If the fractures are small and poorly connected, the reservoir can be modeled as a homogeneous reservoir. However, if the fractures are extensive and well connected, more complex flow models are required.

It is important to note that the presence of natural fractures can completely short-circuit the impact of reservoir architecture. For example, if a reservoir has high connectivity fractures—although it may be a labyrinth in terms of deposits of porous media—it may behave as if a few wells can effectively drain the oil.

Low permeability (tight): Reservoirs can also be classified according to their absolute permeability as shown in Table 1.2.1. We have created a separate classification for low or very low permeability reservoirs because they perform differently than other permeability types. The permeability distribution is often the controlling parameter in low permeability reservoirs. In these reservoirs, a high pressure gradient is required to sustain flow if fractures are not present. Therefore, successful exploitation of tight reservoirs almost always requires either naturally occurring or induced fractures or both to be present. As described above, the fractures change the pressure distribution from radial flow to linear flow. Without fractures, the pressure is low in a

Table 1.2.1 Classification of reservoir type by absolute permeability

Permeability classification	Permeability (mD)
Very low	<0.01
Low	0.01−1
Average	1−100
High	100−10,000
Very high	>10,000

Source: From Golan and Whitson (1991).

region near the wellbore and high a short distance from the well. The low pressure promotes coning if there is a fluid contact near the well. It will also lead to premature evolution of solution gas near the well, further exacerbating the pressure decline. Hence, a well in a low permeability unfractured reservoir can only produce oil from a small area near the well.

1.2.3.3 Flow of Solution Gas

Oil and solution gas fluid flow can be divided into three categories: nonsegregated flow, segregated flow, and gravity drainage.

Nonsegregated (diffuse) flow: Oil and solution gas flow in the same direction, although the relative velocity of each fluid may differ. In other words, the bubbles of evolved solution gas do not rise. Solution gas will not migrate to an existing gas cap or form a secondary gas cap. Nonsegregated flow is most likely to occur in flat reservoirs that are extensive in area, e.g., layer-cake, with high horizontal permeabilities and relatively low vertical-to-horizontal permeability ratios. Figure 1.2.7 shows this type of mechanism.

Segregated flow: Oil and solution gas do not flow in the same direction. Typically, oil flows horizontally, while the liberated gas rises to join an existing gas cap or form a secondary gas cap, as shown in Figure 1.2.8. Segregated flow is likely in pinnacle structures with relatively high vertical-to-horizontal permeability ratios.

Gravity drainage: Gravity drainage is usually associated with reservoirs having a significant amount of structure; the dip angle is at least 15°. The process is similar

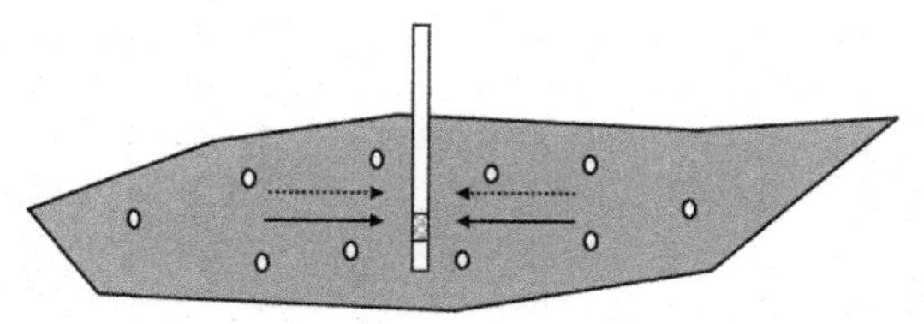

Figure 1.2.7 Nonsegregated gas flow. Gas bubbles are trapped in the oil column of the reservoir, flow horizontally with the oil, and do not rise up and join or form a gas cap.

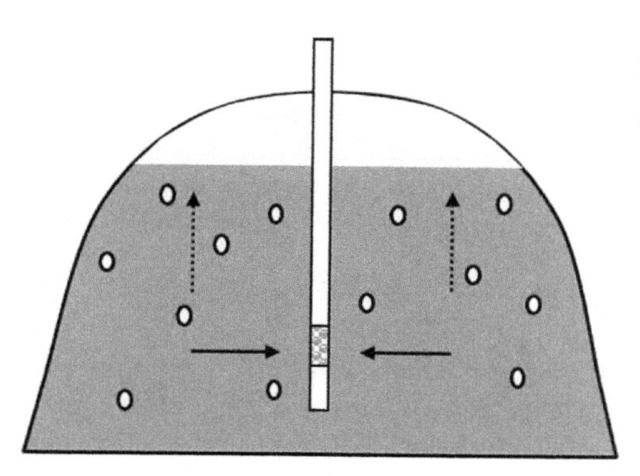

Figure 1.2.8 Segregated gas flow. Gas bubbles do not flow with the oil, but rise up to join or form a gas cap.

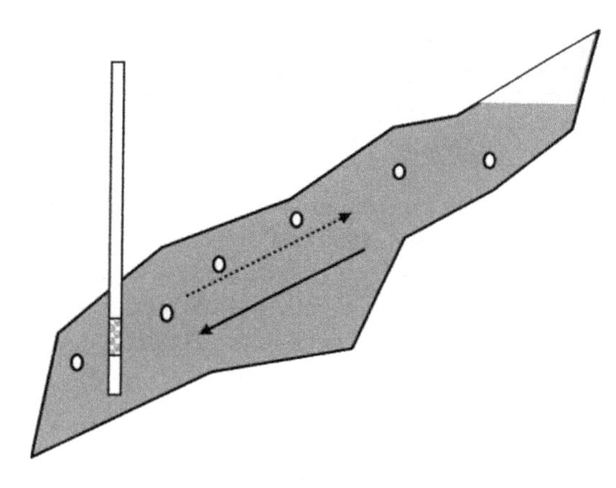

Figure 1.2.9 Gravity drainage. Gas bubbles flow countercurrent to the oil and rise up to join or form a gas cap.

to segregation drive, except the "horizontal" (parallel to formation dip) permeability is used instead of the vertical permeability. Because the lateral permeability is usually higher than the vertical permeability, resistance from the rock to the separation process is typically less than in a segregated drive. The oil and gas flow in opposite directions, causing counterflow. The counterflow occurs in highly permeable, thick formations with vertical fractures or vugs and conditions such as low oil viscosity and a reasonable high degree of dip. Gravity drainage is illustrated in Figure 1.2.9.

1.2.3.4 Reservoir Architecture Limitations of Data and Need for Iteration

It is important to realize that there are considerable limitations to the information about a reservoir. Simply put, core and logs represent only one 10 billionth of the reservoir.

It is no insult to the geologist to say that the subsurface picture is frequently uncertain, for no one knows better than he the difficulties involved in creating this picture and the constant need for revision as new information is obtained from wells that are drilled. The geological picture, however, only depicts the broad features and broad features in themselves are insufficient to explain detailed behavior, for, within the large scale pattern there are much smaller scale variations which have a pronounced effect on observed behavior.

Rowan and Clegg (1963).

Since, 1963 when the above was written, reservoir characterization has improved, particularly with analyzing and integrating sparse data, but there is more to be done.

Reservoir flow, reservoir architecture, fluid properties, and drive mechanism all interact to determine how individual wells and regions will behave. Therefore, an iterative process is often required to complete the reservoir characterization. An initial reservoir architecture is defined and tested for consistency with production and pressure data. If it is not consistent, it must be redefined. As new data are obtained, the reservoir description is modified and refined to better represent the new data.

1.2.4 Near-Wellbore Flow

Multiphase effects such as solution gas evolving from the oil and coning effects as well as turbulence effects in near-wellbore fractures cause nonlinear behavior between drawdown and well flow rate. In other words, the flow rate is not linearly related to pressure drop (see Sections 3.3.4 to 3.3.6). Vogel developed an empirical relationship between flow rate and flowing bottomhole pressure for oil wells that produced below bubble point pressure (Vogel, 1968; Golan and Whitson, 1991).

1.2.5 Using Reservoir Classifications

The reservoir classifications guide the choice of analytical methods that are suitable for a reservoir or part of a reservoir at a given time. The scope of this book is the primary and secondary recovery of oil reservoirs. We exclude those naturally fractured reservoirs that cannot be treated as a homogeneous permeability reservoir. Such reservoirs require a dual porosity model and advanced reservoir engineering methods. We will also focus on black oil reservoirs. Some of the analytical methods available for black oil reservoirs are listed in Table 1.2.2. The reservoir classifications for which each method is applicable are also shown. Often, more than one method is applicable to a given reservoir. For example, in a reservoir with permeability pockets, each pocket may behave differently. Or in a tilted reservoir with a rising water oil contact, both horizontal and vertical flow models may be required.

To develop an analytical approach for a given reservoir, it is convenient to begin with the drive mechanism.

Undersaturated oil reservoirs

Undersaturated oil reservoirs are usually the most straightforward oil reservoirs to analyze. The material balance is simplified to an expansion drive at pressures above

Table 1.2.2 Applicability of analytical methods to black oil reservoir classifications

Method	Applicability
Inflow equations	All
Decline analysis	All
Diagnostic plots	Waterfloods and water drives
Inflow coupled with material balance	Nonsegregated solution gas drive Segregated solution gas drive before breakthrough Gas cap drive before breakthrough
Buckley–Leverett[a]	Horizontal flow in waterflood of homogeneous reservoir
Stiles/Dykstra Parsons[a]	Horizontal flow in waterflood of layered reservoir
Coning models	Vertical flow in water drive, gas cap drive, segregated solution gas drive, and combination drive
Contact movement models[a]	Water drive, gas cap drive, segregated solution gas drive, and combination drive
Aquifer models	Water drive

[a]Displacement models.

the bubble point and solution gas drive when the reservoir pressure is below the bubble point. Both forms provide an accurate estimate of the initial oil volume as long as accurate pressure and production volumes are available.

Because there are no initial fluid contacts, production and injection rates can be modeled with radial or linear inflow equations for any reservoir architecture or permeability distribution. Coning models would only be required if evolved solution gas forms a secondary gas cap. Secondary gas cap formation is unlikely in table-top formations because the gas will flow along the top of the reservoir to the producing wells. However, a secondary gas cap may form in tilted, anticlinal, or dome architectures in which there is enough structure relief for the gas to segregate. If a gas cap forms, coning is only likely to be a significant issue in the latter two categories.

There are analytical models to predict production from homogeneous, high permeability undersaturated reservoirs. For example, there are methods based on material balance described in Dake (1978) and other texts. However, in most cases, production rates are extrapolated using decline analysis. Decline analysis generally applies well to undersaturated oil reservoirs, but the permeability distribution should be taken into account. For example, different adaptations of decline analysis are used for predicting the outcome of infill drilling in low permeability versus high permeability reservoirs. Decline analysis may not be suited to low permeability reservoirs, which take a long time to reach a steady-state condition.

Gas cap reservoirs

The material balance for a gas cap reservoir is dominated by the gas expansion term. If the gas cap is small, it is possible to obtain an accurate estimate of the initial volumes of oil and gas in the reservoir. However, if the gas cap is large, it is usually only possible to determine the volume of gas.

The best approach to modeling production rates depends on the reservoir architecture, oil zone thickness, and permeability distribution. If the oil zone is thick and permeable, production rates can be modeled with radial or linear inflow equations. The lower the thickness and the lower the permeability, the more likely coning will occur. The probability that coning will occur at a given well is dictated by the reservoir architecture and permeability distribution near the well.

As with undersaturated oil reservoirs, production rates can be extrapolated using decline analysis. Different extrapolations will be required for the radial inflow and coning regimes. Decline analysis for coning is less accurate than for radial inflow, because coning is a much less stable process.

Water drive reservoirs

The approach for water drive reservoirs depends on the strength of the aquifer. Weak water drive reservoirs can be treated similarly to undersaturated oil reservoirs, except coning models may be required near the water-oil contact.

For strong water drives, the material balance is dominated by the aquifer influx. It is usually possible to determine the volume of water influx and only rarely the initial volume of oil. Waterfloods are a form of strong water drive. If a waterflood is implemented correctly, there will be no change in reservoir pressure as the oil is produced, and a material balance cannot be performed.

Typically, water drives are subdivided into bottom water drives (vertical displacement of oil) and edge water drive or pattern waterfloods (horizontal displacement of oil). As with gas cap reservoirs, the best approach to modeling production rates depends on the reservoir architecture, oil zone thickness, and permeability distribution. Decline analysis and a variety of diagnostic plots can be applied to water drive reservoirs. There are also a number of analytical methods for pattern waterfloods. The permeability distribution is a critical parameter for many of these analytical methods.

Gas cap and water drive reservoirs

The approach for mixed drive reservoirs again depends on the strength of the aquifer. If the water drive is weak, the reservoirs can be treated similarly to gas cap reservoirs, except coning models may be required near the water-oil contact.

For strong water drives, the material balance is dominated by the aquifer influx. It is usually possible to determine the volume of water influx, but only rarely the initial volume of gas, and almost never the initial volume of oil. Production rates are modeled as with gas cap and water drive reservoirs. Decline analysis can be applied, but with even less certainty than for gas cap or water drive reservoirs. Waterfloods can be implemented on reservoirs with a gas cap depending on the reservoir architecture. The analytical waterflood models do not include a gas cap, and some approximations are required.

The reservoir classification also guides the choice of development strategy. For example, horizontal wells may be selected for a gas cap reservoir with a high potential for coning. Horizontal wells can achieve high production volumes at relatively high wellbore pressures and therefore minimize coning. Infill drilling may be selected for a low permeability reservoir. With low reservoir permeability, infill drilling can improve oil recovery and accelerate production. A waterflood is most likely to give a rapid increase in production for a homogeneous, high permeability, undersaturated oil reservoir. To determine the best strategy, the reservoir engineer must choose the appropriate analytical methods to forecast and compare production rates and recoverable reserves for different development cases.

1.3 General Workflow for Reservoir Characterization

There is no complete recipe for the tasks required to characterize all reservoir types. Often the workflow is dictated by the availability of data, the anticipated size of the resource, and characteristics of the reservoir itself. Sometimes the workflow pathway will seem oblique. This is sometimes troubling for an inexperienced engineer or geoscientist (Kay, 2011).

1.3.1 The Conceptual Model—From the Simple to the Complex

During initial stages of examining a reservoir, we often are interested in pursuing exploratory analysis (Jensen et al., 2000) Rather than imposing our preconceived notions or models, we want to gain insight into the nature of the reservoir. To do this the required data are:

- The reservoir structure and elevation of the fluid contacts (structural reservoir architecture "model");
- The porosity and permeability distribution of the reservoir (reservoir architecture permeability "model");
- Fluid properties (density, viscosity, PVT relationships) productivity indices and/or pressure transient data to determine effective large scale permeability;
- Production and injection rates versus time;
- Reservoir pressure versus time.

The exact spatial distribution of these reservoir parameters (first two items above) are never known exactly because of a lack of spatial sampling (cores and logs sample one ten-billionth of reservoir is sampled). However, as the number of wells, number and types of tests, and time-dependent dynamic data become available, it is often possible to better estimate the size and geometry of the reservoir. It is important to get ranges for important variables, especially for large scale parameters.

The need to integrate and iterate with the dynamic and static (well log) data to populate the geological model is illustrated by Figure 1.2.10, which shows where a hypothetical well has penetrated a number of porous intervals. The log data alone is consistent with a number of geological models. However, with production and reservoir pressure data, we may see large or small differences in rates and pressures, which will cause us to select a particular reservoir architecture type as described in

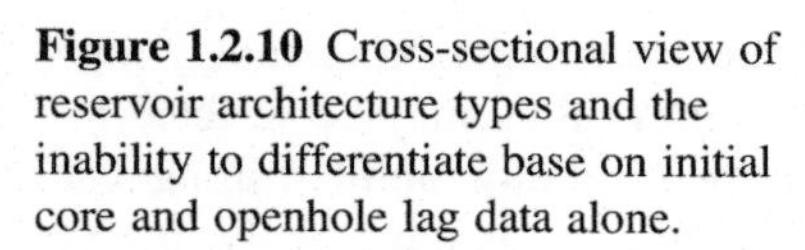

Figure 1.2.10 Cross-sectional view of reservoir architecture types and the inability to differentiate base on initial core and openhole lag data alone.

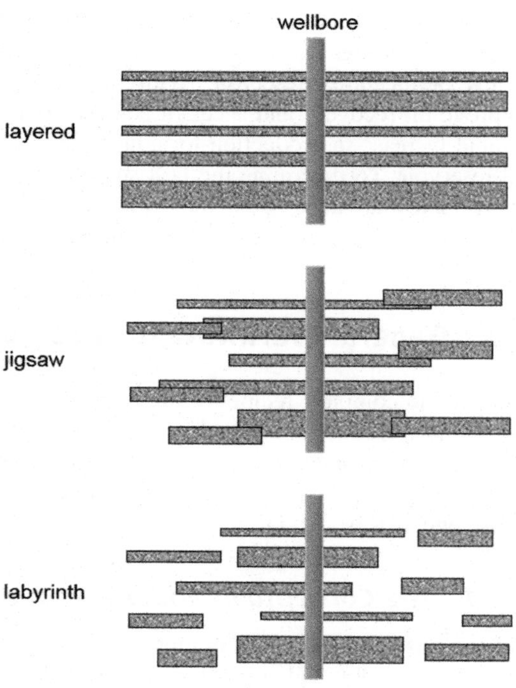

Section 1.2.3. The success of infill drilling and fluid injection projects will be a strong function of the reservoir architecture between wells, so it is critical that we categorize the reservoir architecture type.

To achieve that objective, it is best to start out with approximate simple methods, then move toward more complex methods when data ranges are refined. In practice, there is a critical need to analyze initially the dynamic data with simple decline, analytical and analogous reservoir models. Although the simple models may not be as comprehensive as a flow simulation, the presimulation checks have far fewer unknowns and are great for developing good conceptual models and communicating to nonreservoir engineers about important concepts. In reservoir performance, there are usually a large number of potential parameters that may control the process. In our experience, only a few parameters control the process. Analytical models and simpler models usually help us determine those critical parameters (Dake, 1994).

The most complex and usually final stage of modeling is reservoir simulation. It is sometimes argued that simulation flow models, now widely available, can easily forecast reservoir behavior. While this assertion is partially true, it misses the point. Simulation models *implicitly assume* that we have captured all the relevant physical phenomena and correctly know the spatial distribution of properties. In practical reservoir engineering terms, especially early in the life of a pool, this is a tenuous assumption. Because our log and core sampling is so sparse, it is highly unlikely the initial estimates of parameters are correct.

Even with history matching 30–50 years of dynamic data such as production rates and reservoir pressure, most flow simulation models will still have non-unique solutions. The exercise of history matching individual wells and matching a multitude of variables will usually narrow the range of parameters considerably. For example, by including not only cumulative oil and water or oil and water rates but also reservoir pressure and flowing bottomhole pressures, we can narrow the range of uncertainty substantially. With remote sensing such as four-dimensional seismic, tilt meters, microseismic, and resistivity mapping, it now is possible to narrow the uncertainly to a point at which business decisions can be made.

However, "outside" factors, such as near wellbore phenomena (waxes, particulates, and fractures) and/or unknown reservoir parameters away from the wellbore, may significantly affect reserves from a pool. Therefore it is critical to develop a conceptual model or do "brain simulation" using analytical methods, analogous pools, material balance, or empirical methods first.

For large reservoirs, most reservoir work is aimed at flow simulation, and the processes are shown in Figures 1.3.1 and 1.3.2. The general work flow is to build a static (no flow) model from geological, petrophysical, and geophysical data. The static model is fed into a dynamic model, usually with a reduction in the refinement of the reservoir property distributions. Then, parameters with the greatest uncertainty are adjusted iteratively to history match the dynamic production and pressure data. This approach is termed the shared earth model.

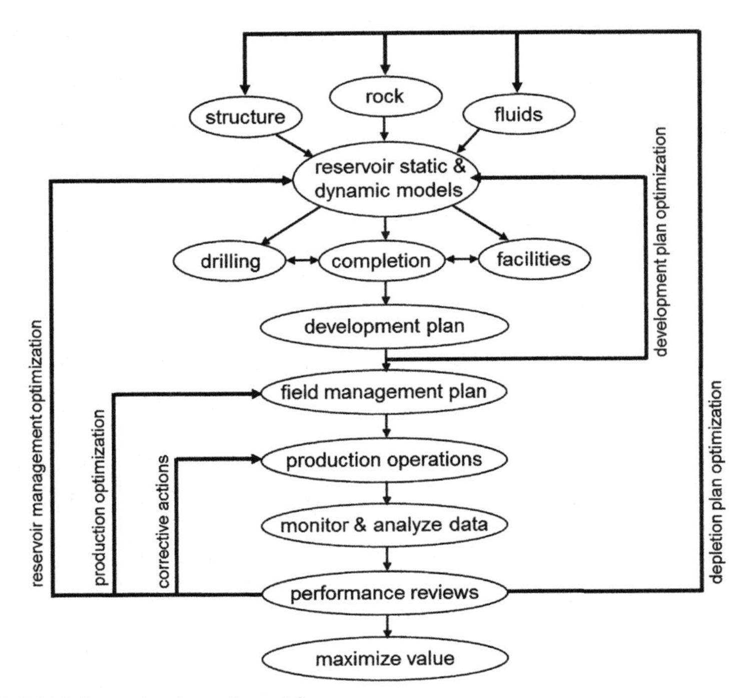

Figure 1.3.1 Schematic view of workflow.

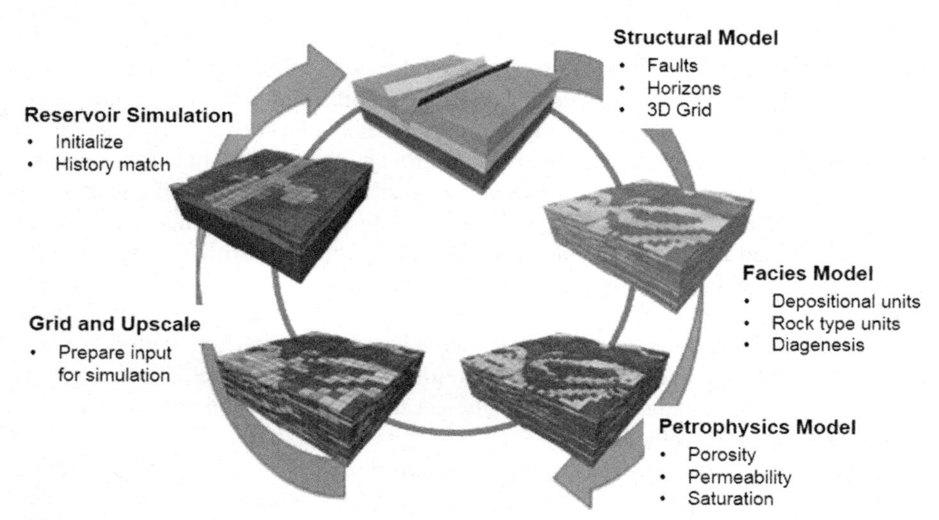

Figure 1.3.2 Schematic view of high level matrix modeling workflow.

The shared earth model concept originated in the early 1990s (Gawith and Gutteridge, 1999). This type of modelling approach had the potential to integrate the hitherto separate views of a reservoir held by reservoir engineers and by geoscientists. It has since been widely taken up by some oil companies and developed further by the industry. However, the shared earth model has substantial drawbacks. Updating the models can be slow, and there is a lot of inertia in the system, making substantial changes difficult to implement. Because the workflow and tasks are in series, it is extremely common to have long delays in project timing.

1.3.2 Recommended Workflow

We strongly recommend a parallel approach in which large flow simulation models are supplemented by simpler flow simulation models and analytical, decline, or material balance models. Saleri coined the term pseudo-models. These pseudo-models test for future problems and give preliminary results such as the range of recovery factor (Saleri, 1998). These initial estimates "clear the way" and give ranges for parameters and targets for more complex models.

Therefore, while in many cases the ultimate goal is to construct the geological and reservoir flow models, we recommend using simpler models as range finders. Also, because our best estimates are not totally correct, we recommend updating both the static and dynamic models regularly. Quality control, the review and screening of the input parameters to the models and forecast, is an important part of the workflow. A typical example is the inspection of core data and the removal of outliers (caused by drilling-induced fractures) from porosity versus permeability plots. Analytical material balance and analytical models are constructed and updated at the same time to support the decision-making process.

We recommend a fit for purpose approach or workflow. Figure 1.3.3 is a workflow for developing an initial conceptual model. This initial mental model becomes the first step of the workflow to characterize a reservoir and select an appropriate analytical model (Figure 1.3.4). The workflow for reservoir simulation proceeds in parallel (Figure 1.3.5). The conceptual model and analytical model guides the initialization and history matching steps. Note that the workflows have a large number of iterative steps, but these are not included in the diagrams for the sake of clarity.

It is important to understand the scale of the data when we are analyzing it. To quote:

> *Reservoir description is the combined effort of discretizing the reservoir into subunits, such as layers and gridblocks, and assigning values of all pertinent physical properties to these blocks. For this task, data from several scales and sources are available, each with a given accuracy and each with a measurement averaging over a different volume (both in terms of size and location) of rock.... This problem ... poses a paramount challenge in arriving at a realistic reservoir description ....*
>
> *Haldorsen (1986).*

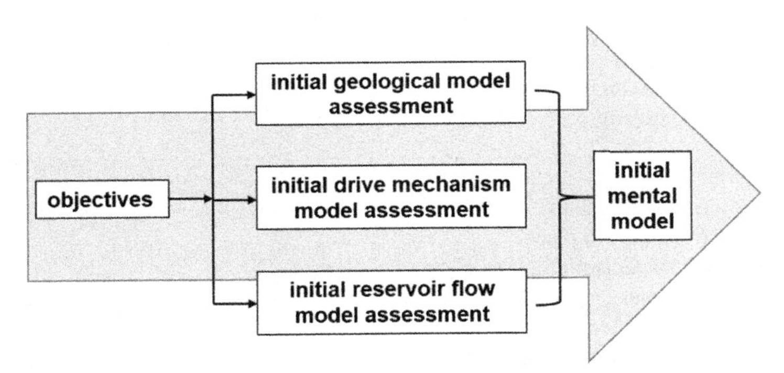

Figure 1.3.3 View of workflow for developing conceptual model part 1: initial assessment.

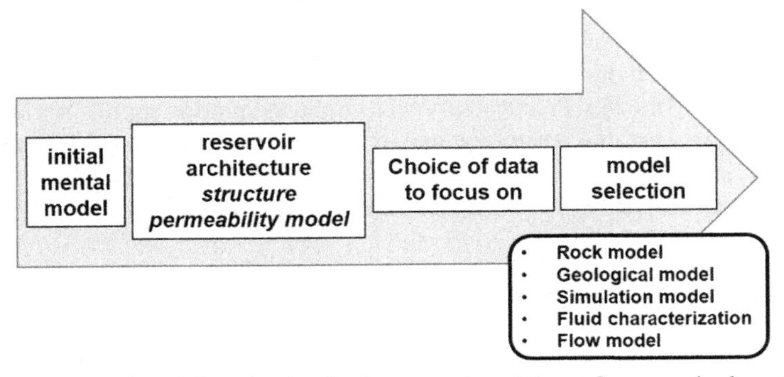

Figure 1.3.4 View of workflow for developing mental model part 2: reservoir characterization and analytical model.

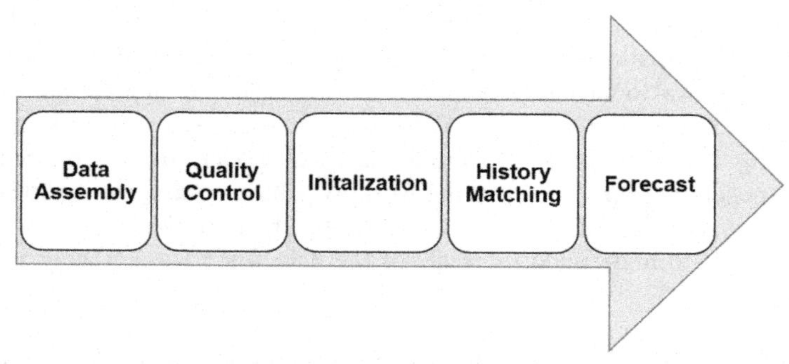

Figure 1.3.5 View of workflow for developing mental model part 3: reservoir simulation.

Real reservoir problems have to be tackled with good laboratory data, but also larger-scale data such as production and pressure (dynamic) data, geological information, and seismic data.

As we will see from the following chapters, there is a large amount of effort, tools/methods, and data needed to classify and optimize reservoir production. There are also a number of questions to address systematically to optimize production. The following questions can be considered as a checklist:

1. What is the reservoir architecture?
   a. Permeability/porosity
   b. Structural
2. What is the fluid type?
3. What is the drive mechanism?
4. What is the original oil in place?
5. What is the current remaining oil in place?
6. Where is the remaining oil in place?
7. How is the oil distributed and what are the pressure and saturation conditions?
8. What do we need to do to optimize recovery?
9. Can oil rates and reserves be economically increased?

Generally, it is recommended to use this checklist to guide the reservoir engineer through the characterization work. We often do not need perfect answers to the questions above, but we need order of magnitude answers, initially $\pm 30\%$, then, with time and data, we can converge to a tighter tolerance, $\pm 10\%$. There is a constant need to iterate on the conceptual (mental) models. Practical surveillance and monitoring approaches are needed for the iteration process. It is a common trait of humans to overestimate their knowledge, and this is also true of reservoir characterization (Chabris and Simons, 2011; Capen, 1976). Unfortunately, this trait sometimes blinds us to alternative explanations.

1.4 Approach and Purpose of This Book

1.4.1 Approach to Reservoir Engineering

As discussed, the two major functions of reservoir engineering are to determine the volume of oil, gas, and water in a reservoir and to predict and optimize production rates and recoverable reserves. To do so, the reservoir engineer must construct a sufficiently accurate (fit-for-purpose) representation of the reservoir, including the reservoir structure, rock properties, and fluid saturations. Depending on the situation, a detailed zero-, one-, two-, or three-dimensional reservoir characterization may be required. Then the engineer can classify the reservoir and apply the appropriate analytical methods.

The challenge facing reservoir engineers is that reservoirs are buried underground, and there is little information available about the reservoir. A large part of the art and science of reservoir engineering is to combine the available fragments of data to construct a model of the reservoir.

Reservoir analysis can be divided into the following phases:

- *Data gathering and reconciliation*: The well logs, core studies, fluid studies, pressure data, and production and injection histories are collected and analyzed. Whenever possible the dataset must be checked for internal consistencies. For example, does the porosity determined from well logs match the core porosity?
- *Reservoir mapping, classification, and material balance*: The structure, thickness, porosity, and permeability of the reservoir are mapped. After mapping, the elevation of the fluid contacts and the volumetric reserves are determined. The reservoir is classified according to fluid type, drive mechanism, and architecture. A material balance is performed if sufficient pressure data are available. The material balance and volumetric reserves must also be reconciled.
- *Diagnostics and forecasting*: The appropriate analytical models are selected. If appropriate, the reservoir history is reviewed using diagnostic plots. Forecasts are made using decline analysis and the appropriate analytical models.
- *Reservoir simulation*: In some cases, analytical models are not adequate to obtain an accurate forecast and a numerical simulation of the reservoir is performed.

Data gathering, mapping, classification, and material balance are all part of reservoir characterization. Reservoir characterization is a vital step in reservoir analysis, because it is not possible to make an accurate forecast based on an inaccurate reservoir model. This argument applies to both analytical modeling and reservoir simulation.

The relative merits of analytical forecasts and reservoir simulation can be debated. Analytical models can be quickly implemented, but often have limited application and cannot represent complex reservoirs. Reservoir simulation is well suited to complex reservoirs, but is time-consuming. It is also easy to fall into the trap of treating the simulator as a black box; that is, inputting reservoir data without adequately checking the data and considering the probable performance of the reservoir. With the advent of portable computers, reservoir simulation has become an almost routine method. However, analytical diagnostic and forecasting methods remain a vital part of reservoir analysis. These methods can be used directly for many reservoirs and as a tool to qualitatively interpret a complex reservoir prior to simulation.

1.4.2 Purpose and Organization of Book

The purpose of the book is to be a guide to reservoir engineers and a resource for asset team members. The data and techniques required to complete most reservoir engineering tasks are reviewed and illustrated through actual reservoir examples and case studies. The aim is to illustrate a systematic approach to characterizing, analyzing, and predicting production from oil reservoirs using analytical methods.

The principles of reservoir engineering are briefly reviewed in Chapter 2. This review is provided for reference and is intended to show the basis of the equations and methods used to analyze reservoirs without considering their application. Chapter 3 involves reservoir calculations for volumetrics, material balance, steady state flow, and transient flow. Chapter 4 discusses how extract information about reservoirs and how to understand production data. The analysis of the various data sources and the use of correlations are discussed. Chapter 5 describes how to use PVT data and how reconcile the PVT data. Chapter 6 describes how to use pressure transient analysis. Chapter 7 describes how to use routine core data for reservoir characterization work. Chapter 8 shows how to use special core analysis. Chapter 9 discusses how to extract information from well logs.

Chapter 10 discusses how to reconcile sometimes contradictory data from cores, well logs, and pressure transient analysis. Also discussed are permeability, flow diagnostics, initialization of the reservoir, mapping, identifying outer boundaries of reservoirs, classifying the reservoir, and constructing a three-dimensional representation of the reservoir. Chapter 11 deals further with how to classify reservoirs, discusses workflows, and provides real field examples of integrating techniques.

This book is intended for both junior and experienced reservoir engineers. Chapters 2–9 are intended as a reference guide for specific data and methods and can be used by any level of engineer or geoscientist. Chapters 10 and 11 will be most useful for experienced geoscientists and novice engineers.

There are a number of outstanding books that cover the basics of reservoir engineering. This book is different in that it stresses three components:

- Using real field data;
- Starting off with simple models, and then building complexity as needed;
- Integrating data to construct a conceptual model.

Often field data are noisy and sometimes have significant measurement errors. An important role of the reservoir engineer is to screen, filter, and integrate that data. In this book, we have for the most part used real unfiltered raw data. Engineers and geoscientists often select a flow simulation as being the most rigorous method but our experience is that simpler models are needed before complex models are constructed. In fact, it is critical to simulate in the brain and build a mental model before going to complex model to understand the basics physics of the problem correctly. Finally, integrating all of the reservoir data, however patchy, is vital to constrain any model to a realistic set of utions.

Part One

Basic Reservoir Engineering Principles

Part 1 is intended as a brief review of the basics of reservoir engineering. It does not deal with practical application but covers the fundamental background required to start analyzing, characterizing, and developing real reservoirs. Chapter 2 focuses on reservoir properties including rock properties (porosity and permeability), reservoir rock-fluid interactions (capillary pressure and relative permeability), and fluid properties. Chapter 3 focuses on basic reservoir engineering calculations including reservoir volumetrics, material balances, and steady state and transient flow through porous media.

The material for Part 1 was drawn from many sources. Sources used for specific issues are cited in the text. More general sources are: Ahmed (2000), Amyx, Bass, and Whiting (1960), Chapman (1983), Craft and Hawkins (1991), Dake (1978), McCain (1990), Selley (1998), Slider (1983), and Towler (2002).

Rock and Fluid Properties

2

Chapter Outline

Reservoirs and reservoir properties are created and shaped through geological processes. Petroleum geology is an important tool for understanding both the architecture and the properties of a reservoir and is the starting point for reservoir engineering.

2.1 Petroleum Geology

Petroleum forms from organic matter that was buried and heated at high pressure over millions of years. At sufficient burial depth, the organic material undergoes thermal cracking to produce petroleum. As shown in Table 2.1.1, there is a window of burial depth that results in the formation of oil. If the burial depth is too low, no cracking occurs. Greater burial depth leads to the formation of gas.

In some instances, geological forces can expose an oil-bearing formation to the surface or to within a few hundred meters of the surface. Microbial reactions can consume the lighter fractions of the crude oil transforming it into a heavy oil or bitumen. The operational definition of conventional oils, heavy oils, and bitumens is based on their density and viscosity, Table 2.1.2.

Petroleum is found in the pore space of rock. The rock in which petroleum is created is termed the source rock. The petroleum almost always migrates upward from the

Practical Reservoir Engineering and Characterization. http://dx.doi.org/10.1016/B978-0-12-801811-8.00002-X

Table 2.1.1 Effect of burial depth on petroleum product

Maturity of source organic matter	Depth (m)	Temperature (°C)	Hydrocarbon product
Immature	0		Biogenic gas
	1000	60	
	Cracking begins		
Mature	1000	60	Heavy oils with gas medium to light oils
	5000	300	Condensate and/or gas
	Degeneration		
Metamorphosed	5000	300	Dry gas
	9000	500	
	Barren		

Table 2.1.2 Unitar definition of crude oils (Gray, 1994)

Classification	Viscosity (mPa s)	Density (kg/m^3)	API gravity
Conventional oil	<100	<934	>20
Heavy oil	100–100,000	934–1000	20–10
Bitumen	>100,000	>1000	<10

source rock until it either reaches the surface or is trapped, creating a petroleum reservoir. An exception is shale oil or gas, in which the source rock is the reservoir. Petroleum geology is concerned with the types of rock and structures in which the petroleum is trapped.

Rock is formed from surface deposits of solid material such as sand, clay, lime, and salt. If the deposits accumulate and are buried to a sufficient depth and for a sufficient time, pressure will consolidate the deposit into rock. Accumulations of deposit are created by the action of wind and water on the surface of the earth. Exposed surfaces are eroded, and the eroded matter is deposited elsewhere. The deposits may be redistributed by later erosion or may lie undisturbed. Undisturbed deposits are gradually buried as more material is laid down above them. The layers of buried deposits are termed sedimentary layers. Areas in which greater accumulation occurs (and thicker sedimentary layers build up) than surrounding areas are termed sedimentary basins. Most significant petroleum deposits are found in sedimentary basins.

Most sediment accumulates at the margin between land and sea. Some examples of such accumulation are delta sand, barrier bars, and mud. Other accumulations found in basins are reefs and lime particles. The reservoir types associated with various accumulations are listed in Table 2.1.3.

Table 2.1.3 **Reservoir deposition types**

1. Clastics (sandstones)
   - **a.** Subaerial—sand dunes and other surface sand deposits
   - **b.** Fluvial—river beds (channels or point bars)
   - **c.** Deltaic—river deltas (channels and mouth bars)
   - **d.** Shoreline—beach sand, barrier bars and tidal channels
   - **e.** Marine—offshore bars, sheet sands, deep sea fans
2. Carbonates (limestone and dolomite)
   - **a.** Lime mud—aragonite and calcite from algae and skeletal debris
   - **b.** Coarse lime particles—skeletal material and grain stones
   - **c.** Carbonate masses—coral reefs
3. Other deposition types
   - **a.** Shale/mudstone—mud
   - **b.** Anhydrite evaporite—salt flats
   - **c.** Coal bed—organic matter

Most productive reservoirs are clastic or carbonate deposits. Shales and evaporites may contain petroleum, but until recently were considered nonproductive. Today, improvements in hydraulic fracturing and multilateral well technology have made shale oil and gas production economically viable. Coal bed methane is another recently developed source of natural gas.

The position of many accumulations depends on their position relative to the continental margin. The position of the continental margin is controlled by the rates of sediment deposition and sea level change. If sediment deposition occurs faster than the sea level rises, the shoreline regresses, that is, migrates seaward. If sea level rise exceeds deposition, then the shoreline transgresses, that is, moves landward. Hence, as the shoreline moves back and forth across one location, layers of different kinds of sediment build up, as shown in Figure 2.1.1. Similarly, carbonate deposits are found near shorelines, and the location of the lagoons and forereefs will shift with the movement of the shoreline. A major component of petroleum exploration is mapping the movement of ancient shorelines to identify likely locations for accumulations of deposits.

After burial, deposits are subject to tectonic forces and may be tilted, folded, fractured, and/or faulted. The type and shape of the network of buried deposits can form various traps for hydrocarbons that are rising from more deeply buried source rock. Traps are usually classified into three main categories: structural—a trap arising from the shape of a deformed deposit; stratigraphic—a trap arising from a change in rock type or porosity; or combination—structural traps enclosing a mixture of reservoir and nonreservoir rock.

The two main types of structural trap are anticlinal folds (Figure 2.1.2(a)) and faults (Figure 2.1.2(b)). There are many varieties of stratigraphic traps, including reefs (Figure 2.1.3(a)), sand bars (Figure 2.1.3(b)), sand lenses, and channel fillings (Figure 2.1.3(c)) as well as erosion and postdepositional chemical alterations. Usually erosion (Figure 2.1.4) and chemical alterations are associated with combination traps.

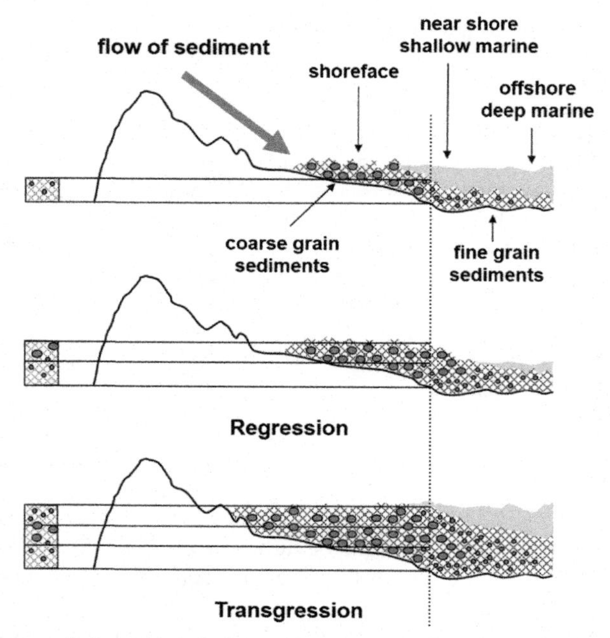

Figure 2.1.1 Accumulation of beach deposits of varying grain size during regression and transgression.

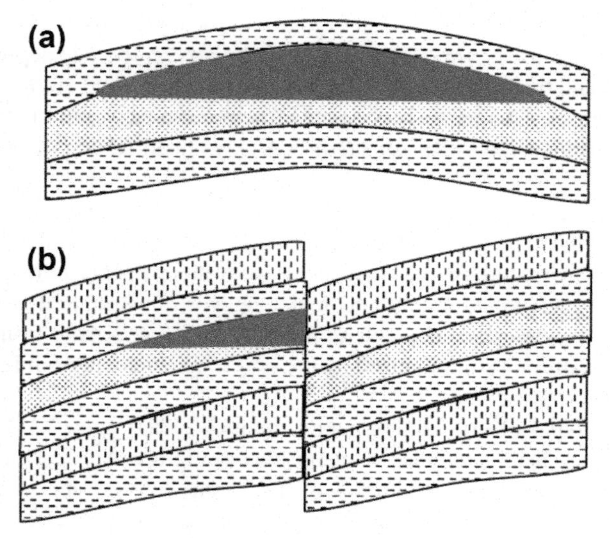

Figure 2.1.2 Examples of structural traps: (a) anticline, (b) fault.

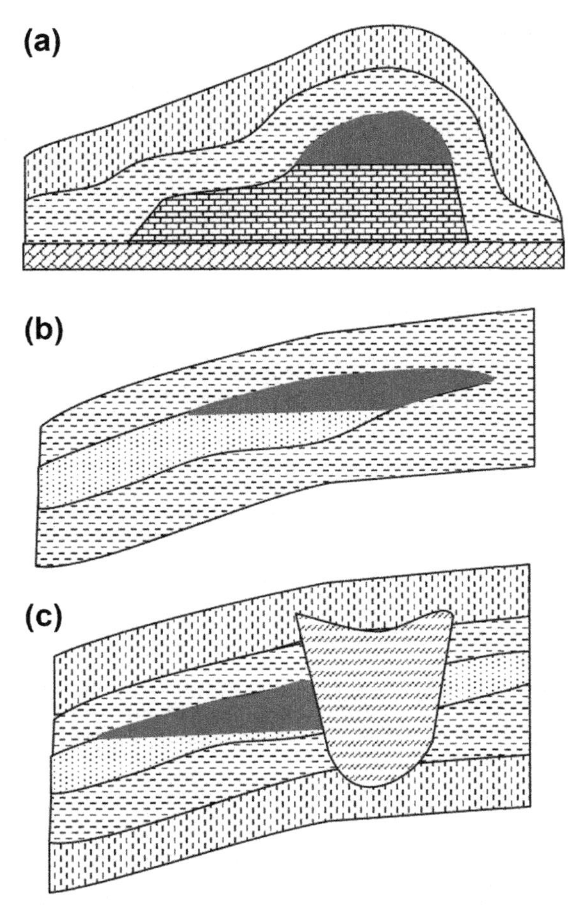

Figure 2.1.3 Examples of stratigraphic traps: (a) reef, (b) sand bar, (c) channel fill.

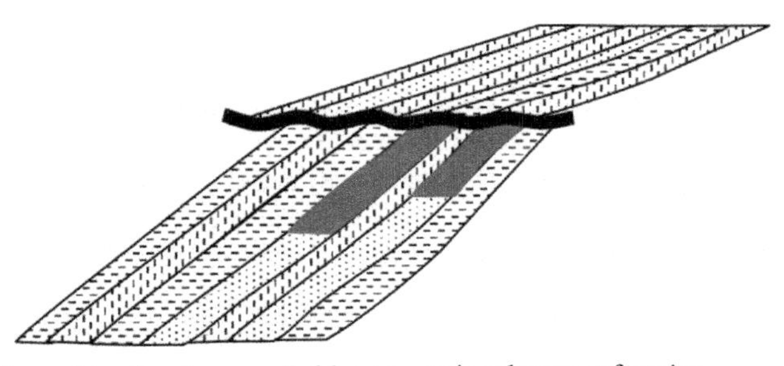

Figure 2.1.4 A combination trap created by an erosional nonconformity.

The existence of a trap is the first criterion for locating a productive petroleum reservoir. Many traps are filled only with water, and the second criterion is that hydrocarbons have migrated into and remain in the trap. Some reservoirs that once held trapped hydrocarbons have lost the mobile hydrocarbons due to subsequent faulting or structural changes, so that only residual hydrocarbon saturations remain. If hydrocarbons remain trapped, there may oil, gas, or both present in the reservoir. If both oil and gas are present, the gas separates under gravity forces into a layer over the oil. In this case, the bottom of the gas layer is termed the gas–oil contact (GOC) and is usually located at a constant elevation. In many reservoirs, a water layer or aquifer remains underneath the trapped hydrocarbons, and a water–oil contact (WOC) or water–gas contact is present. Note that while most WOCs are horizontal, hydrodynamic forces can result in a tilted contact. A schematic of the fluid distributions when all three fluids are present is given in Figure 2.1.5.

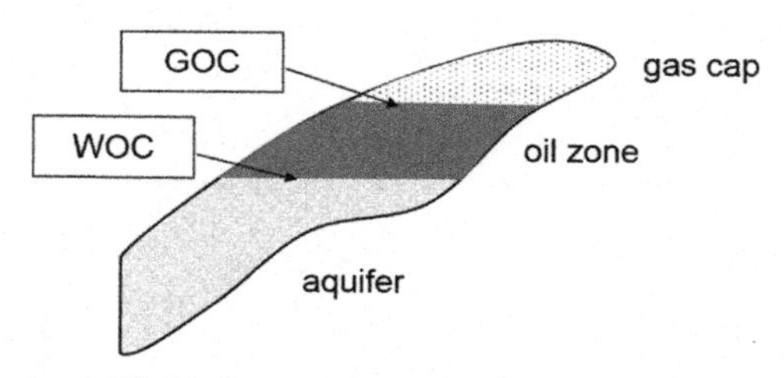

Figure 2.1.5 Distribution of fluids in a gas cap reservoir with an underlying aquifer.

2.2 Rock Properties

The third and final criterion for a productive reservoir is that it contains sufficient petroleum and is capable of sustaining sufficient production rates to be commercially viable. The amount of petroleum in a reservoir and the productivity of a reservoir depend on the shape of the reservoir and its rock properties, including porosity, permeability, fluid saturations, capillary pressure, and relative permeability. Each of these rock and rock–fluid properties are reviewed below.

2.2.1 Diagenesis and Porosity

When a clastic or carbonate is deposited, it is not totally solid. There are microscopic spaces, or pore spaces, between the sand grains or the carbonate crystals that comprise the rock, as shown in Figure 2.2.1. The ratio of the pore space of a rock to its total or bulk volume is defined as the porosity of a rock. The microscopic spaces between sand grains are termed intergranular porosity, and the spaces between carbonate crystals are termed intercrystalline porosity. The original porosity of a deposit is dictated by the size distribution and arrangement of the rock grains and is usually in the order of 40–50% for clastics and 60–80% for carbonates. The effect of the size distribution and sorting is illustrated in Figure 2.2.2, and some examples are provided in

Table 2.2.1. The porosity of a deposit is usually altered through post-burial processes. These changes are termed diagenesis. Sources of diagenesis are compaction, cementation, solution, and fracturing. The magnitude of these effects depends on the type of deposit.

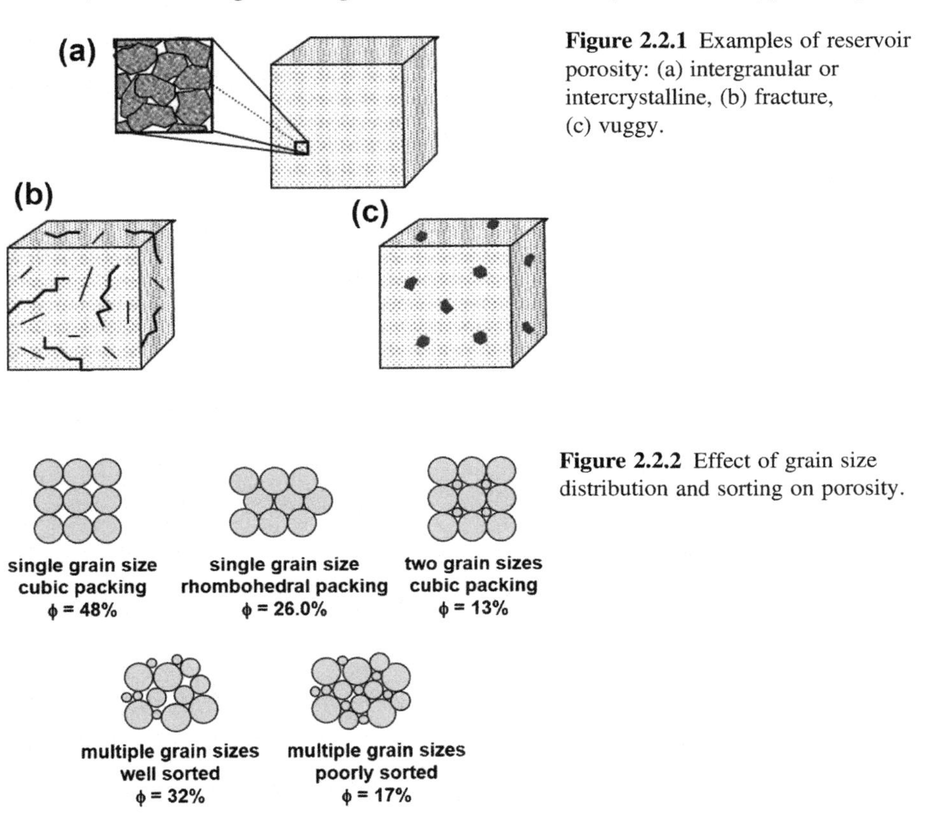

Figure 2.2.1 Examples of reservoir porosity: (a) intergranular or intercrystalline, (b) fracture, (c) vuggy.

Figure 2.2.2 Effect of grain size distribution and sorting on porosity.

Clastics

The effect of diagenesis on clastic porosity is shown in Figure 2.2.3. As the deposit is buried, it is compacted, and the original porosity is reduced to 10–30%. Over geological time scales, cementation may occur due to fines migration and the growth of minerals such as quartz, calcite, and clays in the pore space. Cementation further reduces porosity. In some cases, the cement is leached out of the rock, for example, by dissolution in acidic water. The porosity is then partially or almost wholly restored. The deposit may also fracture when deformed under compaction or tectonic forces. Fracture porosity is usually a small fraction of the total porosity.

Carbonates

The effect of diagenesis on carbonate porosity is usually more significant and more complex than on clastics because the carbonates are more reactive and soluble in water than is silica. Much of the original pore space of carbonates usually disappears after

Table 2.2.1 Typical porosity for different rock types and extent of consolidation

Rock type	Consolidation	Porosity (%)
Clay	Unconsolidated	45–55
Silt	Unconsolidated	40–50
Mixed medium and coarse sand	Unconsolidated	35–40
Mixed fine and medium sand	Unconsolidated	30–35
Gravel	Unconsolidated	30–40
Sandstone	Consolidated	10–30
Shale	Consolidated	1–10
Limestone	Consolidated	1–10
Dolomite	Consolidated	1–30

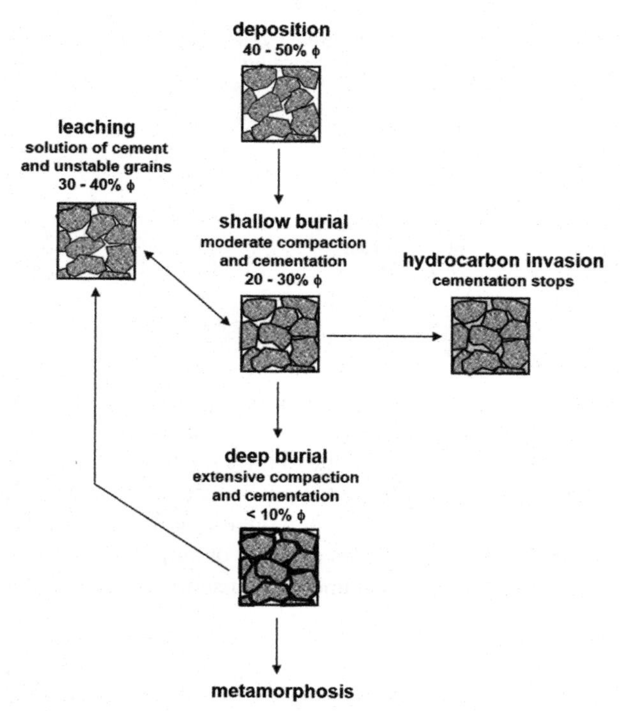

Figure 2.2.3 Schematic of diagenesis pathways in clastics. Adapted from Selley, 1998.

burial due to compaction and cementation. However, minerals are leached out of carbonates as water migrates through the formation. The process of leaching and recrystallization may occur several times before a carbonate deposit is buried in impermeable muds. The end result is a complex mixture of porosities, some of which are defined below:

- *Intercrystalline*: spaces between original or re-precipitated rock crystals
- *Moldic*: created by leaching of either rock grains or matrix
- *Fenestral*: gap in rock fabric greater than intergranular or intercrystalline porosity, usually created by dehydration shrinkage
- *Vuggy*: holes in rock greater than intergranular or intercrystalline porosity, created by dissolution
- *Oolitic*: holes created by dissolution of fossils

Carbonates may also fracture as a result of dehydration, compaction, or tectonic forces.

Carbonates are divided into two groups: limestones (calcite) and dolomites. Limestone reacts reversibly with magnesium to form dolomite:

$$2\,\underset{\text{calcite}}{CaCO_3} + Mg^{2+} \leftrightarrow \underset{\text{dolomite}}{CaCO_3MgCO_3} + Ca^{2+}$$

Limestones are found in reef deposits, lime sands, and lime muds. Primary dolomites are deposited as dolomite and occur in salt marsh sequences, lagoons, and evaporates. Secondary dolomites form in deposited limestones. When calcite reacts to form dolomite, the bulk volume is reduced by approximately 13%. Hence, porosity increases. Although the increased porosity may be reduced through diagenesis, secondary dolomites are often high-quality reservoirs.

Porosity definitions: As we have seen, there are a number of descriptive porosity definitions, such as vuggy and fracture porosity, shown in Figure 2.2.1. Some other commonly used porosity definitions are given below:

- *Primary porosity*: the intergranular or intercrystalline porosity of the original deposit after compaction
- *Secondary porosity*: the porosity developed through diagenesis, including fractures, intergranular and intercrystalline solution porosity, vugs, and caverns
- *Total porosity*: the sum of the primary and secondary porosity; note that not all of the total porosity is necessarily interconnected
- *Effective porosity*: the total interconnected porosity; only hydrocarbons within the effective porosity can be produced

2.2.2 Pore Structure and Permeability

The microscopic and macroscopic porosity forms a partially or wholly interconnected network of passages called the pore structure. The shape of the pore structure governs the flow capacity of a reservoir. Intuitively, we can imagine that fluid flows more readily through a straight large diameter pipe than through a narrow twisting pipe. The narrow pipe has more surface area per flow volume and therefore more drag at a given flow rate. The term permeability is defined as a measure of the flow capacity of a reservoir. Permeability is discussed more formally in Chapter 3. For now, it is sufficient to recognize that a reservoir with narrow twisting passages will have lower permeability than a reservoir with wider, straighter passages.

Figure 2.2.4 is a schematic of a hypothetical pore structure. Note that the pores are irregular in shape and that the connection between the pores, the pore throat, is often very small. In many case, it is the size of the pore throats that controls reservoir permeability. A major concern in production engineering is formation damage, which usually results from a blockage of the pore throats near the wellbore. The pore structure is also an important factor in the water saturation of a reservoir, relative permeability, and capillary pressure.

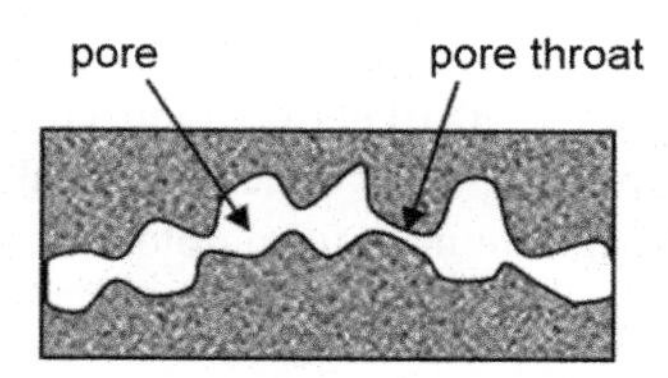

Figure 2.2.4 Schematic of pore structure.

The pore structure and permeability are a strong function of the depositional environment and post-depositional processes. As a result, in most cases, the reservoir permeability is not uniform. Some examples are listed below.

Regressing or transgressing shorelines: A receding (regressing) shoreline deposit may have small, uniform sand grains in the early lower deposits formed in deeper water. As the shoreline regressed, deposits at that location are formed in the wave-swept region and are coarser and more randomly sorted with higher permeability. The resulting rock has progressively coarser grains higher in the structure and is termed a coarsening upward sequence. The permeability is higher at the top of the structure than at the bottom. The opposite trend is observed for transgressing shorelines.

Directional deposition: Many fluvial deposits are affected by the direction of the river flow during deposition. The long axis of the sand grains tends to align in the direction of flow. This orientation is preserved when the deposit becomes rock. Permeability is usually higher along the length of the oriented sand grains (in the direction of the original river flow). Such a reservoir is said to have directional permeability.

Reefs: Carbonate reefs usually exhibit complex geology with several depositional environments such as forereefs, lagoons, and detrital zones. The forereef is the part of the reef exposed to waves and usually has high permeability. The lagoon is a calm region behind the forereef that usually has low permeability. The detrital zone is formed from the debris that collects at the base of the reef front. The detrital zone usually has low to moderate permeability. The location of each of these deposits moves as the shoreline and sea level change over time. Reef permeability is also significantly altered through diagenesis. As a result, the reef may have many pods of high permeability surrounded by less permeable rock.

Fractures: Fracturing can have a significant impact on permeability if a network of connected fractures occurs. The fractures act as flow conduits with a high permeability.

Fracture porosity can be small, 0.01–0.1%, but with vugs the porosity can rise to 2–4%. Many of these same systems will yield 100 mD to tens of Darcies. However, fractures do not necessarily occur uniformly throughout a deposit, and there can be significant local variations in permeability.

Even a brief review of petroleum geology shows that permeability can vary considerably in a given reservoir. However, the distribution of permeability in a reservoir cannot be accurately measured, as will be discussed in Part 2. At best, a small set of local and average permeabilities can be obtained. The estimation of the reservoir permeability distribution from these measurements is perhaps the most challenging aspect of reservoir engineering.

2.3 Rock–Fluid Interactions

The pore space of a reservoir may contain water, oil, and gas either in any combination. (Amyx, 1960) Surface interactions between the rocks and fluids and between the fluids themselves impose restrictions on the fraction of the pore space a given fluid can occupy (the fluid saturation). These interactions also affect the flow capacity of the reservoir to a given fluid; that is, effective permeability. To quantify these effects we begin with the concepts of interfacial and surface tension.

2.3.1 Interfacial Tension, Surface Tension, and Wettability

Consider an oil–water interface, as shown in Figure 2.3.1(a). The water molecules in the bulk aqueous phase are surrounded by other water molecules. Similarly, the oil molecules in the bulk oil phase are surrounded by other oil molecules. However, at the interface, the oil and water molecules are forced into contact with each other. Hence, the molecular interaction energy is higher at the interface. The excess energy per interfacial area is defined as the *interfacial tension*, σ. A similar situation occurs at a solid–liquid interface, as shown in Figure 2.3.1(b). In this case, the excess energy per surface area is defined as the *surface energy*. It is also customary to refer to the excess energy at the gas–oil interface as a *surface tension*.

Figure 2.3.2 shows two fluids (oil and water) in contact with a surface. The angle the fluid–fluid contact forms at the solid surface is defined as the *contact angle*, θ. The

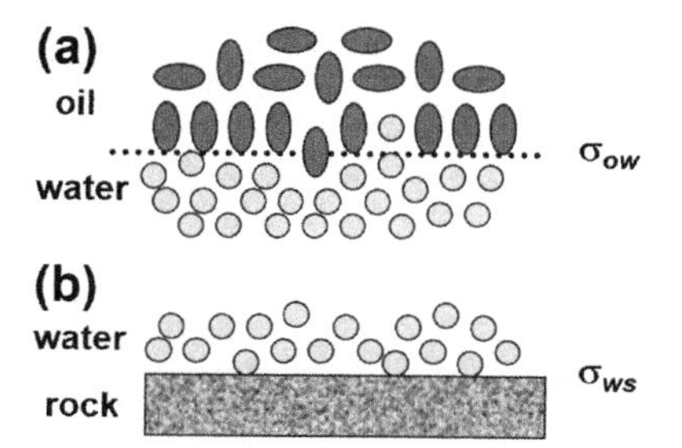

Figure 2.3.1 Schematic of (a) water–oil and (b) water–rock interfaces.

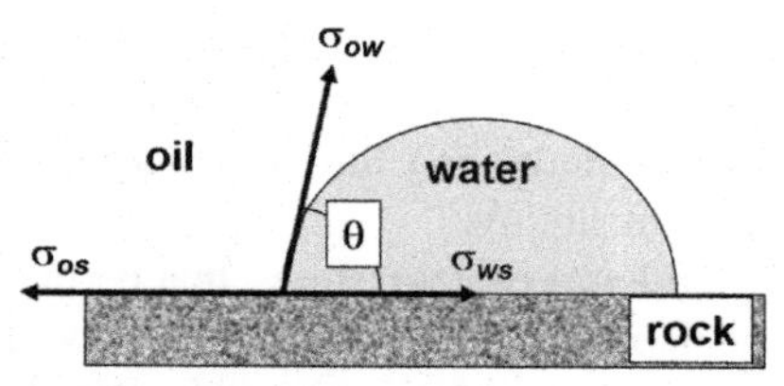

Figure 2.3.2 Contact angle and force balance at the three phase contact line.

contact angle depends on the relative magnitude of the interfacial and surface tensions. A force balance on the line of contact between the three phases gives:

$$\sigma_{os} = \sigma_{ws} + \sigma_{ow}\cos\theta \tag{2.3.1}$$

or

$$\cos\theta = \frac{\sigma_{os} - \sigma_{ws}}{\sigma_{ow}} \tag{2.3.2}$$

in which σ_{os} is the surface tension between the rock and the oil, σ_{ws} is the surface tension between the rock and the water, and σ_{ow} is the interfacial tension between the oil and water.

If the rock–oil surface tension is greater than the rock–water surface tension, the water will tend to spread on the rock to minimize the energy of the system, as shown in Figure 2.3.3. In this case, the contact angle will be less than 90°. The *wettability* of a rock is defined in terms of the contact angle, as follows (Treiber and Owens, 1972):

$0^{\circ} < \theta < 75^{\circ}$ water wet

$75^{\circ} < \theta < 105^{\circ}$ intermediate

$105^{\circ} < \theta < 180^{\circ}$ oil wet

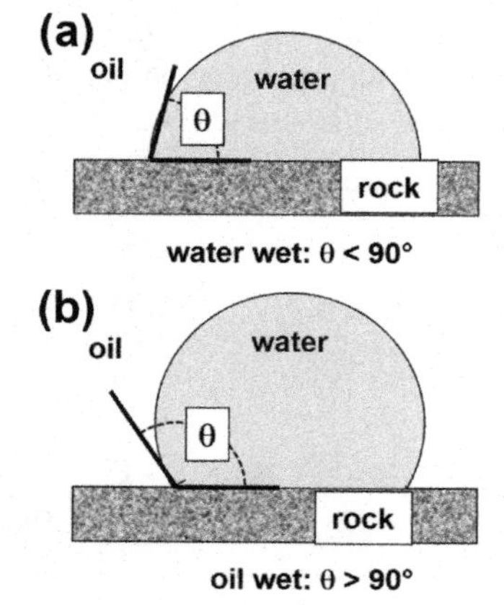

Figure 2.3.3 Schematic of contact angle and wettability (a) water wet: $\theta < 90°$ (b) oil wet: $\theta > 90°$.

The wettability of a reservoir rock has a strong impact on the distribution and movement of fluids within the rock.

2.3.2 Capillary Pressure

The pore structure of a reservoir can be considered as a network of capillaries of different length and diameter. One consequence of interfacial tension and contact angle is that water can be drawn above a WOC through a capillary. This capillary effect can alter the distribution of water in a reservoir.

To understand the capillary effect, first consider a droplet of water in oil, Figure 2.3.4(a). The interfacial energy of the droplet is the product of the interfacial tension and the surface area of the droplet, σA. The energy of the interface is minimized if the droplet shrinks. However, as the droplet shrinks, the internal pressure of the droplet increases and opposes the shrinkage. The equilibrium condition is the balance between the differential change in interfacial energy ($\sigma_{ow} dA$) versus the differential work of shrinkage ($\Delta P dV$):

$$\Delta P dV = \sigma_{ow} dA \tag{2.3.3}$$

in which ΔP is the change in the droplet pressure, dV is the change in drop volume, and dA is the change in interfacial area. For a spherical droplet, the change in internal pressure is given by:

$$\Delta P = \frac{2\sigma_{ow}}{R} \tag{2.3.4}$$

in which R is the radius of the droplet.

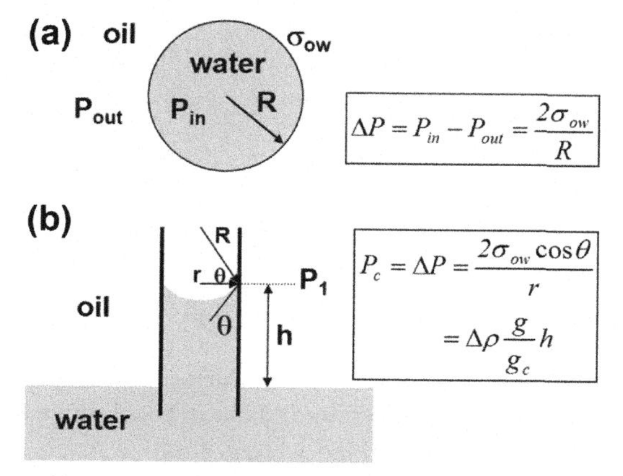

Figure 2.3.4 Effects of interfacial curvature: (a) internal pressure of droplet, (b) capillary pressure and capillary rise.

A similar argument applies to any curved interface between two fluids. Consider water rising in a capillary with oil above, Figure 2.3.4(b). In this case, the water–oil interface is a segment of a sphere with a geometry defined by the contact angle. If r is the radius of the capillary, then, the spherical radius, R, is given by:

$$R = r\cos\theta \tag{2.3.5}$$

This spherical radius expression is substituted into Eqn 2.3.4 to obtain:

$$P_c = \Delta P = \frac{2\sigma_{ow}\cos\theta}{r} \tag{2.3.6}$$

in which P_c is the capillary pressure. Note that the pressure difference across the interface is related to the contact angle and therefore the wettability of that surface. In this example, the contact angle is less than 90°, and the interface is concave downward. The pressure is always higher on the inside of the curved interface, and therefore the pressure is higher in the oil immediately above the interface than in the water immediately below the interface.

Equation 2.3.6 also shows that the capillary pressure increases as the radius of the capillary decreases (Amyx, 1960). Hence, capillary pressure tends to be more significant in narrow pores, that is, lower permeability rock.

Water–oil capillary pressure in water-wet rock

One important consequence of capillary pressure in water-wet rock ($\cos\theta$ is positive) is that water will rise in the capillary to balance hydrostatic forces. A pressure balance at the bulk water-oil interface gives:

$$P_1 - P_c + \rho_w \frac{g}{g_c} h = P_1 + \rho_o \frac{g}{g_c} h \tag{2.3.7}$$

in which P_1 is the pressure above the interface in the capillary, g is gravitational acceleration, g_c is the conversion factor between natural and defined force units, and h is the height of the water in the capillary. Equation 2.3.7 simplifies to:

$$P_c = \Delta\rho \frac{g}{g_c} h \tag{2.3.8}$$

Equating and rearranging Eqns 2.3.6 and 2.3.8 gives the relationship of the height of water in a capillary to the diameter of the capillary and the contact angle:

$$h = \frac{2g_c\sigma_{ow}\cos\theta}{\Delta\rho g r} \tag{2.3.9}$$

As mentioned previously, the reservoir pore structure resembles a network of capillaries of different shape and size. Therefore, there is not a single capillary pressure. Instead, capillary pressure is related to the water saturation in the rock, as shown in Figure 2.3.5. Imagine that a sample of water-wet reservoir rock is saturated with water; that is, the pore spaces are filled with water. Now surround the sample with oil and

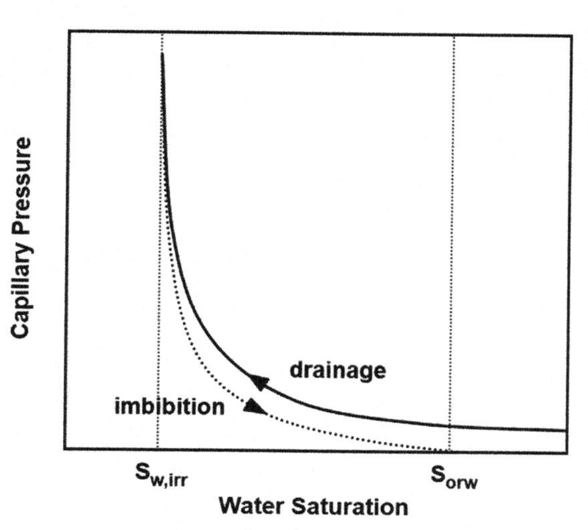

Figure 2.3.5 Typical water–oil capillary pressure drainage and imbibition curves for a water-wet rock.

apply pressure so that the oil displaces water from the sample. This displacement of the wetting phase (water) by the nonwetting phase (oil) is termed *drainage*. As the applied pressure increases, progressively more water is displaced, until a limit is reached and no more water can be displaced. The water saturation at which this limit is reached is the *irreducible water saturation*, $S_{w,irr}$.

Now imagine that the sample is surrounded by water and the pressure is decreased. As pressure decreases, the capillary pressure will draw more and more water into the rock, displacing the oil. The displacement of the nonwetting phase by the wetting phase is termed *imbibition*. When the applied pressure reaches zero, there will still be some oil in the reservoir. This oil saturation is the *residual oil saturation to water*, S_{orw}.

The capillary pressure curve provides several useful pieces of information, including the irreducible water saturation, the residual oil saturation, and the relationship between capillary pressure and water saturation. This relationship is used to determine the distribution of water saturations in the reservoir at the transition zone between the oil and water. The methodology for converting a capillary pressure curve to a water saturation versus depth plot is shown in Figure 2.3.6.

Note that, when there is a transition zone, the irreducible water saturation is not the same as the *initial water saturation*, S_{wi}. However, in many reservoirs, the transition zone is small and the two saturations are the same. In practice, the two terms are often used interchangeably. In this book, the symbol S_{wi} is used to represent both the irreducible and the initial water saturation unless otherwise stated.

Water–oil capillary pressure in oil-wet rock

In practice, it is rare to measure capillary pressures in oil-wet rock. After cleaning and preparation, most core samples are water-wet. Hence, the validity of capillary pressures for originally oil-wet rock is highly questionable. In oil-wet rock, the capillary

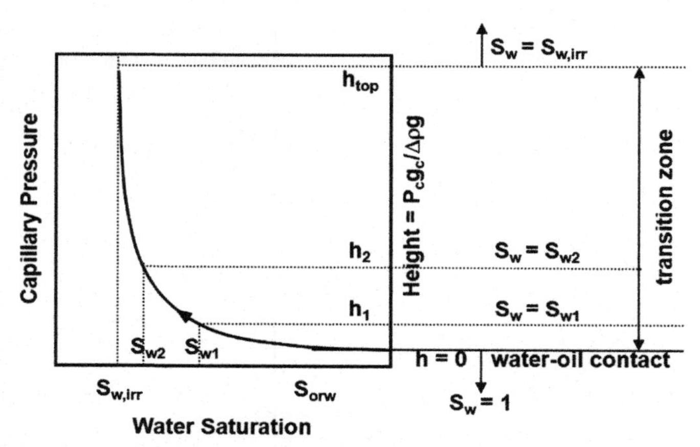

Figure 2.3.6 Methodology for calculating heights and saturations within transition zone between water and oil.

pressures are negative (cos(θ) is less than zero). A positive pressure is required to displace oil with water. A water-saturated sample imbibes oil. The absolute value of the capillary pressure versus water saturation curve has a similar shape to the water-wet capillary pressure curve.

Gas—oil capillary pressure

Gas—oil capillary pressure follows the same principles as water—oil capillary pressure. Gas is always the nonwetting phase. Typically, a sample is saturated with water, then the water is displaced with oil to the irreducible water saturation. Then the oil is displaced with gas (drainage) to the residual oil saturation to gas. The capillary pressure is usually plotted versus gas saturation and has a similar shape to the water—oil capillary pressure curve.

Gas—oil capillary pressure may be of similar magnitude to the water—oil capillary pressure. However, the height of the transition zone is small because the density difference between the gas and oil is so great (Eqn 2.3.10). Therefore, gas—oil capillary pressures are not much used by reservoir engineers.

2.3.3 Relative Permeability

The *absolute permeability* of a porous medium such as reservoir rock is the permeability when the medium is saturated with a single one-phase fluid and only that fluid is flowing through the medium. If a second fluid is present, the permeability will be less than the absolute permeability. A pore containing a single fluid is compared with a pore containing two fluids in Figure 2.3.7. Note that the cross-sectional area for flow for each fluid must be less than the cross-sectional area for flow if only one fluid was present. Also, the presence of a fluid—fluid interface increases the total surface area exposed to fluid flow. Therefore the drag is increased for any given volumetric flow rate, and the effective permeability is decreases. The blockage of pore throats by one fluid further reduces the effective permeability.

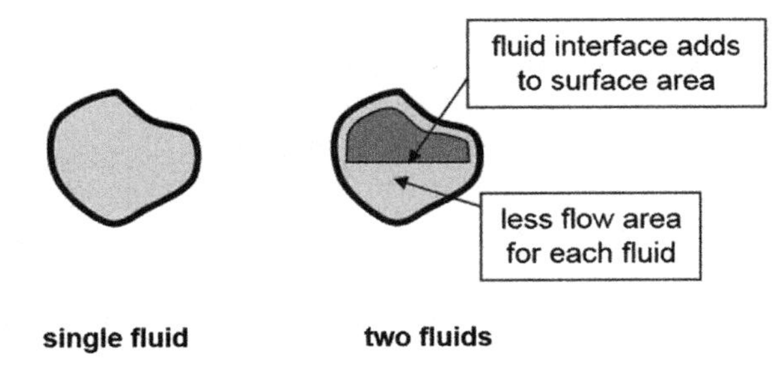

Figure 2.3.7 Cross-section of pore showing effect of adding a fluid on the cross-sectional and surface area for flow through the pore.

To account for the effect of multiple fluids, *relative permeabilities* are defined as follows:

$$\text{oil: } k_{ro} = \frac{k_o}{k} \quad \text{gas: } k_{rg} = \frac{k_g}{k} \quad \text{water: } k_{rw} = \frac{k_w}{k}$$

in which k_r is relative permeability, and k_i (subscript $i = o$ for oil, g for gas, w for water) is the permeability measured for a single fluid in the presence of other fluids. Oil relative permeability can be measured in the presence of water or gas. Therefore, the following definitions are added: k_{row} = relative permeability of oil to water; k_{rog} = relative permeability of oil to gas.

The relative permeability of a given fluid will depend on the saturations of all the fluids in the reservoir, as illustrated in Figure 2.3.8 for water and oil. At the irreducible water saturation, only oil will flow. The effective permeability is still less than the absolute permeability, because water occupies some of the pore space. As the water saturation increases, water begins to flow and the relative permeability to water increases, while the relative permeability to oil decreases. At the residual oil saturation, oil no longer flows and the relative permeability to oil is zero, while the relative permeability to water reaches its maximum.

Water–oil relative permeability

Typical relative permeability curves for oil and water are shown in Figure 2.3.9. Oil permeability decreases monotically from its maximum at the irreducible water

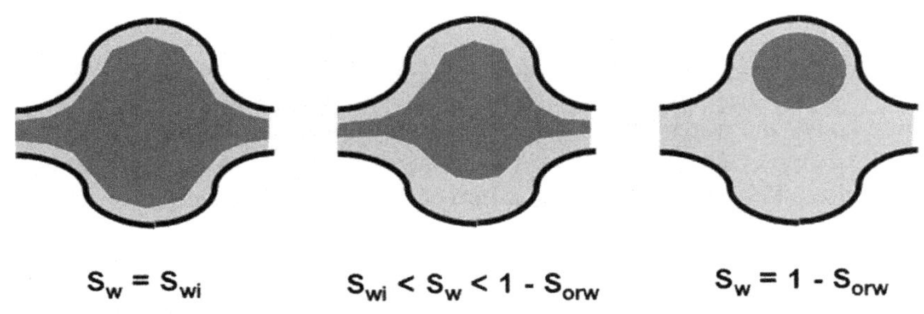

Figure 2.3.8 Effect of increasing water saturation on the distribution and flow of oil through a pore.

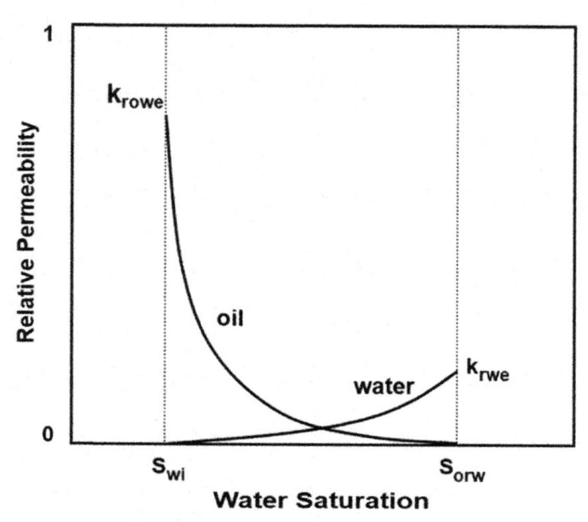

Figure 2.3.9 Typical water–oil relative permeability curves.

saturation, k_{rowe}, to zero at the residual oil saturation to water. Water permeability increases monotonically from zero at the irreducible water saturation to a maximum at the residual oil saturation, k_{rwe}.

Water-wet and oil-wet relative permeability curves are compared in Figure 2.3.10. The end point saturations are strongly dependent on the wettability, as was seen in the capillary pressure discussion. A water-wet rock has a higher irreducible water saturation and a lower residual oil saturation than an oil-wet rock. Oil-wet rocks tend to have a higher relative permeability to water. Some rules of thumb to distinguish water-wet and oil-wet relative permeability curves are given below:

Water wet:	$S_{wi} > 20\%$
	k_{rw} at $S_{orw} < 0.30$
	S_w in which oil and water curves cross over >50%
Oil wet:	$S_{wi} < 15\%$
	k_{rw} at $S_{orw} > 0.50$
	S_w in which oil and water curves cross over <50%

The shape of the water–oil relative permeability curves depends on the rock type and pore structure. Generally, as the permeability of the rock increases, the k_{rw} at S_{orw} end point rises. Relative permeability is further discussed in Chapter 8.

Gas–oil relative permeability

Gas–oil permeabilities are usually measured in samples presaturated with water so that irreducible water is present in the sample as it would be in the reservoir. The relative permeabilities of oil and gas are plotted against either liquid (oil plus water)

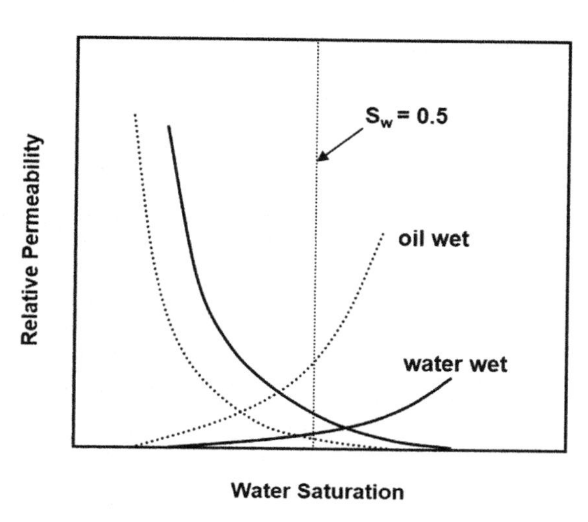

Figure 2.3.10 Comparison of water–oil relative permeability in water-wet and oil-wet rock.

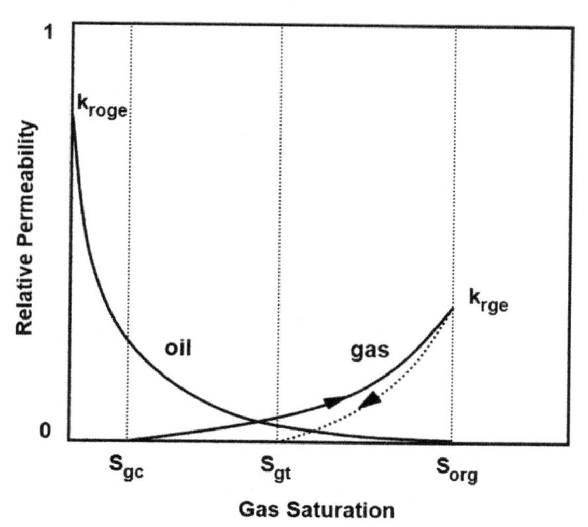

Figure 2.3.11 Typical gas–oil relative permeability curves.

saturation or gas saturation. When plotted against gas saturation as shown in Figure 2.3.11, the gas–oil relative permeability curves are superficially similar to the water–oil relative permeability curves. There is a finite saturation at which oil no longer flows, the residual oil saturation to gas, S_{org}. The relative permeability to gas reaches its maximum at S_{org}.

However, there are some important differences between water–oil and gas–oil relative permeability curves. While there is always at least an irreducible water saturation throughout a reservoir, there is initially no gas in the oil zone of a reservoir. Therefore, the oil relative permeability curve begins at zero gas saturation. As gas evolves in the reservoir or is added to a rock sample in a test, the oil premeability is reduced, but the gas does not necessarily flow until it reaches a *critical gas saturation*, S_{gc}. When oil displaces gas, not all of the gas is displaced. The oil relative permeability to gas reduces to zero at a *trapped gas saturation*, S_{gt}, which is analogous to the irreducible water saturation.

2.4 Types of Reservoir Fluids

Reservoir fluids can exist as a liquid (oil) phase, a gas phase, or both phases together. The conditions in which the two phases appear are readily illustrated on a pressure–temperature (PT) phase diagram, Figure 2.4.1. At relatively high pressures and low temperatures, a liquid phase is formed. At relatively low pressures and high temperatures, a gas phase is formed. At intermediate temperatures and pressures, both liquid and gas phases may be present. The range of pressures and temperatures at which both phases are present defines a two-phase envelope on the PT diagram (McCain, 1990).

The boundary of the two-phase envelope defines the bubble point and dew point curves of the reservoir fluid. The bubble point is the temperature at which the first bubble of gas is formed in a liquid at equilibrium at a given pressure. The dew point is the temperature at which the first drop of liquid is formed again at equilibrium at a given pressure. The dew point and bubble point curves meet at the critical point. The critical point is the temperature and pressure at which the composition of the two phases is the same. Above the critical point, a liquid and gas phase cannot be distinguished, and this single phase is usually termed a fluid phase.

The size and shape of the two-phase envelope depend on the composition of the reservoir fluid. However, reservoir fluids can be classified according to the position of the initial reservoir pressure and temperature relative to the phase envelope.

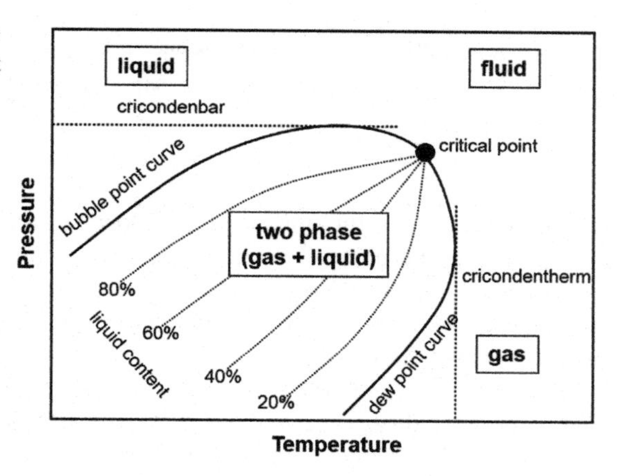

Figure 2.4.1 Pressure–temperature phase diagram of a multicomponent mixture.

To aid in the classification, one more term is defined, the cricondentherm. The cricondentherm is the line of constant temperature on the PT at which the temperature is the maximum two-phase temperature. The various reservoir fluid classifications are: undersaturated oil, bubble point oil, volatile oil, retrograde condensate, and gas.

Undersaturated oil: The reservoir fluid is initially above the bubble point and below the critical temperature. A fluid above the bubble point is an undersaturated liquid, and this type of reservoir fluid is termed an undersaturated oil. The path of an undersaturated oil as the reservoir is produced is shown as line A_1A_2 on Figure 2.4.2. Because the reservoir temperature is maintained by heat transfer from the surrounding rock, reservoirs are assumed to be isothermal. Hence, the reservoir pressure decreases at constant temperature. As pressure decreases, the oil reaches the bubble point, and a gas phase begins to form in the reservoir. The gas is usually produced with the oil, but some gas may remain dispersed in the oil or rise to form a secondary gas cap above the oil.

Saturated (bubble point) oil: In some cases, the reservoir fluid is initially at the bubble point. This is an indication that the reservoir fluid is already in the two-phase region and has separated into an oil layer (oil zone) and a gas layer (primary gas cap). The oil is saturated with gas and is therefore at its bubble point. The gas is saturated with liquid and is therefore at its dew point. The phase envelopes for the oil and gas intersect each other at the bubble point pressure and temperature, as shown on Figure 2.4.3. When a sample of fluid is withdrawn for testing, it is usually taken from either the oil zone or the gas zone and does not represent the total reservoir fluid composition. In most cases, the oil and gas zones are analyzed individually and it is not necessary to obtain a total fluid analysis. As a saturated oil is produced (line B_1B_2 on Figure 2.4.2), gas will evolve immediately from the oil. This gas may be

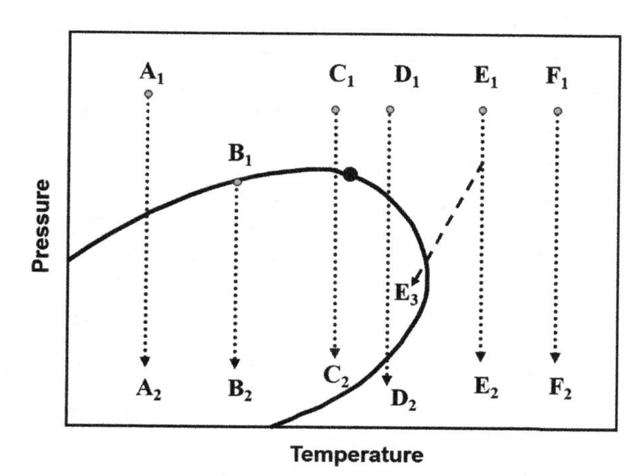

Figure 2.4.2 Location of reservoir fluid types and production paths on a PT diagram. (A_1A_2 = undersaturated oil reservoir; B_1B_2 = saturated oil or gas cap reservoir; C_1C_2 = volatile oil reservoir; D_1D_2 = retrograde condensate reservoir; E_1E_2 = wet gas reservoir, F_1F_2 = dry gas reservoir).

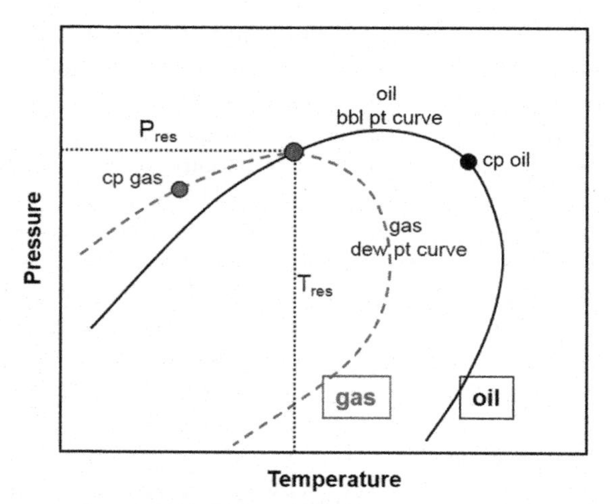

Figure 2.4.3 Phase envelopes of a bubble point oil and its gas cap gas.

produced with the oil, remain dispersed in the oil zone, or rise through the reservoir to join the gas cap. (McCain, 1990)

Volatile oil: If an oil is near its critical point, its properties begin to resemble gas properties, and it is termed a volatile oil. Volatile oils (line C_1C_2 on Figure 2.4.2) evolve a substantial fraction of gas as pressure drops below the bubble point, and the composition and properties of the oil and evolved gas may change significantly. It is sometimes necessary to model the phase behavior of a volatile oil with an equation-of-state-based composition model rather than the black oil model outlined later in this section.

Retrograde condensate: If the initial pressure and temperature lie in the fluid regime above the critical point but the temperature is below the cricondentherm, then the fluid is a retrograde condensate gas. When the pressure is reduced to the dew point, the fluid begins to condense a liquid phase (line D_1D_2 on Figure 2.4.2). As the pressure is reduced further, more liquids condense, but at some point, the liquids will recondense as pressure continues to decrease. The reversal of liquid condensation is termed retrograde condensation. The liquids do not flow as readily as gas and usually inhibit production and reduce fluid recovery from a reservoir. Compositional models are required to evaluate retrograde condensate reservoirs.

Gas: If the initial temperature is above the cricondentherm, the fluid is in the gas phase. As reservoir pressure decreases isothermally, the fluid remains in the single gas phase region (lines E_1E_2 and F_1F_2 on Figure 2.4.2). If the gas is near the two-phase envelope, liquids will condense from a reservoir gas in the wellbore or surface facilities as the temperature decreases (line E_1E_3). A reservoir gas in which condensation occurs on cooling is termed a *wet gas*. Condensation in the wellbore will always occur for gas cap production (unless there has been dry gas injection into the reservoir), because the gas is already at the dew point. If a gas is far enough from the two-phase region, no condensation occurs, even on cooling, and the gas is termed a *dry gas* (line F_1F_2).

In reality, a PT diagram of a given reservoir fluid is almost never available. The type of reservoir fluid is inferred from production history, fluid analyses, and reservoir characteristics, as discussed in Part 3.

2.5 Reservoir Fluid Properties

The fluid properties of interest for reservoir engineering are density, viscosity, and phase behavior, in particular how volume changes with pressure. Let us begin with the most common model for oil phase behavior, the black oil model.

2.5.1 *Black Oil Volume, Pressure, and Temperature Properties*

In many phase behavior applications, such as chemical engineering process calculations, the components of a mixture are known. When gas–liquid phase behavior is examined, the ratio in which each component partitions to each phase can be determined accurately. In most cases, the amount and composition of each phase can be determined exactly for a given set of conditions. In fact, a modification of this approach is used for compositional reservoir fluid modeling. However, the composition of a crude oil is not really known because crude oils consist of hundreds of thousands (or more) of ill-defined chemical species. For most oil reservoirs, the molecular composition is ignored, an isothermal reservoir is assumed, and the volumes of oil and gas are determined as a function of pressure. This approach is termed a black oil model.

To help visualize the black oil model, consider a volume of oil at reservoir pressure, Figure 2.5.1. If the pressure is above the bubble point, the oil volume will increase as the pressure decreases because the oil is slightly compressible. However, if the pressure is at or below the bubble point, the oil will release solution gas as the pressure decreases, and the remaining oil volume will decrease. The solution gas volume will increase significantly as pressure decreases because gas is highly compressible. The black oil model relates the volume of oil and gas volumes at reservoir temperature

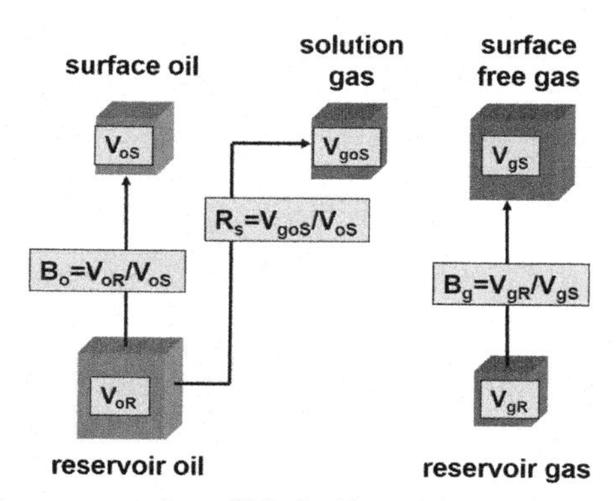

Figure 2.5.1 Schematic representation of black oil model.

and any given pressure to surface volumes usually standard conditions for gas and a specific separator temperature and pressure for oil. The following terms are defined for the black oil model:

B_o = oil formation volume factor $\left(\text{rb/stb or m}^3/\text{scm}\right)$

= the ratio of oil volume at reservoir conditions to the oil volume at surface conditions

R_s = solution gas-oil ratio ratio (SCF/stb or scm/scm)

= the ratio of the standard volume of solution gas dissolved in the oil at a given pressure to the oil volume at surface conditions

B_g = gas formation volume factor $\left(\text{rb/SCF or m}^3/\text{scm}\right)$

= the ratio of gas volume at a reservoir conditions to the gas volume at surface conditions

Because reservoirs also contain water, a water formation volume factor is also introduced:

B_w = water formation volume factor $\left(\text{rb/stb or m}^3/\text{scm}\right)$

= the ratio of water volume at reservoir conditions to the water volume at surface conditions

The relationship between B_o, R_s, and B_g and pressure is shown in Figure 2.5.2. The trends follow from the observations in Figure 2.5.1. Above the bubble point, B_o increases as pressure decreases due to the compressibility of the oil. R_s is constant because no solution gas evolves. Below the bubble point, B_o decreases as pressure decreases and solution gas evolves. R_s decreases with pressure, some solution gas evolves, and less remains dissolved in the oil. B_g increases with a decrease in pressure due to the compressibility of the gas. B_w increases slightly with a decreases in pressure and is often assumed to have a constant value of unity because the compressibility of water is low.

It is simplest to understand the black oil model if we consider a mass of oil initially at its bubble point with a volume, $V_{ob,s}$, measured at surface (or standard) conditions. The oil volume at reservoir conditions is then:

$$V_{ob} = V_{ob,s}B_{ob} \tag{2.5.1}$$

If the oil is initially at its bubble point, then the initial oil volume is the bubble point volume. If the oil is initially above its bubble, the initial oil volume is given by:

$$V_{oi} = V_{ob,s}B_{oi} \tag{2.5.2}$$

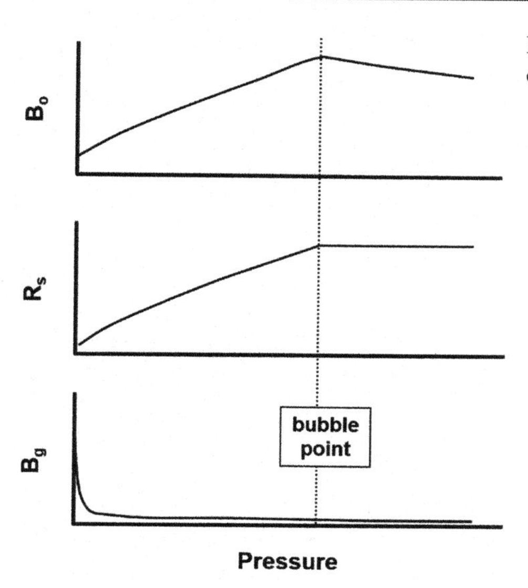

Figure 2.5.2 The effect of pressure on black oil properties.

The initial dissolved solution gas volume is defined as follows:

$$V_{sgi} = V_{ob,s} R_{si} B_{gi} \tag{2.5.3}$$

Note that the volume of dissolved solution gas, V_{sgi}, is included in the reservoir oil volume, V_{oi}, and represents the part of the reservoir oil volume made up of the dissolved solution gas.

Similarly, the initial reservoir volumes for given surface volumes of gas, $V_{gi,s}$, and water, $V_{wi,s}$, are given by:

$$V_{wi} = V_{wi,s} B_{wi} \tag{2.5.4}$$

$$V_{gi} = V_{gi,s} B_{gi} \tag{2.5.5}$$

If the reservoir pressure is decreased for the constant mass of oil, the new volume of the oil is given by:

$$V_o = V_{ob,s} B_o \tag{2.5.6}$$

and the change in volume of the oil at reservoir conditions is then:

$$\Delta V_{oil} = V_{ob,s}(B_o - B_{oi}) \tag{2.5.7}$$

For the same mass of oil and the same pressure change, the volume of solution gas evolved in the reservoir is given by:

$$\Delta V_{sg,s} = V_{ob,s}(R_{si} - R_s) \text{ with units of surface volume} \tag{2.5.8}$$

or

$$\Delta V_{sg} = V_{ob,s}(R_{si} - R_s)B_g \text{ with units of reservoir volume} \tag{2.5.9}$$

Similarly for a mass of gas with a volume of $V_{gi,s}$ at standard conditions, the change in reservoir volume with a decrease in reservoir pressure is given by:

$$\Delta V_{gas} = V_{gi,s}(B_g - B_{gi}) \tag{2.5.10}$$

and for a mass of water with a volume of $V_{wi,s}$ at standard conditions, the change in reservoir volume with a decrease in reservoir pressure is given by:

$$\Delta V_{wtr} = V_{wi,s}(B_w - B_{wi}) \tag{2.5.11}$$

Eqns 2.5.1–2.5.11 are sufficient to determine the volume of oil, gas, and water in any oil reservoir at any reservoir pressure. Note that the oil and solution gas volumes have been related to the bubble point oil volume which can be related to the initial oil volume with Eqn 2.5.2. In effect, all the volume changes have been related to the initial surface volumes of oil, gas, and water in the reservoir. Therefore, the only data required to determine the volume of fluid in the reservoir at any pressure are the initial surface volumes and the black oil volume, pressure, and temperature (PVT) properties, that is, the values of B_o, R_s, B_g, and B_w as a function of reservoir pressure. The PVT properties are measured or determined from correlations as discussed in Section 3.3.3.

Note that above the bubble point, the compressibilities of oil or water are often used instead of the formation volume factors. The isothermal compressibility of oil is defined as:

$$c_o = -\left(\frac{dV_o}{V_o dP}\right)_T \tag{2.5.12}$$

The oil compressibility is related to B_o as follows:

$$c_o \cong \frac{V_o - V_{oi}}{V_{oi}(P_i - P)} = \frac{B_o - B_{oi}}{B_{oi}\Delta P} \tag{2.5.13}$$

Similarly, the isothermal compressibility of water is given by:

$$c_w \cong \frac{V_w - V_{wi}}{V_{wi}(P_i - P)} = \frac{B_w - B_{wi}}{B_{Wi}\Delta P} \tag{2.5.14}$$

2.5.2 Oil Density and Viscosity

Oil density is usually measured at the initial reservoir conditions. For a constant mass, and given that density is inversely proportional to volume, the density at other pressures can be determined from the oil formation volume factor as follows:

$$\rho_o = \rho_{oi} \frac{B_{oi}}{B_o} \tag{2.5.15}$$

For an initially undersaturated oil, the density decreases as the pressure is reduced toward the bubble point. Below the bubble point, the density increases as pressure is reduced and the solution gas evolves.

Alternatively, above the bubble point, the density at any pressure can be expressed as a function of its compressibility as follows:

$$\rho_o = \rho_{o,ref} \exp\{c_o(P - P_{ref})\} \tag{2.5.16}$$

in which ρ_o is the oil density, and P_{ref} is a reference pressure such as the bubble point or initial pressure. Equation 2.5.16 is valid as long as the compressibility does not vary significantly with pressure.

Oil viscosity as a function of pressure is measured directly or obtained from correlations (see Chapter 5). For an undersaturated oil, the oil viscosity decreases as pressure is reduced toward the bubble point because the fluid expands. Below the bubble point, the viscosity increases as the solution gas evolves from the oil and the oil becomes denser.

2.5.3 Gas Volume, Pressure, and Temperature Properties

The relationship between pressure and volume for a gas is usually expressed as the real gas law:

$$Pv = zRT \tag{2.5.17}$$

in which v is the molar volume, z is the gas compressibility factor, R is the universal gas constant, and T is temperature. The gas compressibility factor is a function of the reduced temperature and pressure for which correlations are provided in Chapter 5. Gas volume changes can be calculated directly from the real gas law or using the gas formation volume faction, which is related to the real gas law as follows:

$$B_g = \frac{V_g}{V_{g,s}} = \frac{zT_r P_s}{T_s P_r} \tag{2.5.18}$$

in which the subscripts r and s indicate reservoir and surface conditions, respectively. The z-factor is calculated at reservoir conditions; the z-factor at surface conditions is approximately unity.

2.5.4 Heavy Oil Volume, Pressure, and Temperature Properties

Heavy oil can be treated as a black oil, except for foamy oil production. Foamy oil occurs when the oil is so viscous that evolved solution gas remains dispersed as small bubbles. In this case, the oil may retain much of its original compressibility; that is, B_o does not decrease with pressure as much as expected with a black oil. In addition, foams are not thermodynamically stable, and therefore the formation volume factors may change even at constant pressure. Foamy oil viscosity is equally uncertain and likely depends on the volume fraction of evolved gas that remains dispersed in the oil. Therefore, foamy oil properties are usually obtained by history matching reservoir production data.

Laboratory tests on heavy oil show that foams retard the formation of continous gas saturation, resulting in lower gas relative permeability and higher trapped or critical gas saturation for gas-oil relative permeability curves than are observed for conventional oils (Shen and Batycky, 1999; Zhang et al., 1999; Tang and Firoozabadi, 2003). In laboratory tests, the rate of pressure depletion is rapid (100−1000 psi/d), which favors foam formation; therefore, foamy effects have a large effect on laboratory measurements of relative permeability. The effects on field scale relative permeability are harder to estimate because the pressure depletion rates are much lower. Normally, the foamy oil effect is expected to occur only near the wellbore. However, in unconsolidated heavy oil reservoirs, sand production enhances permeability through wormhole formation, which causes large localized spatial pressure gradients. Hence, there may be a foamy oil effect near these wormoles. Note, the foamy oil effect decreases as the viscosity of the fluid decreases, for example, with increasing temperature.

To produce many heavy oil reservoirs, the oil viscosity must be reduced either by heating or by dilution with a solvent. Both methods have a dramatic effect of viscosity. Heating and dilution will also alter the PVT properties. For example, density decreases linearly with temperature. The bubble point pressure decreases with increasing temperature and will usually decrease with solvent addition. The bubble point depends largely on the methane content of the fluid and solvent addition reduces the methane concentration. The effect of temperature on PVT properties is usually determined experimentally. Heavy oil properties are discussed in more detail in Chapter 5.

2.5.5 Compositional Fluid Models

Black oil models are suitable for oils far from the critical point, and the real gas law applies outside the two-phase envelope. For volatile oils, retrograde condensates, and miscible-enhanced oil recovery, a more rigorous fluid model is required because the phase compositions and properties change significantly with small changes in pressure and temperature. There are two main approaches for these types of fluid: K-value models and equations-of-state.

Both approaches are used to solve a flash calculation to determine the amount and composition of each phase. A flash calculation starts with a material balance for each component of fluid which can be expressed the Rachford−Rice equation:

$$\Psi = \sum_{i=1}^{n} \frac{z_i(K_i - 1)}{1 + \alpha(K_i - 1)} = 0 \tag{2.5.19}$$

in which n is the number of components, α is the molar ratio of the vapor to the feed, $K_i = y_i/x_i$ is the K-value of component i, and x, y, and z are the liquid, vapor, and feed mole fractions, respectively. A flash calculation involves the iterative solution of Eqn 2.5.19 to find the unknown α. Once α and the K-values are determined, all the phase volumes and compositions can be determined from the material balances as follows:

$$x_i = \frac{z_i}{1 + \alpha(K_i - 1)} \quad \text{and} \quad y_i = K_i x_i \tag{2.5.20}$$

$$\dot{n}_L = (1 - \alpha)\dot{n}_F \tag{2.5.21}$$

in which $\dot{n}$ is the molar flow rate.

2.5.5.1 K-Value Model

The K-value is also known as the equilibrium ratio. By definition, when a vapor and a liquid are at equilibrium, the fugacities of the two phases are equal:

$$f_i^L = f_i^V \tag{2.5.22}$$

which is related to the phase compositions as follows:

$$x\varphi_i^L P = y_i \phi_i^V P \tag{2.5.23}$$

in which f is the fugacity, ϕ is the fugacity coefficient, and the superscripts L and V denote the liquid and vapor states. Equation 2.5.21 can be rearranged to obtain:

$$K_i = \frac{y_i}{x_i} = \frac{\phi_i^L}{\phi_i^V} \tag{2.5.24}$$

The K-value model uses tabulated or correlated K-values in the flash calculation (see Chapter 5). This approach is convenient when the K-values do not vary significantly with composition, for example, with natural gases (GPSA Handbook, 1980) and for some enhanced oil recovery calculations (Whitsen and Brulé, 2000). K-value models are less prevalent today because increased computational speed now allows rapid and robust equation-of-state (EoS) calculations in which the K-values can depend on composition.

2.5.5.2 Equation-of-State Models

EoSs are designed to provide accurate vapor—liquid equilibrium calculations for the entire two-phase envelope, including near the critical point. An EoS defines the relationship between pressure temperature and volume based on a balance of the attractive and repulsive forces between the molecules in the fluid. The attractive forces are expressed via an attractive parameter, a, and the repulsive forces via a molecular co-volume, b. There are many EoSs, most of which are cubic. The most commonly used for hydrocarbon fluids is the Peng and Robinson EoS:

$$P = \frac{RT}{v - b} - \frac{a}{v^2 + 2bv - b^2} \tag{2.5.25}$$

The attractive term is a function of temperature given by:

$$a = a_c \alpha(T) \tag{2.5.26}$$

in which a_c is the attractive term calculated in such way as to obey the van der Waals conditions at the critical point, and $\alpha(T)$ is an empirical function of temperature. This empirical function of temperature can be correlated in terms of acentric factor for nonpolar or slightly pure components such as hydrocarbons. The $\alpha(T)$ function used for hydrocarbons is given by:

$$\sqrt{\alpha_i} = 1 + f_{wi}\left(1 - \sqrt{T_{ri}}\right) \tag{2.5.27}$$

in which T_{ri} is the reduced temperature of component i and f_{wi} is the correlation based on acentric factor given by:

$$\text{for } \omega_i < 0.5: \quad f_{wi} = 0.37464 + 1.54226 w_i - 0.26992 w_i^2 \tag{2.5.28a}$$

$$\text{for } \omega_i \geq 0.5: \quad f_{wi} = 0.3796 + 1.4850 w_i - 0.1644 w_i^2 + 0.01666 w_i^3 \tag{2.5.28b}$$

For mixtures, the mixing rules are required to determine a and b. A commonly used set of mixing rules is given by:

$$a = \sum_{i=1}^{N} \sum_{j-1}^{N} (1 - k_{ij}) \sqrt{a_i a_j} x_i x_j \tag{2.5.29}$$

$$b = \sum_{i=1}^{N} x_i b_i \tag{2.5.30}$$

$$k_{ij} = k_{ij}^o + \frac{k_{ij}^1}{T} + k_{ij}^2 \ln T \tag{2.5.31}$$

in which N is the number of components, x_i is the mole fraction of component i, and k_{ij} is the interaction parameter between components i and j and determined based on experimental vapor–liquid equilibrium data.

Depending on the pressure, temperature, and composition of a fluid, the EoS may have one, two, or three specific volume roots. A single root corresponds to a single phase, and two roots to a two-phase mixture. When there are three roots, the roots with the highest and lowest specific volume correspond to two different phases, while the middle root is nonphysical.

Calculation of K-values

With an EoS, the K-values used in the flash are calculated from the fugacity coefficients of each component in each phase and Eqn 2.5.24. The fugacity coefficient of a component is calculated from the Peng–Robinson equation as follows:

$$\ln \phi_i = \frac{B_i}{B}(Z-1) - \ln(Z-B) + \frac{A}{2\sqrt{2}B}\left(\frac{B_i}{B} - \frac{2}{A}\sum_{j=1}^{N} x_i(1-k_{ij})\sqrt{A_iA_j}\right) \times \ln\left\{\frac{Z+(1+\sqrt{2})B}{Z-(1-\sqrt{2})B}\right\} \tag{2.5.32}$$

in which:

$$Z = \frac{Pv}{RT} \quad A_i = a_i\frac{P}{(RT)^2} \quad B_i = b_i\frac{P}{RT},$$

$$A = \sum_{i=1}^{N}\sum_{j-1}^{N}(1-k_{ij})\sqrt{A_iA_j}x_ix_j \quad B = \sum_{i=1}^{N}x_iB_i$$

The Zs are another form of the roots of the EoS, and each phase has its own Z. Because the K-value is not necessarily constant but depends on the composition, an iterative flash calculation is used to solve for the roots of the EoS and the phase amounts and compositions, Figure 2.5.3.

Volume translation

Simple cubic equations of state are not usually able to calculate liquid densities accurately. A simple way to correct this is through the use of a volume translation factor (Peneloux et al., 1982):

$$v^T = v - \sum_{i=1}^{N} x_ic_i \tag{2.5.33}$$

in which c_i is the volume translation factor for the ith component. The beauty of the volume translation is that it corrects the density without altering the phase equilibrium calculations. Values of the volume translation factor are tabulated or correlated for petroleum components.

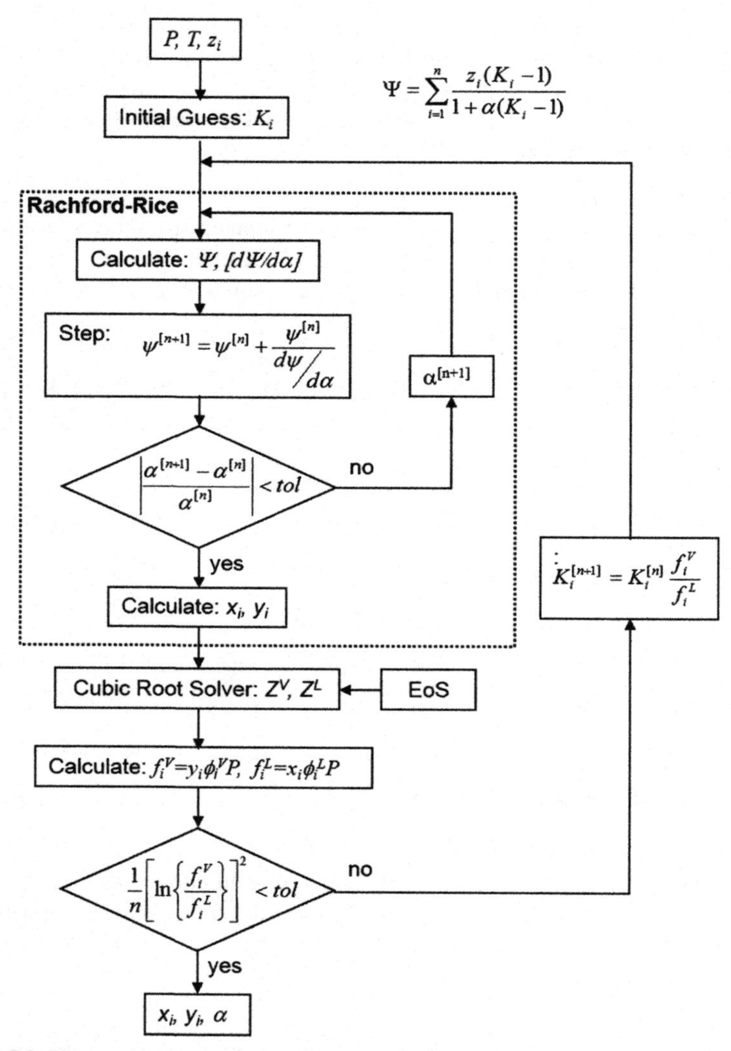

Figure 2.5.3 Schematic of a flash algorithm for liquid–vapor equilibrium based on an EoS.

Compositional fluid modeling of petroleum with equations of state or K-value models is beyond the scope of this book, and the interested reader is referred to Whitsen and Brulé (2000) and Pedersen et al. (1989).

Basic Reservoir Engineering Calculations

Chapter Outline

Arguably the two most important reservoir calculations are to determine how much hydrocarbon is in the reservoir and to predict fluid flow rates in the reservoir and wells. The amount of hydrocarbons in the reservoir (in place) can be determined from the shape and porosity of the reservoir (volumetrics) and from a material balance. Fluid flow is usually determined from applications of Darcy's law and the radial inflow equation.

3.1 Reservoir Volumetrics

Petroleum is found in the pore space of a hydrocarbon-bearing rock formation. At discovery, it shares the pore space with the initial (connate) water present in the

Practical Reservoir Engineering and Characterization. http://dx.doi.org/10.1016/B978-0-12-801811-8.00003-1

hydrocarbon-bearing zone. Hence, for an oil-bearing zone, the volume of oil in the reservoir at initial reservoir conditions is given by:

$$OOIP = \frac{PV_o(1 - S_{wi})}{B_o} \tag{3.1.1}$$

in which *OOIP* is the original oil in place in surface volume units, *PV* is the pore volume of the oil-bearing rock, and S_{wi} is the initial water saturation in the oil-bearing zone. The pore volume is the product of the bulk reservoir volume and the average porosity of the reservoir rock. Equation (2.1.1) can then be expressed as:

$$OOIP = \frac{BV_o\phi(1 - S_{wi})}{B_o} \tag{3.1.2}$$

The bulk volume is often expressed as the product of the outside area and the average thickness of the oil-bearing zone:

$$OOIP = \frac{h_oA_o\phi(1 - S_{wi})}{B_o} \tag{3.1.3}$$

In some cases, the oil-bearing zone is a mixture of productive and unproductive rock, in which a productive rock has sufficient permeability to sustain economic production rates. In these reservoirs, a ratio of net productive thickness to total or gross reservoir thickness, r_{ng}, is defined, and Eqn (3.1.3) is modified as follows:

$$OOIP = \frac{r_{ng}h_oA_o\phi(1 - S_{wi})}{B_o} \tag{3.1.4}$$

Similarly, for a gas reservoir or gas cap of an oil reservoir, the volume of free gas in place at surface conditions is given by:

$$OGIP = \frac{r_{ng}h_gA_g\phi(1 - S_{wi})}{B_g} \tag{3.1.5}$$

Note that the net-to-gross pay ratio, average porosity, and initial water saturation of an oil zone and its gas cap are not necessarily the same.

Sometimes it is desired to determine the amount of solution gas in an oil reservoir, for example, when booking reserves. The volume of solution gas in place in a reservoir at surface conditions is given by:

$$OSGIP = \left(R_{si} - R_{ssep}\right)OOIP \tag{3.1.6}$$

in which R_{ssep} is the solution gas-oil ratio at the separator conditions. If the volume, pressure, and temperature (PVT) data match the current operating conditions, R_{ssep} is zero. Note the solution gas is dissolved in the oil in the reservoir and is included in the reservoir volume of the original oil in place (OOIP), not as part of the gas cap. The original solution gas in place (OSGIP) only has meaning as a surface quantity and is not part of any material balance calculations.

3.2 Reservoir Material Balance

When a reservoir is produced, the pressure drops, and the rock and fluids remaining in the reservoir expand. The reservoir volume (defined as the hydrocarbon occupied volume above the *initial* water-oil contact) is constant, so the volume of expansion must equal the volume of fluid withdrawn from the reservoir:

$$withdrawal = expansion \tag{3.2.1}$$

in which the volumes of withdrawal and expansion are determined at the current reservoir conditions. Note expansion is defined to include water influx from an underlying aquifer into the defined reservoir volume. Equation (3.2.1) is the general form of the material balance for petroleum reservoirs. To apply Eqn (3.2.1), it is necessary to relate the withdrawal and expansion terms to measurable quantities.

3.2.1 Oil Reservoir Material Balance

3.2.1.1 Withdrawal

Because the material balance is applied on a cumulative basis, the following cumulative production and injection data are required:

N_p = cumulative oil production
R_p = cumulative producing gas-oil ratio
= cumulative gas production/cumulative oil production
G_i = cumulative gas injection
W_p = cumulative water production
W_i = cumulative gas injection

All the production data are reported at surface conditions because they are measured at the surface facilities. If other fluids are injected, the appropriate terms must be included. Because the material balance is applied at reservoir conditions, the produced volumes are converted to reservoir volumes using the formation volume factors. The oil withdrawal is:

$$oil\ withdrawal = N_pB_o \tag{3.2.2}$$

This term includes the reservoir volume of the produced solution gas because the oil formation volume factor accounts for this volume. The reservoir volume of all the produced gas is:

$$total\ gas\ withdrawal = N_pR_pB_g \tag{3.2.3}$$

However, this expression includes the solution gas that was produced. If Eqns (3.2.2) and (3.2.3) are combined, the reservoir volume of the produced solution gas will be counted twice. Therefore, the produced solution gas is subtracted from the gas withdrawal to obtain:

$$\textit{net gas withdrawal} = N_p(R_p - R_s)B_g \qquad (3.2.4)$$

The reservoir volume of the water withdrawal is given by:

$$\textit{water withdrawal} = W_p B_w \qquad (3.2.5)$$

The reservoir volumes of the injected fluids are given by:

$$\textit{injection} = G_i B_{gi} + W_i B_{wi} + \ldots \qquad (3.2.6)$$

Note that each injection fluid will have a formation volume factor that depends on the composition of the fluid. The combined withdrawal is given by:

$$\textit{withdrawal} = N_p\left(B_o + (R_p - R_s)B_g\right) + W_p B_w - G_i B_{gi} - W_i B_{wi} - \ldots \qquad (3.2.7)$$

The injection withdrawals are negative because injection adds material to the reservoir. All the withdrawal terms are determined at a given reservoir pressure. Hence, the withdrawal represents the volume that all of the produced and injected fluids would occupy in the reservoir at the given reservoir pressure.

3.2.1.2 Expansion

The expansion in a reservoir is a combination of rock compaction, and water, oil, and gas expansion. Oil expansion is determined from the oil formation volume factor:

$$\textit{oil expansion} = N(B_o - B_{oi}) \qquad (3.2.8)$$

in which N is the original oil in place in surface volume units. N and $OOIP$ are identical, but we have used different nomenclature to distinguish values determined from the volumetric and material balance methods. Note that oil expansion is negative because B_o decreases as pressure decreases and solution gas is released. The solution gas evolution and expansion are found as follows:

$$\textit{solution gas expansion} = N(R_{si} - R_s)B_g \qquad (3.2.9)$$

If there is a gas cap, the expansion of the free (gas cap) gas is given by:

$$\textit{gas cap expansion} = \frac{mNB_{oi}}{B_{gi}}\left(B_g - B_{gi}\right) \qquad (3.2.10)$$

in which m is the ratio of the initial reservoir volume of free gas to initial reservoir volume of oil. The term mNB_{oi}/B_{gi} is the initial standard volume of the free gas.

The contribution of rock compaction and water expansion are determined from the compressibility of the formation and water as follows:

$$formation\ expansion = V_f c_f \Delta P + V_w c_w \Delta P \tag{3.2.11}$$

in which V_f is the bulk volume of the hydrocarbon bearing zone (the formation), V_w is the water volume in the hydrocarbon bearing zone, c_f and c_w are the respective compressibilities of the formation and water, and ΔP is the change in average reservoir pressure. The water compressibility was defined in Eqn (2.5.14). The isothermal rock compressibility is defined as:

$$c_f = -\left(\frac{dPV_f}{PV_f dP}\right)_T = \left(\frac{dV_f}{V_f dP}\right)_T \tag{3.2.12}$$

in which PV_f is the pore volume of the formation. The volume of water in the oil-bearing zone is related to the formation volume through the initial water saturation:

$$V_w = S_{wi} V_f \tag{3.2.13}$$

The formation volume is related to the initial hydrocarbon volume as follows:

$$V_f = \frac{NB_{oi} + mNB_{oi}}{1 - S_{wi}} \tag{3.2.14}$$

Combining Eqns (3.2.11), (3.2.13) and (3.2.14), the formation expansion can be expressed as:

$$formation\ expansion = \left[\frac{(1+m)NB_{oi}}{1 - S_{wi}}\right](c_f + S_{wi} c_w)\Delta P \tag{3.2.15}$$

If there is an aquifer adjoining the reservoir, then the expansion of the aquifer must also be accounted for. There are several approaches to predict aquifer expansion or influx (see Chapter 10). For now, let us simply define the volume of aquifer influx as:

$$aquifer\ expansion = W_e B_w \tag{3.2.16}$$

The combined expansion is the sum of Eqns (3.2.8), (3.3.3), (3.2.10), (3.2.15), and (3.2.16):

$$expansion = N\left[\begin{array}{l}(B_o - B_{oi}) + (R_{si} - R_s)B_g + \dfrac{mB_{oi}}{B_{gi}}(B_g - B_{gi}) \\ \\ + \dfrac{(1+m)B_{oi}}{1 - S_{wi}}(c_f + S_{wi} c_w)\Delta P\end{array}\right] + W_e B_w \tag{3.2.17}$$

3.2.1.3 General Oil Reservoir Material Balance

To obtain the general material balance for an oil reservoir, the expansion, Eqn (3.2.17), is equated with the withdrawal, Eqn (3.2.7):

$$N_p\left(B_o+\left(R_p-R_s\right)B_g\right)+W_pB_w-G_iB_{gi}-W_iB_{wi}$$
$$=N\left[\begin{array}{l}(B_o-B_{oi})+(R_{si}-R_s)B_g+\dfrac{mB_{oi}}{B_{gi}}(B_g-B_{gi})\\ \\ +\dfrac{(1+m)B_{oi}}{1-S_{wi}}(c_f+S_{wi}c_w)\Delta P\end{array}\right]+W_eB_w \qquad (3.2.18)$$

The total volume formation factor is defined as:

$$\begin{aligned}B_t&=B_o+(R_{si}-R_s)B_g\\ B_{ti}&=B_{oi}\end{aligned} \qquad (3.2.19)$$

and is sometimes introduced into Eqn (3.2.18) to obtain the following more concise expression of the general material balance:

$$N_p\left(B_t+\left(R_p-R_{si}\right)B_g\right)+W_pB_w-G_iB_{gi}-W_iB_{wi}$$
$$=N\left[(B_t-B_{ti})+\frac{mB_{oi}}{B_{gi}}\left(B_g-B_{gi}\right)+\frac{(1+m)B_{ti}}{1-S_{wi}}(c_f+S_{wi}c_w)\Delta P\right]+W_eB_w \qquad (3.2.20)$$

The application of the general material balance to different types of reservoir is discussed in Chapter 10. The material balance is used to determine the volumes of oil and gas in a reservoir and the volume of aquifer influx. Aquifer influx models are also presented in Chapter 10.

3.2.2 Gas Reservoir Material Balance

Except when connected to a strong aquifer, the material balance for a gas reservoir is completely dominated by the gas expansion; rock compaction and connate water expansion are negligible. The material balance in this situation is identical to blowing down a tank of gas:

$$\frac{dN_g}{dt}=-\dot{n}_g \qquad (3.2.21)$$

in which N_g is moles of gas in the reservoir, $\dot{n}_g$ is the molar gas flow rate out of the reservoir, and t is time. Using the real gas law, the moles of gas can be expressed in terms of pressures and volumes to obtain:

$$d\left(\frac{P}{z}\right) = -\frac{T_r P_s}{T_s V} q_g dt \tag{3.2.22}$$

in which V is the initial volume of gas in the reservoir at reservoir conditions, and q_g is the gas production rate in surface volumes. Equation (3.2.22) is integrated to obtain:

$$\frac{P}{z} = \left(\frac{P}{z}\right)_o - \frac{T_r P_s}{T_s V} Q_g \tag{3.2.23}$$

in which $Q_g = \int_o^t q_g dt$ is simply the cumulative gas production in surface volumes. Note that when P/z goes to zero, Eqn (3.2.23) becomes:

$$Q_g = \frac{P_{ro} V}{z_o T_r} \frac{T_s}{P_s} = OGIP \tag{3.2.24}$$

Equation (3.2.23) is the gas material balance, and it shows that a plot of P/z versus gas production is linear, Figure 3.2.1. Equation (3.2.24) shows that the x-intercept of a P/z plot is equal to the original gas in place in surface volumes.

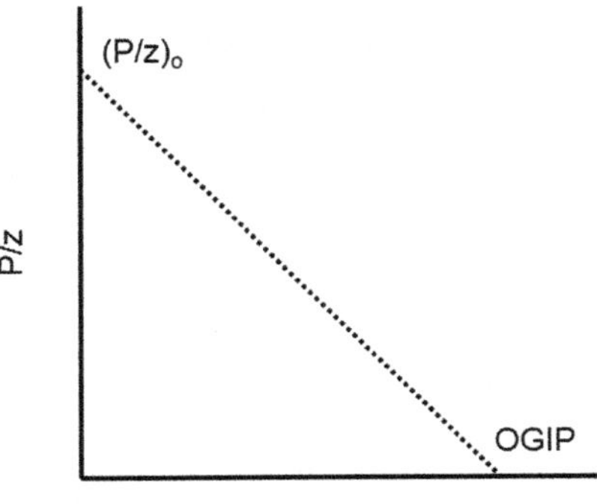

Figure 3.2.1 P/z plot—a gas reservoir material balance.

3.3 Steady-State Flow Through Porous Media

A porous media such as a reservoir can be considered as a collection of capillaries of different length and diameter. Because flow in an oil reservoir is typically horizontal and laminar, a logical starting point for modeling reservoir flow is to consider laminar flow through a horizontal capillary. We then extend this approach to linear and radial flow through porous media and to reservoir fluids.

3.3.1 Horizontal Laminar Flow Through a Capillary

The Hagen–Poiseiulle equation describes steady-state horizontal laminar flow of a single-phase fluid through a cylindrical segment. This equation can be derived from a force balance around a disk-shaped element, as shown in Figure 3.3.1. The net force on the disc is given by:

$$F = \pi r^2 P - \pi r^2 (P + dP) - 2\pi r dL\tau = 0 \tag{3.3.1}$$

in which F is the net force, r is the radius of the disc, P is the upstream pressure, dP is the change in pressure across the disc, dL is the thickness of the disc, and τ is the shear stress acting on the edge of the disc.

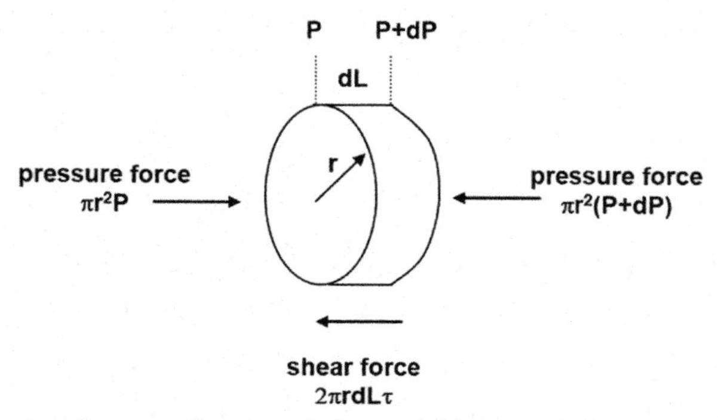

Figure 3.3.1 Force balance on disc shaped element of horizontal flow.

Equation (3.3.1) can be expressed as:

$$\frac{dP}{dL} + \frac{2\tau}{r} = 0 \tag{3.3.2}$$

If the fluid is Newtonian, the shear stress is related to the velocity of the fluid as follows:

$$\tau = -\mu \frac{du}{dr} \tag{3.3.3}$$

Substituting Eqn (3.3.3) into Eqn (3.3.2) gives:

$$\frac{dP}{dL} - \frac{2\mu du}{rdr} = 0 \tag{3.3.4}$$

It can be shown from Eqn (3.3.4) that the average velocity of laminar horizontal flow in a cylindrical segment is given by:

$$u = -\frac{d^2}{32\mu} \frac{dP}{dL} \tag{3.3.5}$$

in which d is the diameter of the cylindrical segment. The square diameter arises from the integration of the ratio of the shear stress to the area of flow. In effect, the larger the diameter, the less drag per area of flow and the greater the velocity. The negative sign in Eqn (3.3.5) indicates that pressure decreases in the direction of flow.

3.3.2 Darcy's Law for Linear Flow

To apply Eqn (3.3.5) directly to laminar flow through a porous media, it would be necessary to determine the geometry of all the capillaries in the media. Instead, Darcy defined an empirical factor, permeability, to account for the cross-sectional and surface areas of a network of capillaries in a porous media. Darcy's expression for laminar flow through a porous media is:

$$u = -\frac{k}{\mu}\frac{dP}{dL} \tag{3.3.6}$$

in which:

u = the apparent velocity, cm/s
k = permeability, D
μ = viscosity, cp
P = pressure, atm
L = length of cylinder, cm

Comparing Eqns (3.3.5) and (3.3.6), it is clear that the dimension of permeability is length squared. It is equivalent to the integrated ratio of shear stress to flow area for the porous media. The higher the permeability, the less drag per flow area and the greater the velocity. In a sense, the permeability is a measure of the flow capacity of the porous media. This is best illustrated by modifying Darcy's law to determine the volumetric flow rate:

$$q = -\frac{kA}{\mu}\frac{dP}{dL} \tag{3.3.7}$$

in which q is the volumetric flow rate in cm^3/s, and A is the cross-sectional area in cm^2. Note that for a reservoir problem, the volumetric flow rate in Eqn (3.3.7) would have reservoir units, not surface units. Equation (3.3.7) is analogous to Fourier's law, the equation that describes heat conduction across a temperature gradient. Darcy's law describes steady-state volumetric flow across a pressure gradient. The ratio of permeability to viscosity (the mobility ratio) is the conductivity to flow and is analogous to thermal conductivity.

The unit of permeability is the Darcy. From Eqn (3.3.7), it is apparent that a rock of one Darcy will flow 1 cm/s of a 1 cp fluid under a pressure gradient of 1 atm/cm. It can be shown that one Darcy is equivalent to $1.0133 \times 10^{-8}\ cm^2$. Most reservoirs have permeabilities lower than one Darcy, and units of millidarcy (mD) are frequently used. Permeability must be determined experimentally for any given porous medium.

3.3.3 Nonhorizontal Linear Flow

Equation (3.3.7) is valid for any horizontal geometry of constant cross-sectional area and permeability. If the porous medium is tilted, the effect of gravity force on the pressure gradient must be considered. Figure 3.3.2 shows a tilted porous medium. The gravity force acts over the height of the medium and is given by:

$$dP_g = \rho \frac{g}{g_c} dh \tag{3.3.8}$$

in which ρ is the density of the fluid, g is gravitational acceleration, and g_c is the conversion factor between natural and defined force units. The height is related to the length of the bed through the angle on inclination, α, in which $dh = dLcos\,\alpha$. Hence, the pressure gradient arising from the gravity force is:

$$\frac{dP_g}{dL} = \rho \frac{g}{g_c} \cos\alpha \tag{3.3.9}$$

When the gravity contribution is included, Darcy's law becomes:

$$q = -\frac{kA}{\mu}\left(\frac{dP}{dL} - \rho \frac{g}{g_c} \cos\alpha\right) \tag{3.3.10}$$

The units of the gravity term must of course be consistent with the units of the pressure gradient.

3.3.4 Darcy's Law for Radial Flow

Many reservoirs are produced from vertical wells that have not been hydraulically fractured. In this case, the reservoir fluid flows radially toward the well, as shown in Figure 3.3.3. Darcy's law for horizontal radial flow for a single-phase fluid can be expressed as:

$$q = \frac{kA}{\mu} \frac{dP}{dr} \tag{3.3.11}$$

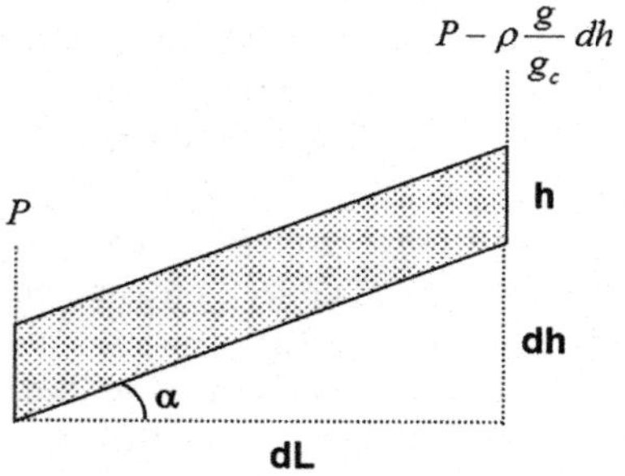

Figure 3.3.2 Hydrostatic pressure in a tilted reservoir.

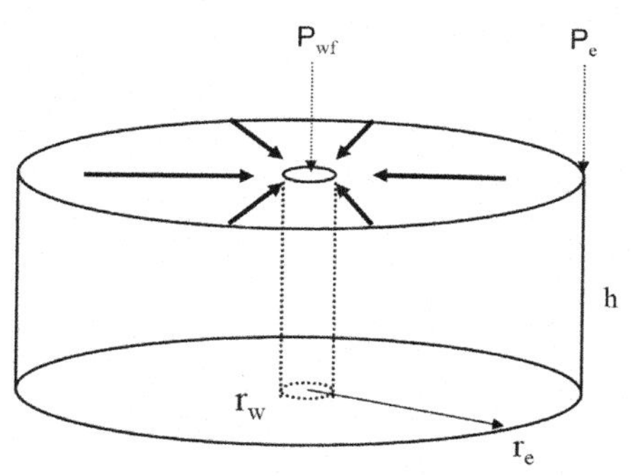

Figure 3.3.3 Schematic of radial inflow to a well.

The sign is positive because the radial coordinate is measured from the wellbore outward, that is, opposite to the direction of flow. The cross-sectional area for flow is the edge of a disk around the wellbore, $A = 2\pi rh$. Therefore, Eqn (3.3.11) becomes:

$$q = \frac{2\pi rkh}{\mu}\frac{dP}{dr} \tag{3.3.12}$$

To solve Eqn (3.3.12), we need to examine the pressure profile around a vertical well under steady-state radial flow, Figure 3.3.4. The flow across any given radius is constant. Nearer the wellbore, the area for flow decreases rapidly, and therefore the fluid velocity and pressure drop increase rapidly. Conversely, the reservoir pressure increases gradually further from the wellbore, and eventually a point is reached where there is no change in pressure as the radius increases. This hypothetical radius is termed the drainage radius. It may correspond to the outer edge of the reservoir or to the point where the pressure profiles of adjacent wells overlap. To obtain a solution for Eqn (3.3.12), it is integrated from the wellbore radius to the drainage radius:

$$q = \frac{2\pi kh}{\mu}\frac{P_e - P_{wf}}{\ln\left\{\frac{r_e}{r_w}\right\}} \tag{3.3.13}$$

in which r_w is the wellbore radius, r_e is the drainage radius, P_{wf} is the flowing pressure at the wellbore, and P_e is the pressure at the boundary of the drainage radius. Note that the flow rate and viscosity are assumed to be independent of the pressure and radius.

In some cases, the formation near the wellbore is damaged during drilling, and the near wellbore permeability is reduced. There is in an increased pressure drop through the damaged zone. van Everdingen (1953) defined a skin factor, S, to empirically account for the increased pressure drop:

$$\Delta P_{skin} = \frac{q\mu}{2\pi kh}S \tag{3.3.14}$$

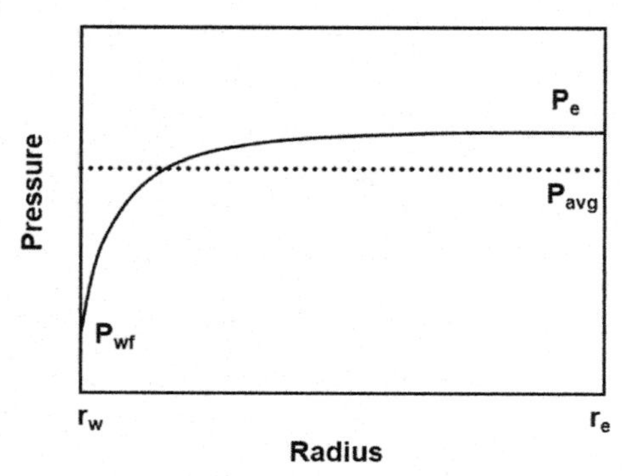

Figure 3.3.4 Radial pressure gradient around a producing well.

When Eqn (3.3.14) is introduced to Eqn (3.3.13), Darcy's law becomes:

$$q = \frac{2\pi kh}{\mu} \frac{P_e - P_{wf}}{\ln\left\{\frac{r_e}{r_w}\right\} + S} \tag{3.3.15}$$

Note that when the well is undamaged, the skin factor is zero, and the original radial flow equation is obtained. If the well is stimulated beyond its native capacity, the skin factor is negative.

It is often more convenient to use the average reservoir pressure, P_r, rather than the pressure at the drainage radius. If it is assumed that there is no flow across the outer boundary, r_e, the "average pressure" solution to Eqn (3.3.12) is given by:

$$q = \frac{2\pi kh}{\mu} \frac{P_r - P_{wf}}{\ln\left\{\frac{r_e}{r_w}\right\} - \frac{1}{2} + S} = \frac{2\pi kh}{\mu} \frac{P_r - P_{wf}}{\ln\left\{0.607\frac{r_e}{r_w}\right\} + S} \tag{3.3.16}$$

3.3.5 Darcy's Law for Reservoir Fluids

For reservoir calculations, the volumetric flow rate is desired in surface units, while Darcy's law is based on the actual flow rate, that is, reservoir units. If the flowing fluid is single-phase oil, the oil formation volume factor is introduced to obtain:

$$q_o = \frac{2\pi k_o h}{\mu_o B_o} \frac{P_e - P_{wf}}{\ln\left\{\frac{r_e}{r_w}\right\} + S} \tag{3.3.17}$$

in which the subscript o indicates an oil property. An equivalent expression applies to laminar water flow.

Gas flow is more complicated because gas is highly compressible, and gas properties vary significantly with pressure. It cannot be assumed that flow rate and viscosity are independent of pressure. Therefore the gas formation factor must be introduced to Eqn (3.3.12) prior to the integration. The gas formation factor is related to pressure through the nonideal gas law and the compressibility factor as follows:

$$B_g = \frac{P_s T z}{T_s P} \tag{3.3.18}$$

in which T_s and P_s are standard temperature and pressure, respectively, and z is the gas compressibility factor. When Eqn (3.3.18) is introduced into Eqn (3.3.12), the following expression is obtained:

$$q_g \frac{dr}{r} = \frac{2\pi k_g h T_s}{P_s T} \frac{PdP}{\mu_g z} \tag{3.3.19}$$

Wattenbarger and Ramey (1968) observed that at pressures below 2000 psia, $\mu_g z$ is approximately constant and that at pressures above 2000 psia, $\mu_g z/P$ is approximately constant. Using these assumptions, two pseudo steady-state solutions of Eqn (3.3.19) are found:

Below 2000 psia:

$$q_g = \frac{\pi T_s k_g h}{P_s T\left(z\mu_g\right)} \frac{P_e^2 - P_{wf}^2}{\ln\left\{\frac{r_e}{r_w}\right\} + S} \tag{3.3.20a}$$

Above 2000 psia:

$$q_g = \frac{2\pi T_s k_g h}{P_s T\left(z\mu_g/P\right)} \frac{P_e - P_{wf}}{\ln\left\{\frac{r_e}{r_w}\right\} + S} \tag{3.3.20b}$$

When using Eqn (3.3.20b), the term $z\mu_g/P$ should be evaluated at the average pressure between P_r and P_{wf}.

Note that in Eqns (3.3.16) and (3.3.20) the permeability is the absolute permeability. The use of absolute permeability is valid if the reservoir rock contains only one single-phase fluid. If more than one fluid is present, the relative permeability must be introduced as follows:

$$q_o = \frac{2\pi k_{ro} k h}{\mu_o B_o} \frac{P_e - P_{wf}}{\ln\left\{\frac{r_e}{r_w}\right\} + S} \tag{3.3.21}$$

in which k is the average absolute permeability within the drainage radius, and k_{ro} is the relative permeability to oil. Also note that a constant absolute permeability has been assumed so far in the development of the steady-state flow equations. For heterogeneous reservoirs, the average absolute permeability can be determined using the techniques described in Chapters 6 and 7.

3.3.6 Summary of Darcy's Law

Darcy's law applies to steady-state flow in which the flow rate across the outer boundary of a given volume is the same as the flow rate across the inner boundary. Equation (3.3.21) gives Darcy's law for radial inflow. In practice, this situation occurs when pressure is maintained through natural water influx or through injection. Another common reservoir situation occurs when the drainage radius reaches the boundaries of the reservoir or the drainage area of surrounding wells. In this case, as will be shown in Section 3.4.2, the "pseudo steady state" radial inflow equation for a fluid with constant compressibility and viscosity, such as an undersaturated black oil, is given by:

$$q = \frac{2\pi k_{ro}kh}{\mu}\frac{P_e - P_{wf}}{\ln\left\{\frac{r_e}{r_w}\right\} - \frac{1}{2} + S} = \frac{2\pi k_{ro}kh}{\mu}\frac{P_e - P_{wf}}{\ln\left\{0.607\frac{r_e}{r_w}\right\} + S} \tag{3.3.22}$$

and the average pressure solution is:

$$q = \frac{2\pi k_{ro}kh}{\mu}\frac{P_r - P_{wf}}{\ln\left\{\frac{r_e}{r_w}\right\} - \frac{3}{4} + S} = \frac{2\pi k_{ro}kh}{\mu}\frac{P_r - P_{wf}}{\ln\left\{0.472\frac{r_e}{r_w}\right\} + S} \tag{3.3.23}$$

The steady-state and pseudo steady-state radial inflow equations are summarized in Insert 1. The steady solutions for different fluids are summarized in Field and SI units in Insert 2. For any given reservoir analysis, the reservoir or production engineer must choose the most appropriate inflow equation for the given reservoir and fluid. Transient equations are discussed in Section 3.4.

Insert 1 Summary of Radial Inflow Equations

Steady-state and semi-steady-state formulations for a constant compressibility and viscosity fluid, such as an undersaturated black oil, are listed. The equations are presented using either the outer boundary pressure or the average pressure. The steady-state equations apply when reservoir pressure is maintained through influx or injection. The semi-steady-state equations apply when the drainage radius has reached a boundary (physical or other drainage areas). The adjustments required for different fluids or for Field units are presented in Insert 2.

Steady-state, boundary pressure:

$$q_o = 0.0005355\frac{k_{ro}kh}{\mu_o B_o}\frac{P_e - P_{wf}}{\ln\left\{\frac{r_e}{r_w}\right\} + S}$$

Insert 1 Summary of Radial Inflow Equations—cont'd

Steady-state, average pressure:

$$q_o = 0.0005355\frac{k_{ro}kh}{\mu_o B_o}\frac{P_e - P_{wf}}{\ln\left\{\frac{r_e}{r_w}\right\} - \frac{1}{2} + S}$$

Pseudo steady-state, boundary pressure:

$$q_o = 0.0005355\frac{k_{ro}kh}{\mu_o B_o}\frac{P_e - P_{wf}}{\ln\left\{\frac{r_e}{r_w}\right\} - \frac{1}{2} + S}$$

Pseudo steady-state, average pressure:

$$q_o = 0.0005355\frac{k_{ro}kh}{\mu_o B_o}\frac{P_e - P_{wf}}{\ln\left\{\frac{r_e}{r_w}\right\} - \frac{3}{4} + S}$$

in which q is in sm^3/d, k in mD, μ in mPa s, h in m, r in m, T in K, and P in kPa.

Insert 2 Summary of Darcy's Law—Steady-State

Linear and radial inflow equations for oil, gas, and water are presented in Field and SI units. All equations are steady-state formulations expressed in terms of the outer boundary pressure. Pseudo steady-state or average pressure formulations can be substituted as required.

Linear flow: Darcy's law for horizontal linear flow is used for determining permeability from core samples and for analyzing reservoir flow to a hydraulically fractured well. The expression in Field units is given by:

oil $$q_o = -0.001127\frac{k_{ro}kA}{\mu_o L}(P_2 - P_1)$$

water $$q_w = -0.001127\frac{k_{rw}kA}{\mu_w L}(P_2 - P_1)$$

gas below 2000 psia $$q_g = -0.003164\frac{T_s k_{rg}kA}{P_s T(z\mu_g)L}\left(P_2^2 - P_1^2\right)$$

gas above 2000 psia $$q_g = -0.006328\frac{T_s k_{rg}kA}{P_s T(z\mu_g/P)L}(P_2 - P_1)$$

Continued

Insert 2 Summary of Darcy's Law—Steady-State—cont'd

in which q_o and q_w are in bbl/d, q_g in SCF/d, k in mD, μ in cp, A in ft^2, L in ft, T in R, and P in psia.

Darcy's law for linear flow in SI units is given by:

oil $$q_o = -8.523 \times 10^{-5} \frac{k_{ro} k A}{\mu_o L} (P_2 - P_1)$$

water $$q_w = -8.523 \times 10^{-5} \frac{k_{rw} k A}{\mu_w L} (P_2 - P_1)$$

gas below 14 MPa $$q_g = -4.261 \times 10^{-5} \frac{T_s k_{rg} k A}{P_s T (z\mu_g) L} (P_2^2 - P_1^2)$$

gas above 14 MPa $$q_g = -8.523 \times 10^{-5} \frac{T_s k_{rg} k A}{P_s T (z\mu_g / P) L} (P_2 - P_1)$$

in which q is in sm^3/d, k in mD, μ in mPa s, A in m^2, L in m, T in K, and P in kPa.

Radial flow: Darcy's law for radial flow is used to analyze reservoir flow to vertical wells without hydraulic fractures and to hydraulically fractured wells in which the drainage radius is much larger than the fracture length. The expression in Field units is given by:

oil $$q_o = 0.00708 \frac{k_{ro} k h}{\mu_o B_o} \frac{P_r - P_{wf}}{\ln\left\{\frac{r_e}{r_w}\right\} + S}$$

water $$q_w = 0.00708 \frac{k_{rw} k h}{\mu_w B_w} \frac{P_e - P_{wf}}{\ln\left\{\frac{r_e}{r_w}\right\} + S}$$

gas below 2000 psia $$q_g = 0.01988 \frac{T_s k_{rg} k h}{P_s T (z\mu_g)} \frac{P_e^2 - P_{wf}^2}{\ln\left\{\frac{r_e}{r_w}\right\} + S}$$

gas above 2000 psia $$q_g = 0.03976 \frac{T_s k_{rg} k h}{P_s T (z\mu_g / P)} \frac{P_e - P_{wf}}{\ln\left\{\frac{r_e}{r_w}\right\} + S}$$

in which q_o and q_w are in bbl/d, q_g in SCF/d, k in mD, μ in cp, h in ft, r in ft, T in R, and P in psia.

Insert 2 Summary of Darcy's Law—Steady-State—cont'd

Darcy's law for radial flow in SI units is given by:

oil
$$q_o = 0.0005355\frac{k_{ro}kh}{\mu_o B_o}\frac{P_e - P_{wf}}{\ln\left\{\frac{r_e}{r_w}\right\} + S}$$

water
$$q_w = 0.0005355\frac{k_{rw}kh}{\mu_w B_w}\frac{P_e - P_{wf}}{\ln\left\{\frac{r_e}{r_w}\right\} + S}$$

gas below 14 MPa
$$q_g = 0.0002678\frac{T_s k_{rg}kh}{P_s T(z\mu_g)}\frac{P_e^2 - P_{wf}^2}{\ln\left\{\frac{r_e}{r_w}\right\} + S}$$

gas above 14 MPa
$$q_g = 0.0005355\frac{T_s k_{rg}kh}{P_s T(z\mu_g/P)}\frac{P_e - P_{wf}}{\ln\left\{\frac{r_e}{r_w}\right\} + S}$$

in which q is in sm^3/d, k in mD, μ in mPa s, h in m, r in m, T in K, and P in kPa.

3.4 Transient Flow Through Porous Media

Darcy's law is derived assuming steady-state flow. In practice, flow in reservoirs is often transient, for example, when wells are activated, shut-in, or experience a change in bottom hole flowing pressure. An important discipline in reservoir engineering is pressure transient analysis; that is, the analysis of the pressure changes on shutting in well production or injection. Pressure transient analysis (PTA) can provide a great deal of information about a reservoir including: an estimate of the average reservoir permeability near the well; an estimate of the extent of well damage or stimulation; and some indications of the permeability distribution near the well. Before discussing PTA, let us derive the equation for transient radial flow, the radial diffusivity equation.

3.4.1 Radial Diffusivity Equation

Consider a radial volume element in a reservoir, Figure 3.4.1. Note that the radius increases outward, while flow is inward. Therefore, the mass flow rate out of the

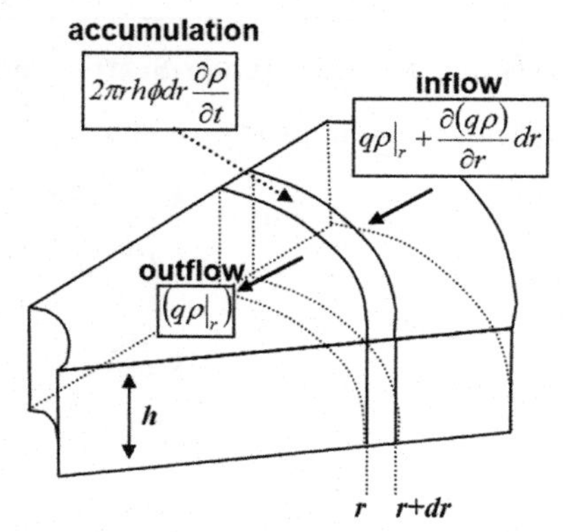

Figure 3.4.1 Radial volume element.

element at the inner radius is taken as the basis of the derivation. A mass balance on the element gives:

$$\underbrace{\left(q\rho|_r + \frac{\partial(q\rho)}{\partial r}dr\right)}_{\substack{\text{mass rate in} \\ \text{at } r+dr}} - \underbrace{(q\rho|_r)}_{\substack{\text{mass rate out} \\ \text{at } r}} = \underbrace{2\pi rh\phi dr\frac{\partial\rho}{\partial t}}_{\text{mass accumulation}} \tag{3.4.1}$$

in which q is the flow rate, ρ is density, r is radius, h is height, ϕ is porosity, and t is time. Equation (3.4.1) simplifies to:

$$\frac{\partial(q\rho)}{\partial r} = 2\pi rh\phi\frac{\partial\rho}{\partial t} \tag{3.4.2}$$

Both density and flow rate are functions of pressure. Flow rate is related to pressure through Darcy's law for radial flow, Eqn (3.3.12). Equation (3.3.12) (now a partial derivative) is substituted into Eqn (3.4.2) to obtain:

$$\frac{\partial}{\partial r}\left(\frac{2\pi rkh}{\mu}\rho\frac{\partial P}{\partial r}\right) = 2\pi rh\phi\frac{\partial\rho}{\partial t} \tag{3.4.3}$$

which simplifies to:

$$\frac{1}{r}\frac{\partial}{\partial r}\left(\frac{k\rho}{\mu}r\frac{\partial P}{\partial r}\right) = \phi\frac{\partial\rho}{\partial t} \tag{3.4.4}$$

Density is related to pressure through the compressibility as follows:

$$c = -\frac{1}{V}\frac{\partial V}{\partial P} = +\frac{1}{\rho}\frac{\partial \rho}{\partial P} \tag{3.4.5}$$

The derivative of density with time is then:

$$\frac{\partial \rho}{\partial t} = c\rho\frac{\partial P}{\partial t} \tag{3.4.6}$$

Equation (3.4.6) is substituted into Eqn (3.4.4) to obtain the radial diffusivity equation:

$$\frac{1}{r}\frac{\partial}{\partial r}\left(\frac{k\rho}{\mu}r\frac{\partial P}{\partial r}\right) = \phi c\rho\frac{\partial P}{\partial t} \tag{3.4.7}$$

The radial diffusivity equation describes the radial flow of any single-phase fluid through a porous medium. Pressure appears explicitly in the partial derivative and implicitly in the density, viscosity, and compressibility terms. Hence, the radial diffusivity equation is a nonlinear partial differential equation.

3.4.2 Radial Diffusivity Equation Applied to Fluids with Constant Compressibility

It is necessary to linearize the radial diffusivity equation to obtain a solution. For undersaturated black oils, the oil compressibility and viscosity are both nearly independent of pressure. Therefore, it is assumed that the oil compressibility and viscosity are constant. If we further assume that $\partial P/\partial r$ is small and that terms of the order $(\partial P/\partial r)^2$ can be neglected, the radial diffusivity equation can be linearized into the following form:

$$\frac{\partial^2 P}{\partial r^2} + \frac{1}{r}\frac{\partial P}{\partial r} = \frac{\phi\mu c}{k}\frac{\partial P}{\partial t} \tag{3.4.8a}$$

or

$$\frac{1}{r}\frac{\partial}{\partial r}\left(r\frac{\partial P}{\partial r}\right) = \frac{\phi\mu c}{k}\frac{\partial P}{\partial t} \tag{3.4.8b}$$

With the above assumptions, the coefficient $\phi\mu c/k$ is a constant. The reciprocal of this constant, $k/\phi\mu c$, is defined as the diffusivity constant. Note that this solution of the radial diffusivity equation is mathematically identical to the radial heat conduction equation:

$$\frac{1}{r}\frac{\partial}{\partial r}\left(r\frac{\partial T}{\partial r}\right) = \frac{1}{K}\frac{\partial T}{\partial t} \tag{3.4.9}$$

in which K is the thermal diffusivity constant. Hence, the integrated solutions have the same mathematical form.

To integrate Eqn (3.4.8), an initial condition and two boundary conditions are required. The initial condition is that the initial pressure is a constant, P_i. Also, if constant flow rate is assumed, Darcy's equation applies at the wellbore:

$$q = 2\pi r \frac{kh}{\mu} \left(\frac{dP}{dr} \right)_{r=r_w} \tag{3.4.10}$$

Four sets of the remaining boundary conditions are considered:

- Steady−state,
- Stabilized rate−pseudo steady−state,
- Constant terminal rate−pseudo steady−state,
- Constant terminal rate−transient.

The boundary conditions and solutions are described below.

3.4.2.1 Steady-State Conditions

The steady-state condition means that the flow rate into and out of any radial element is the same. In this case, there is no accumulation, and the pressure is uniform. The boundary conditions are expressed as follows:

$$P = P_e = constant, \quad \text{at } r = r_e \tag{3.4.11a}$$

$$\frac{\partial P}{\partial t} = 0, \quad \text{at all } r \text{ and } t \tag{3.4.11b}$$

These conditions apply when reservoir pressure is maintained, for example, through natural water influx or water injection. For stabilized inflow, the solution to Eqn (3.4.8) with these boundary conditions is the expected steady-state solution, Darcy's law:

$$q = \frac{2\pi kh}{\mu} \frac{P_e - P_{wf}}{\ln\left\{\frac{r_e}{r_w}\right\}} \tag{3.4.12}$$

As discussed previously, the solution when an average pressure is used in place of the boundary pressure is given by:

$$q = \frac{2\pi kh}{\mu} \frac{P_r - P_{wf}}{\ln\left\{\frac{r_e}{r_w}\right\} - \frac{1}{2}} = \frac{2\pi kh}{\mu} \frac{P_r - P_{wf}}{\ln\left\{0.606\frac{r_e}{r_w}\right\}} \tag{3.4.13}$$

The effect of wellbore damage or stimulation is accounted for using a skin factor, as was described in Section 3.3.4. The expression for Darcy's radial flow equation including the skin factor is as follows:

$$q = \frac{2\pi kh}{\mu} \frac{P_r - P_{wf}}{\ln\left\{\frac{r_e}{r_w}\right\} - \frac{1}{2} + S} \tag{3.4.14}$$

3.4.2.2 Stabilized Rate—Pseudo Steady-State Conditions

Stabilized rate presumes that the wellbore flowing pressure is constant. The pseudo steady-state boundary condition applies when the pool pressure is not constant and the drainage radius has reached its boundaries. In this case, the boundary conditions are:

$$\frac{\partial P}{\partial r} = 0, \text{ at } r = r_e \tag{3.4.15a}$$

$$\frac{\partial P}{\partial t} = constant, \text{ at all } r \text{ and } t \tag{3.4.15b}$$

These conditions apply to a single well that has drained a reservoir to the point at which the drainage radius has reached the pool boundaries. They also apply to a well with a drainage radius bounded by the drainage radii of other wells, as long as the boundaries do not move over time. These conditions are usually satisfied when the dimensionless group $kt/\phi\mu cA$ is greater than 0.1, in which t is the time the well flowed at a steady rate.

The solution to Eqn (3.4.8) with these boundary conditions and including skin factor is given by:

$$q = \frac{2\pi kh}{\mu} \frac{P_e - P_{wf}}{\ln\left\{\frac{r_e}{r_w}\right\} - \frac{1}{2} + S} \tag{3.4.16}$$

The solution using the average pressure instead of the boundary pressure is given by:

$$q = \frac{2\pi kh}{\mu} \frac{P_r - P_{wf}}{\ln\left\{\frac{r_e}{r_w}\right\} - \frac{3}{4} + S} = \frac{2\pi kh}{\mu} \frac{P_r - P_{wf}}{\ln\left\{0.472\frac{r_e}{r_w}\right\} + S} \tag{3.4.17}$$

3.4.2.3 Constant Terminal Rate—Pseudo Steady-State Conditions

The constant terminal rate solution applies when the flow rate at the wellbore is held constant, but the flowing pressure changes with time, as shown in Figure 3.4.2. As before, the pseudo steady-state condition implies that the drainage radius has reached

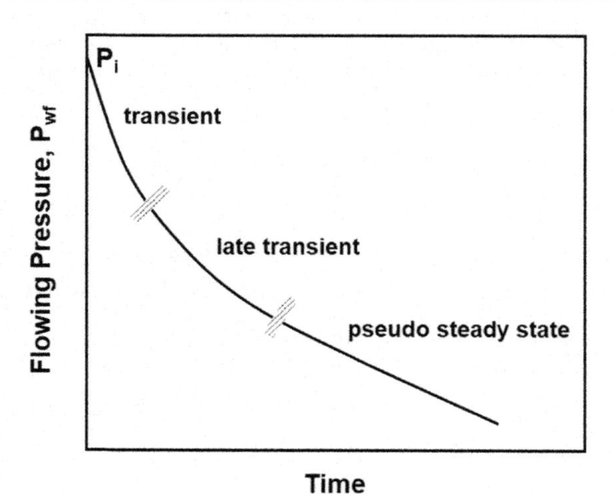

Figure 3.4.2 Typical wellbore flowing pressure profile at a constant terminal rate.

its boundaries. The stabilized rate, pseudo steady-state boundary conditions, Eqn (3.4.15a) and (3.4.15b), still apply and the same solution, Eqn (3.4.17), is obtained. Now, however, the change in reservoir pressure with time must be accounted for. When a pseudo steady-state condition is reached, a material balance can be performed on the drainage area to obtain:

$$cAh\phi(P_i - P_r) = qt \tag{3.4.18}$$

The material balance is introduced into the pseudo steady-state inflow equation, Eqn (3.4.14), to obtain the constant terminal rate solution:

$$P_{wf}(t) = P_i - \frac{q\mu}{2\pi kh}\left(\ln\left\{0.472\frac{r_e}{r_w}\right\} + 2\pi\frac{kt}{\phi\mu cA} + S\right) \tag{3.4.19}$$

Equations (3.4.19) can be expressed as:

1. in Field units:

$$P_{wf}(t) = P_i - \frac{162.6 q_o\mu_o B_o}{k_{ro}kh}\left[\log\left\{0.223\frac{r_e^2}{r_w^2}\right\} + 0.87S\right] - \frac{0.2339 q_o B_o t}{Ah\phi c_t} \tag{3.4.20a}$$

in which P in psia, q_o is in bbl/d, k in mD, μ in cp, A in ft^2, h in ft, r in ft, t in hours, and c_t in psia^{-1}.

2. in SI units:

$$P_{wf}(t) = P_i - \frac{2149 q_o\mu_o B_o}{k_{ro}kh}\left[\log\left\{0.223\frac{r_e^2}{r_w^2}\right\} + 0.87S\right] - \frac{0.04166 q_o B_o t}{Ah\phi c_t} \tag{3.4.20b}$$

in which P in kPa, q_o is in m^3/d, k in mD, μ in mPa.s, A in m^2, h in m, r in m, t in hours, and c_t in kPa^{-1}.

3.4.2.4 Pseudo Steady-State Solutions for Other Geometries

Stabilized rate solution: So far, we have assumed a radial geometry with the well in the center of the drainage area. In practice, the drainage area may assume any shape depending on the shape of the reservoir and the pattern of drainage areas established by the wells in the reservoir. Dietz (1965) developed a generalized pseudo steady-state equation based on a shape factor, C_A. To develop the generalized equation, the stabilized rate solution at pseudo steady-state conditions, Eqn (3.4.17), is recast as follows:

$$\begin{aligned} P_r - P_{wf} &= \frac{q\mu}{2\pi kh}\left(\frac{1}{2}\ln\left\{\frac{4\pi r_e^2}{4\pi r_w^2 e^{3/2}}\right\} + S\right) \\ &= \frac{q\mu}{2\pi kh}\left(\frac{1}{2}\ln\left\{\frac{4A}{1.781(31.6)r_w^2}\right\} + S\right) \end{aligned} \tag{3.4.21}$$

in which A is the drainage area, πr_e^2. The constant 31.6 is defined as the shape factor, C_A, so that Eqn (3.4.21) becomes:

$$P_r - P_{wf} = \frac{q\mu}{2\pi kh}\left(\frac{1}{2}\ln\left\{\frac{4A}{1.781 C_A r_w^2}\right\} + S\right) \tag{3.4.22}$$

in which $C_A = 31.6$ for a well in the center of a circular drainage area. Dietz was able to show that Eqn (3.4.22) applies for all drainage geometries, but that each geometry has a specific value of the shape factor. Some shape factors are shown in Figure 3.4.3.

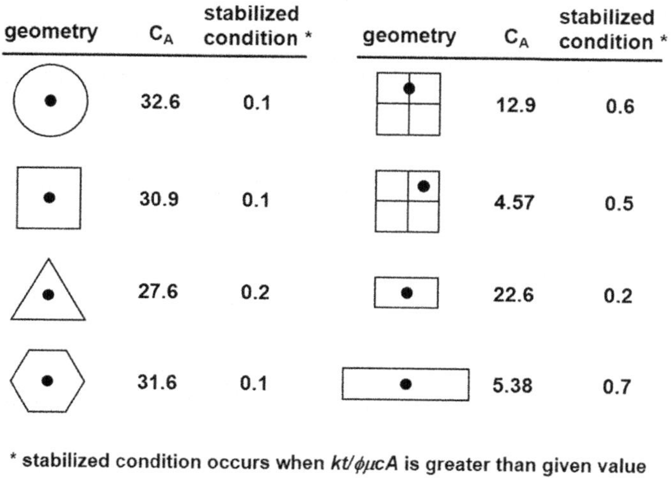

geometry	C_A	stabilized condition *	geometry	C_A	stabilized condition *
circle, well at center	32.6	0.1	square quadrants, well at top edge	12.9	0.6
square, well at center	30.9	0.1	square quadrants, well in upper-right quadrant	4.57	0.5
triangle, well at center	27.6	0.2	2:1 rectangle, well at center	22.6	0.2
hexagon, well at center	31.6	0.1	4:1 rectangle, well at center	5.38	0.7

* stabilized condition occurs when $kt/\phi\mu cA$ is greater than given value

Figure 3.4.3 Examples of Deitz shape factors.

Equation (3.4.22) can be expressed as:

1. in Field units:

$$P_{wf}(t) = P_r - \frac{162.6 q_o \mu_o B_o}{k_{ro} k h}\left[\log\left\{\frac{4A}{1.781 C_A r_w^2}\right\} + 0.87S\right] \tag{3.4.23a}$$

in which P in psia, q_o is in bbl/d, k in mD, μ in cp, A in ft^2, h in ft, r in ft, t in hours, and c_t in psia^{-1}.

2. in SI units:

$$P_{wf}(t) = P_r - \frac{2149 q_o \mu_o B_o}{k_{ro} k h}\left[\log\left\{\frac{4A}{1.781 C_A r_w^2}\right\} + 0.87S\right] \tag{3.4.23b}$$

in which P in kPa, q_o is in m^3/d, k in mD, μ in mPa.s, A in m^2, h in m, r in m, t in hours, and c_t in kPa^{-1}.

Constant terminal rate solution: Using the same arguments, the generalized terminal rate solution at pseudo steady-state conditions is given by:

$$P_{wf}(t) = P_i - \frac{q\mu}{2\pi k h}\left(\frac{1}{2}\ln\left\{\frac{4A}{1.781 C_A r_w^2}\right\} + 2\pi\frac{kt}{\phi\mu c A} + S\right) \tag{3.4.24}$$

Equation (3.4.25) can be expressed as:

1. in Field units:

$$P_{wf}(t) = P_i - \frac{162.6 q_o \mu_o B_o}{k_{ro} k h}\left[\log\left\{\frac{4A}{1.781 C_A r_w^2}\right\} + 0.87S\right] - \frac{0.2339 q_o B_o t}{A h \phi c_t} \tag{3.4.25a}$$

in which P in psia, q_o is in bbl/d, k in mD, μ in cp, A in ft^2, h in ft, r in ft, t in hours, and c_t in psia^{-1}.

2. in SI units:

$$P_{wf}(t) = P_i - \frac{2149 q_o \mu_o B_o}{k_{ro} k h}\left[\log\left\{\frac{4A}{1.781 C_A r_w^2}\right\} + 0.87S\right] - \frac{0.04166 q_o B_o t}{A h \phi c_t} \tag{3.4.25b}$$

in which P in kPa, q_o is in m^3/d, k in mD, μ in mPa s, A in m^2, h in m, r in m, t in hours, and c_t in kPa^{-1}.

3.4.2.5 Constant Terminal Rate, Transient Conditions

Transient conditions apply for a relatively short time after a disturbance at the well, such as a rate change. The pressure response is considered transient until the effect of the disturbance reaches the boundary of the drainage area. In this case, the reservoir appears to be infinite in extent, and the following boundary conditions apply:

$$P = P_i, \text{ as } r \to \infty \text{ for all } t \tag{3.4.26a}$$

$$P = P_i, \text{ at } t = 0 \text{ for all } r \tag{3.4.26b}$$

The transient, constant terminal rate solution of the radial diffusivity equation is given by:

$$P(r,t) = P_i - \frac{q\mu}{4\pi kh}\left[-E_i\left(-\frac{\phi\mu c r^2}{4kt}\right)\right] \tag{3.4.27}$$

in which E_i is the exponential integral, defined as:

$$E_i(-x) = -\int_x^{\infty} \frac{e^{-u}du}{u} \tag{3.4.28}$$

When $x < 0.01$, the approximate solution of $E_i(x)$ is given by:

$$E_i(-x) = -\ln(x) - 0.5772 = -\ln(1.781x) \tag{3.4.29}$$

Typically in well test analysis, a solution is sought at the wellbore. At the wellbore, r is small, and $x = \phi\mu c r_w^2/4kt$ is usually less than 0.01. The approximate solution of Eqn (3.4.27) is then given by:

$$P_{wf}(t) = P(r_w,t) = P_i - \frac{q\mu}{4\pi kh}\ln\left\{\frac{4kt}{1.781\phi\mu c r_w^2}\right\} \tag{3.4.30}$$

For ease of graphical analysis, Eqn (3.4.30) is usually recast in the form of a Base 10 logarithm:

$$P_{wf}(t) = P_i - \frac{2.302q\mu}{4\pi kh}\log\left\{\frac{4kt}{1.781\phi\mu c r_w^2}\right\} \tag{3.4.31}$$

When applied to oil reservoirs and the pressure drop due to skin is included, Eqn (3.4.31) becomes:

1. in Field units:

$$P_{wf}(t) = P_i - \frac{162.6 q_o \mu_o B_o}{k_{ro} k h}\left[\log\left\{\frac{k_{ro} k t}{1698 \phi \mu_o c_t r_w^2}\right\} + 0.87S\right] \quad (3.4.32a)$$

in which P in psia, q_o is in bbl/d, k in mD, μ in cp, h in ft, r in ft, t in hours, and c_t in psia^{-1}.

2. in SI units:

$$P_{wf}(t) = P_i - \frac{2149 q_o \mu_o B_o}{k_{ro} k h}\left[\log\left\{\frac{k_{ro} k t}{126000 \phi \mu_o c_t r_w^2}\right\} + 0.87S\right] \quad (3.4.32b)$$

in which P in kPa, q_o is in m^3/d, k in mD, μ in mPa s, h in m, r in m, t in hours, and c_t in kPa^{-1}.

Equation (3.4.32) includes the oil formation volume factor. Also, note that the compressibility includes the contribution of the formation and connate water compressibility. The total compressibility, c_t, is given by:

$$c_t = c_o S_o + c_w S_w + c_f \quad (3.4.33)$$

in which S is saturation, c is compressibility, and the subscripts o, w, and f denote oil, water, and formation, respectively. Equation (3.4.32) is the basis for pressure transient (well test) analysis of oil wells. Pressure transient analysis is discussed in Chapter 6.

Part Two

Reservoir Data Analysis

Part 2 focuses on the data a reservoir engineer must work from, to characterize a reservoir. Table 4.0.1 lists the data required to characterize, forecast, and simulate a reservoir. Possible sources for the data are also provided in the table. The data sources are discussed in detail in the following chapters, including measurement methods, correlations, sources of error, and accuracy checks.

Table 4.0.1 **Types and sources of reservoir data**

Required data	Source
Drilling, completion, and workover history	Company records Scout reports
Production history	Company records Public data base
Pressure history	Static pressure tests Pressure transient analysis
Fluid properties	Oil, gas, and water analyses Fluid study Correlations
Reservoir structure	Seismic data Well logs Core data
Pay thickness	Well logs Core data
Porosity	Core data Well logs
Water saturation	Special core data Well logs
Permeability	Core data Pressure transient analysis Inflow tests
Relative permeability	Special core data Correlations
Capillary pressure	Special core data

The material for Part 2 was drawn from many sources. Sources used for specific issues are cited in the text. More general sources are: Allen and Roberts (1982), Arnold and Stewart (1986), Baker Hughes (1995), Bass (1987), Butler (1987), Chilingarian, Robertson, and Kumar (1987), Core Laboratories (1973), Dresser Atlas (1995), Earlogher (1977), Golan and Whitson (1991), Lee (1982), McCain (1990), Pedersen, Fredenslund, and Thomassen (1989), Riazi (2005), Rose (1987), Schlumberger Educational Services (1987), Welex (1987), and Whitson and Brulé (2000).

Pool History

4

Chapter Outline

4.1 Well History

Well history involves the physical description of the wells in a pool including: the casing and production tubing size and layout; the type of completion; the type and timing of workovers; and the type and timing of artificial lift. This information is used in some reservoir and production engineering calculations and aids in the interpretation of production and pressure data. It is also required for well and reservoir simulation.

4.1.1 Review of Drilling and Completions

Vertical wells

Let us first briefly review how a "typical" vertical well is drilled and completed. The first step is to drill an initial large diameter hole to a depth below any surface water-bearing formations. A relatively large diameter pipe is then cemented into place so that the annular space between the pipe and the surrounding earth is full of cement. This pipe is called the surface casing. Its purpose is to protect ground water from drilling, completion, and production fluids.

The well is then drilled to or below the depth of some target formation. While drilling, the rock is cut with a drill bit on the end of piping (the drill string) suspended from the drilling rig, as shown in Figure 4.1.1. The drill string consists of a series of pipe sections (joints) that are screwed together as they are lowered into the wellbore. A fluid (drilling mud) is circulated down the drill string and up the annulus between the drill string and the surrounding rock. The drilling mud cools the drill bit, assists in cutting through the rock, and carries rock cuttings away from the drill bit and up to the surface. There are many types of drilling mud that can be broadly classified as either oil-based or water-based muds. Once the target depth is reached, the drill string is pulled out of

Practical Reservoir Engineering and Characterization. http://dx.doi.org/10.1016/B978-0-12-801811-8.00004-3

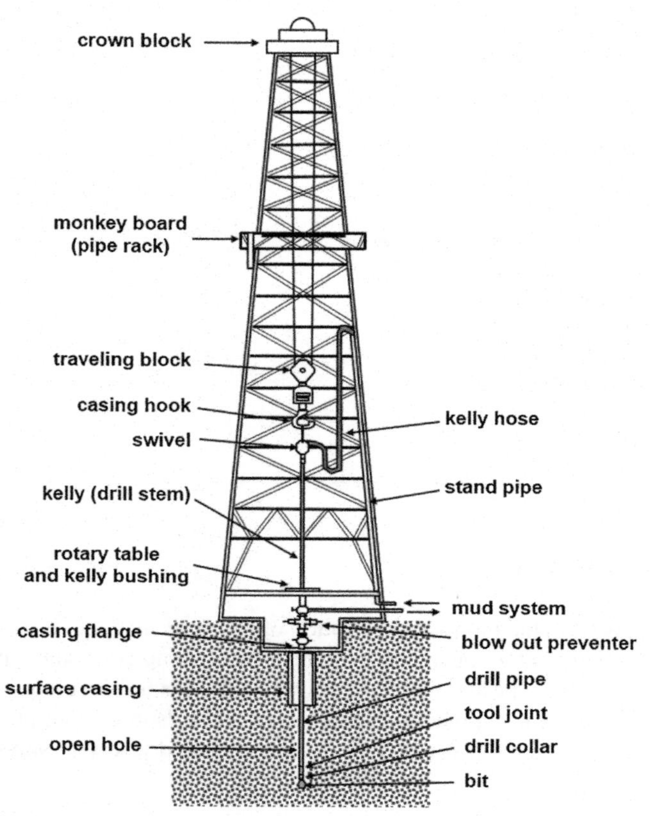

Figure 4.1.1 Schematic of drilling rig and drill string.

the hole, and another length of pipe (the production casing) is cemented in place. The purpose of the production casing is to prevent hole collapse and to isolate productive formations from one another.

The next step is to complete the well, that is, to open a pathway from the formation of interest to the wellbore. Figure 4.1.2 shows the three main types of completion: open hole, liner, and perforations. The most common form of completion is perforations, which involves firing a series of bullets or explosive charges through the casing and cement into the formation. To perform an open hole completion, the production casing is set at a depth above the formation of interest. Then the well is drilled below the casing into the formation of interest. While this method provides the most complete contact between the formation and the wellbore, there is nothing to prevent the collapse of the hole below the production casing. In some cases, the open hole interval is packed with gravel, but it is more common to set a liner (usually a section of slotted pipe) into the open hole interval. If hole sloughing still occurs, the liner may be cemented in place and then perforated. Some of the pros and cons of each completion technique from the perspective of a reservoir engineer are summarized in Table 4.1.1.

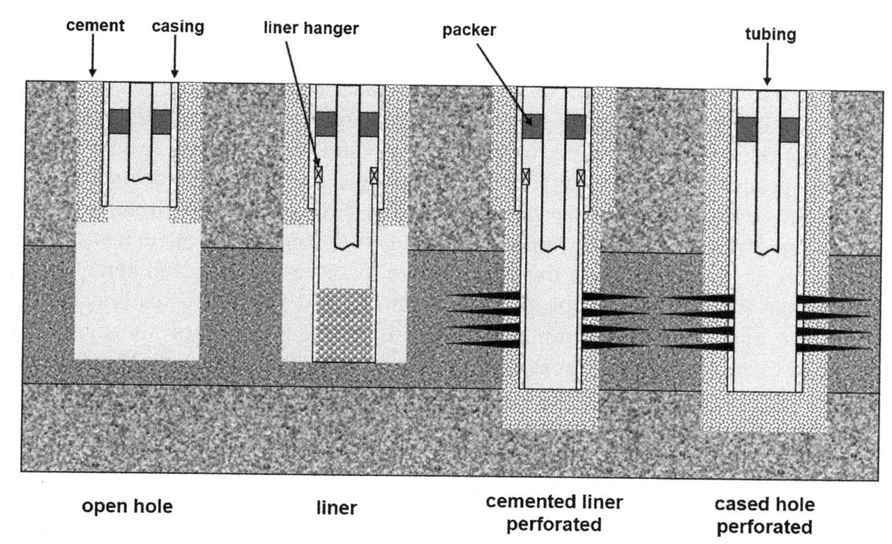

Figure 4.1.2 Common types of well completion.

Table 4.1.1 Pros and cons of different completion techniques from a reservoir engineering perspective

Technique	Pros	Cons
Open hole	No barrier between formation and wellbore	Sloughing No control over water or gas coning
Open hole—packed or lined	No barrier between formation and wellbore inhibits sand production	Sand or fines fill No control over water or gas coning
Cemented liner, perforated	Will fit in small diameter hole Allows control of height and location of fluid entry	Requires perforation to connect to formation More potential for formation damage Restricted flow in small diameter liner Little room to perform workover operations
Cased hole, perforated	Allows control of height and location of fluid entry	Requires perforation to connect to formation More potential for formation damage

In many cases, the action of drilling and completing a well damages the formation around the wellbore. For example, the drilling mud may cause fines in the formation to swell or move. These fines can block the pore throats in the reservoir so that permeability is reduced. Also, when perforating, the disintegrated cement and rock may plug the formation around the perforations. As a result, the well may not produce at all or may produce at lower rates than the undamaged formation could sustain.

There are several methods to remove damage and stimulate the well, but the most common are acidization and hydraulic fracturing. It is common to inject a small volume of acid (acid squeeze) into the formation to dissolve the fines and cement fragments around the wellbore. Typically, hydrochloric acid is used for carbonates and hydrofluoric acid (mud acid) for sandstones. Large acid squeezes are sometimes used in carbonate reservoirs to create deeper pathways for fluid flow and stimulate production above what even the undamaged well would be capable of. It is important to choose the acid type carefully, especially with sandstones, because an incompatible acid will further damage the formation.

Hydraulic fracturing is most commonly used for sandstones. Hydraulic fracturing (fracing) involves injecting fluid at high pressure to crack open the formation. Solid particles, "proppants," are injected into the open fractures to hold the fracture open after the pressure is reduced. The propped fractures create two planar conduits for fluid flow, as shown in Figure 4.1.3. Small fracs are used to bypass near-wellbore damage. Large fracs are used to stimulate low permeability wells to productivity above their undamaged potential. Acid fracs are sometimes used to stimulate carbonates.

Once the well is completed and if necessary stimulated, production tubing is lowered into the well and suspended from a wellhead. A schematic of a completed well is provided in Figure 4.1.4. It is sometimes desired to isolate the formation from the casing. In this case, a packer is set above the perforations or open hole interval. In other cases, the well may be incapable of flowing against the hydrostatic pressure in the wellbore. A down hole pump (artificial lift) may be installed to overcome this pressure. In unconsolidated formations, sand can be produced with the oil and can damage tubing and lift equipment and even fill up the wellbore. To control the sand, a gravel pack (coarse sand filling in a milled out larger diameter hole) or a slotted liner may be required.

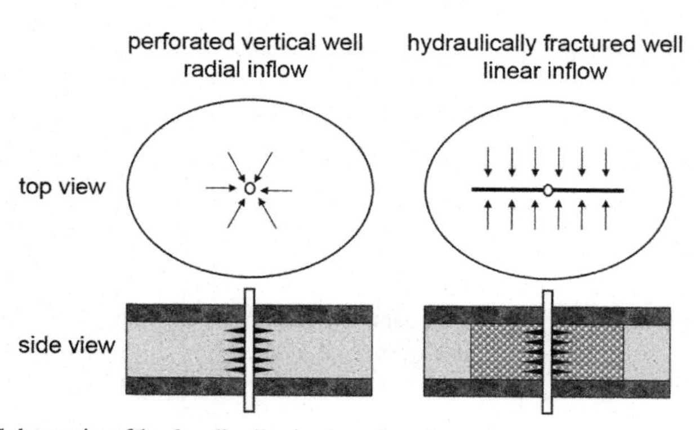

Figure 4.1.3 Schematic of hydraulically fractured well.

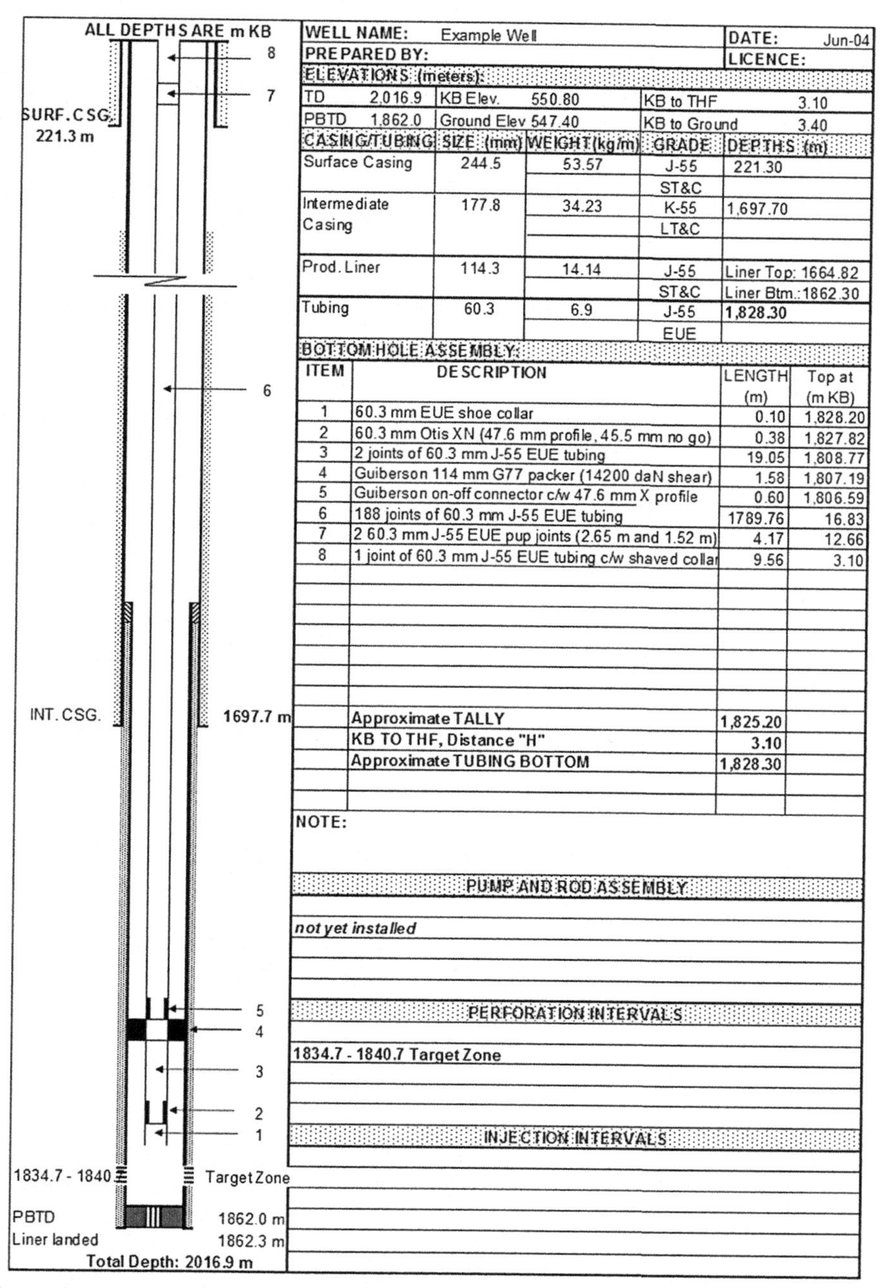

WELL NAME:	Example Well			DATE:	Jun-04
PREPARED BY:				LICENCE:	
ELEVATIONS (meters):					
TD 2,016.9	KB Elev. 550.80		KB to THF		3.10
PBTD 1,862.0	Ground Elev 547.40		KB to Ground		3.40

CASING/TUBING	SIZE (mm)	WEIGHT (kg/m)	GRADE	DEPTHS (m)
Surface Casing	244.5	53.57	J-55	221.30
			ST&C	
Intermediate Casing	177.8	34.23	K-55	1,697.70
			LT&C	
Prod. Liner	114.3	14.14	J-55	Liner Top: 1664.82
			ST&C	Liner Btm.: 1862.30
Tubing	60.3	6.9	J-55	**1,828.30**
			EUE	

BOTTOM HOLE ASSEMBLY:

ITEM	DESCRIPTION	LENGTH (m)	Top at (m KB)
1	60.3 mm EUE shoe collar	0.10	1,828.20
2	60.3 mm Otis XN (47.6 mm profile, 45.5 mm no go)	0.38	1,827.82
3	2 joints of 60.3 mm J-55 EUE tubing	19.05	1,808.77
4	Guiberson 114 mm G77 packer (14200 daN shear)	1.58	1,807.19
5	Guiberson on-off connector c/w 47.6 mm X profile	0.60	1,806.59
6	188 joints of 60.3 mm J-55 EUE tubing	1789.76	16.83
7	2 60.3 mm J-55 EUE pup joints (2.65 m and 1.52 m)	4.17	12.66
8	1 joint of 60.3 mm J-55 EUE tubing c/w shaved collar	9.56	3.10
	Approximate TALLY	**1,825.20**	
	KB TO THF, Distance "H"	**3.10**	
	Approximate TUBING BOTTOM	**1,828.30**	

NOTE:

PUMP AND ROD ASSEMBLY

not yet installed

PERFORATION INTERVALS

1834.7 - 1840.7 Target Zone

INJECTION INTERVALS

Figure 4.1.4 Schematic of a completed well.

Work on a well does not necessarily stop after the well is first put on production. Sometimes, fluid flow and pressure drop in the reservoir causes near-wellbore damage, for example, if fines are mobilized. A stimulation may be performed to regain the lost productivity. Or a decision may be made after reviewing a well's production history to stimulate the well above its undamaged potential, for example, to frac a low-permeability well. These stimulations that occur after initial production are called workovers. Workovers do not always involve the formation itself. Sometimes there is a problem in the wellbore or facilities. For example, wax build up may occur in the tubulars and the well may be treated with hot oil to remove the wax. Finally, as reservoir pressure decreases, the well may not be able to produce against hydrostatic pressure. In this case, an artificial lift may be installed.

Directional, horizontal, and multilateral wells

Directional wells are drilled at an angle from the surface in situations in which it is not practical to drill straight down to the desired location for the perforations, for example, when the target zone is under a lake. Directional wells are similar to vertical wells in terms of their intersection with the reservoir and production performance.

Horizontal wells are drilled vertically or directionally to the zone of interest and then drilled sideways into the formation. Horizontal well technology gained prominence in the late 1980s and has had a dramatic impact on the rate profiles and recovery factor of fields. There are four main reasons to select horizontal over vertical wells (Joshi, 1991; Springer et al., 2002):

- Horizontal wells dramatically increase the "contact" between the wellbore and reservoir, providing a larger drainage area and therefore fewer wells for the same production, Figure 4.1.5, (Joshi, 1988).

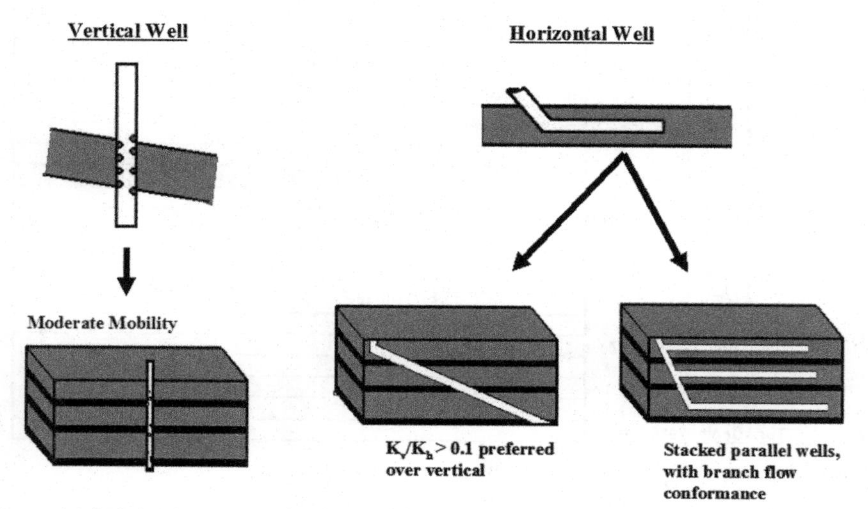

Figure 4.1.5 Schematic of vertical, horizontal, and directional drilled well applications: tight and laminated reservoirs.

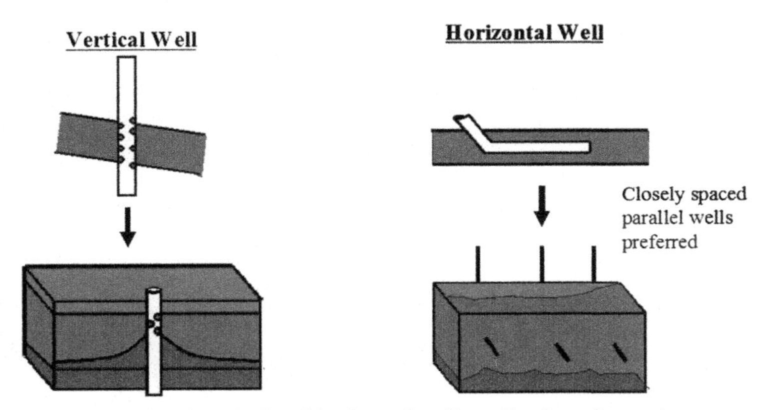

Figure 4.1.6 Schematic of vertical and horizontal well applications for coning.

- Horizontal wells can lower drawdown pressures for the same production rate compared to vertical wells and therefore are able to delay the onset of coning or cresting (Figure 4.1.6) and to control gas or water production from gas caps or waterleg better (Berge et al., 2006; Chaperon, 1986).
- In naturally fractured reservoirs, horizontal wells drilled perpendicular to open fracture systems can give large improvements in oil or fluid productivity, Figure 4.1.7 (Beliveau, 1995).
- In injection processes, horizontal wells may provide better volumetric sweep efficiency.

Horizontal wells are particular useful for low permeability reservoirs, because the contact area is so much larger; often horizontal wells will allow the development of a field, whereas vertical wells may be subeconomic. For the same reason, multilateral wells are now used for very low permeability fields. Multilateral wells are horizontal wells with more than one horizontal leg, Figure 4.1.8. A similar strategy is to hydraulically fracture a horizontal well at several intervals along its length (multifracing).

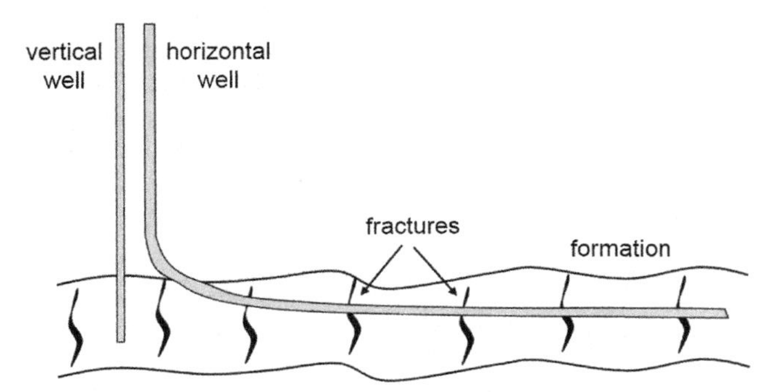

Figure 4.1.7 Schematic diagram of a horizontal well application. Note how the horizontal well intersects fractures in a naturally fractured reservoir, while the vertical well does not.

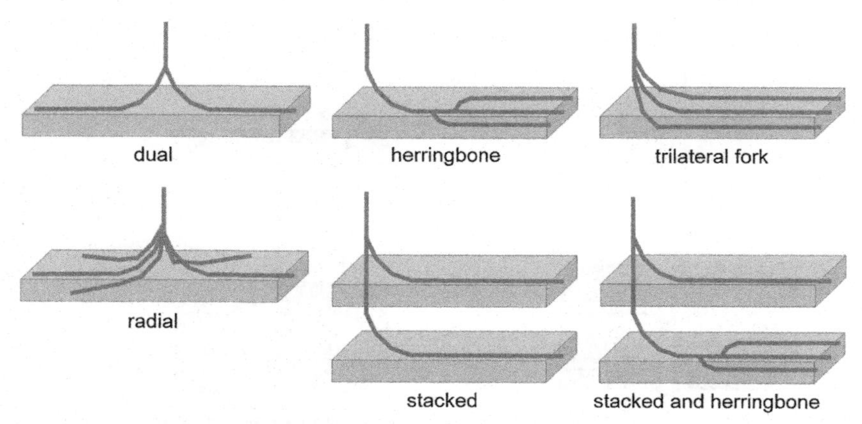

Figure 4.1.8 Horizontal multilateral well configurations.

4.1.2 Reservoir Engineering Issues

It is usually the role of a production engineer to ensure that a well is producing at its capability and that there are no unnecessary restrictions to its production. However, a reservoir engineer needs to be aware of these factors when assessing reservoir performance. *It is vital to distinguish between production behavior controlled by the reservoir and behavior controlled by the well and/or facilities.* There are also some well data that the reservoir engineer requires for analytical calculations and for reservoir simulation.

When interpreting the production history of a well or pool, it is necessary to identify well locations, the dates and types of well completions, workovers, and artificial lift. Many trends or spikes in production are a result of work on a well rather than reservoir dynamics. These human-caused effects must be identified before the reservoir performance can be assessed. If there are reservoir effects on well production, such as water breakthrough, they must be located correctly in the reservoir from the well location.

For reservoirs with gas caps or aquifers, the type of completion is important. An open hole completion usually provides little opportunity to mitigate coning. However, cased hole completions with perforations can be cement squeezed, and the perforations can be relocated when coning occurs. It is useful to assess the quality of the cement near the perforations using a cement bond log. If the cement is poor, water or gas may flow along channels in the cement, as shown in Figure 4.1.9. In this case, water or gas breakthrough may occur well before coning would occur.

Other useful information is the bit size used to drill the well and the wellbore configuration. The bit size is helpful in interpreting well logs (see Chapter 9). The wellbore configuration refers to the type and lengths of the casing, tubing, and down hole equipment in a well. The wellbore configuration is required when estimating pressure drop in a well. Wellbore pressure drop is not only useful for identifying production bottle necks, but also for relating wellhead to bottom hole pressures. Often only wellhead

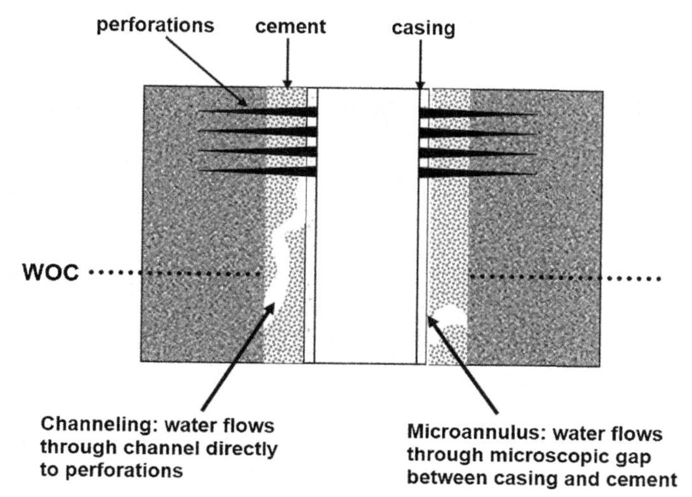

Figure 4.1.9 Schematic of water channeling behind cemented casing.

pressures are available, and an estimate of the bottom hole pressure can be made if the wellbore pressure drop is known.

Well history data are usually reliable, but not always available. Drilling, completion, and some artificial lift data are usually available from scout reports. Detailed descriptions of workovers and the wellbore configuration are usually only available in company well files.

4.1.3 Scout Reports

When well files are not available, the main and often only source of data is scout reports. Scout reports are sometimes available from commercial databases, and an example is given in Figure 4.1.10. A scout report is rather cryptic at first glance, so let us review some of the items applicable to reservoir engineering in detail. Note, this example is from Alberta, Canada, and details will differ in other countries.

Well location
The header begins with the well location (100/10-07-050-07W5/00 [AB]). The AB indicates that the well is in Alberta (Canada) and the numbers inside the slash marks (10-07-050-07W5) locate the well in the Western Canada grid of townships, ranges, sections, and legal survey descriptions (LSDs), as shown in Figure 4.1.11. This well location is LSD 10, section 7, township 50, range 7 west of the 5th meridian. The first three numbers (100) are used to distinguish between wells drilled on the same LSD. The last two numbers (00) are used to distinguish between different producing formations in the same well. In this case, the well is the only well on the LSD and is producing from only one formation.

The report provides enough information to locate the well in space as required for accurate mapping and initialization of reservoir simulations. Surface elevations are

UWID: 100/10-07-050-07W5/00 [AB] STATUS: Pumping OIL
PRD/INJ FORMATION: card_ss
NAME: AMOCO PEM 10-7-50-7 LAHEE: DEV

CURR OPER: Penn West Petrl Ltd [0BP8] LIC#: 0018607
ORIG OPER: Cnsld Mic Mac Oils Lmtd [#220] Date: 1960/01/29

GOVT KB: 832.1m GL: 827.8m MD: TVD: 1388.7m
LOGS KB: 832.1m GL: 827.8m FM@TD: card_ss

BH COORDS: S 603.5m NE 07-050-07W5 BH LAT: 53.30317 deg N
: W 603.5m BH LON: 115.00913 deg W

SPUD: 1960/02/19 CMPL DRL: 1960/02/23 DAYS: 5 RR:

STAT: 1960/01/29 Location; 1960/02/23 Drilled & Cased;
1960/03/07 Flowing; 1988/09/06 Pumping OIL
CASG: 219.1mm SRF @ 146.0m; 114.3mm PRD @ 1388.4m
LOGS: (WLX, Rm=3.020@41C) NEUT (1325.9 ~ 1382.3m),
LL (145.4 ~ 1380.4m), PERF (1367.3 ~ 1376.8m)

FORMATION TOPS
geoSCOUT REF ELEV: +832.1m

FORMATION	TVD	ELEV	FORMATION	TVD	ELEV
edmonton	260.6	+571.5	colorado	1192.7	-360.6
knhl_tuf	311.5	+520.6	cardium	1346.0	-513.9
belly_rv	794.9	+37.2	card_ss	1366.1	-534.0
lea_park	1067.7	-235.6			

PERFS/TREATMENTS

#01.	1960/03/04	Jet	1367.3 ~ 1367.9m	@ 7sh/m	card_ss
#02.	1960/03/04	Jet	1369.5 ~ 1370.1m	@ 7sh/m	" "
#03.	1960/03/04	Jet	1371.3 ~ 1371.6m	@ 7sh/m	" "
#04.	1960/03/04	Jet	1374.3 ~ 1375.0m	@ 7sh/m	" "
#05.	1960/03/04	Jet	1375.6 ~ 1375.9m	@ 7sh/m	" "
#06.	1960/03/04	Jet	1376.5 ~ 1376.8m	@ 7sh/m	" "
#07.	1960/03/05	AcidSq	1367.3 ~ 1376.8m		" "
#08.	1960/03/05	Fracd	1367.3 ~ 1376.8m		" "
#09.	1974/03/18	Fracd	1367.3 ~ 1376.8m		" "
#10.	1974/03/18	AcidSq	1367.3 ~ 1376.8m		" "
#11.	1978/07/20	Fracd	1367.3 ~ 1376.8m		" "

PRODUCTION SUMMARY
FIELD: PEMBINA[0685] ST: Pumping OIL card_ss
POOL: CARDIUM[0176000] ON: 1960/03
UNIT: Lobstick Cardium Unit [65623]
BATTERY: 6850680 Operator: Penn West Petrl Ltd

No: IP Tests, Cores, Oil Zone Pressure Tests, AOFPs/Pressure Tests or Formation Tests Reported.

Figure 4.1.10 Scout card from Geoscout™.

provided in the header. The ground level (GL) elevation at the well head is 827.8 m above sea level. Two elevations for the kelly bushing on the drilling rig floor (KB) are listed: GOVT KB from provincial government records, and LOGS KB from the header of the wireline logs run on this well. In this case, both reported elevations

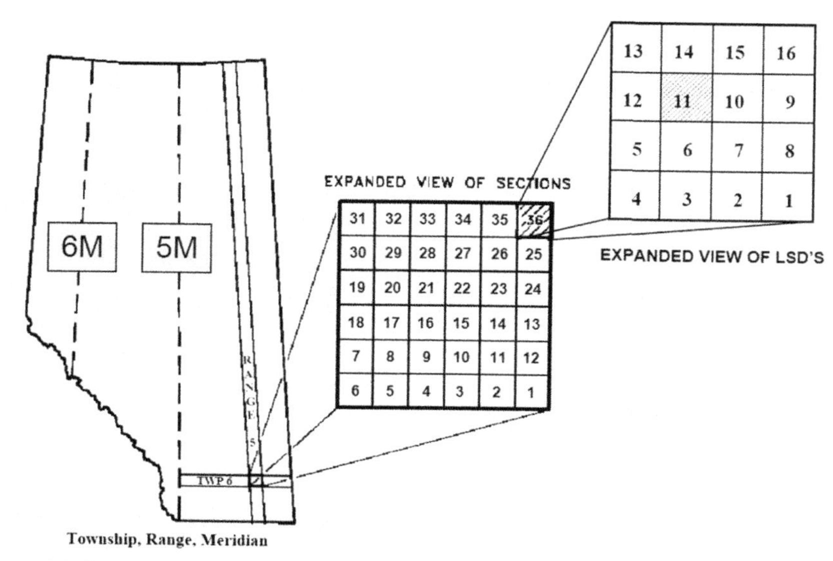

Figure 4.1.11 Reading locations in Western Canada. The dashed lines are meridians of longitude. The first expanded square is one township, which contains 36 sections. Each section contains 16 LSDs. The highlighted location is 11-36-06-05W4 (LSD 11, section 36, township 6, range 5, west of the 4th meridian).

are the same, 832.1 m above sea level. *All measured depths are reported relative to the kelly bushing.* The elevation and subsea depth are determined from the measured depth as follows:

Elevation = KB Elevation − Measured Depth
Subsea Depth = −Elevation = Measured Depth − KB Elevation

The measured depth (MD) and true vertical depth (TVD) at the bottom hole are reported in the second section of the scout report. In this case, the TVD and MD are the same, and only the TVD is listed, 1388.7 mKB (meters below kelly bushing). For deviated and horizontal wells, both MD and TVD are reported, and a deviation survey may also be available. The bottom hole formation (FM@TD) is listed and is the Cardium SS formation for this well.

The bottom hole coordinates are listed relative to the section boundaries and as a latitude and longitude. Here, under the BH COORDS heading, the bottom hole location is given as 603.4 m south and 603.5 m west of the NE corner of section 7-50-07W5. The latitude and longitude are self-evident.

Production status and wellbore configuration

The header also includes the current production status of the well and the current producing formation. In this case, the well is producing oil from the Cardium SS formation and is on artificial lift (pumping). Details of the historical well status are provided in the second section of the scout report under the SPUD and STAT headings. Here, the well location was approved for drilling in 1960/01/29, and the well was spudded (rig moved on site for drilling) on 1960/02/18 and cased by 1960/02/23. The rig release (RR) date is not provided for this well. The well was put on production in 1960/03/07 as a flowing well, and artificial lift was installed in 1988/09/06.

Some information of the wellbore configuration is provided under the CASG heading. This well has 219.1 mm surface casing to a depth of 146.0 mKB and 114.3 mm production casing to a depth of 1388.4 mKB.

Geological formations
The tops of distinct geological formations are summarized under the FORMATION TOPS heading. The tops are listed as a true vertical depth and as a subsea elevation. The reference elevation is the KB elevation of 832.1 m above sea level. This well encountered seven geological formations. The Cardium SS was the bottom formation at an elevation of −534 m subsea. The tops of the geological formations are a useful aid to log interpretation and area-wide mapping.

Perforations and stimulations
The perforation and treatment history of the well is summarized under the PERFS/TREATMENTS heading. In this case, the well was initially completed in the Cardium SS formation on 1960/03/04 with a series of jet perforations from 1367.3 to 1376.8 mKB with a perforation density of 7 shots per meter. This interval was acid squeezed and hydraulically fractured on the next day. No further details of the acid squeeze and frac are typically available. The well was worked over twice: an acid squeeze and frac in 1974/03/18 and another acid squeeze and frac in 1978/07/20.

Logs and cores
The intervals logged with open hole wireline instruments are also reported under the LOGS heading. Here, a Neutron log was run from 1325.9 to 1382.3 mKB, a Laterolog was run from 145.4 to 1380.4 mKB, and a perforation log (a Gamma Ray log) was run from 1367.3 to 1376.8 mKB. If the well had been cored, the cored interval would also be reported. This information is useful for quickly identifying what log and core data may be available for a given pool. Sometimes electronic copies of logs and core reports are also available.

Flow and pressure tests
An example of the pressure and formation test data from a scout report is provided in Figure 4.1.12. Pressure data are given under the OIL ZONE PRESSURE TEST SUMMARY heading. The field (Pembina), pool (Viking B), perforated interval (1994.2−1996.2 mKB) and the original pressure at a datum depth (18894 kPa at 1977.1 mKB) are summarized at the beginning of the table. Then, pressure test data for the particular well are listed. Each pressure test lists the test date, the pressure at the datum depth, the shut-in time, and the test type. For example, the first pressure test was a bottom hole static gradient (BHStGr) run on 1983/03/25 with a shut-in time of 826 h and a pressure of 18,580 kPa. More detailed reports can be obtained that also include the gauge depth, gauge pressure, fluid gradients, and the pressure at the midpoint of the perforations (MPP). Build-up and fall-off test data are not usually available through scout services.

```
~~~~~~~~~~~~~~~~~~~~~~~~~~~~~~~~~~~~~~~~~~~~~~~~~~~~~~~~~~~~~~~~~~~~~~~~~~~~
UWID: 100/16-23-045-07W5/00 [AB]        STATUS: Flowing OIL
                                        PRD/INJ FORMATION: vik_ss
NAME: PENN WEST ET AL PEMBINA 16-23-45-7         LAHEE: DEV
~~~~~~~~~~~~~~~~~~~~~~~~~~~~~~~~~~~~~~~~~~~~~~~~~~~~~~~~~~~~~~~~~~~~~~~~~~~~
PERFS/TREATMENTS
#01. 1983/01/07 Jet           1994.5 ~ 1996.5m @ 13sh/m          vik_ss
#02. 1983/02/05 AcidSq        1994.5 ~ 1996.5m                    "  "
#03. 1983/02/06 Fracd         1994.5 ~ 1996.5m                    "  "
~~~~~~~~~~~~~~~~~~~~~~~~~~~~~~~~~~~~~~~~~~~~~~~~~~~~~~~~~~~~~~~~~~~~~~~~~~~~
CORES
#01. 1989.0 ~ 2001.0m vik_ss   Rec 12.0m
~~~~~~~~~~~~~~~~~~~~~~~~~~~~~~~~~~~~~~~~~~~~~~~~~~~~~~~~~~~~~~~~~~~~~~~~~~~~
OIL ZONE PRESSURE TEST SUMMARY
Field : PEMBINA[0685]            Pool: VIKING B[0218002]
Intvl : 1994.2 ~ 1996.2mKB vik_ss   Porg:18894kPa @ 1977.1mKB Datum Depth
First : Pres #01.  18580kPa 1983/03/25  SI=     826Hr  Srvy=BHStGr
Last  : Pres #06.  6951kPa 1989/10/13  SI=    3072Hr  Srvy=BHStGr
~~~~~~~~~~~~~~~~~~~~~~~~~~~~~~~~~~~~~~~~~~~~~~~~~~~~~~~~~~~~~~~~~~~~~~~~~~~~
FORMATION TESTS
DST # 1. 1992.0 ~ 2000.0m  Fm: vik_ss       MD:  2030.0m  Date: 1982/12/16
 Source:  geoLOGIC
  VO: 15/60    SI: 30/180   Inflate Strad
  SFC: WAB < SAB, GTS 15 mn / SGB, GTS imm.
  GTS:  15 mn  Flo:Max     1.19  Final  1.19e3m3/d
  REC: 110.0m=100.0m cl OIL, 10.0m oc MUD
  HP: 21949/21949   FP: 625/896              SIP: 19004/19083
  P Un: kPa   RCD DPTH:  1993.0m              Slope: 87,05/88,02
  BHT: 59.8C                           XP: /19098
  OIL: 0.820SpGr      VISC: 1.9227mm2/s @ 59.8C
  RMK: High Perm indicated.
~~~~~~~~~~~~~~~~~~~~~~~~~~~~~~~~~~~~~~~~~~~~~~~~~~~~~~~~~~~~~~~~~~~~~~~~~~~~
```

Figure 4.1.12 Excerpted pressure and formation tests from scout report for well 16-23-45-7W5 from Geoscout™.

Formation test data are listed under the FORMATION TESTS heading. In Figure 4.1.12, the results of a drill stem test (DST) are provided. This DST includes a short build-up test and fluid sample data. The test was conducted in 1982/12/16 on the Viking SS formation over the interval 1992.0 to 2000.0 mKB. Let us consider the build up test first. The relevant symbols are:

VO = valve open, indicates the flowing time;
SI = shut-in time;
FP = flowing pressure;
SIP = shut-in pressure.

In this case, the well was flowed and shut-in twice during the test with the following results:

1. flow 15 min, shut-in 30 min, shut-in pressure = 19,004 kPa;
2. flow 60 min, shut-in 180 min, shut-in pressure = 19,083 kPa.

The build-up data were analyzed, and the extrapolated reservoir pressure was 19,098 kPa. The remarks (RMK) state that the test indicated high permeability. The

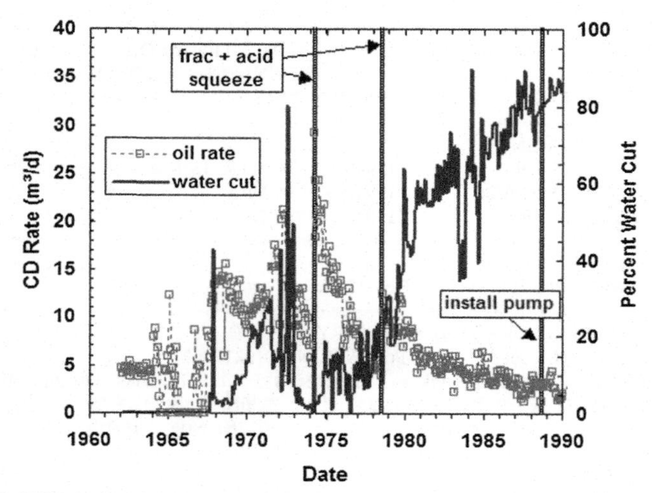

Figure 4.1.13 Effect of workovers on oil rate and water cut of well 10-07-50-07W5.

bottom hole temperature (BHT) was 59.8 °C, and the hydrostatic pressure was 21,949 kPa. Note that DST pressures are reported at the recorder depth (REC DPTH) of 1993.0 mKB.

A sample was also recovered (REC), which consisted of 100 m of clean oil and 10.0 m of oil cut mud. The oil had a specific gravity of 0.820 (OIL) and a kinematic viscosity (VISC) of 1.9227 mm^2/s at 59.8 °C. Observations made at the surface during the DST are summarized under the SFC heading. In this case, during the first flow period, there was a weak air blow (WAB) to strong air blow (SAB) and then gas to surface (GTS) after 15 min of flow. During the second flow period, there was a strong gas blow (SGB) and gas to surface immediately. The maximum (Flo:Max) and final (Final) gas flow rates were both $1.19 \cdot 10^3$ m^3/d (Figure 4.1.12).

Using scout data to interpret production data

The oil rate and water cut from the 10-07-50-7W5 well discussed above (Figure 4.1.6) are shown in Figure 4.1.13. There are several spikes in the water cut and oil production that may be a result of fluid flow in the reservoir or from workovers. Two workovers were performed on this well, the first in 1974 and the second in 1978. There is a clear increase in oil production after each workover. Artificial lift was reported in 1988. Based on the increase in oil rate and water cut, the actual implementation of artificial lift may have been in 1987.

4.1.4 Sources of Error

The two pieces of well data that are most subject to error are the bottom hole well location and the depth of the perforation interval. The surface location and elevation of a well are surveyed and are usually accurate. However, many wells are not perfectly vertical, but are deviated from vertical. Today, this deviation is monitored during drilling, and the areal location and elevation of any point in the well bore can be determined from the deviation survey. Errors are rare, but can arise when an angle is calculated or reported incorrectly. Older wells may not have a deviation survey, and there may be some error in assuming that the bottom hole location is the same as the surface location. Note that it is important to use the bottom hole location rather than the surface location for all reservoir mapping, interpretation, and simulation.

Depth (or elevation) can also be checked from a tally of the production casing or tubing because the number and length of each joint is recorded. This method is subject to human error, and depth is usually determined from a wireline measurement during well logging, as is discussed in Chapter 9. These depths are typically accurate to within 2 m.

It must also be kept in mind that scout reports are not always complete. In particular, treatments and pressure tests are often not reported. The installation of artificial lift may not be reported, and well status dates may reflect the time the well status was reported rather than the date the status change was implemented.

4.2 Production History

Production history refers to the recorded volumes of produced and injected oil, gas, water, or other injection fluids over time. These volumes are measured in surface facilities at the operating temperature and pressure of those facilities. The volumes are corrected to standard oilfield conditions (60 °F, 14.7 psia; 15.6 °C, 101 kPa). Production is reported as monthly volumes of oil, gas, and water. The number of hours a well was on production is also reported. These data are manipulated into a variety of amounts and ratios, some of which are listed in Table 4.2.1.

4.2.1 Surface Facilities

A simplified schematic of a typical oilfield battery is provided in Figure 4.2.1. The produced fluids from each well are gathered in individual flow lines. The flow lines converge on a production header at the inlet to the oil battery. At the header, the flow lines can be closed off, sent to a test separator, or sent to a group separator. The primary function of the separators is to remove the gas from the production fluids. The liquid stream from the group separator usually flows to a treater, where the oil and water are separated.

The production from an individual well is measured at the test separator, while the production from the group of wells is measured after the group separator. Each well is tested individually in the test separator for 2–3 days. The test time and frequency for each well in each field may vary dramatically, depending on the number of wells attached to a header and the flow rates into the test seperators. Wells with high flow rates tend to have better tests because there are shorter purge times and larger volumes associated with these wells. Group production is reported on a monthly basis. Group production volumes are assumed to be accurate. Production is allocated to each well based on prorated test separator production rates.

Table 4.2.1 Selected production and injection variables

Variable	Description
Monthly hours	Number of hours a well was on production or injection
Well count	Number of wells producing and/or injecting
Producing day oil rate	Monthly oil volume divided by hours on production times 24 h per day
Calendar day oil rate	Monthly oil volume divided by number of days in the month
Monthly oil rate	Monthly oil volume
Cumulative oil production	Sum of monthly oil volumes to date
Producing day gas rate	Monthly gas production volume divided by hours on production times 24 h per day
Producing day water rate	Monthly water production volume divided by hours on production times 24 h per day
Producing day fluid rate	Monthly oil and water production volumes divided by hours on production times 24 h per day
Injecting day water rate	Monthly water injection volume divided by hours on production times 24 h per day
Injecting day gas rate	Monthly gas injection volume divided by hours on production times 24 h per day
GOR (gas-oil ratio)	Ratio of monthly gas production volume to monthly oil volume
WOR (water-oil ratio)	Ratio of monthly water production volume to monthly oil volume
Water cut	Ratio of monthly water production volume to monthly fluid volume

At the test separator, gas is separated and its flow rate measured using an orifice meter. The fluid rate (oil plus water) is also measured. The water–oil ratio is measured with a spin test. A sample is withdrawn and centrifuged, and the volume of water and oil is measured.

After the group separator and treater, gas, oil, and water are metered individually before they leave the facility. The produced gas is first treated to control the dew point. For example, water may be condensed out of the gas in a glycol dehydration unit. The gas is then compressed, metered, and pipelined to a gas plant. Liquids such as condensate, butane, and propane may be recovered at the gas plant and allocated to individual wells based on their gas analysis and flow rate.

Oil is metered, for example, with a positive displacement meter and either pumped to storage tanks or a transmission pipeline. If the oil is sent to storage, any additional

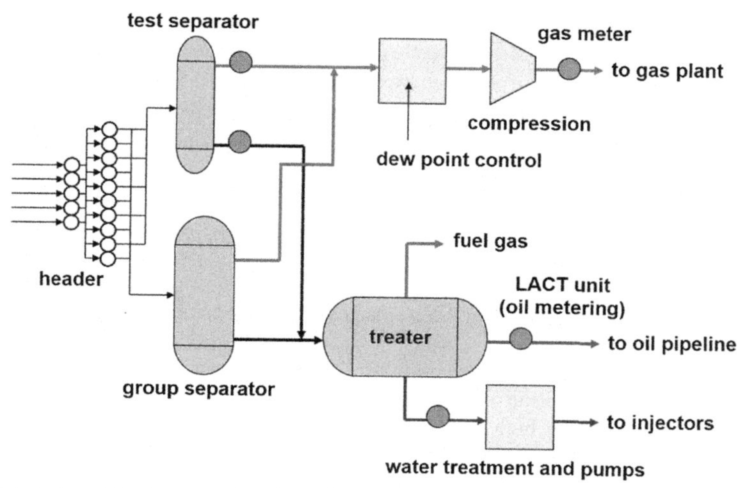

Figure 4.2.1 Simplified schematic of an oilfield battery.

gas that evolves is recaptured and used as fuel gas or returned to the treater. Water is metered, pumped to storage tanks, and allowed to settle. Oil is skimmed from the surface and returned to the treater. The water is then trucked or pipelined to a water handling facility, where it is usually filtered, sometimes treated to remove calcium and other scale forming components, and then pumped to an injector.

Oilfield batteries may be more complex than the simple battery shown in Figure 4.2.1. For example, there may be a series of group separators, a free water knock out, more elaborate water treatment, and so on. Also, oil production may be tested in satellites that feed a main battery where the group separation occurs. In other cases, when an oil well is geographically isolated or newly on production, separation may occur in a single well battery. Nonetheless, Figure 4.2.1 is sufficient to explain typical oilfield separation and metering, as well as potential sources of error.

Gas field facilities are more straightforward than oil facilities. Gas is tied directly into a pipeline to a gas plant. It will usually be treated with methanol, heated, or passed through a dehydration unit to avoid hydrate formation. Compression may be added at the well site, distributed within a gathering system, and/or located at the gas plant inlet. Gas flow rates are metered at the well site and the plant inlet.

4.2.2 Sources of Error

The most obvious source of error is metering error; for example, an incorrect calibration in an orifice meter. Metering error is occasionally a problem with historical gas production, particularly if the gas was flared and not recovered for sales. In this case, there was no incentive to accurately monitor the gas. Similarly, water production is not always monitored carefully. Oil production volumes are usually the most reliable. For example, Alberta's measurement standards for group production from an oil battery are ±0.5% for oil and ±3.0% for gas and water (Alberta Oil and Gas Conservation Regulations, 2004).

Another source of error is in the allocation of production to individual wells. Production is prorated based on a two- to three-day test. The production during this test may not be representative of the average monthly production, particularly if the well exhibits intermittent slugs of gas or water production (spikes in the GOR or water cut). Furthermore, the individual well water cut is based on centrifuge tests from a small number of samples. Sampling error can be high. Also, if the water is emulsified in the oil, the water cut from a spin test is unreliable.

Production volumes are usually the most accurate data available from a reservoir. If production volume errors are suspected:

1. Check for step changes in the GOR. Step changes occur when the orifice meter calibration is checked and corrected.
2. Examine individual well production profiles. An inconsistent profile that has no reservoir-based explanation could result from an allocation error; for example, if a high water cut is reported for a structurally high well, while a low water cut is reported for a structurally low well.
3. Whenever possible, talk to the operators, learn the history of the field operations, and identify any potential sources of error. In fact, a good engineer should always discuss the field with the operators.

Production volumes are surface volumes and must be converted to reservoir conditions for reservoir calculations. This conversion will be discussed in Chapter 6 and can also be a source of error.

4.2.3 Types of Production Plots

A summary of some commonly used primary production plots is provided in Table 4.2.2. Each type of plot is reviewed below.

Composite plots: Production plots are used for single wells, groups of wells, and for entire pools. The most common format for viewing oil production data is a composite plot, that is, combined plots of oil rate, GOR, and water cut versus time or cumulative production. The data can be plotted on Cartesian or semilog coordinates. Example pool production plots versus time are provided in Figures 4.2.2 and 4.2.3 on semilog and Cartesian coordinates, respectively.

Daily production rates are a useful diagnostic tool. There are two forms of daily rate: calendar day (CD) rate and producing day (PD) rate. CD rate is defined as:

$$q_{CD} = \frac{q_{month}}{days} \tag{4.2.1}$$

while PD rate is defined as:

$$q_{PD} = q_{month}\frac{op\ hours}{hours} \tag{4.2.2}$$

in which q_{month} is the monthly production volume, *days* is the number of days in the month, *op hours* is the number of hours the well was on production, and *hours* is the number of hours in the month. If a well is on continuous production, $q_{CD} = q_{PD}$. If a

Table 4.2.2 **Types of production plots for oil reservoirs**

Plot	Y-axis	X-axis	Purpose
Composite	Cartesian • Oil rate • Fluid rate • Water injection rate • Gas injection rate • GOR • Water cut • Operating hours • Well count	Log or Cartesian Time or cumulative oil production	• Assess drive mechanism • Assess response to injection • Find breakthrough time • Decline analysis • Estimate recovery factor • Identify patterns in well performance • Identify workover candidates
Cumulative	Cartesian • Cumulative oil production • Cumulative gas production • Cumulative water production • Cumulative gas injection • Cumulative water injection	Cartesian Time or cumulative oil production	• Relate oil decline with fluid withdrawals
Cumulative gas–cumulative oil	Log • Cumulative gas production	Log or Cartesian Cumulative oil production	Solution gas drives • Forecast GOR • Forecast ultimate oil recovery
Purvis	Log • Cumulative fluid production • Daily fluid rate • Daily oil rate • WOR+1	Cartesian Cumulative oil production	Water drives/ waterfloods • Forecast production • Assess waterflood and waterflood performance Lo et al. (1990) and Startzman and Wu (1984)
Ershaghi (Ershaghi and Omoregie, 1978)	$(1/f_w - \ln(1 - f_w) - 1)$ f_w = water cut	Cumulative oil production	Water drives/ waterfloods • Forecast production • Assess waterflood and waterflood performance

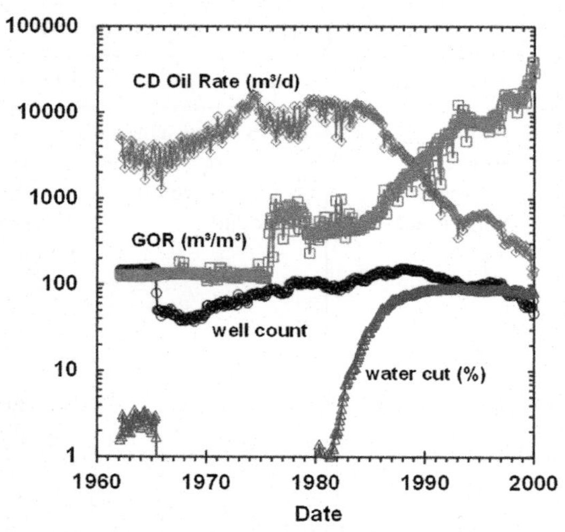

Figure 4.2.2 Semilog composite plot of pool production versus time for a combination drive reservoir.

well is produced intermittently, then $q_{CD} < q_{PD}$. PD rate is a better indication of the well's productivity, as long as the well is produced for sufficient hours to achieve a stabilized rate. CD rate is a better measure of the average producing rate of the well.

The rate versus time plots, Figures 4.2.2 and 4.2.3, are useful for identifying the oil decline rate and the time of gas and/or water breakthrough. The rate versus time plot is also useful when attempting to assess the effect of workovers or infill drilling. For this reason, the number of producing and/or injecting wells is usually included, too. The use of composite plots is discussed in more detail in Part 3.

The rate versus cumulative oil production plot, Figure 4.2.4, is also used for identifying the oil decline rate, but also for material balances and some reservoir forecasting models. It can be helpful in assessing the reservoir drive mechanism and for estimating pool oil recovery. Cumulative production of oil, water, and gas can also be plotted versus time or cumulative oil production. These plots are often used when assessing history matches from analytical or numerical models.

Cumulative gas–cumulative oil plot: When assessing solution gas drive reservoirs, a log–log or semilog plot of cumulative gas production (Q_g) versus cumulative oil production (Q_o) can be used as a forecast tool (Cronquist, 2001). For reservoirs with nonsegregated gas flow, there is often a linear trend of Q_g versus Q_o. The trend can be used to forecast cumulative gas production and can be extrapolated to determine the ultimate recoverable oil (Figure 4.2.5).

To forecast GOR, it is sufficient to calculate cumulative gas production as a function of cumulative oil production from the linear trend (usually on a monthly basis). The GOR is simply:

$$\text{GOR} = \frac{\Delta Q_g}{\Delta Q_o} \tag{4.2.3}$$

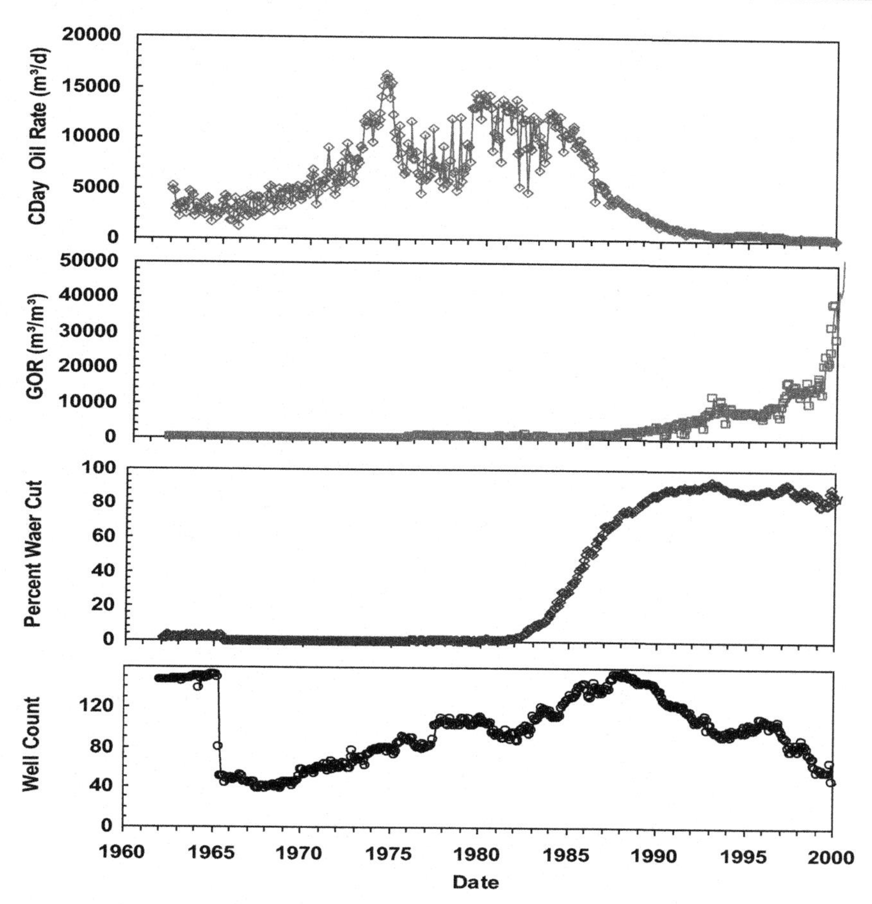

Figure 4.2.3 Cartesian composite plot of pool production versus time for a combination drive reservoir.

in which the incremental oil (ΔQ_o) and gas production (ΔQ_g) values are recalculated at each month of the forecast.

To extrapolate to the ultimate oil recovery, it is assumed that the solution gas in place is known and that the ultimate solution gas recovery factor is approximately 90%. Typical solution gas recoveries range from 60% to 95%. The linear trend is then extrapolated to a Q_g equal to the recoverable solution gas in place. The underlying concept in the plot is that most of the drive energy in a solution gas drive reservoir comes from gas compressibility, so when most solution gas is produced, the reservoir pressure will be small. The cumulative oil production at that point is the recoverable oil, as shown in Figure 4.2.5.

Purvis plot: When assessing water drives or waterfloods, the water–oil ratio (WOR) is sometimes used instead of the water cut (Purvis, 1987). A semilog plot of

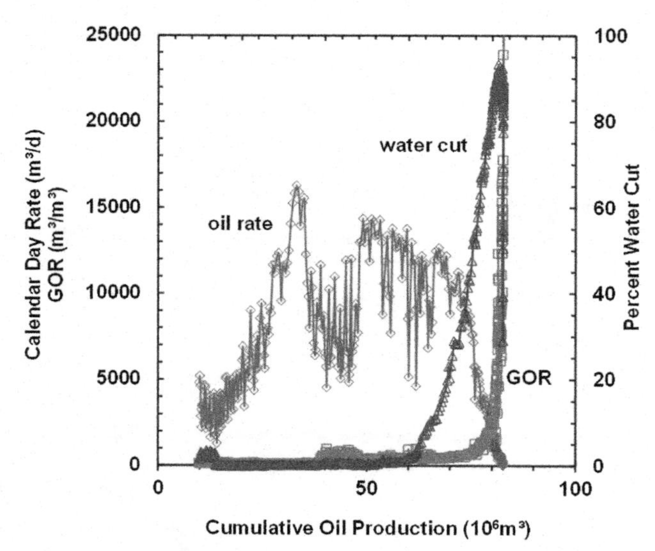

Figure 4.2.4 Cartesian plot of pool production versus cumulative oil production for a combination drive reservoir.

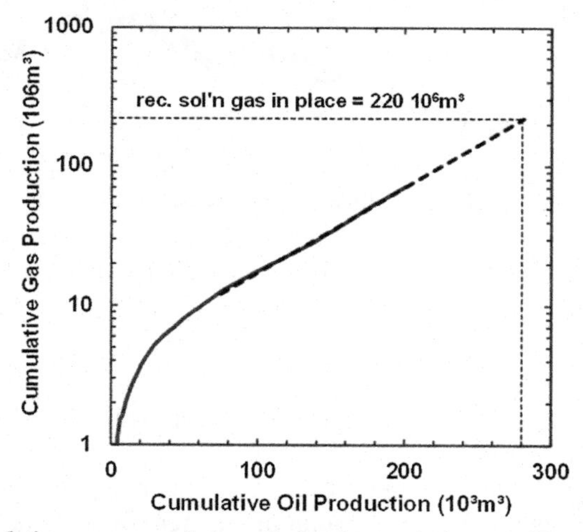

Figure 4.2.5 Cumulative gas–cumulative oil plot for a solution gas drive reservoir.

WOR or (WOR + 1) versus cumulative oil production can sometimes be used to forecast production in conjunction with a plot of fluid rate versus time or cumulative oil production. A Purvis plot combines on a single semilog plot cumulative fluid production, daily fluid rate, daily oil rate, and WOR + 1, all versus cumulative oil production. Figure 4.2.6 is an example of a nearly linear trend in WOR + 1 on a Purvis plot.

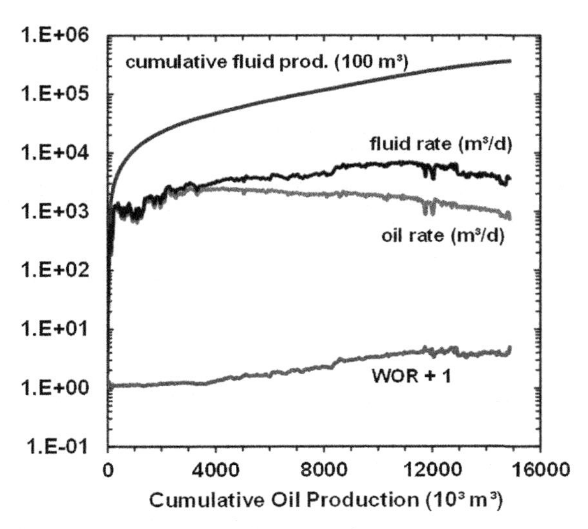

Figure 4.2.6 Purvis plot of for a water drive reservoir.

To make a forecast, the fluid rate is usually assumed to be constant, and the WOR is predicted from the linear trend. The oil rate is then given by:

$$q_o = \frac{1}{1 + \mathrm{WOR}} q_f \tag{4.2.4}$$

in which q_f is the daily fluid rate.

If there is injection, a plot of injection rate versus time is included in the composite plot. A comparison of oil rates, GORs, and water cuts with injection rates can provide a qualitative assessment of the reservoir response to the injection.

Production plots for gas reservoirs: Composite and cumulative production plots are also used for gas reservoirs. In this case, gas rate, water rate, cumulative gas production, cumulative water production, water–gas ratio (WGR), and well count can be plotted versus time or cumulative gas production. If there is injection, such as a retrograde condensate reservoir with dry gas injection, then the injection rates are included in the composite and cumulative production plots. Note that reservoir gas is saturated with water, and therefore the WGR is always greater than zero. A WGR rising above the saturation value is an indication of water production from an underlying water zone or aquifer.

One thing to note when we are interpreting WGR is that sometimes high liquid production will "kill" a well. In other words, the combined effect of vapor plus condensed or free liquids in the reservoir means that the hydrostatic gradient in the wellbore results in a low tubing head pressure, and thus low and unstable flow rates. In these cases, there may not be long-term production trends for WGR or condensate–gas ratio.

Another useful plot when examining gas wells is a plot of flowing wellhead pressure versus time or cumulative gas (Mattar and McNeil, 1998). If the reservoir has reasonably good permeability and is in transient flow only for a short time, then the decline rate of the flowing wellhead pressure can be used as a proxy for the decline in reservoir pressure. In shallower dry gas reservoirs, there is often not a substantial difference between wellhead pressure and bottom hole flowing pressures. Therefore, reservoir pressure trends can be estimated from the wellhead pressure. In this case, the amount of gas in place can be approximated from wellhead flowing pressure data converted to reservoir pressure and used in a gas material balance (Chapter 3). A preliminary plot of wellhead pressure versus time or cumulative gas will identify whether there is a suitable pressure trend to perform the material balance.

Fluid Properties (PVT Data)

Chapter Outline

Fluid properties are required for almost all reservoir calculations, including volumetrics, material balances, and flow equations. In this chapter, we review fluid data sources, the analysis of fluid studies, fluid property correlations, and how to check and correct fluid data. Heavy oil fluid properties and specialized fluid studies are also reviewed.

Practical Reservoir Engineering and Characterization. http://dx.doi.org/10.1016/B978-0-12-801811-8.00005-5

5.1 Fluid Property Assays and Studies

5.1.1 Fluid Analyses

Gas, condensate, oil, and water analysis can be conducted on samples collected after separation in a battery or from flow testing equipment at a well site. Samples can also be obtained downhole using a wireline sampling chamber. Each type of analysis is discussed below.

Gas analysis

A gas analysis, Figure 5.1.1, includes a molar compositional analysis, usually of hydrocarbons from methane to hexane or heptane as well as hydrogen, helium, hydrogen sulfide, nitrogen, and carbon dioxide. Any higher hydrocarbons are lumped into a single fraction, such as a C_{7+} (heptane and higher hydrocarbon numbers) fraction. The molar mass, density, critical properties, and heating value of the gas are

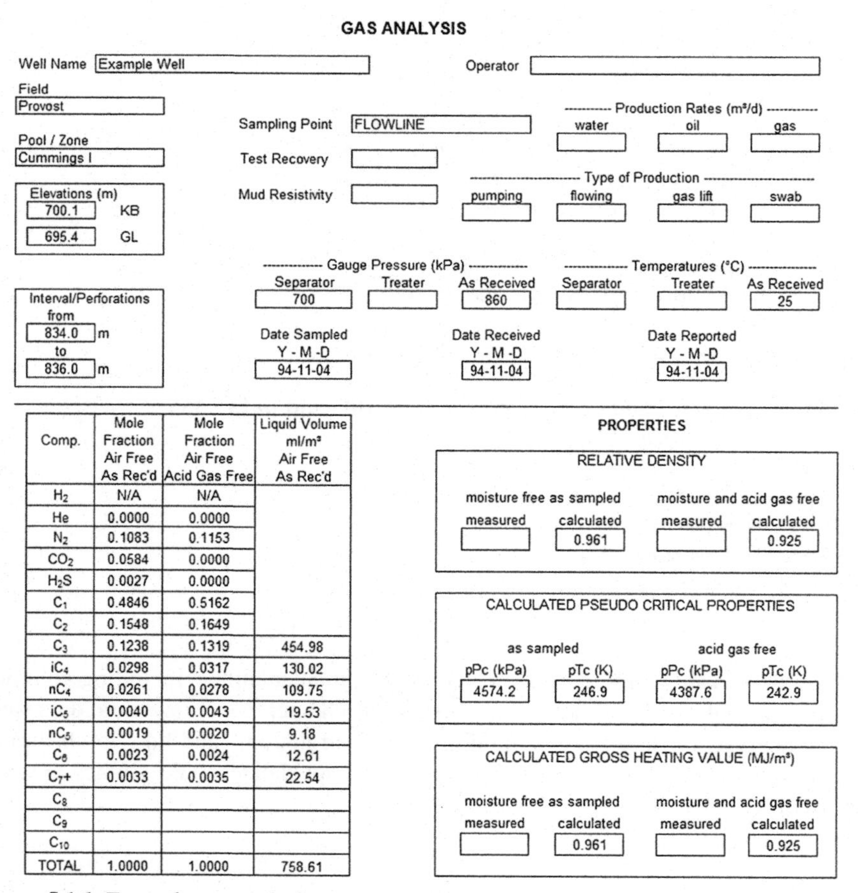

GAS ANALYSIS

Well Name: Example Well
Operator:
Field: Provost
Pool / Zone: Cummings I
Sampling Point: FLOWLINE
Test Recovery:
Mud Resistivity:
Elevations (m): 700.1 KB; 695.4 GL
Production Rates (m³/d): water | oil | gas
Type of Production: pumping | flowing | gas lift | swab
Interval/Perforations: from 834.0 m to 836.0 m

Gauge Pressure (kPa)			Temperatures (°C)		
Separator	Treater	As Received	Separator	Treater	As Received
700		860			25

Date Sampled Y-M-D	Date Received Y-M-D	Date Reported Y-M-D
94-11-04	94-11-04	94-11-04

Comp.	Mole Fraction Air Free As Rec'd	Mole Fraction Air Free Acid Gas Free	Liquid Volume ml/m³ Air Free As Rec'd
H_2	N/A	N/A	
He	0.0000	0.0000	
N_2	0.1083	0.1153	
CO_2	0.0584	0.0000	
H_2S	0.0027	0.0000	
C_1	0.4846	0.5162	
C_2	0.1548	0.1649	
C_3	0.1238	0.1319	454.98
iC_4	0.0298	0.0317	130.02
nC_4	0.0261	0.0278	109.75
iC_5	0.0040	0.0043	19.53
nC_5	0.0019	0.0020	9.18
C_6	0.0023	0.0024	12.61
C_7+	0.0033	0.0035	22.54
C_8			
C_9			
C_{10}			
TOTAL	1.0000	1.0000	758.61

PROPERTIES

RELATIVE DENSITY

moisture free as sampled		moisture and acid gas free	
measured	calculated	measured	calculated
	0.961		0.925

CALCULATED PSEUDO CRITICAL PROPERTIES

as sampled		acid gas free	
pPc (kPa)	pTc (K)	pPc (kPa)	pTc (K)
4574.2	246.9	4387.6	242.9

CALCULATED GROSS HEATING VALUE (MJ/m³)

moisture free as sampled		moisture and acid gas free	
measured	calculated	measured	calculated
	0.961		0.925

Figure 5.1.1 Example gas analysis.

determined. Properties are often also reported on an acid gas (H_2S and CO_2)-free basis, because these components are removed during processing to obtain a sales gas.

Gas density and viscosity are required for reservoir calculations. Gas density can be determined at any temperature and pressure using the nonideal gas law:

$$\rho_g = \frac{MP}{zRT} \tag{5.1.1}$$

in which ρ_g is the gas density, M is the molar mass, z is the compressibility factor, and R is the universal gas constant. Gas viscosity is usually determined from correlations. More details are provided in Section 5.3.

Condensate analysis

Often a gas has associated liquids (condensate) that drop out of the gas as the pressure decreases from the reservoir to the surface. For gas reservoirs and in some cases for gas-cap gas, the condensate is analyzed in a similar manner to a gas analysis. A summary of the lighter fractions usually up to hexane or heptane is provided, as shown in Figure 5.1.2. The density and molar mass of the lumped higher hydrocarbon fractions are also listed in the summary. A more detailed compositional analysis is usually performed as well, as shown in Figure 5.1.3.

Gas-condensate recombination

If the surface volumes of the gas and liquids are known, the gas and liquid analyses can be recombined into a single compositional analysis of the original single phase fluid, as shown in the following example. The composition and properties of the recombined analysis are used for reservoir calculations.

Example gas-condensate recombination: Assume that the gas and condensate analyses given in Figures 5.1.1 and 5.1.2 were obtained from a well flowing at 120×10^3 m^3/day gas (at standard conditions of 15 °C and 101 kPa) and 35 m^3/day condensate at 15 °C. Find the recombined gas analysis.

The first step is to determine the molar flow rates of gas and condensate. At standard conditions, the molar flow rate of the gas, n_g, is determined using the ideal gas law:

$$n_g = \frac{q_g P_s}{RT_s} = \frac{(120\ \text{m}^3/\text{day})(101.325\ \text{kPa})}{(8.314\ \text{kPa}\cdot\text{m}^3/\ \text{kmol}\cdot\text{K})(273.15 + 15\ \text{K})} = 5075\ \text{kmol/day}$$

The molar flow rate of the condensate, n_{liq}, is determined from its density and molar mass as follows:

$$n_{liq} = \frac{q_{liq}\rho_{liq}}{M_{liq}} = \frac{(5\ \text{m}^3/\text{day})(908\ \text{kg/m}^3)}{268\ \text{kg/kmol}} = 119\ \text{kmol/day}$$

in which the density and molar mass are taken from the condensate analysis.

The second step is to find the molar flow rate of each component in both the gas and liquid streams, sum the flow rate for each component, and calculate the composition of the combined stream. The calculations for this example are summarized in Table 5.1.1.

CONDENSATE ANALYSIS

Well Name: Example Well
Operator:
Field: Provost
Pool / Zone: Cummings I
Sampling Point: FLOWLINE
Test Recovery:
Mud Resistivity:

Elevations (m): 700.1 KB; 695.4 GL

Interval/Perforations: from 834.0 m to 836.0 m

Production Rates (m^3/d)		
water	oil	gas

Type of Production			
pumping	flowing	gas lift	swab

Gauge Pressure (kPa)			Temperatures (°C)		
Separator	Treater	As Received	Separator	Treater	As Received
700		860			25

Date Sampled (Y - M - D)	Date Received (Y - M - D)	Date Reported (Y - M - D)
94-11-04	94-11-04	94-11-04

Comp.	Mole Fraction	Mass Fraction	Volume Fraction
N_2	0.0003	TRACE	TRACE
CO_2	0.0115	0.0019	0.0021
H_2S	0.0025	0.0003	0.0004
C_1	0.0330	0.0020	0.0060
C_2	0.0219	0.0025	0.0063
C_3	0.0344	0.0057	0.0101
iC_4	0.0188	0.0041	0.0066
nC_4	0.0223	0.0048	0.0075
iC_5	0.0087	0.0023	0.0034
nC_5	0.0054	0.0015	0.0021
C_6	0.0121	0.0039	0.0051
C_7+	0.8291	0.9710	0.9504
C_8			
C_9			
C_{10}			
C_{11}			
C_{12}			
TOTAL	1.0000	1.0000	758.61

Residue	Relative Density @ 15°C		Relative Molar Mass	
	Observed	Calculated	Observed	Calculated
C5+		0.925		307
C6+		0.927		311
C7+	0.928		314	
C10+				
C12+				
Total Sample		0.908		268

Residue	DATA SUMMARY		
	Mole Fraction	Mass Fraction	Vol. Fraction
C5+	0.8553	0.9787	0.9610
C6+	0.8412	0.9749	0.9555
C7+	0.8291	0.9710	0.9504
C10+	0.7116	0.9028	0.8802
C12+	0.6439	0.8498	0.8258

Absolute density of stabilized condensate @ 15°C: 923 kg/m^3

Total sulfur of stabilized condensate (mass fraction): 0.0212

Figure 5.1.2 Summary of condensate analysis.

Once the composition is known, the molar mass and critical properties of the recombined stream can be determined using molar averages of the component properties. The density can then be determined at any temperature and pressure using the nonideal gas law.

Oil analysis

An oil analysis usually includes density, viscosity, pour point, and distillation data, as shown in Figure 5.1.4. The water, solids (BS), and sulfur contents are also included. This information is required to determine the quality and value of the oil and to ensure pipeline and refinery specifications are met. An oil analysis may also include a detailed compositional analysis similar to the condensate analysis presented in Figure 5.1.3. This detailed analysis is required for composition fluid models.

Probably the most commonly used oil fluid property is simply the oil density. The oil density is an important parameter for many reservoir calculations and is an input for

CONDENSATE ANALYSIS

DETAILED COMPONENT SUMMARY

well name Example Well
sample point FLOWLINE

COMPONENT	BOILING POINT (°C)	MOLE FRACTION	MASS FRACTION	VOLUME FRACTION
NITROGEN	-196	0.0003	TRACE	TRACE
CARBON DIOXIDE	-79	0.0115	0.0019	0.0021
HYDROGEN SULPHIDE	-60	0.0025	0.0003	0.0004
METHANE	-162	0.0330	0.0020	0.0060
ETHANE	-89	0.0219	0.0025	0.0063
PROPANE	-42	0.0344	0.0057	0.0101
ISO-BUTANE.	-12	0.0188	0.0041	0.0066
N-BUTANE	-1	0.0223	0.0048	0.0075
ISO-PENTANE	28	0.0087	0.0023	0.0034
N-PENTANE	36	0.0054	0.0015	0.0021
HEXANES	69	0.0110	0.0036	0.0048
HEPTANES	98	0.0178	0.0099	0.0111
OCTANES	126	0.0260	0.0162	0.0176
NONANES	151	0.0195	0.0136	0.0144
DECANES	174	0.0318	0.0242	0.0253
UNDECANES	196	0.0312	0.0256	0.0263
DODECANES	216	0.0411	0.0363	0.0369
TRIDECANES	236	0.0480	0.0451	0.0454
TETRADECANES	253	0.0444	0.0441	0.0440
PENTADECANES	271	0.0513	0.0535	0.0530
HEXADECANES	287	0.0452	0.0494	0.0487
HEPTADECANES	302	0.0535	0.0616	0.0609
OCTADECANES	317	0.0420	0.0517	0.0503
NONADECANES	331	0.0324	0.0421	0.0408
EICOSANES	343	0.0342	0.0480	0.0464
HENEICOSANES	357	0.0318	0.0469	0.0453
DOCOSANE'S	369	0.0305	0.0458	0.0441
TRICOSANES	380	0.0296	0.0460	0.0443
TETRACOSANES	391	0.0275	0.0440	0.0420
PENTACOSANES	402	0.0261	0.0431	0.0412
HEXACOSANES	412	0.0201	0.0341	0.0326
HEPTACOSANES	422	0.0197	0.0344	0.0326
OCTACOSANES	432	0.0183	0.0330	0.0313
NONACOSANES	441	0.0165	0.0305	0.0289
TRIACONTANES+	449	0.0317	0.0602	0.0571
BENZENE	80	0.0000	0.0000	0.0000
TOLUENE	111	0.0040	0.0020	0.0018
XYLENE	139	0.0139	0.0082	0.0071
1,2,4-TRIMETHYLBENZENE	169	0.0047	0.0032	0.0028
CYCLOPENTANE	49	0.0011	0.0003	0.0003
METHYL CYCLOPENTANE	72	0.0092	0.0043	0.0043
CYCLOHEXANE	81	0.0085	0.0039	0.0039
METHYL CYCLOHEXANE	101	0.0186	0.0101	0.0100
TOTALS		1.0000	1.0000	1.0000

Figure 5.1.3 Detailed compositional analysis of a condensate.

Table 5.1.1 Gas–liquid recombination calculations

Component	Gas mole fraction	Liquid mole fraction	Gas flow rate (kmol/day)	Liquid flow rate (kmol/day)	Recombined flow rate (kmol/day)	Recombined mole fraction
N_2	0.1083	0.0003	549.7	0.0	549.7	0.1058
CO_2	0.0584	0.0115	296.4	1.4	297.8	0.0573
H_2S	0.0027	0.0025	13.7	0.3	14.0	0.0027
C_1	0.4846	0.0330	2459.5	3.9	2463.4	0.4743
C_2	0.1548	0.0219	785.7	2.6	788.3	0.1518
C_3	0.1238	0.0344	628.3	4.1	632.4	0.1218
iC_4	0.0298	0.0188	151.2	2.2	153.5	0.0295
nC_4	0.0261	0.0223	132.5	2.6	135.1	0.0260
iC_5	0.0040	0.0087	20.3	1.0	21.3	0.0041
nC_5	0.0019	0.0054	9.6	0.6	10.3	0.0020
C_6	0.0023	0.0121	11.7	1.4	13.1	0.0025
C_{7+}	0.0033	0.8291	16.7	98.3	115.1	0.0222
Total	1.0000	1.0000	5075.4	118.6	5194.0	1.0000

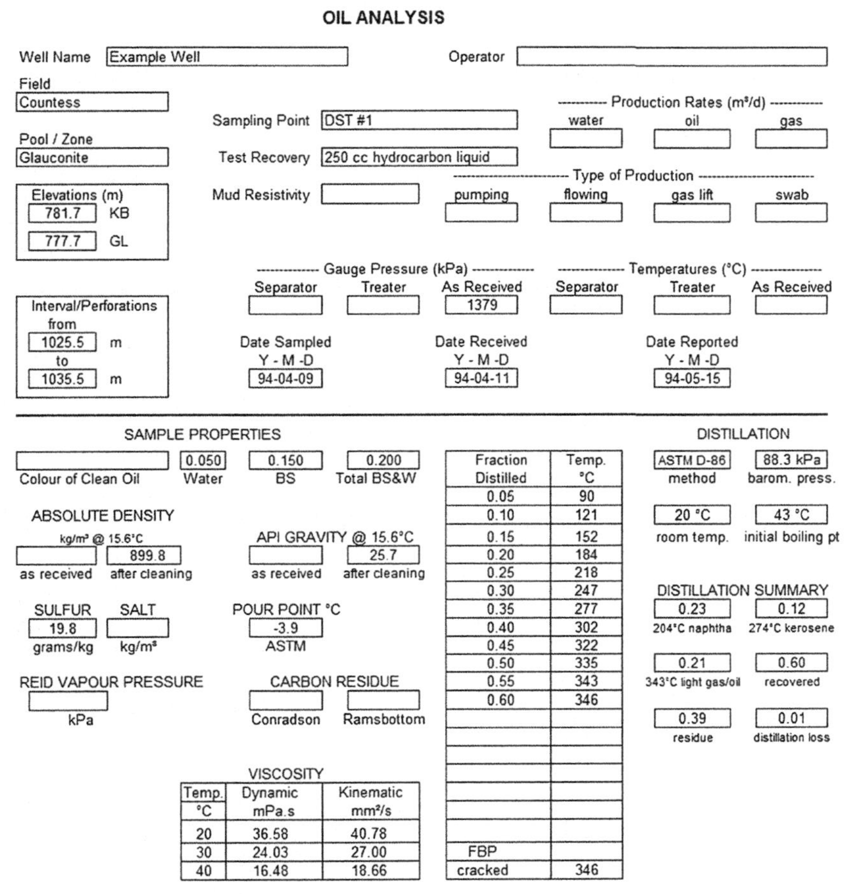

OIL ANALYSIS

Well Name: Example Well
Operator:
Field: Countess
Pool / Zone: Glauconite
Sampling Point: DST #1
Test Recovery: 250 cc hydrocarbon liquid
Mud Resistivity:

Production Rates (m³/d): water | oil | gas

Type of Production: pumping | flowing | gas lift | swab

Elevations (m): 781.7 KB; 777.7 GL

Interval/Perforations: from 1025.5 m to 1035.5 m

	Separator	Treater	As Received
Gauge Pressure (kPa)			1379
Temperatures (°C)			

Date Sampled (Y-M-D): 94-04-09
Date Received (Y-M-D): 94-04-11
Date Reported (Y-M-D): 94-05-15

SAMPLE PROPERTIES

Colour of Clean Oil: ; Water: 0.050; BS: 0.150; Total BS&W: 0.200

ABSOLUTE DENSITY kg/m³ @ 15.6°C: as received: ; after cleaning: 899.8

API GRAVITY @ 15.6°C: as received: ; after cleaning: 25.7

SULFUR: 19.8 grams/kg; SALT: kg/m³

POUR POINT °C: -3.9 ASTM

REID VAPOUR PRESSURE: kPa

CARBON RESIDUE: Conradson: ; Ramsbottom:

VISCOSITY

Temp. °C	Dynamic mPa.s	Kinematic mm²/s
20	36.58	40.78
30	24.03	27.00
40	16.48	18.66

Fraction Distilled	Temp. °C
0.05	90
0.10	121
0.15	152
0.20	184
0.25	218
0.30	247
0.35	277
0.40	302
0.45	322
0.50	335
0.55	343
0.60	346
FBP	
cracked	346

DISTILLATION

method: ASTM D-86; barom. press.: 88.3 kPa
room temp.: 20 °C; initial boiling pt: 43 °C

DISTILLATION SUMMARY

204°C naphtha: 0.23; 274°C kerosene: 0.12
343°C light gas/oil: 0.21; recovered: 0.60
residue: 0.39; distillation loss: 0.01

Figure 5.1.4 Oil analysis.

all oil property correlations. Density is sometimes presented as a specific gravity, defined as follows:

$$SG = \frac{\rho}{\rho_{w(25\ ^\circ\mathrm{C},101\ \mathrm{kPa})}} \tag{5.1.2}$$

in which ρ is the fluid density at a given temperature and pressure, and $\rho_{w(25\ ^\circ\mathrm{C},\ 101\ \mathrm{kPa})}$ is the density of water at standard conditions. Density can also be presented as an API (American Petroleum Institute) gravity. API gravity is related to specific gravity as follows:

$$API = \frac{141.5}{SG} - 131.5 \tag{5.1.3}$$

The API gravity scales oil densities from approximately 10–50 API, in which the lower the API gravity, the denser the oil.

The measurement of oil properties is discussed in more detail in Section 5.1.2. The analysis of oil property data is considered in Section 5.2 and correlations for oil properties are reviewed in Section 5.3.

Water analysis

A water analysis includes the concentration of commonly occurring positive and negative ions in the water, as shown in Figure 5.1.5. Dissolved solids, water density, pH, and water resistivity are also reported. Water analyses are used to:

- Assess fluid compatibility, for example, when determining completion fluids;
- Prepare test brines for compatibility studies of injection water in a particular formation;
- Identify the corrosion and scaling potential of the water;
- Assist in log interpretation (see Chapter 9).

Reservoir calculations require the water density, viscosity, and compressibility at reservoir conditions. The determination of these properties is discussed in Section 5.3.3.

WATER ANALYSIS

Well Name: Example Well
Operator:
Field: Boundary Lakes
Pool / Zone: Halfway
Sampling Point: TESTER
Test Recovery:
Mud Resistivity:

Production Rates (m^3/d): water | oil | gas

Type of Production: pumping | flowing | gas lift | swab

Elevations (m): 751.9 KB; 747.6 GL

Gauge Pressure (kPa): Separator | Treater | As Received

Temperatures (°C): Separator | Treater | As Received

Interval/Perforations: from 1388.0 m; to m

Date Sampled (Y-M-D)	Date Received (Y-M-D)	Date Reported (Y-M-D)
97-04-03	97-04-03	97-04-14

ION	mg/L	mass fraction	mmol/L
Na	7820.0	0.335	340.0
K	285.0	0.012	7.3
Ca	26.8	0.001	0.7
Mg	31.7	0.001	1.3
Ba	N.A.	N.A.	N.A.
Sr	N.A.	N.A.	N.A.
Fe	0.1	TRACE	TRACE

ION	mg/L	mass fraction	mmol/L
Cl	10400.0	0.446	295.0
Br	N.A.	N.A.	N.A.
I	N.A.	N.A.	N.A.
HCO_3	2970.0	0.127	48.6
SO_4	692.0	0.030	7.2
CO_3	300.0	0.013	5.0
OH	NIL	NIL	NIL
H_2S	818.0	0.035	24.0

DISSOLVED Total Solids (mg/L): evap. @110°C: ; evap. @180°C: ; at ignition: ; calculated: 23300

Relative Density: 1.009 @25C; Refractive Index: N/A

Observed pH: 8.65 @25C; Resistivity: 0.33 @25C Ohm.m

SALINITY: 1.87%

Figure 5.1.5 Water analysis.

5.1.2 Black Oil Fluid (PVT) Study

For black oil reservoir fluids, the oil and gas formation volume factors, the solution gas-oil ratio, oil and gas densities, and the oil and gas viscosities are required at reservoir temperature as a function of pressure. These properties can be determined in a fluid study consisting of the following tests:

- Flash expansion test: used to determine the bubble point pressure and the compressibility of the undersaturated oil as a function of pressure.
- Differential liberation test: used to determine the relationship of the oil and gas formation volume factors and the solution gas-oil ratio to pressure. Oil density, oil viscosity, and gas density are also measured at each pressure. Gas viscosity is calculated at each pressure.
- Separator test: used to determine the bubble point formation volume factors and solution gas-oil ratio at separator conditions.

These tests are performed in a pressure, volume, and temperature (PVT) cell, as shown in Figure 5.1.6. A PVT cell is a high-pressure chamber, usually equipped with a site glass. The chamber resides in an air bath in which a constant temperature is maintained. A fluid sample is pumped into the cell, and the volume and therefore pressure of the sample chamber is controlled with a piston. Most sample chambers also are equipped with a magnetic stirrer to ensure that the system is well mixed. When two phases are present, the interface between the faces can be detected through the site glass and the volume of each phase measured. When required, a volume of the lighter phase can be displaced from the sample chamber. Procedures for the fluid tests are described below.

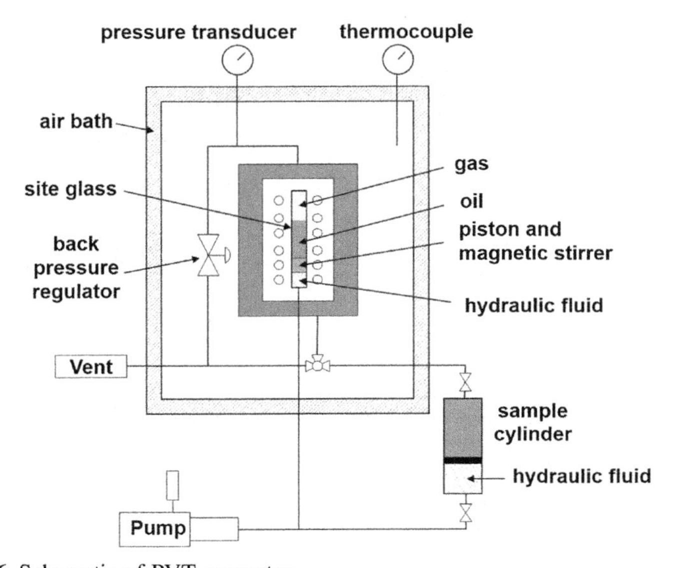

Figure 5.1.6 Schematic of PVT apparatus.

Sample preparation: The recovered oil and gas samples are recombined at a pressure far above the bubble point so that a single phase fluid is obtained. It is critical to recombine at the correct ratio to recreate an accurate bubble point pressure and solution gas-oil ratio. Subsurface samples are recombined at their measured gas-oil ratio. Surface samples are recombined at the gas-oil ratio in the separator vessel from which the sample was obtained, $R_{s,SEP}$. Because the gas-oil ratio is not measured directly, the ratio is calculated from the measured gas-oil ratio at stock tank conditions, $R_{s,ST}$ as follows:

$$R_{s,SEP} = S \cdot R_{s,ST} \tag{5.1.4}$$

in which S is the ratio of the stock tank oil volume to the separator oil volume. The value of S is determined as part of the fluid study by flashing a sample of the separator oil to stock tank conditions. As is discussed in Section 5.4, it is important to check bubble point pressure and solution gas-oil ratio against field pressure and production data.

Flash expansion test: A sample of recombined fluid is compressed to a pressure well above the bubble point and maintained at reservoir temperature. The volume of the sample chamber is increased incrementally, and hence the pressure is reduced incrementally, as shown in Figure 5.1.7(a). The volume of the fluid in the cell and the pressure are recorded at each step after some time for equilibration. The data are usually scaled to the bubble point oil volume, as shown in Table 5.1.2. The system compressibility changes dramatically when gas is liberated at and below the bubble

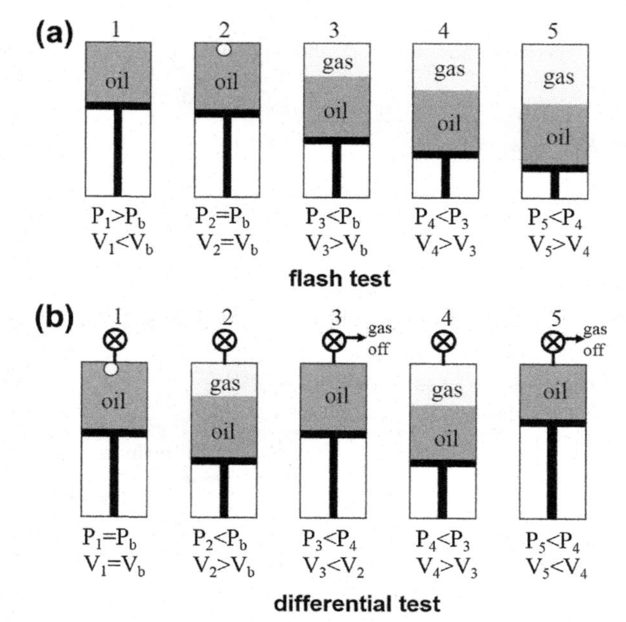

Figure 5.1.7 Flash (a) and differential (b) liberation tests.

Table 5.1.2 Flash expansion test results from Pembina Viking B Pool fluid study

Pressure (kPag)	V/V_{sat}[a]	Viscosity (mPa·s)
34,474	0.9670	0.361
31,026	0.9729	0.349
27,579	0.9796	0.337
24,132	0.9868	0.324
20,684	0.9948	0.312
19,995	0.9965	–
19,305	0.9983	–
18,685 (P_{bp})	1.0000	–
18,140	1.0107	–
17,065	1.0347	–
15,658	1.0742	–
14,058	1.1336	–
12,473	1.2132	–
10,894	1.3240	–
9460	1.4670	–
8039	1.6727	–
7412	1.7934	–

[a]V_{sat} is the volume of oil at the saturation pressure (bubble point).

point. Therefore, the bubble point pressure is determined from a change in slope of the measured pressure versus volume, as shown in Figure 5.1.8. Oil viscosity is also measured above the bubble point. The calculation of the oil compressibility above the bubble point is discussed in Section 5.1.2.

Differential liberation test: A sample of recombined fluid is compressed to the bubble point pressure and maintained at reservoir temperature. As with the flash expansion test, the volume of the sample chamber is increased incrementally; however, in this case, the evolved solution gas is withdrawn from the sample chamber after each incremental expansion, as shown in Figure 5.1.7(b). After each step, the evolved gas is expanded to standard conditions, and the expansion factor is recorded. Also, the pressure, the total volume of the fluid in the cell, the volume of the liquid phase, the viscosity of the liquid phase, the volume of evolved gas, and the density of the evolved gas are measured at each step. The viscosity of the evolved gas is usually calculated rather than measured.

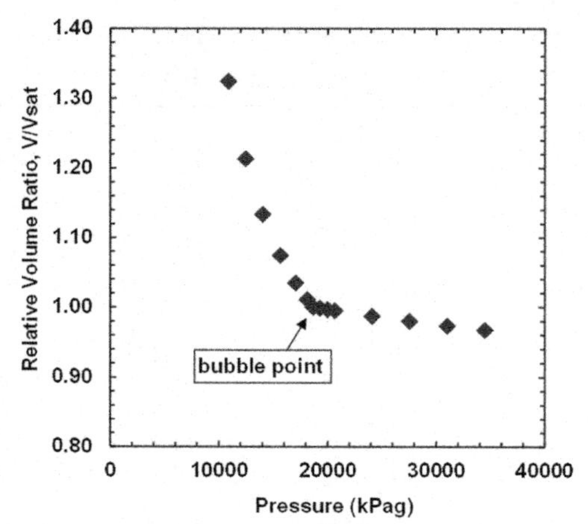

Figure 5.1.8 Bubble point determination from flash expansion test. Data from Pembina Viking B Pool fluid study (Table 5.1.2).

The oil volume data are sometimes scaled to the residual oil volume at atmospheric pressure and 15 °C, as shown in Table 5.1.3. The residual oil volume is determined in the final step of a differential test when the oil at atmospheric pressure and reservoir temperature is flashed to 15 °C. In other cases, the data may be scaled to the bubble point oil volume, as shown in Table 5.1.4. In many fluid studies, the solution gas-oil ratio calculated from the evolved gas volume and the relative oil volume is reported instead of the evolved gas volume.

The data in Table 5.1.3 or Table 5.1.4 are sufficient to determine oil and gas PVT properties for a differential liberation. Oil density is used directly as reported. The other oil properties of interest are the differential liberation oil formation volume factor, B_{od} (oil volume at a give pressure per volume of residual oil at 15 °C and atmospheric pressure) and the differential liberation solution gas-oil ratio, R_{sd} (gas volume at standard conditions per volume of residual oil at 15 °C and atmospheric pressure). These properties are calculated as follows:

If the measurements are scaled to bubble point oil volume:

$$B_{od} = V_{or} \tag{5.1.5}$$

$$R_{sd} = \left(F_{r,residual} - F_r\right) \tag{5.1.6}$$

If the measurements are scaled to the residual oil volume:

$$B_{od} = \frac{V_{ob}}{V_{ob,residual}} \tag{5.1.7}$$

$$R_{sd} = \frac{\left(F_{b,residual} - F_b\right)}{V_{ob,residual}} \tag{5.1.8}$$

Table 5.1.3 Differential liberation test volume and density measurements at 82.2 °C from Pembina Viking B Pool fluid study

Pressure (kPag)	Oil density (kg/m^3)	Relative oil volume[a] (V_{or})	Relative total volume[a]	Relative evolved gas volume[b] (F_r)	Gas-specific gravity[c]	Gas expansion factor[d] (E_g)
18,685	644.6	1.642	1.642	0.00	–	–
16,547	655.5	1.579	1.723	23.86	0.814	166.11
14,486	666.1	1.525	1.833	44.13	0.773	143.06
12,376	677.0	1.473	2.008	63.87	0.765	119.47
10,321	687.7	1.426	2.266	82.01	0.753	97.66
8239	698.0	1.382	2.685	98.91	0.769	75.93
6191	708.6	1.338	3.426	115.81	0.789	55.46
4137	719.5	1.295	4.941	131.99	0.820	36.21
1931	733.3	1.238	10.307	151.91	0.983	16.75
979	741.6	1.204	19.747	162.78	1.131	8.78
0	776.3	1.061	241.079	194.24	1.797	0.81

Scaled to residual oil volume.
Density of residual oil = 823.0 kg/m^3 at 15.56 °C.
[a]Oil and gas at indicated temperature and pressure relative to residual oil volume at 15 °C.
[b]Gas at 101 kPa and 15 °C relative to residual oil volume.
[c]Relative to air at 101 kPa and 15 °C.
[d]Gas volume at 101 kPa and 15 °C relative to volume at given pressure and temperature.

Table 5.1.4 **Differential liberation test volume and density measurements at 82.2 °C from Pembina Viking B Pool fluid study**

Pressure (kPag)	Oil density (kg/m^3)	Relative oil volume[a] (V_{or})	Relative total volume[a]	Relative evolved gas volume[b] (F_r)	Gas-specific gravity[c]	Gas expansion factor[d] (E_g)
18,685	644.6	1.0000	1.000		–	–
16,547	655.5	0.9616	1.049	14.53	0.814	166.11
14,486	666.1	0.9287	1.116	26.88	0.773	143.06
12,376	677.0	0.8971	1.223	38.90	0.765	119.47
10,321	687.7	0.8685	1.380	49.95	0.753	97.66
8239	698.0	0.8417	1.635	60.24	0.769	75.93
6191	708.6	0.8149	2.086	70.53	0.789	55.46
4137	719.5	0.7887	3.009	80.38	0.820	36.21
1931	733.3	0.7540	6.277	92.52	0.983	16.75
979	741.6	0.7333	12.03	99.14	1.131	8.78
0	776.3	0.6462	146.8	118.29	1.797	0.81
0 (15 °C)	823.0	0.6090	–	118.29	–	1.00

Scaled to bubble point oil volume.
[a]Oil and gas at indicated temperature and pressure relative to residual oil volume at 15 °C.
[b]Gas at 101 kPa and 15 °C relative to residual oil volume.
[c]Relative to air at 101 kPa and 15 °C.
[d]Gas volume at 101 kPa and 15 °C relative to volume at given pressure and temperature.

Table 5.1.5 **Pembina Viking B Pool oil properties from differential test**

Pressure (kPag)	Oil density (kg/m^3)	Oil viscosity (mPa·s)	Oil volume factor, B_{od} (m^3/scm)	Solution gas-oil ratio, R_s (scm/scm)
18,685	644.6	0.304	1.642	194.24
16,547	655.5	0.330	1.579	170.38
14,486	666.1	0.360	1.525	150.11
12,376	677.0	0.394	1.473	130.37
10,321	687.7	0.432	1.426	112.23
8239	698.0	0.474	1.382	95.33
6191	708.6	0.506	1.338	78.43
4137	719.5	0.613	1.295	62.25
1931	733.3	0.759	1.238	42.33
979	741.6	0.866	1.204	31.46
0	776.3	1.579	1.061	0

in which V_o is the relative oil volume, and F is the evolved gas volume. The subscripts are defined as follows: subscript *r*—the property is relative to the residual oil volume; subscript *b*—the property is relative to bubble point oil; subscript *residual*—residual oil property. Oil properties calculated from the data in Table 5.1.3 are provided in Table 5.1.5. Note that oil viscosity is also measured in a differential liberation test and is included in Table 5.1.5.

The gas density is used directly as reported. The gas formation volume factor, B_g, is the reciprocal of the expansion factor, E_g. The gas compressibility factor (z-factor) can be calculated from the expansion factor as follows:

$$z = \frac{PT_s}{P_s T E_g} \tag{5.1.9}$$

Gas properties calculated from the data in Table 5.1.3 are provided in Table 5.1.6. Gas viscosity is calculated from a correlation and was included in Table 5.1.3.

Separator test: A sample of recombined fluid is compressed to the bubble point pressure at the reservoir temperature. The sample is then flashed to the temperature and pressure of the field separator. The volume of oil and gas are measured. If the field has a second stage separator, the sample is then flashed to the temperature and pressure of the second separator. The oil and gas volumes are measured again. After all of the separation stages have been tested, the sample is flashed to stock tank conditions, and the stock tank oil and gas volumes are measured. The results of a separator test are usually reported in terms of oil volume factors and gas-oil ratios, as shown in Table 5.1.7.

Table 5.1.6 Pembina Viking B Pool gas properties

Pressure (kPag)	Gas-specific gravity	Gas viscosity (mPa·s)	Compressibility factor, z	Gas expansion factor, E_g (scm/m^3)	Gas volume factor, B_g (m^3/scm)
18,685	–	–	–	–	–
16,547	0.814	0.0204	0.802	166.11	0.00602
14,486	0.773	0.0183	0.816	143.06	0.00699
12,376	0.765	0.0169	0.836	119.47	0.00837
10,321	0.753	0.0157	0.854	97.66	0.01024
8239	0.769	0.0147	0.879	75.93	0.01317
6191	0.789	0.0138	0.908	55.46	0.01803
4137	0.820	0.0130	0.937	36.21	0.02762
1931	0.983	0.0118	0.971	16.75	0.05970
979	1.131	0.0111	0.985	8.78	0.1139
0	1.797	0.0089	0.999	0.81	1.236

The two properties of most interest are: the total gas-oil ratio, which is termed the separator flash bubble point solution gas-oil ratio, R_{sfb}; and the oil formation volume factor, which is termed separator flash bubble point volume factor, B_{ofb}.

5.1.3 Specialized Pressure, Volume, and Temperature Tests

Specialized fluid data are required for fluids exhibiting more complex phase behavior, including retrograde condensates, volatile oils, and miscible fluids. Fluid studies performed for volatile fluids include constant mass expansion, constant volume depletion, swelling, and slim tube miscibility tests. Each is reviewed briefly below.

Constant mass expansion: A constant mass expansion test is essentially the same as the flash expansion test described for a black oil study in Section 5.1.2. However, for a gas-condensate system, the objective is to determine the dew point rather than the bubble point. The test apparatus is designed so that the small volumes of liquid that drop out can be collected at the end of each pressure step. A volatile fluid is near the critical point, and either a bubble point or dew point may be measured. The test apparatus usually has a long site glass so that the gas–liquid interface can be observed even with large changes in the phase volumes.

Constant volume depletion: A constant volume depletion test is essentially the same as the differential liberation test described in Section 5.1.2 with modifications as described for the constant mass expansion test. For a gas-condensate, the volume of liquid that drops out below the dew point is measured. Liquid volumes and gas phase

Table 5.1.7 **Separator test results from Pembina Viking B Pool fluid study**

Seperator pressure (kPag)	Seperator temperature (°C)	Gas-oil ratio[a]	Gas-oil ratio[b] (R_{sf})	Oil API gravity (15.6 °C)	Oil volume factor[c] (B_{ofb})	Separator volume factor[d]	Gas-specific gravity[e]
545	16	127.25	133.71	–	–	1.051	0.753
0	16	15.39	15.40	44.0	1.474	1.001	1.123
Total			149.11				

[a]Gas at 101 kPa and 15 °C relative to oil volume at given pressure and temperature.
[b]Gas at 101 kPa and 15 °C relative to stock tank oil volume at 15 °C.
[c]Volume of saturated oil at bubble point relative to volume of stock tank oil at 15 °C.
[d]Volume of oil at given pressure and temperature relative to volume of stock tank oil at 15 °C.
[e]Relative to air at *101* kPa and *15* °C.

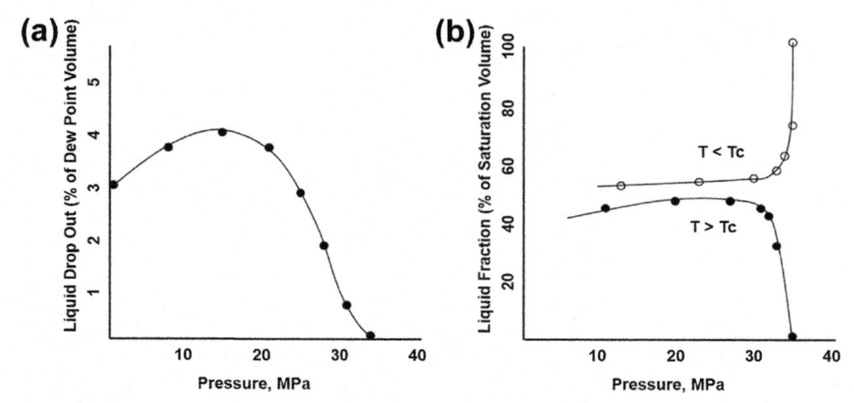

Figure 5.1.9 Representations of (a) liquid drop out from a gas condensate and (b) liquid volume fractions in a volatile fluid.

compositions are usually measured after each pressure step, Figure 5.1.9(a). Volatile fluids are near their critical point, and significant changes in phase behavior occur over relatively small changes in temperature and pressure. The depletion experiments may be repeated at different temperatures to map out the phase boundary, liquid drop out at temperatures above the critical point, and gas evolution at temperatures below the critical point, Figure 5.1.9(b).

Swelling test: Swelling tests are used to determine the change in live oil volume when it is saturated with an injection gas. A series of flash expansion tests are performed on the oil with increasing volumes of injection gas, and the change in saturation pressure is measured, Figure 5.1.10. The data are used to relate the "swollen" oil volume to gas injection pressure.

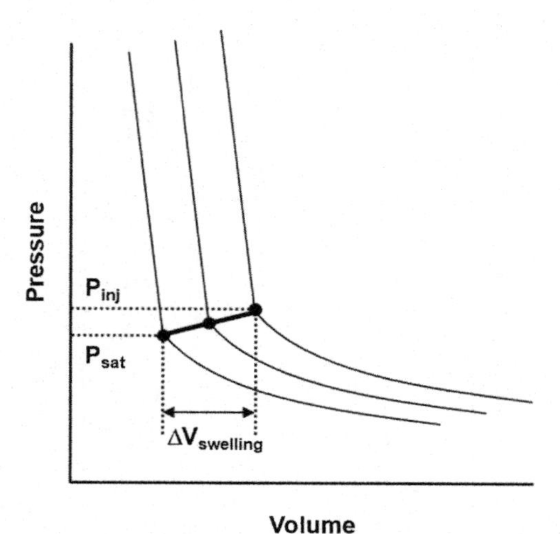

Figure 5.1.10 Representation of a swelling test.
Adapted from Pedersen et al. (1989).

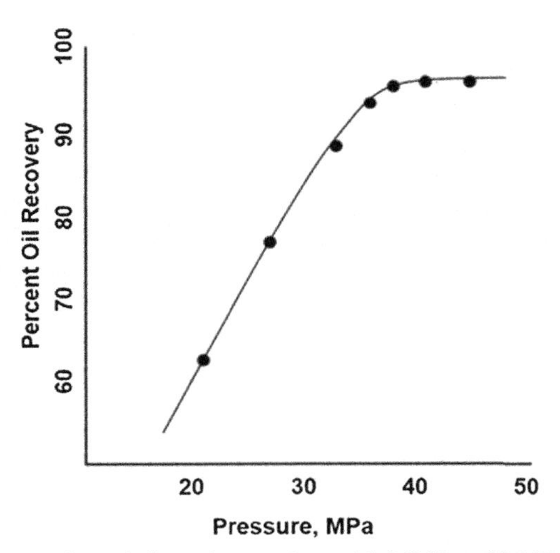

Figure 5.1.11 Representation of slim tube test data with MMP = 38 MPa.

Slim tube test: Slim tube tests are used to determine the minimum miscibility pressure, the lowest pressure at which an injection gas is fully miscible in a crude oil. The test involves displacing the crude oil from a sand pack with the injection gas and is based on a simple principle: oil recovery reaches a limit near 100% when the pressure is increased to the point of miscibility. The sand is packed in a long, slim tube (e.g., 10 m length and 3.5 cm diameter steel tubing), hence the name of the test. The gas is injected at a relatively low pressure, and the volumes and compositions of the produced oil and gas are measured. When no more oil is recovered, the pressure is increased and the procedure repeated. The point at which increasing the pressure does not increase the oil recovery (usually between 90% and 100% recovery) is the minimum miscibility pressure, Figure 5.1.11.

5.2 Analysis of a Black Oil Dataset

5.2.1 Determination of B_o and R_s

As we saw in the previous section, B_o and R_s can be measured using a differential liberation or a flash separation test. The two tests will give slightly different results, because in a flash test the overall fluid composition is constant, while in the differential liberation test the overall fluid composition changes as the gas phase is removed. The phase equilibrium changes when the composition changes. The question is which test best represents the phase behavior of the reservoir fluid as it flows from the reservoir to the wellbore to the surface facilities?

As pressure is reduced in the reservoir, a gas phase evolves and usual flows as a separate continuous phase. Because gas mobility is greater than oil mobility, the evolved gas usually separates and flows away from its source oil. This process more closely resembles differential liberation. On the other hand, once the fluid reaches the wellbore and flows to the surface facilities, it usually undergoes a rapid depressurization before there is an opportunity for fluid separation. This process more closely resembles the flash separator test.

To approximate the behavior of the reservoir fluid, the differential data are used to determine the relative change of oil volume and solution gas-oil ratio with pressure. However, the differential data are adjusted so that the absolute values at the bubble point correspond to the separator flash test values. In this way, the effect of oil and gas separation in the reservoir is accounted for, while the surface volumes obtained for a given separation scheme are also honored. It must be emphasized that this approach is an approximation of the complex, nonequilibrium phase separations that really occur. In general, the approximation is reasonable for low-volatility oils. It may be necessary to use a compositional model for high-volatility oils.

Below the bubble point, the adjusted B_o is determined as follows:

$$B_o = \frac{B_{od}}{B_{odb}} \tag{5.2.1}$$

Above the bubble point, B_o is determined as follows:

$$B_o = \frac{V_o}{V_{sat}} B_{ofb} \tag{5.2.2}$$

Below the bubble point, the adjusted R_s is given by:

$$R_s = R_{sfb} - \frac{B_{ofb}}{B_{odb}}(R_{sdb} - R_{sd}) \tag{5.2.3}$$

Above the bubble point, R_s is constant and is given by:

$$R_s = R_{sfb} \tag{5.2.4}$$

The gas volume factor, B_g, is taken directly from the differential liberation test without adjustments. In this case, the gas volume is relative to the gas volume at standard conditions and is independent of the separator conditions.

The complete set of Pembina Viking B Pool PVT data after the adjustment to separator conditions is provided in Table 5.2.1. Plots of B_o, R_s, B_g, μ_o, and μ_g versus pressure are given in Figures 5.2.1–5.2.5.

5.2.2 Calculation of Oil Compressibility

In most oil fluid studies, the oil compressibility is reported above the bubble point. Recall the definition of oil compressibility:

$$c_o = -\frac{1}{v_o}\frac{dv_o}{dP} = -\frac{1}{B_o}\frac{dB_o}{dP} \tag{5.2.5}$$

Table 5.2.1 Pembina Viking B pool PVT data

Pressure (kPag)	B_o (m³/STm³)	R_s (scm/STm³)	B_g (m³/scm)	μ_o (cp)	μ_g (cp)
34,474	1.425	149.1		0.361	
31,026	1.434	149.1		0.349	
27,579	1.444	149.1		0.337	
24,132	1.455	149.1		0.324	
20,684	1.466	149.1		0.312	
19,995	1.469	149.1		–	
19,305	1.471	149.1		–	
18,685[a]	1.474	149.1		0.304	
16,547	1.417	127.7	0.00602	0.33	0.0204
14,486	1.369	109.5	0.00699	0.36	0.0183
12,376	1.322	91.8	0.00837	0.394	0.0169
10,321	1.280	75.5	0.01024	0.432	0.0157
8239	1.241	60.3	0.01317	0.474	0.0147
6191	1.201	45.1	0.01803	0.506	0.0138
4137	1.163	30.6	0.02762	0.613	0.013
1931	1.111	12.7	0.05970	0.759	0.0118
979	1.081	3.0	0.1139	0.866	0.0111
0	0.952	0	1.236	1.579	0.0089

[a]Bubble point.

The compressibility is usually approximated from the oil formation volume factors above the bubble point as follows:

$$c_o = \frac{B_{o1} - B_{o2}}{B_{o2}(P_2 - P_1)} \tag{5.2.6}$$

Compressibilities calculated from the Pembina Viking B Pool fluid study are listed in Table 5.2.2.

Note that if the oil volume ratio above the bubble point (V/V_{sat}) is curve fit, a more exact determination of c_o can be obtained from the derivative of the curve fit equation. For example, the oil volume ratio data of Table 5.2.2 can be fit with the following polynomial:

$$\frac{V}{V_{sat}} = 3.04 \times 10^{-11} P^2 - 3.96 \times 10^{-6} P + 1.0583 \tag{5.2.7}$$

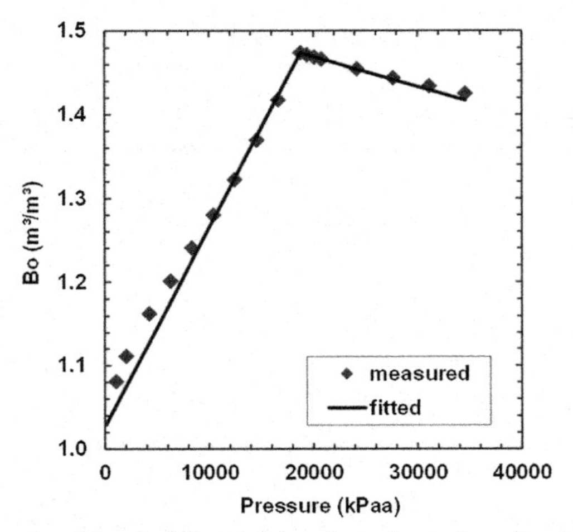

Figure 5.2.1 Measured and fitted oil formation volume factor from Pembina Viking B Pool fluid study.

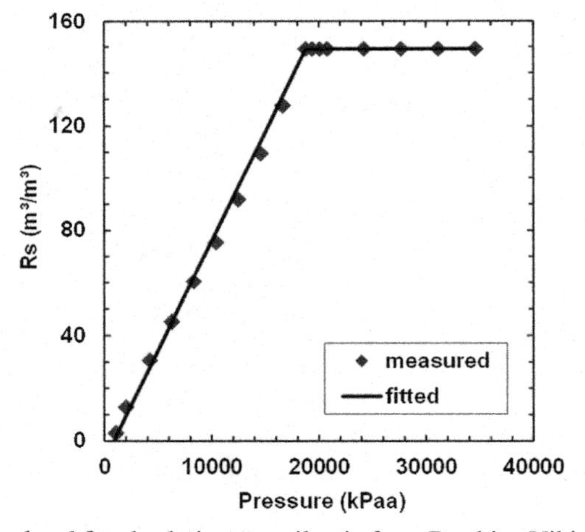

Figure 5.2.2 Measured and fitted solution gas-oil ratio from Pembina Viking B Pool fluid study.

in which P is in kPa. The compressibility is then given by:

$$c_o = \frac{6.08 \times 10^{-11}P - 3.96 \times 10^{-6}}{3.04 \times 10^{-11}P^2 - 3.96 \times 10^{-6}P + 1.0583} \tag{5.2.8}$$

The compressibilities determined from the curve fit method are compared with the approximate values in Table 5.2.2. The compressibilities from each method are within 10% of each other. The approximate method is simpler, but the curve fit method is more easily adapted for use in a spreadsheet. Therefore, the choice of method depends on the required application.

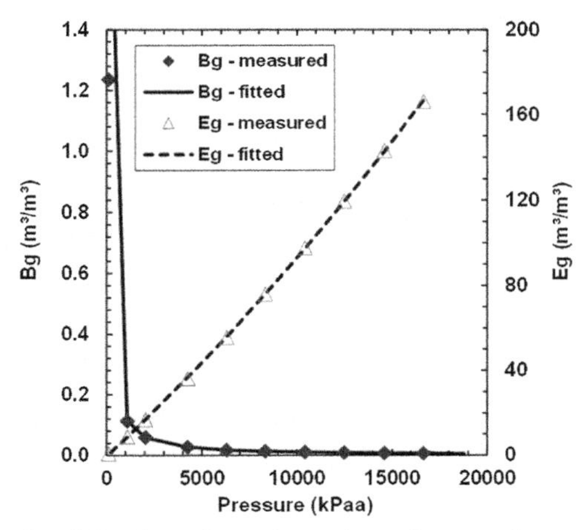

Figure 5.2.3 Measured and fitted gas formation volume factor and gas expansion factor from Pembina Viking B Pool Fluid Study.

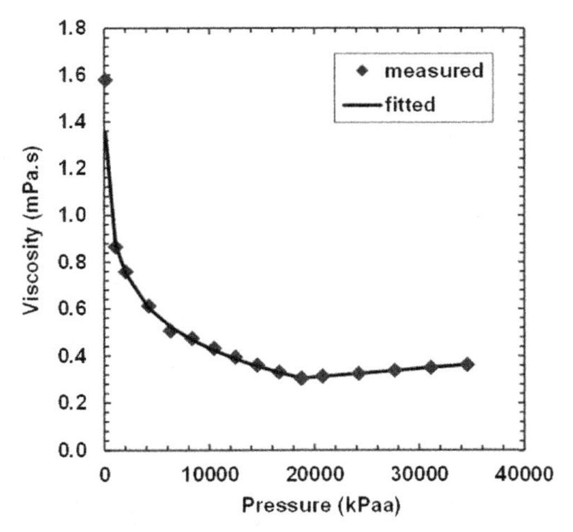

Figure 5.2.4 Measured and fitted oil viscosity from Pembina Viking B Pool Fluid Study.

5.2.3 Curve Fitting Black Oil Fluid Data

It is often convenient to use curve fits for the PVT data, particularly for use in a spreadsheet. The following forms of equations can usually provide reasonably good fits of oil properties. Note that the equations are structured to fit the bubble point properties exactly.

Above bubble point:

$$B_o = B_{ob} \exp\{-c_o(P - P_b)\} \tag{5.2.9}$$

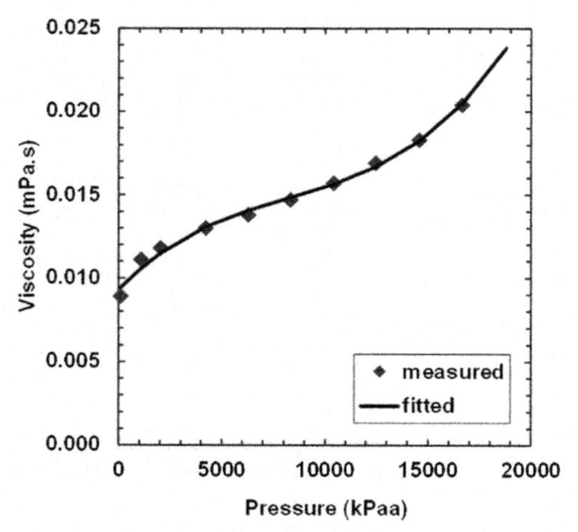

Figure 5.2.5 Measured and fitted gas viscosity from Pembina Viking B Pool Fluid Study.

Table 5.2.2 Oil compressibility determined from tabulated B_o values and from curve fit equation for V/V_{sat}

P_1 kPa	P_2 kPa	c_o from B_o kPa^{-1}	c_o from V/V_{sat} at P_2 kPa^{-1}
–	18,685	–	2.554×10^{-6}
18,685	20,684	2.615×10^{-6}	2.445×10^{-6}
20,684	24,132	2.351×10^{-6}	2.253×10^{-6}
24,132	27,579	2.133×10^{-6}	2.055×10^{-6}
27,579	31,026	1.998×10^{-6}	1.853×10^{-6}
31,026	34,474	1.769×10^{-6}	1.648×10^{-6}

$$R_s = R_{sb} \tag{5.2.10}$$

$$\mu_o = \mu_{ob} - a_{vo1}(P - P_b) \tag{5.2.11}$$

Below bubble point:

$$B_o = B_{ob} - a_b(P_b - P) \tag{5.2.12}$$

$$R_s = R_{sb} - a_r(P_b - P) \tag{5.2.13}$$

$$\mu_o = \mu_{ob} - a_{vo2} \ln\left\{\frac{P}{P_b}\right\} \tag{5.2.14}$$

Table 5.2.3 Parameters for curve fit equations for Pembina Viking B Pool PVT data

Property	a	b	c	d
B_o below P_b	2.380×10^{-5}	—	—	—
R_s below P_b	8.322×10^{-3}	—	—	—
μ_o above P_b	3.643×10^{-6}	—	—	—
μ_o above P_b	0.2006	—	—	—
E_g	−0.3475	8.264×10^{-3}	1.058×10^{-7}	—
μ_g	9.314×10^{-3}	1.277×10^{-3}	-1.108×10^{-10}	4.471×10^{-15}

Units are kPa and cp.

in which a_i is the fitting coefficient. Note that it is usually sufficient to assume a constant compressibility to calculate B_o above the bubble point. The compressibility curve fit equation can be used for greater accuracy.

The following equations can be used to fit gas properties:

$$E_g = a_{e1} + b_{e1}P + c_{e1}P^2 \tag{5.2.15}$$

$$B_g = \frac{1}{E_g} \tag{5.2.16}$$

$$\mu_g = a_{vg} + b_{vg}P + c_{vg}P^2 + d_{vg}P^3 \tag{5.2.17}$$

in which a_i, b_i, c_i, and d_i are the fit coefficients.

The above equations were used to fit the Pembina Viking B Pool PVT data, as shown in Figures 5.2.1–5.2.5. The bubble point properties were given in Table 5.2.1, and the fitting parameters for pressure in units of kPa absolute are given in Table 5.2.3. Note that B_g and E_g can also be determined analytically based on a gas analysis, and μ_g can be found directly from the viscosity correlation originally used to generate the data. Property correlations are discussed in Section 5.3.

5.3 Correlations for Fluid Data

Fluid data can be modeled with an equation of state. Components and pseudocomponent composition is initialized based on a recombined C30 hydrocarbon analysis. The model is usually tuned to a flash test. Such a tuned model is able to predict differential liberation data to within the accuracy of measured data. Therefore, many modern fluid studies do not measure differential liberation data, but generate it from the tuned

Table 5.3.1 Commonly used fluid property correlations

Property	Correlation	Applicability
P_b, B_o, R_s	Vasquez–Beggs (1980)	
R_s	Lasater (1958)	15 < API < 30
P_b, B_o, R_s	Standing (1947)	API < 15
μ_o	Beggs–Robinson (1975)	16 < API < 58
B_g	Corresponding states	Non-polar hydrocarbons
μ_g	Lee et al. (1966)	$P < 8000$ psia, $100 < T < 340$ K
B_w	McCain (1988)	–
R_{sw}	McCain (1988)	$1000 < P < 10{,}000$ psia, $100 < T < 250$ F, $S < 30\%$
c_w	Osif and McCain (1984)	$1000 < P < 20{,}000$ psia, $200 < T < 270$ F, $C_{NaCl} < 200$ g/L
μ_w	McCain (1973)	$P < 15{,}000$ psia, $100 < T < 400$ F, $S < 26\%$

equation of state. Equation of state models are beyond the scope of this book, but are discussed elsewhere (Whitson and Brulé, 2000; Pedersen et al., 1989).

There are several correlations available for oil and gas properties. Some of the more commonly used correlations are listed in Table 5.3.1. Each of these correlations is discussed below.

5.3.1 Correlations for Oil Properties

5.3.1.1 Vasquez–Beggs Correlation for P_b, B_o, and R_s

The Vasquez–Beggs (1980) correlation is based on the data from more than 600 fluid studies from all over the world. The Vasquez–Beggs correlation is best suited for oils with gravity greater than 30 API. The input data for the Vasquez–Beggs correlation are the oil gravity, the gas gravity, the reservoir temperature, the separator temperature, and the separator pressure. Either the bubble point or the initial producing gas-oil ratio also must be specified.

The first step is to calculate a corrected gas gravity as follows:

$$\gamma_{gc} = \gamma_g\left(1.0 + 5.912 \times 10^{-5}\gamma_o T_{sep} \log\left\{\frac{P_{sep}}{114.7}\right\}\right) \tag{5.3.1}$$

in which γ_{gc} is the corrected gas gravity, γ_g is the actual gas gravity, γ_o is the oil API gravity, T_{sep} is the separator temperature (°F), and P_{sep} is the separator pressure (psia).

Next, the bubble point is determined. If the bubble point is known, then the actual bubble point is used. If the bubble point is not known, the bubble point is estimated as follows:

$$P_b = \left[\frac{R_p}{c_1 \gamma_{gc} \exp\{c_3 \gamma_o/(T+460)\}}\right]^{1/c_2} \tag{5.3.2}$$

in which P_b is the bubble point pressure (psia), R_p is the initial producing gas-oil ratio (SCF/stb), T is the reservoir temperature (°F), and c_1, c_2, and c_3 are constants. The values of the constants are given in Table 5.3.2.

If the bubble point is not known or not selected as an input parameter, the solution gas-oil ratio at the bubble point is set equal to the initial producing gas-oil ratio. Otherwise, the solution gas-oil ratio (SCF/stb) is calculated as follows:

$$R_{sb} = c_1 \gamma_{gc} P_b^{c_2} \exp\{c_3 \gamma_o/(T+460)\} \tag{5.3.3}$$

The final step is to determine the solution gas-oil ratio and the oil formation volume factor as a function of pressure. The procedure is given below.

Above the bubble point:

$$R_s = R_{sb} \tag{5.3.4}$$

$$B_o = B_{ob} \exp\{-c_o(P - P_b)\} \tag{5.3.5}$$

in which c_o, the oil compressibility (psi^{-1}) is given by:

$$c_o = \frac{5R_s + 17.2T - 1180\gamma_{gc} + 12.61\gamma_o - 1433}{10^5 P} \tag{5.3.6}$$

Table 5.3.2 Constants for the Vasquez–Beggs correlation

Constant	γ_o <= 30° API	γ_o > 30° API
c_1	0.0362	0.0178
c_2	1.0937	1.1870
c_3	25.7240	23.9310
c_4	4.677×10^{-4}	4.670×10^{-4}
c_5	1.751×10^{-5}	1.100×10^{-5}
c_6	-1.811×10^{-8}	1.337×10^{-9}

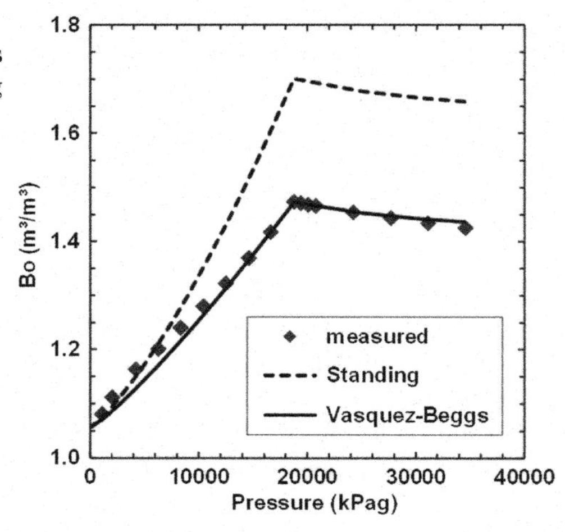

Figure 5.3.1 Comparison of Vasquez–Beggs and Standing correlations against measured B_o from Pembina Viking B Pool fluid study.

At and below the bubble point:

$$R_s = c_1 \gamma_{gc} P^{c_2} \exp\{c_3 \gamma_o/(T + 460)\} \tag{5.3.7}$$

$$B_o = 1 + c_4 R_s + (c_5 + c_6 R_s)(T - 60)\frac{\gamma_o}{\gamma_{gc}} \tag{5.3.8}$$

The values of the constants c_4, c_5, and c_6 are listed in Table 5.3.2. The Vasquez–Beggs correlations are compared with the Pembina Viking B fluid data in Figures 5.3.1 and 5.3.2. The inputs to the correlation were a reservoir temperature of 180 F (82.2 °C), bubble point pressure of 2727 psia (18,785 kPaa), an oil gravity

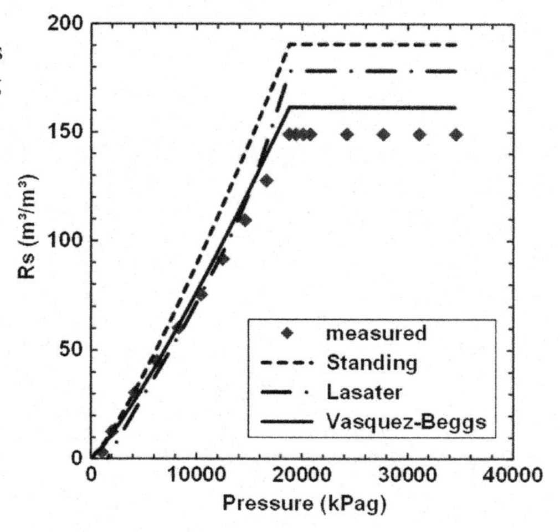

Figure 5.3.2 Comparison of Vasquez–Beggs and Standing correlations against measured R_s from Pembina Viking B Pool fluid study.

of 40.3 API, a gas gravity of 0.989, a separator pressure of 60 psia, and a separator temperature of 70 °F. The Vasquez–Beggs correlation provides a good prediction of the PVT properties of this high API oil.

5.3.1.2 Lasater Correlation for R_s

The Lasater (1958) correlation is based on 137 black oils from Canada, the United States, and South America. It is best suited for oils with gravity between 15 and 30 API. The input data for the Lasater correlation are the oil gravity, the gas gravity, and the reservoir temperature. The bubble point must be specified or determined from another correlation. The oil volume formation factor must be determined from another correlation.

The solution gas-oil ratio at and below the bubble point is given by:

$$R_s = 132755 \frac{\gamma_o}{M_{eff}} \frac{x_g}{1 - x_g} \tag{5.3.9}$$

in which γ_o is the oil API gravity, M_{eff} is the effective molar mass of the stock tank oil, and x_g is the mole fraction of gas. The effective molar mass is determined as follows: for $\gamma_o > 40$ API:

$$M_{eff} = 73110 \gamma_o^{-1.562} \tag{5.3.10a}$$

for $\gamma_o \leq 40$ API:

$$M_{eff} = 630 - 10\gamma_o \tag{5.3.10b}$$

To determine the gas mole fraction, a bubble point pressure factor, P_{bpf}, is introduced:

$$P_{bpf} = \frac{P\gamma_g}{T + 460} \tag{5.3.11}$$

in which γ_g is the gas gravity, P is pressure in psia, and T is temperature in °F. The gas mole fraction is then given by:

for $P_{bpf} \geq 3.29$:

$$x_g = \left[0.121 P_{bpf} - 0.236\right]^{0.281} \tag{5.3.12a}$$

for $P_{bpf} < 3.29$:

$$x_g = 0.359 \ln\left\{1.473 P_{bpf} + 0.476\right\} \tag{5.3.12b}$$

The Lasater correlation is compared with the Pembina Viking B Pool fluid properties in Figure 5.3.2. The correlation overpredicts the R_s of this 40 API oil. As mentioned previously, the Lasater correlation is better suited to medium API oils.

5.3.1.3 Standing Correlation for P_b, B_o, and R_s

The Standing (1947) correlation is based on a laboratory study of 22 different crude oils from California. It is best suited for oils with gravity less than 15 API. The input data for the Standing correlation are the oil gravity, the gas gravity, and the reservoir temperature. Either the bubble point or the initial producing gas-oil ratio must also be specified.

First, the bubble point is determined. If the bubble point is known, then the actual bubble point is used. If the bubble point is not known, the bubble point is estimated as follows:

$$P_b = 18\left[\frac{R_p}{\gamma_g}\right]^{0.83} 10^Y \tag{5.3.13}$$

in which the exponent Y is given by:

$$Y = 0.00091T - 0.0125\gamma_o \tag{5.3.14}$$

in which P_b is the bubble point pressure (psia), R_p is the initial producing gas-oil ratio (SCF/stb), and T is the reservoir temperature (°F).

The solution gas-oil ratio and the oil formation volume factor are determined as a function of pressure as outlined below.

At and below the bubble point:

$$R_s = \gamma_g\left[\frac{P}{18 \times 10^Y}\right]^{1.204} \tag{5.3.15}$$

$$B_o = 0.972 + 0.000147\left[R_s\left(\frac{\gamma_g}{\gamma_o^*}\right)^{0.5} + 1.25T\right]^{1.175} \tag{5.3.16}$$

in which γ_o* is the specific gravity of the oil relative to water. Properties above the bubble point can be found using the procedure described in the Vasquez–Beggs method.

The Standing correlation is compared with the Pembina Viking B Pool fluid properties in Figures 5.3.1 and 5.3.2. The correlation overpredicts both the B_o and R_s of this 40 API oil. As mentioned previously, the Standing correlation is better suited to low API oils.

5.3.1.4 Beggs–Robinson Correlation for μ_o

The Beggs–Robinson (1975) correlation is based on more than 2000 oil samples covering the following ranges:

- Pressures up to 5250 psig;
- Temperatures from 70 to 295 °F;
- Solution gas-oil ratios from 20 to 2070 SCF/stb;
- Oil densities from 16 to 58 API.

The input data for the Beggs–Robinson correlation are the API oil gravity, the solution gas-oil ratio, and the reservoir temperature. The first step is to calculate the dead oil viscosity as follows:

$$\mu_{od} = 10^X - 1.0 \tag{5.3.17a}$$

in which μ_{od} is the dead oil viscosity in cp, and X is given by:

$$X = 10^{(3.0324-0.0203\gamma_o)}T^{-1.163} \tag{5.3.17b}$$

in which T is in °F. The oil viscosity below the bubble point is given by:

$$\mu_o = A\mu_{od}^B \tag{5.3.18a}$$

in which:

$$A = 10.715[R_s + 100]^{-0.515} \tag{5.3.18b}$$

$$B = 5.44[R_s + 150]^{-0.338} \tag{5.3.18c}$$

The oil viscosity above the bubble point is given by:

$$\mu_o = \mu_{ob}\left[\frac{P}{P_b}\right]^m \tag{5.3.19a}$$

in which:

$$m = 2.6P^{1.187}\exp\{-11.513 - 8.98 \times 10^{-5}P\} \tag{5.3.19b}$$

The Beggs–Robinson correlation is compared with the Pembina Viking B Pool measured oil viscosity in Figure 5.3.3.

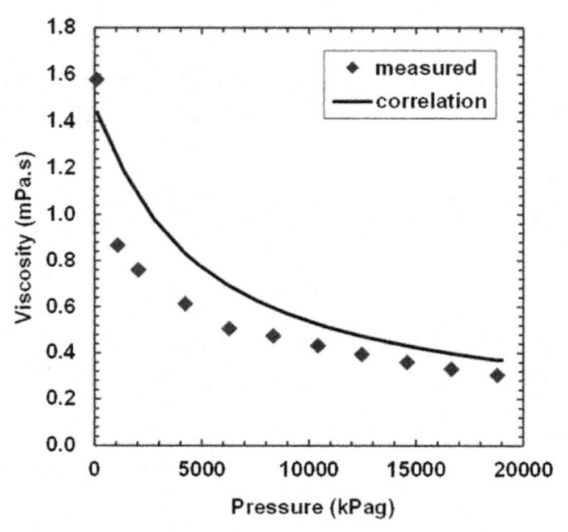

Figure 5.3.3 Comparison of Beggs–Robinson correlation against measured oil viscosity from Pembina Viking B Pool fluid study.

5.3.2 Correlations for Gas Properties

5.3.2.1 Corresponding States for B_g

The data required to use the law of corresponding states are the temperature, pressure, and either the critical properties or the density of the gas. The gas formation factor is related to the gas compressibility factor as follows:

$$B_g = 5.037\frac{z(T+460)}{P} \tag{5.3.20}$$

in which B_g has units of bbl/MSCF, T is the temperature (°F), and P is the pressure (psia). The factor of 460 converts the temperature from °F to R. The following equation for the compressibility factor was fitted to compressibility chart data:

$$z = A + (1-A)\exp\{-B\} + CP_r^D \tag{5.3.21a}$$

in which:

$$A = 1.39[T_r - 0.92]^{0.5} - 0.36T_r - 0.101 \tag{5.3.21b}$$

$$B = (0.62 - 0.23T_r)P_r + \left(\frac{0.066}{T_r - 0.86} - 0.037\right)P_r^2 + 0.32\frac{P_r^6}{10^{9(T_r-1)}} \tag{5.3.21c}$$

$$C = 0.132 - 0.32\log\{T_r\} \tag{5.3.21d}$$

$$D = 10^{\left(0.3106-0.49T_r+0.1824T_r^2\right)} \tag{5.3.21e}$$

in which T_r and P_r are the reduced temperature and pressure, respectively, defined as follows:

$$T_r = \frac{T+460}{T_c} \tag{5.3.22}$$

$$P_r = \frac{P}{P_c} \tag{5.3.23}$$

and T_c and P_c are the critical temperature and pressure, respectively. The critical properties can be obtained from a gas analysis. If a gas analysis is not available, they can be estimated from the gas gravity as follows:

$$T_c = 168 + 325\gamma_g - 12.5\gamma_g^2 \tag{5.3.24}$$

$$P_c = 677 + 15.0\gamma_g - 37.5\gamma_g^2 \tag{5.3.25}$$

in which the units of T_c and P_c are R and psia, respectively.

The corresponding states calculation is compared versus the Pembina Viking B Pool measured gas formation and expansion factors in Figure 5.3.4. The calculated expansion factor deviates from the measured values because a constant gas density was assumed for the calculation. In fact, the density of the evolved solution gas changes with pressure as more and richer solution gas evolves.

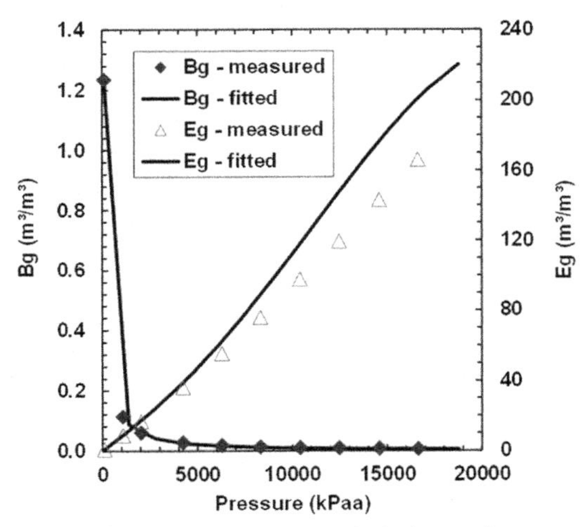

Figure 5.3.4 Comparison of corresponding states calculation against measured B_g and E_g from Pembina Viking B Pool fluid study.

5.3.2.2 Lee et al. correlation for μ_g

The Lee et al. (1966) correlation applies to the following conditions:

- Pressures from 100 to 8000 psia;
- Temperatures from 100 to 340 K;
- Carbon dioxide contents from 0.90 to 3.20 mol%.

The input data for the Lee et al. correlation are the gas gravity and the temperature. The gas viscosity is given by:

$$\mu_g = 10^{-4} K \exp\left\{X \rho_g^{(2.4-0.2X)}\right\} \tag{5.3.26a}$$

in which μ_g is the gas viscosity in cp and K and X are given by:

$$K = \frac{(9.4 + 0.02M)[T + 460]^{1.5}}{209 + 19M + T + 460} \tag{5.3.26b}$$

$$X = 3.5 + \frac{986}{T + 460} + 0.01M \tag{5.3.26c}$$

The gas density is given by:

$$\rho_g = 0.0014935 \frac{PM}{z(T + 460)} \tag{5.3.27}$$

in which M is the molar mass of the gas and is related to the specific gravity of the gas as follows:

$$M = 28.964 \gamma_g \tag{5.3.28}$$

Note that the Lee et al. correlation was used to generate the Pembina Viking B Pool gas viscosity data.

5.3.3 Correlations for Water Properties

5.3.3.1 McCain Correlation for B_w

The input data for the McCain (1988) correlation are the temperature and pressure of the water. The water formation volume factor is given by:

$$B_w = (1 + \Delta V_{wt})(1 + \Delta V_{wp}) \tag{5.3.29a}$$

in which:

$$\Delta V_{wt} = -0.010001 + 1.33391 \times 10^{-4}T + 5.50654 \times 10^{-7}T^2 \quad (5.3.29b)$$

$$\Delta V_{wp} = -1.95301 \times 10^{-9}PT - 1.72834 \times 10^{-13}P^2T - 3.58922 \times 10^{-7}P - 2.25341 \times 10^{-10}P^2 \quad (5.3.29c)$$

and T is in °F, and P is in psia. This correlation does not account for the salinity of the water. However, variations in salinity were observed to cause offsetting errors in ΔV_{wt} and ΔV_{wp}.

5.3.2.2 McCain Correlation for R_{sw}

The McCain (1988) correlation for the solution gas-water ratio, R_{sw}, is applicable at the following conditions:

- Pressures from 1000 to 10,000 psia;
- Temperatures from 100 to 250 °F;
- Salinity up to 30%.

The input data for the correlation are temperature, pressure, and salinity. The solution gas—water ratio is given by:

$$R_{sw} = R_{swp}10^X \quad (5.3.30a)$$

in which:

$$X = -0.0840655 \cdot S \cdot T^{-0.285854} \quad (5.3.30b)$$

and T is in °F, S is the salinity in wt%, and R_{swp} is given by:

$$R_{swp} = A + BP + CP^2 \quad (5.3.31a)$$

in which:

$$A = 8.15839 - 0.0612265T + 1.91663 \times 10^{-4}T^2 - 2.1654 \times 10^{-7}T^2 \quad (5.3.31b)$$

$$B = 1.01021 \times 10^{-2} - 7.44241 \times 10^{-5}T + 3.05553 \times 10^{-7}T^2 - 2.94883 \times 10^{-10}T^3 \quad (5.3.31c)$$

$$C = -9.02505 \times 10^{-7} + 1.30237 \times 10^{-8}T - 8.53425 \times 10^{-11}T^2 + 2.34122 \times 10^{-13}T^3 - 2.37049 \times 10^{-16}T^4 \quad (5.3.31d)$$

5.3.3.3 Osif and McCain Correlations for c_w

The Osif (1984) correlation is for the isothermal compressibility of water, c_w, above the bubble point. The correlation is applicable at the following conditions:

- Pressures from 1000 to 20,000 psia;
- Temperatures from 200 to 270 °F;
- NaCl concentrations up to 200 g/L.

The input data for the correlation are temperature, pressure, and salt concentration. The isothermal water compressibility is given by:

$$c_w = -\frac{1}{B_w}\left(\frac{\partial B_w}{\partial P}\right)_T = \frac{1}{7.033P + 541.5C_{\mathrm{NaCl}} - 537.0T + 403000} \quad (5.3.32)$$

in which P is pressure in psia, T is temperature in °F, and C_{NaCl} is the salt concentration in g/L.

The water compressibility is strongly affected by free gas. McCain (1988) proposed the following relationship for the isothermal compressibility of water below the bubble point:

$$c_w = -\frac{1}{B_w}\left(\frac{\partial B_w}{\partial P}\right)_T + \frac{B_g}{B_w}\left(\frac{\partial R_{swp}}{\partial P}\right)_T \quad (5.3.33)$$

The first term in Eqn (5.3.33) is solved using the Osif correlation. The second term is the derivative of Eqn (5.3.31a) and is given by:

$$\left(\frac{\partial R_{swp}}{\partial P}\right)_T = B + 2CP \quad (5.3.34)$$

in which B and C are obtained from Eqn (5.3.31c) and (5.3.31d), respectively. Equation (5.3.33) was not tested against experimental data and is recommended for estimation purposes only.

5.3.3.4 McCain Correlation for μ_w

The McCain (1988) correlation is for brine viscosity, μ_w, and is applicable at the following conditions:

- Pressures up to 15,000 psia;
- Temperatures from 90 to 170 °F;
- Salinity up to 15 wt%.

The input data for the correlation are temperature, pressure, and salinity. The water viscosity in cp is given by:

$$\mu_w = \mu_w^o\left[0.9994 + 4.0295 \times 10^{-5}P + 3.1062 \times 10^{-9}P^2\right] \quad (5.3.35a)$$

in which μ_w^o is the water viscosity at atmospheric pressure in cp given by:

$$\mu_w^o = AT^B \quad (5.3.35b)$$

in which:

$$A = 109.574 - 8.40564S + 0.313314S^2 + 0.00872213S^3 \quad (5.3.35c)$$

$$B = -1.12166 + 0.0263951S - 6.79461 \times 10^{-4}S^2 - 5.47119 \times 10^{-5}S^3 + 1.55586 \times 10^{-6}S^4 \quad (5.3.35d)$$

and S is salinity in wt%.

5.4 Sources of Error and Corrections for Black Oil Fluid Data

5.4.1 Sources of Error

The main source of error for fluid data is in obtaining a sample that accurately represents the reservoir fluid composition. Ideally, the samples should be collected at near initial reservoir conditions before solution gas evolves from the oil in the reservoir. If solution gas has evolved, the producing gas-oil ratio (GOR) may not be the same as the initial solution gas-oil ratio. If the gas saturation in the reservoir is less than the critical gas saturation, the evolved solution gas may not flow to the wellbore, and the GOR will be less than R_s. If the gas saturation exceeds the critical gas saturation, the GOR may be higher than R_s, because the gas is more mobile than the oil.

When samples are collected at surface, it is vital to measure correct flow rates for both oil and gas so that an accurate recombined fluid analysis is obtained. Another potential source of error is the determination of the shrinkage of surface samples, that is, the difference between stock tank and separator oil volumes. When samples are collected downhole, the main source of error is the small size of the sample. Typically, only a few liters are collected, which may not be sufficient to ensure that the sample is representative.

In many lower permeability reservoirs in which the bubble point pressure is near initial reservoir pressure, it is very difficult to get a representative sample. In these cases, it is important to obtain reliable field tests of producing GORs and recombine gas and oil in the laboratory to those GORs.

5.4.2 Checking and Correcting Fluid Data

Given the potential for sampling and recombination errors, fluid data should always be checked against production data. Compare the bubble point solution gas-oil ratio, R_{sb}, with the initial producing GOR of individual wells and for the entire reservoir. At constant separator conditions, the R_{sb} is expected to be the same as the initial producing GOR, except for wells near a gas-oil contact or wells with a high drawdown.

For reservoirs with a gas cap, the bubble point pressure must be the same as the initial reservoir pressure, by definition. For reservoirs without a gas cap, it is sometimes possible to check the bubble point using the pool production profiles, as long as pressure data are available. When the reservoir pressure drops below the bubble point, gas evolves, but remains in the reservoir. As more gas evolves and the critical gas saturation is exceeded, gas flows to the wellbore, and the producing GOR fluctuates, but generally increases. Therefore, on a plot of GOR versus time, the date at which the reservoir reached the bubble point can be estimated from the time when the GOR dips and then increases. In practice, it is often only possible to detect the increase in GOR. A change in the slope of a pressure versus cumulative oil production plot also is expected at the bubble point. Determination of the bubble point from production data is discussed in more detail in Part 3.

When the fluid data cannot be reconciled with the production data, the fluid data must be corrected or replaced. The appropriate action depends on the nature of the discrepancy. Some examples are discussed below:

1. Incorrect gas-oil recombination: If the fluid sample was recombined at too low a gas-oil ratio, the bubble point, the R_{sb}, and B_{ob} will be too low. The converse is also true. In this case, it may be possible to extrapolate the trends in the PVT data to a new bubble point that better matches the production data, as shown in Figure 5.4.1. If the fluid data were curve fit, the fit equations can be used with the same coefficients with the extrapolated bubble point properties. This method is only suitable if the actual GOR is within approximately 20% of the recombined GOR. If the fluid data were fit with a compositional model, a more realistic recombined fluid analysis can be input and used to regenerate the fluid data.
2. Similar but nonidentical fluid study from analogous pool: For many small reservoirs, a fluid study was not undertaken, and the reservoir engineer is forced to rely on a fluid study from a nearby analogous reservoir. If the differences between the fluid data and the production data are less than 10%, the bubble point properties can be corrected and the trends of the fluid data nearly preserved, as shown in Figure 5.4.2. These corrections are only approximations, and if the discrepancies are greater than 10%, it is best to seek another analogous fluid study or turn to correlations.

5.5 Properties of Unconventional Fluids

5.5.1 Heavy Oil

A number of unique issues arise in heavy oil property measurement because heavy oils are such viscous fluids. Some key points are reviewed briefly below including

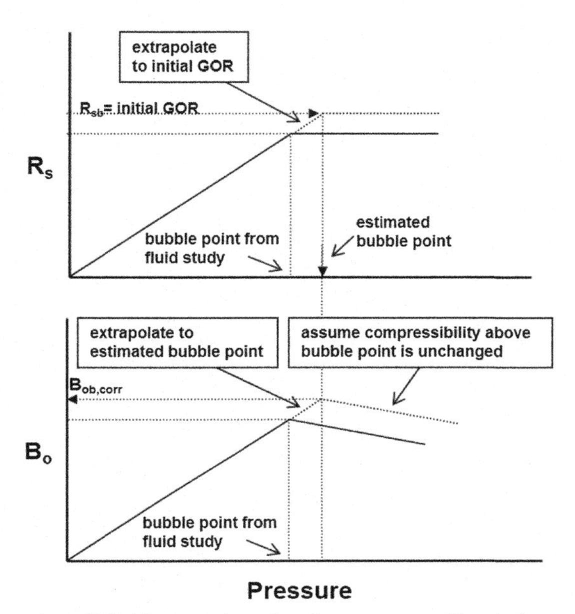

Figure 5.4.1 Correcting oil fluid properties when incorrect gas-oil ratio is used to generate fluid data.

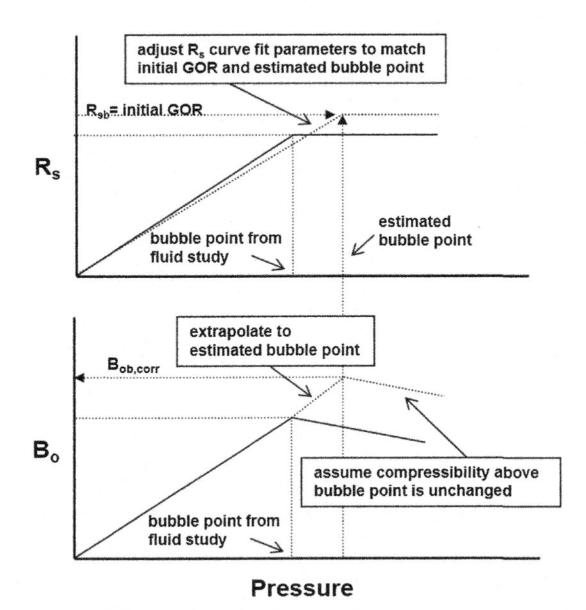

Figure 5.4.2 Correcting oil fluid properties from analogous reservoir to solution gas-oil ratio and bubble point estimated from production data.

representative sampling, sample preparation, PVT measurement, GOR estimation, and the effect of temperature on density and viscosity. More details can be found in Butler (1997) and the AOSTRA Technical Handbook on Oil Sands, Bitumens, and Heavy Oils (1989).

Representative sampling: Unlike most conventional reservoirs, there is evidence that heavy oil properties vary spatially, both horizontally and vertically, within the same reservoir (Larter et al., 2008). Generally, viscosity is highest near water zones that provide access for microbes and subsequent oil degradation. Because heavy oil density is similar to water density, the water zones can be below, above, or dispersed throughout the heavy oil zone. Hence, the distribution of viscosity is not straightforward, and a measurement program may be required to map the viscosities.

A second problem is recovering or preparing a representative live oil sample. If foamy oil flow occurs, the amount of gas in the recovered heavy oil sample may not match the true solution gas content of the heavy oil. Field measurements of the producing GOR are also notoriously unreliable, partly because the gas volumes are small and partly from the difficulty in separating the gas from the oil. An alternative to determine the solution gas-oil ratio is to calculate it based on the solubility of methane in heavy oil, as will be discussed later.

Sample preparation: The high viscosity of heavy oil also makes it challenging to separate water and solids from a heavy oil sample (Memon et al., 2010). Typical methods are heating with centrifugation, solvent dilution with centrifugation, and distillation.

- Heating and centrifugation has the least impact on the oil sample, as long as the temperature does not approach reaction conditions. Temperatures below 50−100 °C are usually recommended, but often reduce the oil viscosity too little to remove the water and solids effectively. Some loss of light ends can also occur, particularly at higher temperatures.
- Solvent dilution and centrifugation is the most effective method to remove water and solids. However, it is almost impossible to remove all the solvent from the oil, and therefore the oil properties, particularly viscosity, are altered.
- Distillation can remove much of the water from the heavy oil, but also removes light ends. The light ends can be condensed and recombined with the oil, but usually some are lost, and the oil properties are altered.

Although heating and centrifugation and distillation are less likely to alter oil properties than solvent dilution, there is simply no method guaranteed to provide a clean heavy oil sample that retains its original properties. For this reason, heavy oil property data can be unreliable, particularly viscosity data, which are sensitive to small amounts of contaminants (Miller et al., 2006).

PVT measurement: Recall that bubble points are usually determined from a pressure-volume isotherm, Figure 5.5.1. Each data point in the isotherm is measured after a finite equilibration time. In heavy oils, the equilibration times can be long (days or weeks), because the oil viscosity is high and therefore gas diffusivity in the oil is low. Foamy oil effects will also hinder the release of evolved solution gas. Therefore, if insufficient equilibration time is allowed, the oil may become supersaturated, and the bubble point determined from the test will be too low, Figure 5.5.1. In other

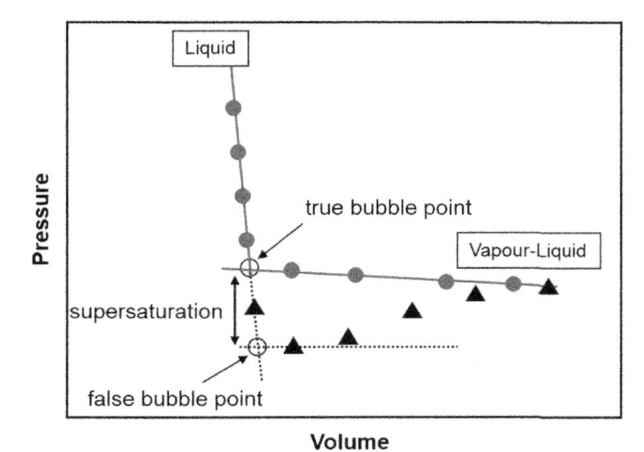

Figure 5.5.1 Determination of bubble point of heavy oil from pressure–volume isotherm (circles = equilibrated data; triangles = insufficient equilibration time).

respects, heavy oils are similar to conventional black oils, and the PVT data are well represented by a black oil model.

GOR estimation: If reliable PVT data are not available, the initial solution gas-oil ratio can be estimated from the solubility of methane in heavy oil. This approximation is based on the observation that the solution gases from heavy oil consist primarily of methane. Also, over geological time, the heavy oil is expected to be saturated with the gas. Butler (1997) found the following correlation fit methane solubilities in bitumen within 20% of experimental data:

$$R_{si,CH_4} = 4.66 P_b \exp\left\{\frac{343}{T_r}\right\} \tag{5.5.1}$$

in which R_{si,CH_4} is the initial solution gas-oil ratio in SCF/stb if the solution gas were composed only of methane, P_b is the bubble point pressure in MPa, and T_r is the reservoir temperature in K. Once the solution gas-oil ratio and bubble point are established, the oil volume factor can be determined from black oil correlations.

Effect of temperature on density: While conventional oil reservoirs are generally produced isothermally within the reservoir, heavy oil reservoirs are often heated with steam to reduce the oil viscosity. Therefore, it is necessary to determine the effect of temperature on fluid properties including density. The density of heavy oils generally follows a linear trend with temperature:

$$\rho_o = \mathring{\rho}_o - A(T - T_{ref}) \tag{5.5.2}$$

in which ρ_o is the oil density in kg/m^3 at T in °C, and $\mathring{\rho}_o$ is the oil density at a reference temperature, T_{ref}. A value for A of 0.62 has been reported for topped Athabasca bitumen (AOSTRA Handbook, 1989). Other values for heavy oils and bitumens generally fall within 0.63 and 0.66.

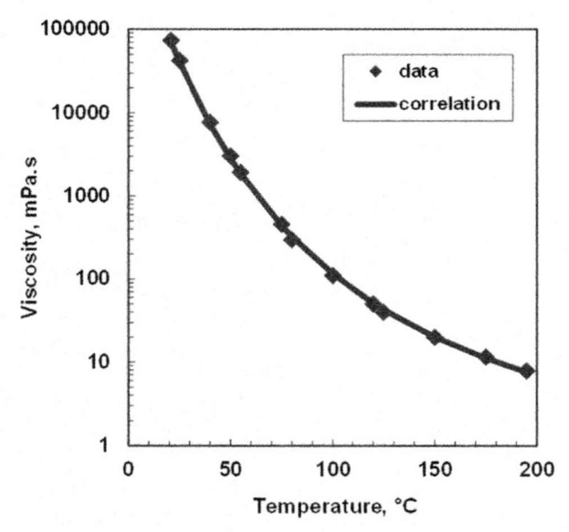

Figure 5.5.2 Effect of temperature on viscosity of an Alberta bitumen at 2.5 MPa. Data from Motahhari et al. (2013) are fitted with the Modified Walther correlation.

Effect of temperature on viscosity: The relationship between viscosity and temperature is arguably the most important property data for heavy oil processes. The viscosity of any fluid is most sensitive to temperature near the melting point or glass transition temperature. Heavy oils are near a glass transition at ambient temperatures and therefore are expected to have highly temperature-sensitive viscosities, as is confirmed in Figure 5.5.2. In this example, the viscosity decreases almost four orders of magnitude as the temperature is raised from 20 to 195 °C.

There are a number of correlations to fit and smooth viscosity-temperature data, a few of which are given below:

Modified Walther (Twu, 1985; Yarranton et al., 2013):

$$\log(\log(\mu + 0.7)) = A + B\log(T) \tag{5.5.3}$$

in which T is in K, and A and B are constants. The value added to the viscosity inside the brackets ranges from 0.7 to 1, depending on the version of the correlation.

Bergman and Sutton (2007):

$$\ln(\ln(\mu + 1)) = A + B\log(T^* + 310) \tag{5.5.4}$$

in which T^* is in °F.

Vogel (1921):

$$\ln(\mu) = A + \frac{B}{T + C} \tag{5.5.5}$$

The modified Walther correlation fit the data from the example bitumen data with an average absolute relative deviation of 5%, Figure 5.5.2. The other two correlations provide a similar quality fit to the data; however, the Vogel correlation requires an extra parameter.

5.5.2 Gas Condensates

The phase behavior of gas condensates was illustrated in Figure 2.4.2. Gas-condensate fluids are found in deeper hot reservoirs. Typically, these reservoirs have a much higher fraction of propane or higher fraction (C_3-C_{7+}) in the total recombined sample. To identify gas condensate reservoirs, engineers use a number of indicators, including gas-oil ratios, C_{7+} mole fraction, and API of separator liquids.

Usually, the initial separator GORs are in the range of 3000 to 25,000 SCF/bbl, with separator API gravities of 45–55. The C_{7+} mole fraction is typically greater than 3–12.5 mol%. However, it is not the surface conditions that define whether the fluid is a gas condensate; rather, it is whether or not liquid drops out in the reservoir during pressure depletion. Using rule-of-thumb criteria based on surface conditions is good for a first pass, but it is not a substitute for a full analysis.

The techniques used for a gas condensate analysis are different than those used for dry gas or black oil systems. A standard analysis for gas condensates is:

- Pressure–volume pressure analysis;
- Constant volume depletion test;
- Analysis of well stream effluent at various stages of depletion.

The pressure–volume relationship for a gas condensate is very different than that of a conventional oil. Unlike a conventional oil, an abrupt change in slope does not occur at the dew point of a gas condensate. A gradual transition is characteristic of a fluid near its critical point, Figure 5.5.3. Generally, gas condensates should be surface-sampled at the surface, rather than downhole, at close to initial reservoir pressure on a well with a low stabilized GOR (Gilchrist, 1993).

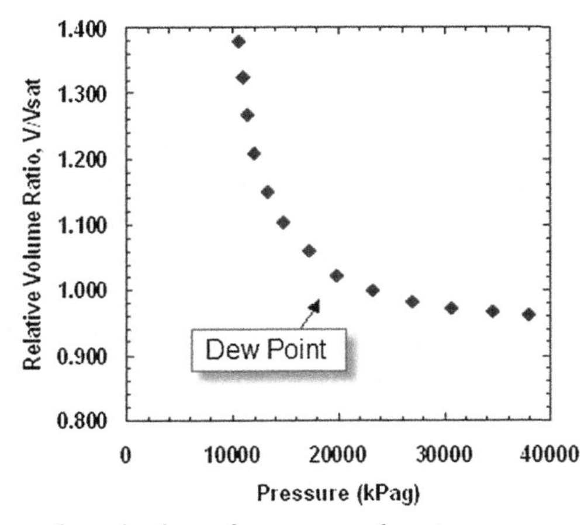

Figure 5.5.3 Pressure volume isotherm for a gas condensate.

In black oil PVT, PVT properties are assumed to be primarily a function of pressure, but with gas condensate and volatile oils, PVT properties are a function of pressure as well as composition. When analyzing gas condensates, we assume that gas is produced as the pressure depletes, and therefore a constant volume depletion (CVD) test at reservoir temperature is required. The initial part of the CVD report is a constant composition expansion analysis, which gives the density above dew point and the compressibility of the fluid above and below the dew point pressure. The second part of the report consists of compositional analyses of the fluid at various pressure levels. The pressure levels typically consist of the dew point pressure plus five lower pressures ($P < P_{\mathrm{dew}}$). The composition Z factors and the gallons per million SCF of C_{3+}, C_{4+} and C_{5+} are reported at each pressure. The third part of the report provides the calculated gas and liquid recoveries (bbl/Mscf) at each pressure level based on the Z-factors and the measured gas compositions. These calculated recoveries are the only recovery estimate available in the early life of a reservoir (Gilchrist, 1993).

Gas-condensate fluid properties are usually modeled with an equation of state approach. Because gas condensates are multicomponent mixtures, laboratory data are required to tune the EOS. The compositions of the produced fluids over time can also be used to constrain the fluid model. The interested reader is referred to Whitson and Brulé (2000) and Pedersen et al. (1989).

Pressure and Flow Test Data

Chapter Outline

Pressure and flow tests provide a great deal of information useful to production and reservoir engineers. As will be seen in Chapter 10, accurate pressure data are required to perform a material balance on a reservoir. Pressure transient tests can provide information on reservoir permeability, continuity, and formation damage. Flow tests are used to predict the performance of a well under different pressure drawdowns. In this chapter, the different types of pressure and flow tests and the analysis of test data are discussed.

6.1 Pressure Measurements

There are a number of techniques to measure bottom hole pressure. Pressures can be measured in the open hole before a well is cased (drill stem tests, wireline formation tests) or in a completed interval (wireline bottom hole recorders, acoustic well sounder tests). In this section, the different measurement techniques are reviewed. Pressure transient analysis (PTA) is discussed in Section 6.2, and the analysis of pool pressure data is considered in Section 6.3.

6.1.1 Open Hole Measurements

Drill Stem Test

Drill stem tests (DSTs) are a type of bottom hole build-up test. They are run during or after the drilling of a well, but before casing installation. A conventional DST is run on the end of the drill string immediately after a zone of interest has been drilled through.

Practical Reservoir Engineering and Characterization. http://dx.doi.org/10.1016/B978-0-12-801811-8.00006-7

In this way, the contact time between the reservoir and the drilling mud is minimized. To perform the test, the zone of interest is isolated with packers set above that zone. A straddle test is run after the well is logged, and zones of interest are identified with greater confidence. In this case, the test interval is isolated with packers set above and below the interval.

A schematic of a conventional DST string is provided in Figure 6.1.1. During a test, the packer is set above the zone of interest, and fluid enters the drill string through the perforated anchor. Pressure is recorded continuously throughout the test using two pressure transducers. A pair of transducers is used both as an accuracy check and for backup in case one transducer fails. Flow through the test string is controlled using the tester valve.

A typical test involves the following steps:

- Running the string into the well;
- Setting the packer(s);
- Initial flow: open tester valve for 15 min;
- Initial shut-in: close tester valve for 30 min;
- Final flow period: open tester valve for 30 min;
- Final shut-in: close the tester valve for 60 min;
- Unset the packer and pull out of hole.

The amount and rate of any flows to surface are recorded at surface. As the test string is withdrawn from the well, any fluid produced in the test that did not reach surface is held in the pipe because the tester valve is closed. The heights of the different

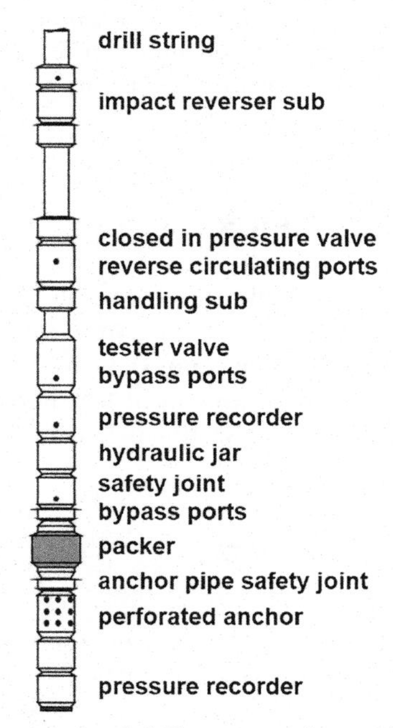

Figure 6.1.1 Schematic of conventional drill stem test pipe configuration.

fluids in the pipe string are recorded. Typical fluids encountered are: water, mud, oil, oil-cut (OC) mud, and gas-cut (GC) mud. The fluid recovery is sometimes the first indication of hydrocarbons if oil or gas is present or of well damage or low permeability if recovery is low.

A diagram of the pressure recorded during a typical test of an undamaged, moderately permeable zone is shown in Figure 6.1.2. Note that the time scale increases from left to right, and the pressure scale increases from top to bottom on the pressure recording. The measured pressure increases as the test string is run in the hole and stabilizes as the packer is set. There is a decrease in pressure when the tester valve is opened, and pressure remains constant or increases slightly during the flow period. There is a rapid increase in pressure when the well is shut-in until the pressure begins to stabilize as it approaches a built-up reservoir pressure. A similar pattern is usually observed in the second flow and build-up period and then pressure decreases as the test string is pulled out of the hole. The pressure data from a drill stem test can be used to determine the average reservoir pressure around the well, the extent of damage to the formation at the wellbore, and sometimes information about reservoir heterogeneity and continuity, as discussed in Section 6.2.

Wireline Formation Tester

A wireline formation tester is an open hole wireline tool. The key components are:

- A donut-shaped packer;
- Two setting pistons to hold the packer against the wellbore;
- A piston drive probe with a filter screen;
- A pressure transducer;
- Two sample chambers.

The operation of the tool is illustrated in Figure 6.1.3. After the packer is set in place, the probe is driven through the center of the packer, through any filter cake

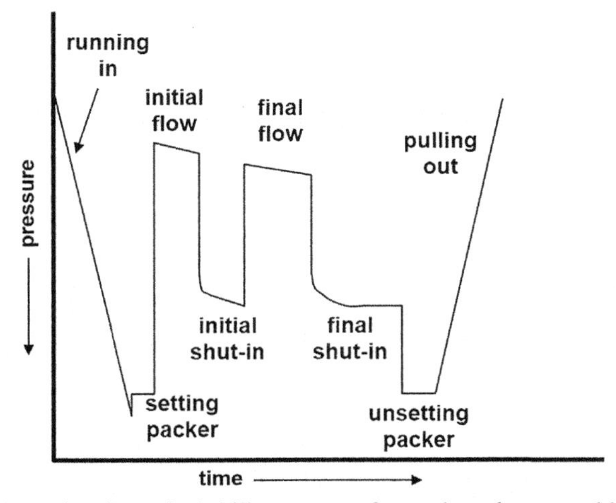

Figure 6.1.2 Schematic of a typical drill stem test of a moderately permeable well.

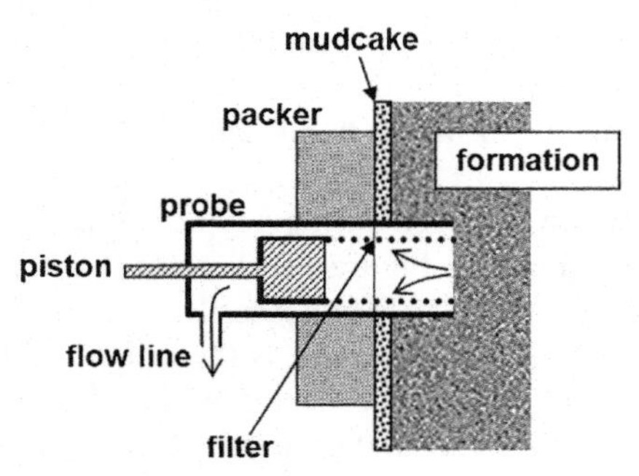

Figure 6.1.3 Schematic of a wireline formation test.

on the wellbore, and into the formation of interest. The pressure is measured continuously, and the signal from the transducer is sent through the wireline to surface. The tool can be reset to measure as many intervals as desired. A single formation can be tested at different depths to determine fluid gradients and contacts in the reservoir. Formation tests can also be conducted in cased hole; a shaped charge is fired through the casing for the test. The methodology is the same as described for a static gradient (SG) test, Section 6.1.2. Build-up tests can also be recorded, and two fluid samples can be obtained. As with a DST, the duration of any flow time is limited due to the expense of maintaining a rig on the well.

6.1.2 Cased Hole Measurements

Direct Bottom Hole Pressure Measurement

The most accurate pressure test is a direct measurement of pressure at the perforated interval. This can be accomplished with a pressure transducer lowered on wireline or a transmitter installed in the tubing string. In some cases, usually for offshore wells, a pressure transmitter is installed downhole permanently, and pressure is recorded continuously. A wireline tool is used for most other applications.

To perform a wireline test, any installed artificial lift, such as a pump and rods, must be removed prior to the pressure test so that the tool can be run in the hole. Removing and reinstalling artificial lift significantly adds to the cost of the test; hence, wireline tests are usually only practical for flowing wells. To perform a bottom hole pressure test, the well is first shut-in for the desired time, and then a pressure measurement tool is lowered into the well on wireline. In the early days, a pressure "bomb" was used because pressure transducers were too large to fit in the production tubing. Once the bomb was in place, the valve was opened until a bottom hole sample was obtained. The pressure was measured at surface by attached a gauge to the bomb valve. Nowadays, pressures are recorded with a pair of transducers lowered into the well. The transducer signals may be recorded in the bottom hole tool or sent through the wireline and recorded at surface.

The most common wireline pressure test is the SG test. An example SG is provided in Figure 6.1.4. The SG gets its name because pressures are recorded at several

SUBSURFACE PRESSURE MEASUREMENT

Company	XXXXX	Well Name	XXXXX
Field and Pool	Owl	Status	flowing oil
Test Type	Static Gradient	Date	Dec. 19/ 95
Prod. Int. (mCF)	1279.7 - 1279.85	CF Elev	719.1
MPP (mCF)	1279.78	KB Elev	723.4

Tubing Pressure (kPag)	6550	Shut-In Date	Oct. 30/ 95 @ 08:00
Casing Pressure (kPag)	5805	Date On Bottom	Dec. 19/ 95 @ 13:03
B.H. Temp. (°C)	55.1	Date Off Bottom	Dec. 19/ 95 @ 13:23

	Gauge 1	Gauge 2
Run Depth (mCF)	1278.8	1279.8
Run Depth Pressure (kPag)	8291	8304
Run Depth Gradient (kPag)	7.4723	7.4372
MPP Pressure (kPag)	8298	8304

Time (hr-min)	Gauge 1 Depth (m CF)	Gauge 1 Pressure (kPag)	Gauge 1 Gradient (kpa/m)	Gauge 2 Depth (m CF)	Gauge 2 Pressure (kPag)	Gauge 2 Gradient (kpa/m)
11:21 - 11:31	SURFACE	6446.6		SURFACE	6458.8	
11:36 - 11:46	299.0	6648.7	0.6759	300.0	6656.6	0.6593
11:50 - 12:00	599.0	6844.7	0.6533	600.0	6847.9	0.6377
12:04 - 12:14	899.0	7032.9	0.6273	900.0	7037.4	0.6317
12:16 - 12:26	999.0	7093.6	0.6070	1000.0	7097.4	0.6000
12:28 - 12:38	1099.0	7153.1	0.5950	1100.0	7158.1	0.6070
12:40 - 12:50	1199.0	7694.4	5.4130	1200.0	7710.2	5.5210
12:51 - 13:01	1239.0	7993.2	7.4700	1240.0	8007.8	7.4400
13:03 - 13:23	1278.8	8290.6	7.4723	1279.8	8303.8	7.4372

Figure 6.1.4 Example of a static gradient test.

different depths as the tool is lowered into the well. The density of the fluids in the wellbore can then be determined from the pressure gradient as follows:

$$\rho = \frac{g\Delta P}{g_c \Delta h} \tag{6.1.1}$$

Recall that g is the gravitational constant and g_c is a units conversion factor. The fluid contacts also can be determined from changes in the pressure gradient, as shown in Figure 6.1.5. Wireline gauges are also used for transient tests such as build-up (BU) or fall-off (FO) tests. Pressure transient tests are discussed in Section 6.2.

Usually, the pressure of interest is the pressure at the midpoint of the perforations (MPP) or at a given datum depth. However, for all direct pressure measurements, the pressure is measured at the depth of the pressure transducers. This depth usually does not coincide with the MPP or the datum depth, as shown in Figure 6.1.6. The pressure at MPP or at the datum depth must be calculated based on the fluid levels and fluid gradients in the well. For example, the MPP pressure for the situation in Figure 6.1.6 is calculated as follows:

$$\begin{aligned} P_{MPP} &= P_{gauge} + \rho_g \frac{g}{g_c}(h_{gauge} - h_{GOC}) + \rho_o \frac{g}{g_c}(h_{GOC} - h_{WOC}) \\ &\quad + \rho_w \frac{g}{g_c}(h_{WOC} - h_{MPP}) \end{aligned} \tag{6.1.2}$$

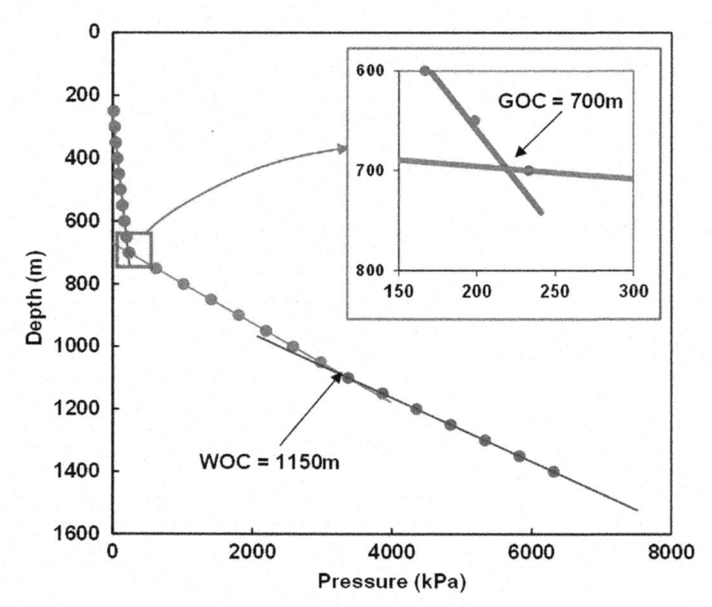

Figure 6.1.5 Determination of fluid levels in well from static gradient pressures.

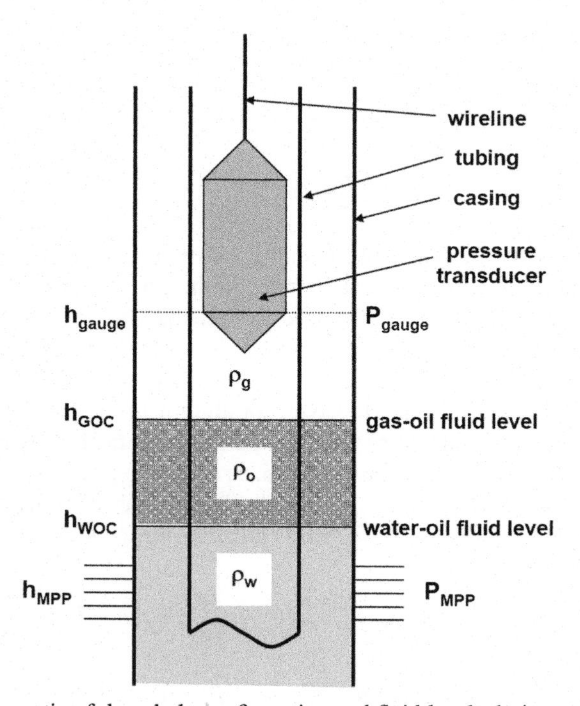

Figure 6.1.6 Schematic of downhole configuration and fluid levels during a pressure test.

in which P_{MPP} is the pressure at the MPP, P_{gauge} is the measured pressure, h_{gauge} is the depth of the pressure transducer, h_{GOC} is the depth of the gas-oil fluid level in the wellbore, h_{WOC} is the depth of the water–oil fluid level in the wellbore, and h_{MPP} is the depth of the MPP.

Acoustic Pressure Measurement

Wireline pressure measurements are usually impractical for wells on artificial lift. Instead, the casing head pressure is recorded at surface and the bottom hole pressure is calculated from the fluid levels in the well and the pressure gradients of the fluids (Figure 6.1.7). The depth of the fluid levels is determined acoustically. A shot or charge is fired at the surface, and the echoes from tubing collars and fluid interfaces are recorded. The depth to the fluid level is calculated from the tubing tally. The acoustic well sounder (AWS), depthograph, echometer, and sonoloy tests all operate on this principle. An example acoustic test is given in Figure 6.1.8.

Acoustic tests can be used for single pressure measurements or for transient tests. However, the calculated pressures are less accurate than direct pressure measurements because the pressure gradient in each fluid is not measured and must be estimated either from surface measurements or from a prior or offset SG test. Also, the measured fluid level may be inaccurate if a gas-oil foam or a water–oil emulsion is present.

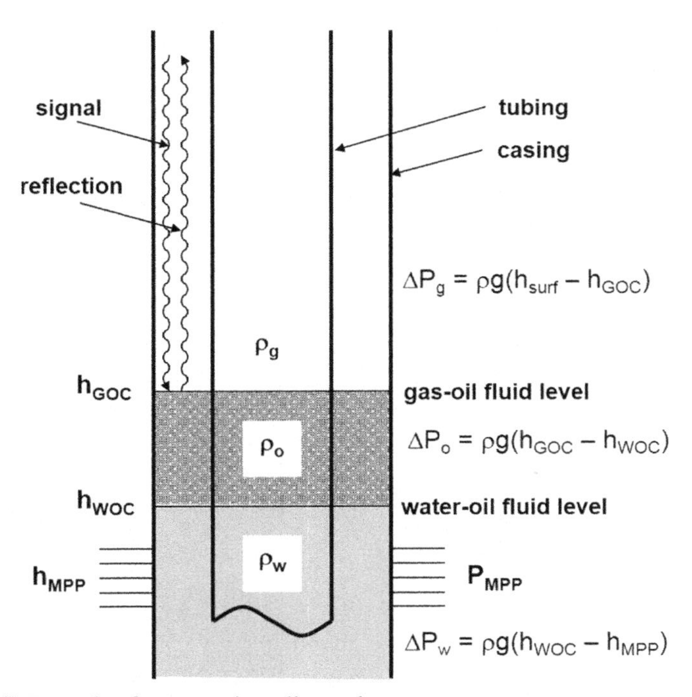

Figure 6.1.7 Example of an acoustic well sounder test.

ACOUSTIC TEST

Well Name	XXXX	Formation	Glauconite	Status	Flowing Gas		
Field Name	Hussar	License #	XXX	Test Type	Build Up		
KB (m)	897.00	KB to CF (m)	4.40	Oil Rate (m^3/d)	0.08	Gas Gravity	0.69
Tubing Bottom (mKB)	1424.79	Surface Temp (°C)	2.2	Water Rate (m^3/d)	0.00	Oil API	66.1
Number of Joints	149	Reservoir Temp (°C)	48.0	Gas Rate ($10^3m^3/d$)	13.00	Water Gravity	1.01

	top of perfs	bottom of perfs	mid-point
Measured Depth (mKB)	1396.00	1408.00	1402.00
True Vertical Depth (mKB)	1396.00	1408.00	1402.00

Date	Time	Shut-In Time (hr)	Joints to Fluid	Column Heights (mCF)			Gradients (kPa/m)			Pressure (kPaa)				
				Gas	Oil	Water	Gas	Oil	Water	Surface	Gas	Oil	Water	MPP
06/12/2004	10:00:00	0	145	1382.3	15.34	0	0.0247	6.583	-	293.00	34.18	100.98	0.00	428.16
06/18/2004	14:31:19	148.52	145	1382.3	15.34	0	0.0884	6.486	-	1022.84	122.13	99.50	0.00	1244.46
06/18/2004	14:46:19	148.77	145	1382.3	15.34	0	0.0884	6.486	-	1023.10	122.16	99.50	0.00	1244.75
06/18/2004	15:01:19	149.02	145	1382.3	15.34	0	0.0884	6.486	-	1023.04	122.15	99.50	0.00	1244.69
06/18/2004	15:16:19	149.27	145	1382.3	15.34	0	0.0883	6.486	-	1022.63	122.10	99.50	0.00	1244.22
06/18/2004	15:31:19	149.52	145	1382.3	15.34	0	0.0883	6.486	-	1023.05	122.11	99.50	0.00	1244.66

Figure 6.1.8 Example results from an acoustic test.

6.1.3 Sources of Error

The main source of error in any single point pressure measurement is insufficient shut-in time for the reservoir to build up to the average reservoir pressure. It is not usually economically feasible to shut-in a producing well for more than a few hours or days. Hence, the shut-in time is usually limited. If the build-up time is longer than the shut-in time, the measured pressure will be less than the average reservoir pressure. The most accurate pressure data are usually obtained from observation wells that have been shut-in for extended periods.

With direct pressure measurements, there may also be errors in the pressure transducer readings. Usually two transducers are run so that a measurement error is apparent when the two pressure measurements are compared. In rare cases, the depth of the transducers is measured incorrectly, and the calculated pressure at the perforations will then be incorrect.

With acoustic tests, errors arise from incorrect fluid gradients or indistinct fluid contacts, as discussed in Section 6.1.2.

6.2 Pressure Transient Tests of Oil Wells

PTA can be used to evaluate the following:

- Initial pressure, P_i;
- Average reservoir pressure within the drainage area, P_r;
- The size of the drainage area, A;
- The permeability-pay thickness, kh;
- The skin factor, S;
- The Dietz shape factor, C_A.

Exactly what information can be obtained depends on the type of test, the test conditions, and the reservoir.

There are several types of pressure transient tests, including pressure drawdown, build-up and fall-off tests. The analysis of each test type is briefly described below, based on the following assumptions:

- Single-phase, laminar, horizontal flow;
- Homogeneous reservoir;
- Circular drainage area with central well (unless otherwise specified);
- Reservoir and fluid properties k, ϕ, h, c, μ, and B_o are independent of pressure;
- Flow is in the transient regime (unless otherwise specified).

6.2.1 Pressure Drawdown Test

A drawdown test involves shutting-in a well until a stabilized pressure is achieved, and then flowing the well at a constant rate while recording the bottom hole flowing pressure. A typical pressure profile during a drawdown test is provided in Figure 6.2.1.

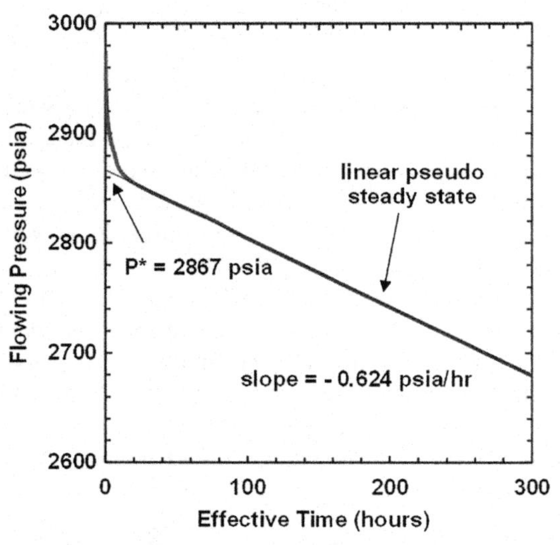

Figure 6.2.1 Pressure response over time during a drawdown test.

Transient Regime

When the pressure is plotted on semilog coordinates, as shown in Figure 6.2.2, a linear trend is apparent. This linear semilog trend is the transient regime, and the terminal rate solution derived in Chapter 3 applies:

$$P_{wf} = P_i - \frac{162.6 q_o \mu_o B_o}{k_o h}\left[\log\left\{\frac{k_{ro} k t}{1698 \phi \mu_o c_t r_w^2}\right\} + 0.87S\right] \tag{6.2.1}$$

in which P is in psia, q_o in bbl/d, k in mD, μ_o in cp, h in ft, r_w in ft, t in hours, and c_t in psia^{-1}.

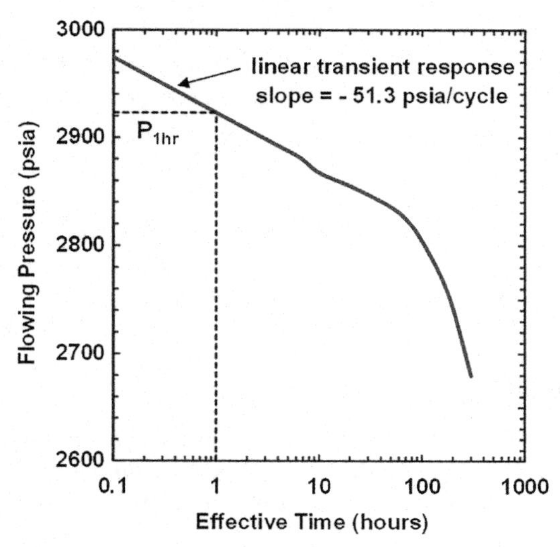

Figure 6.2.2 Semilog plot of pressure response during a drawdown test.

To analyze a drawdown test, all the constants in Eqn (6.2.1) are grouped, and the equation is simplified to:

$$P_{wf} = b + m\log(t) \tag{6.2.2}$$

in which b contains the constants from the log term, and m is given by:

$$m = -\frac{162.6 q_o \mu_o B_o}{k_o h} \tag{6.2.3}$$

The average permeability to oil within the drainage area can be obtained from a simple rearrangement of Eqn (6.2.3):

$$k_o = -\frac{162.6 q_o \mu_o B_o}{mh} \tag{6.2.4}$$

Once m is known, Eqn (6.2.1) can be rearranged to solve for the skin factor:

$$S = 1.151\left[\frac{P_i - P_{wf}}{-m} - \log\left\{\frac{k_o t}{1698\phi\mu_o c_t r_w^2}\right\}\right] \tag{6.2.5}$$

Equation (6.2.5) is solved at 1 h to eliminate the time term and obtain:

$$S = 1.151\left[\frac{P_i - P_{1hr}}{-m} - \log\left\{\frac{k_o}{1698\phi\mu_o c_t r_w^2}\right\}\right] \tag{6.2.6}$$

in which P_{1hr} is the theoretical wellbore flowing pressure 1 h after the start of flow. It is calculated from an interpolation or extrapolation of the linear semilog transient pressure response, as shown in Figure 6.2.2.

Pseudo Steady-State Regime

If the drawdown test is conducted long enough, the pseudo steady-state condition is reached. In this regime, pressure decreases linearly with time, as shown on Figure 6.2.1. The terminal rate solution at pseudo steady state was derived in Chapter 3 and is given by:

$$P_{wf}(t) = P_i - \frac{162.6 q_o \mu_o B_o}{k_o h}\left[\log\left\{\frac{4A}{1.781 C_A r_w^2}\right\} + 0.87S\right] - \frac{0.2339 q_o B_o t}{Ah\phi c_t} \tag{6.2.7}$$

in which P is in psia, q_o is in bbl/d, k_o is in mD, μ_o is in cp, A is in ft^2, h is in ft, r is in ft, t is in hours, and c_t is in psia^{-1}. C_A is the Dietz shape factor (see Figure 3.4.3).

Following a similar procedure as for the transient drawdown, all the constants in Eqn (6.2.7) are grouped, and the equation simplifies to:

$$P_{wf} = b' + m't \tag{6.2.8}$$

in which b' is a constant, and m' is given by:

$$m' = -\frac{0.2339 q_o B_o}{Ah\phi c_t} \tag{6.2.9}$$

The drainage area can be obtained from a simple rearrangement of Eqn (6.2.9):

$$A = -\frac{0.2339 q_o B_o}{m' c_t h\phi} \tag{6.2.10}$$

The linear steady state response can also be extrapolated to zero time to obtain an extrapolated flowing pressure, P^* (Figure 6.2.1). Hence, at $t = 0$, Eqn (6.2.7) becomes:

$$\begin{aligned} P_i - P^* &= \frac{162.6 q_o \mu_o B_o}{k_o h}\left[\log\left\{\frac{4A}{1.781 C_A r_w^2}\right\} + 0.87S\right] \\ &= -m\left[\log\left\{\frac{4A}{1.781 r_w^2}\right\} - \log\{C_A\} + 0.87S\right] \end{aligned} \tag{6.2.11}$$

If m, S, and A are known, the value of C_A can be determined from Eqn (6.2.11). The geometry of the drainage area can then be estimated from a table of Dietz values.

Analysis of Drawdown Test

An examination of Eqns (6.2.4), (6.2.5), and (6.2.10) indicates that the following information is required to interpret a drawdown test:

- Wellbore radius, r_w;
- Flow rate, q_o;
- Oil formation volume factor, B_o;
- Oil viscosity at formation conditions, μ_o;
- Porosity, ϕ;
- Pay thickness, h;
- Total compressibility, c_t;
- Initial pressure, P_i (if not measured during test).

The flow rate is assumed to be constant. When the flow rate experiences minor random variations during a test, the final flow rate is used, and an effective flowing time, t_{eff}, is determined as follows:

$$t_{eff} = \frac{Q_o}{q_{o,final}} \tag{6.2.12}$$

in which Q_o is the cumulative oil production at a given time, and $q_{o,final}$ is the final flow rate. If the flow rate varies significantly or experiences step changes, superposition is required for the analysis, as described in Section 6.2.2.

The wellbore radius is usually based on the casing size. For example, the radius for 7.5-inch casing is approximately 0.3 ft. A rule of thumb for SI units is to use an r_w of 0.1 m. The oil formation volume factor and oil viscosity are found from a fluid study or estimated from correlations (Chapter 5). Porosity is determined from well logs and core data (Chapters 7 and 9). The total compressibility, c_t, is defined as follows:

$$c_t = c_o S_o + c_w S_w + c_f \tag{6.2.13}$$

in which S is saturation, c is compressibility, and the subscripts o, w, and f denote oil, water, and formation, respectively. The saturations are determined from well logs and special core data. Oil and water compressibility are determined from fluid data or correlations. Rock compressibility is determined from special core data or correlations.

Example: Figure 6.2.1 presents a continuous pressure recording taken during a drawdown test, corrected to absolute pressure at a datum depth, and plotted against effective time. The operator wants to determine the permeability, skin factor, and tested drainage area from the pressure data. The following data are available for the well and the reservoir:

$r_w = 0.333$ ft	$B_o = 1.2$ bbl/stb
$P_i = 3300$ psia	$\mu_o = 1.5$ cp
$q_o = 300$ stb/d	$h = 15$ ft
$c_t = 3.0 \cdot 10^{-5}$ psi^{-1}	$\phi = 0.25$

First consider the linear transient response on a semilog plot, Figure 6.2.2. The slope in the transient region is −51.3 psia/hr. From Eqn (6.2.4), the permeability to oil is:

$$k_o = -\frac{162.6 q_o \mu_o B_o}{mh} = \frac{162.6(300\ \text{stb/d})(1.5\ \text{cp})(1.2\ \text{bbl/stb})}{(51.3\ \text{psia/cycle})(15\ \text{ft})} = 114\ \text{mD}$$

Figure 6.2.2 shows that the flowing pressure at 1 h falls within the linear transient region and is 2923 psia. From Eqn (6.2.5), the skin factor is:

$$S = 1.151\left[\frac{P_i - P_{wf}}{-m} - \log\left\{\frac{k_o t}{1698\phi\mu_o c_t r_w^2}\right\}\right]$$

$$= 1.151\left[\frac{3300 - 2923}{51.3} - \log\left\{\frac{114}{1698(0.25)(1.5)(3\cdot 10^{-5})(0.333)^2}\right\}\right] = 3.0$$

Now consider the pseudo steady-state linear response shown on Figure 6.2.1. The slope is −0.624 psia/hr. From Eqn (6.2.10), the drainage area of the test is:

$$A = -\frac{0.2339 q_o B_o}{m' c_t h \phi} = \frac{0.2339(300\ \text{stb/d})(1.2\ \text{bbl/stb})}{(0.624\ \text{psia/hr})(3.0 \cdot 10^{-5}\text{psi}^{-1})(15\ \text{ft})(0.25)}$$
$$= 1.2 \cdot 10^6 \text{ft}^2$$

An area of 1.2 million square feet is 27.5 acres and is equivalent to a drainage radius of 618 ft. This is the area sampled by the test. The permeability of 114 mD is the average permeability to oil within this drainage area.

To determine the shape of the drainage area, the linear steady state response is extrapolated to zero time to obtain $P^* = 2867$ psia. The shape of the drainage area is determined from Eqn (6.2.11) as follows:

$$\log\{C_A\} = \log\left\{\frac{4A}{1.781 r_w^2}\right\} + 0.87S + \frac{P_i - P^*}{m}$$

$$= \log\left\{\frac{4(1.2 \cdot 10^6)}{1.781(0.33)^2}\right\} + 0.87(3.0) - \frac{3300 - 2867}{51.3} = 1.562$$

Hence, C_A is 36.5, consistent with a circular drainage area with a central well ($C_A = 36.1$). The shape factor calculation is sensitive because an antilog is involved. It is not always possible to determine the shape factor accurately when there is uncertainty in the skin factor or if other factors such as permeability variations or well interference influence the pressure response.

6.2.2 Build-up and Fall-off Tests

A build-up test involves producing a well at a constant rate and then shutting-in the well while recording the bottom hole pressure. A typical pressure profile during a build-up test is provided in Figure 6.2.3. A FO test is the same as a build-up test, except that fluid is injected into the well instead of produced from the well. Because the principles of analysis are the same, only build-up test analysis is discussed here. Before examining the equations used to analyze build-up data, it is necessary to review the principle of superposition.

Superposition

The principle of superposition states that the solutions to the individual linear differential equations arising from point sources are additive. In other words, the pressure drop contributions at a point in space and time from a number of wells can be summed to obtain the overall pressure drop. For example, consider the three wells shown in Figure 6.2.4. Let us assume that two wells have just been put on production, and it is desired to monitor the pressure response at the third well. The transient

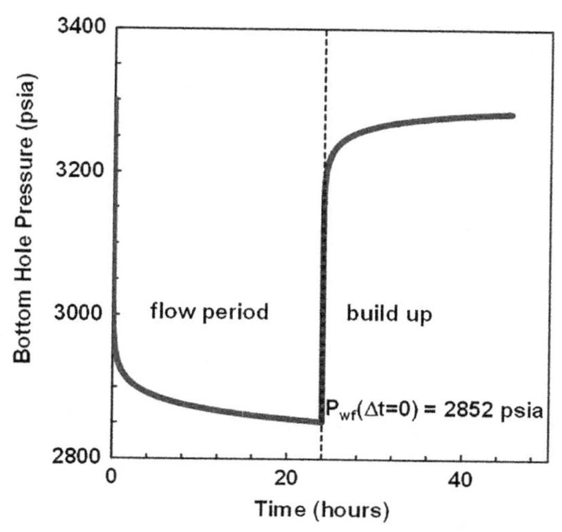

Figure 6.2.3 Pressure profile during a flow and build-up test.

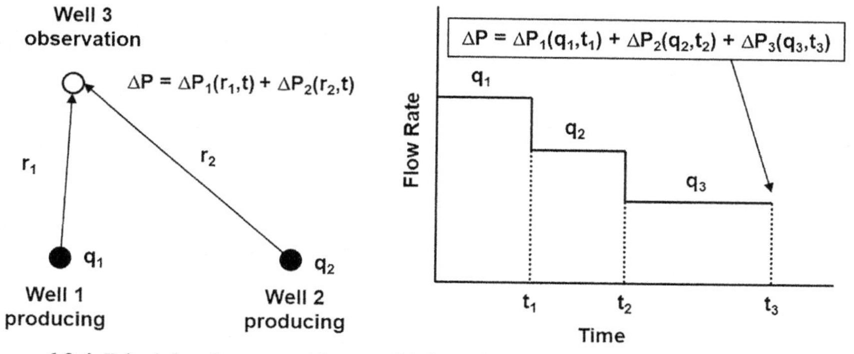

Figure 6.2.4 Principle of superposition: multiple wells (left); multiple rates (right).

pressure response, ΔP_j, from each producing well was derived in Chapter 3 and is given by:

$$\Delta P_j = P_i - P_j(r,t) = \frac{q_j \mu}{4\pi kh}\left[-E_i\left(-\frac{\phi \mu c r_j^2}{4\,kt}\right)\right] \tag{6.2.14}$$

The principle of superposition states that the pressure drop at the nonproducing well (observation well) is simply the sum of the pressure drops from the other wells:

$$\Delta P = \sum_j^n \Delta P_j = \sum_j^n \frac{q_j \mu}{4\pi kh}\left[-E_i\left(-\frac{\phi \mu c r_j^2}{4\,kt}\right)\right] \tag{6.2.15}$$

and the pressure at the observation well is:

$$P = P_i - \sum_{j}^{n} \Delta P_j \tag{6.2.16}$$

The principle of superposition can also be used when the flow rate of the well changes during a drawdown test. In this case, the pressure change contribution from each flow period is accounted for as follows:

$$\Delta P = \sum_{j}^{n} \Delta P_j = \sum_{j}^{n} \frac{162.6\left(q_j - q_{j-1}\right)\mu_o B_o}{k_o h}\left[\log\left\{\frac{1698\phi\mu_o c_t r_w^2}{k_o\left(t - t_{j-1}\right)}\right\} - 0.87S\right] \tag{6.2.17}$$

in which j indicates a flow period from time t_{j-1} to t_j at rate q_j.

Horner Plot

A build-up test is represented mathematically using superposition. At the start of the flowing period, the pressure change is represented by the equation of a single well flowing at a constant rate, q_o. At the end of the flow period, time $= t_p$, a second equation is added representing a well flowing at a constant rate of $-q_o$. The superposition of the equations gives no flow after the end of the flow period, when the well is shut-in. The pressure response is the sum of the predicted pressure change from each equation, given in Field units by:

$$\begin{aligned} P_{ws} &= P_i - \frac{162.6 q_o \mu_o B_o}{k_o h}\log\left\{\frac{k_o\left(t_p + \Delta t\right)}{1698\phi\mu_o c_t r_w^2}\right\} \\ &\quad - \frac{162.6\left(-q_o\right)\mu_o B_o}{k_o h}\log\left\{\frac{k_o \Delta t}{1698\phi\mu_o c_t r_w^2}\right\} \end{aligned} \tag{6.2.18}$$

which simplifies to:

$$P_{ws} = P_i - \frac{162.6 q_o \mu_o B_o}{k_o h}\log\left\{\frac{\left(t_p + \Delta t\right)}{\Delta t}\right\} \tag{6.2.19}$$

in which P_{ws} is the bottom hole shut-in pressure, Δt is the time after shutting-in the well, and $t_p + \Delta t$ is the total time from the start of flow. Note that the flow time is calculated in the same manner as the effective time of a drawdown test:

$$t_p = \frac{Q_o}{q_{o,final}} \tag{6.2.20}$$

in which Q_o is the cumulative oil production at the end of the flow period, and $q_{o,final}$ is the final flow rate.

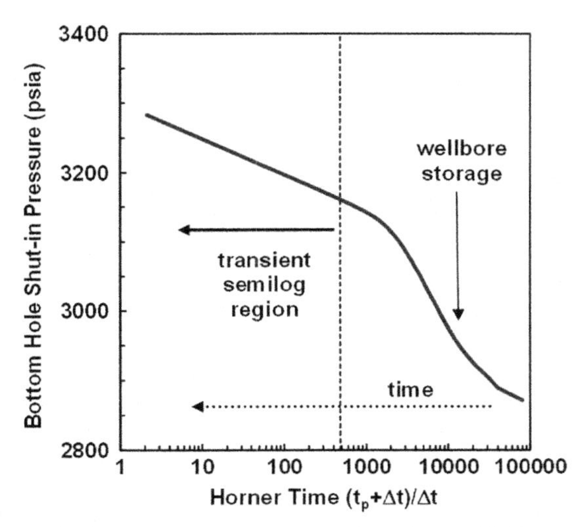

Figure 6.2.5 Horner plot for a homogeneous reservoir with radial inflow.

Equation (6.2.19) indicates that the transient pressure response is linear on a plot of shut-in pressure versus "Horner" time, in which Horner time is defined as $(t_p + \Delta t)/\Delta t$. This type of plot is referred to as a Horner plot and is shown in Figure 6.2.5 (Horner, 1951).

The slope, m, of a Horner plot in the transient regime is:

$$m = -\frac{162.6 q_o \mu_o B_o}{k_o h} \tag{6.2.21}$$

The average permeability within the drainage area can be obtained from a simple rearrangement of Eqn (6.2.21):

$$k_o = -\frac{162.6 q_o \mu_o B_o}{m h} \tag{6.2.22}$$

Once m is known, Eqns (6.2.5) and (6.2.17) can be rearranged to solve for the skin factor:

$$S = 1.151 \left[\frac{P_{ws} - P_{wf}(\Delta t = 0)}{-m} - \log\left\{ \frac{k_o t_p \Delta t}{1698 \phi \mu_o c_t r_w^2 (t_p + \Delta t)} \right\} \right] \tag{6.2.23}$$

Equation (6.2.23) is solved at 1 h to eliminate the Δt term and obtain:

$$S = 1.151 \left[\frac{P_{1hr} - P_{wf}(\Delta t = 0)}{-m} - \log\left\{ \frac{k_o t_p}{1698 \phi \mu_o c_t r_w^2 (t_p + 1)} \right\} \right] \tag{6.2.24}$$

If $t_p >> 1$, then Eqn (6.2.24) reduces to:

$$S = 1.151\left[\frac{P_{1hr} - P_{wf}(\Delta t = 0)}{-m} - \log\left\{\frac{k_o}{1698\phi\mu_o c_t r_w^2}\right\}\right] \tag{6.2.25}$$

in which P_{1hr} is the theoretical wellbore flowing pressure 1 h after shut-in. It is calculated from an interpolation or extrapolation of the linear semilog transient pressure response.

Analysis of Build-Up Test

An examination of Eqns (6.2.19) and (6.2.24) indicates that the following information is required to interpret a build up test:

- Wellbore radius, r_w;
- Flow rate, q_o;
- Oil formation volume factor, B_o;
- Oil viscosity at formation conditions, μ_o;
- Porosity, ϕ;
- Pay thickness, h;
- Total compressibility, c_t.

These parameters are determined as described in the drawdown test discussion.

The first step in analyzing a build up test is to identify the linear region on the Horner plot. In theory, the lower limit of the linear region extends to time zero. However, in practice, wellbore storage obscures the early transient response. Wellbore storage refers to the movement of fluid in the well back into the formation after the well is shut-in. This movement influences the early pressure data. The linear or "semilog" region is shown on the Horner plot in Figure 6.2.5.

It is not always straightforward to precisely identify the semilog region on a Horner plot. A more precise determination can be made from a "derivative" plot. As its name suggests, a derivative plot is a plot of the derivative of a Horner plot versus time, in which the derivative is given by:

$$\frac{dP_{ws}}{d\log\left\{t_p + \Delta t/_{\Delta t}\right\}} \tag{6.2.26}$$

An example derivative plot is provided in Figure 6.2.6. The linear transient region is easily observed as a horizontal line where the derivative is constant.

Example: Determine permeability and skin factor from the build-up test data of Figure 6.2.5. The following data are available for the well and the reservoir:

$r_w = 0.333$ ft	$B_o = 1.2$ bbl/stb
$P_i = 3300$ psia	$\mu_o = 1.5$ cp
$q_o = 300$ stb/d	$h = 15$ ft
$c_t = 3.0 \cdot 10^{-5}$ psi^{-1}	$\phi = 0.25$

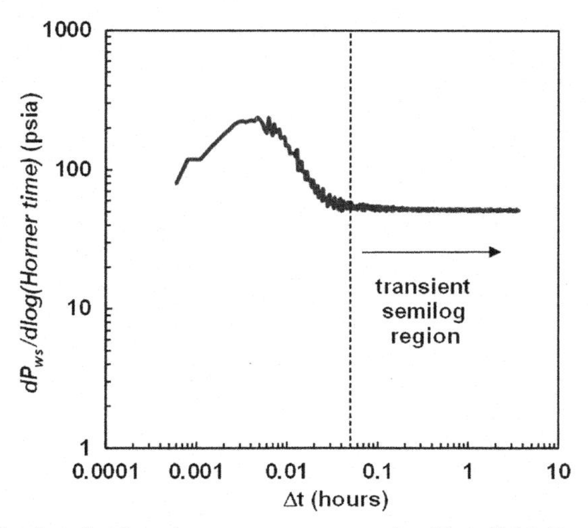

Figure 6.2.6 Derivative plot for a homogeneous reservoir with radial inflow.

The analyzed Horner plot is provided in Figure 6.2.7.

Based on the derivative plot (Figure 6.2.6), the linear transient region begins at 0.05 h or a Horner time of 481. The slope in the transient region on the Horner plot after a Horner time of 481 is −51.8 psia/cycle. From Eqn (6.2.21), the permeability to oil is:

$$k_o = -\frac{162.6 q_o \mu_o B_o}{mh} = \frac{162.6(300\ \text{stb/d})(1.5\ \text{cp})(1.2\ \text{bbl/stb})}{(51.8\ \text{psia/cycle})(15\ \text{ft})} = 113\ \text{mD}$$

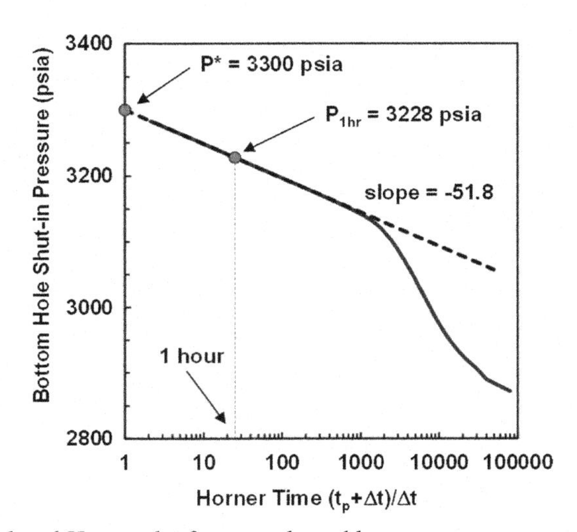

Figure 6.2.7 Analyzed Horner plot for example problem.

Figure 6.2.7 shows that the shut-in pressure at 1 h falls within the linear transient region and is interpolated to be 3228 psia. Figure 6.2.3 shows that the final flowing pressure was 2852 psia. From Eqn (6.2.24), the skin factor is:

$$S = 1.151\left[\frac{P_{1hr} - P_{wf}(\Delta t = 0)}{-m} - \log\left\{\frac{k_o}{1698\phi\mu_o c_t r_w^2}\right\}\right]$$

$$= 1.151\left[\frac{3228 - 2852}{51.8} - \log\left\{\frac{113}{1698(0.25)(1.5)(3\cdot10^{-5})(0.333)^2}\right\}\right] = 2.9$$

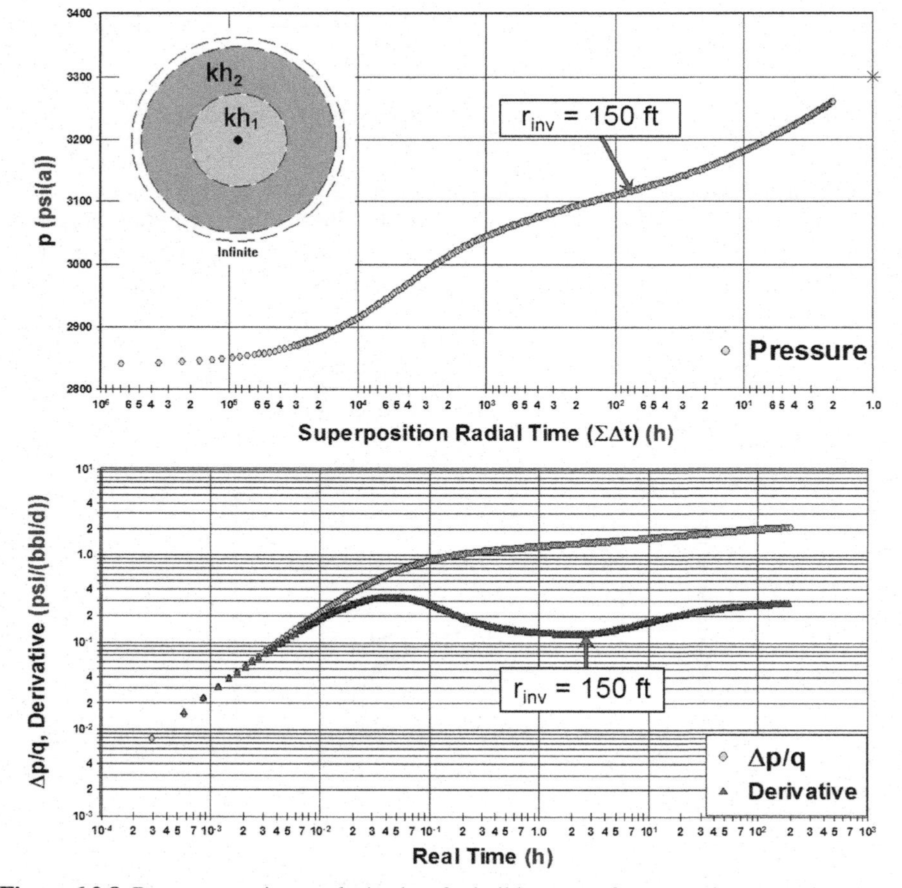

Figure 6.2.8 Pressure transient analysis plots for build up test of a composite reservoir: Horner plot (above) and derivative plot (below). r_{inv} is the radius of investigation at the $kh_1 - kh_2$ boundary.
Courtesy of Fekete & Associates (Ewens, 2012).

Finally, note that the extrapolated pressure is the same as the initial pressure. This is an indication that the pressure transient response has not reached any boundaries during the test.

Advanced Well Test Analysis

So far, we have presented only the most straightforward application of PTA: single-phase radial flow with constant compressibility in a homogeneous reservoir. PTA has been adapted for compressible flow, spherical, linear and bilinear flow patterns, and horizontal wells. PTA is also capable of detecting a variety of flow boundaries, permeability variations, interference from offsetting wells, and fracture systems. Some representative build up responses are shown in Figures 6.2.8–6.2.11. The details of advanced PTA are beyond the scope of this book. The interested reader is referred to Earlogher (1977) and Lee (1982).

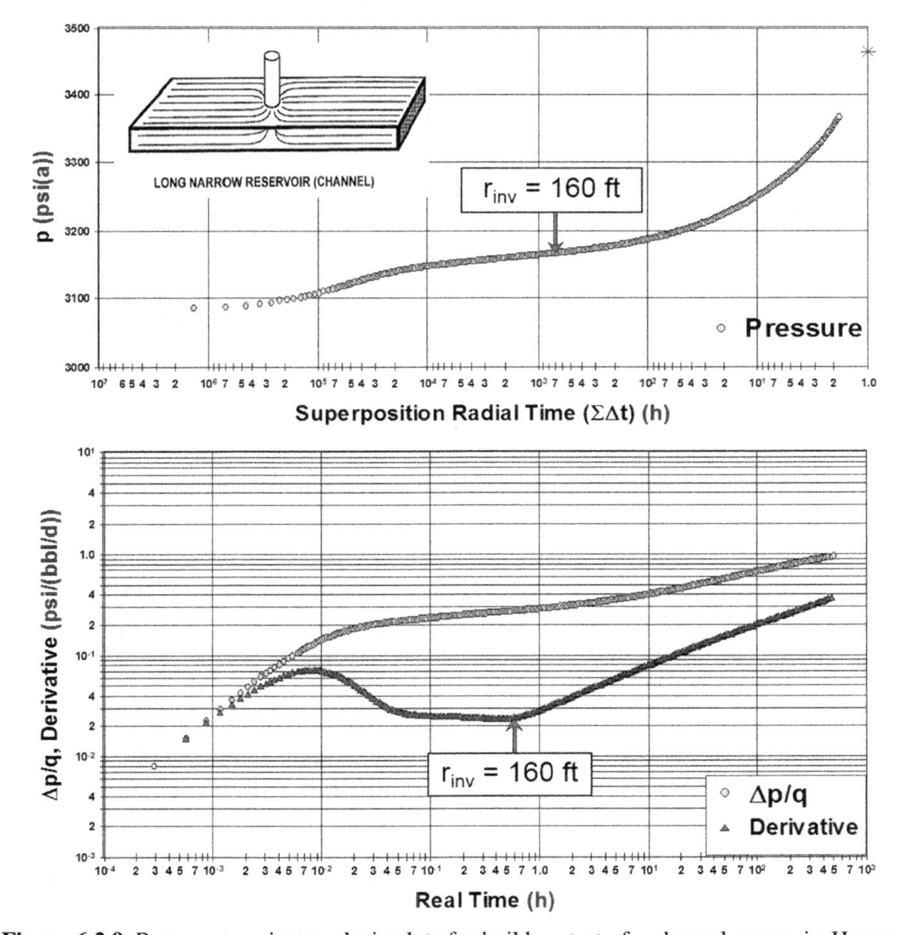

Figure 6.2.9 Pressure transient analysis plots for build up test of a channel reservoir: Horner plot (above) and derivative plot (below). r_{inv} is the radius of investigation, where the disturbance has reached the nearest channel boundaries.
Courtesy of Fekete & Associates (Ewens, 2012).

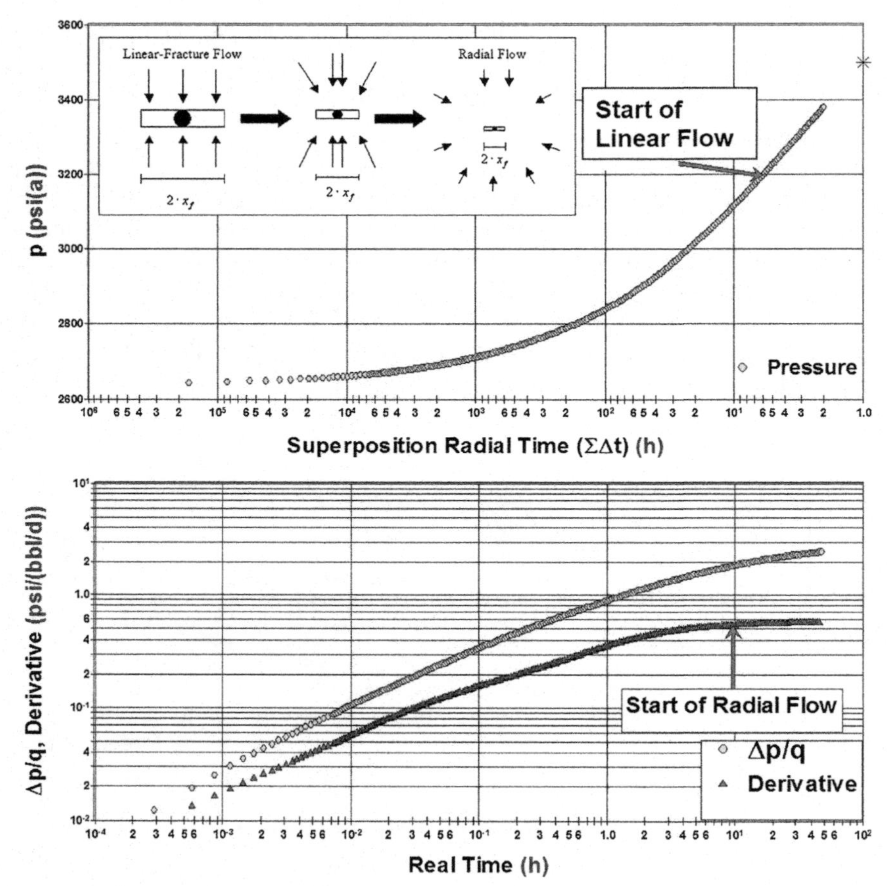

Figure 6.2.10 Pressure transient analysis plots for build up test of a hydraulically fractured reservoir: Horner plot (above) and derivative plot (below). Linear (1/2 slope), bi-linear (1/4 slope), and radial (0 slope) flow regimes are present.
Courtesy of Fekete & Associates (Ewens, 2012).

6.3 Preparation of Pool Pressure History

To interpret reservoir performance and to apply a material balance, it is necessary to collate all the individual pressure points. Erroneous pressure points must be identified, and the remaining pressure points must be assessed to determine average pool pressure or to identify pressure gradients in the reservoir. A methodology for analyzing pressure data is outlined below.

The first step is to adjust all of the pressure data to a common depth, the datum depth. If the reservoir has a gas-oil contact (GOC), the datum depth is usually set to the depth of the GOC. In this way, the pressure, volume, and temperature properties are determined at the pressure at the gas-oil interface, where the reservoir most closely

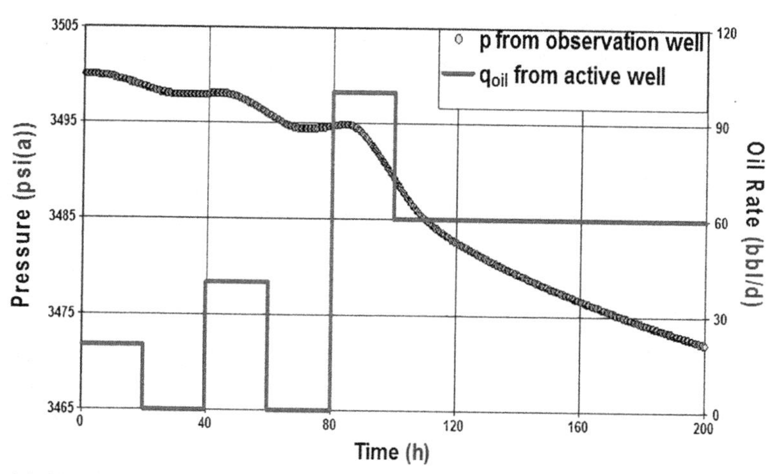

Figure 6.2.11 Pressure transient response for a pulse interference test. Courtesy of Fekete & Associates (Ewens, 2012).

approaches an equilibrium condition. If there is no gas-oil contact, the top of structure, the midpoint of the structure, or the water–oil contact may be used. MPPs are adjusted to the datum depth based on the reservoir fluid contacts. For example, if the MPP pressure is measured in the water zone and the datum depth is the oil zone, the datum pressure is calculated as follows:

$$P_{datum} = P_{MPP} + \rho_o \frac{g}{g_c}(h_{datum} - h_{WOC}) + \rho_w \frac{g}{g_c}(h_{WOC} - h_{MPP}) \tag{6.3.1}$$

in which P_{datum} is the datum pressure, P_{MPP} is the pressure at the MPP, h_{datum} is the datum depth, h_{WOC} is the depth of the water–oil contact in the reservoir, and h_{MPP} is the depth of the midpoint of the perforations. See also Eqn. (6.1.2).

Once the pressure data have been adjusted to the datum depth, the next step is to determine pressure trends and identify outliers in the dataset. Trends can be identified from one of the following plots:

- Pressure versus time;
- Pressure versus cumulative oil production;
- Pressure versus net withdrawal.

It is recommended to make two plots. On one, mark the data from each well; for example, use a different symbol for each well. On the second, mark the date for each test type (SG = static gradient, AWS = acoustic test, BU = build up test, ABU = acoustic build up test, FO = fall-off test, etc.). The pressure plot marked for wells can be used to determine whether:

- Pressure is uniform across the reservoir;
- Pressure gradients exist in the reservoir;
- Wells from different reservoirs have been incorrectly lumped into a single reservoir.

At the same time, the pressure plot marked for test type can be used to identify suspect pressure data. In general, bottom hole build-up and fall-off tests provide the most accurate pressure points. SG tests are accurate given sufficient time to build up pressure. Acoustic tests are more prone to error and, if possible, should be compared with bottom hole pressure data before inclusion in a pool pressure plot.

There are several methods for determining an adequate shut-in time:

1. Find the time required to build up to reservoir pressure in a build up test. For example, the pressure test given in Figure 6.2.3 showed that for the given flow rate, the pressure built up to within 99% of the initial pressure in 6 h.
2. A rule of thumb for undersaturated black oils at 40 acre spacing is:

$$t_{SI} > \frac{2000}{k} \tag{6.3.2}$$

in which t_{SI} is the recommended shut-in time in hours, and k is permeability in mD. For example, the recommended shut-in time for the pressure test of Figure 6.2.3, in which the permeability was 113 mD, would be 17 h.
3. Use a dimensionless shut-in time of 0.1 or:

$$t_{DA,SI} = \frac{0.000264 k_o t}{\phi \mu_o c_t A} > 0.1 \tag{6.3.3}$$

in which $t_{DA,SI}$ is the dimensionless shut-in time, and the equation is given in Field units. For the pressure test of Figure 6.2.3, the recommended shut-in time would be 45 h. Note that this shut-in time assumes that the pressure gradient has reached the boundaries of the drainage area and therefore applies best to tests after extended flow periods.
4. Use a pressure plot. Mark each data point with its shut-in time and identify at what shut-in time the measured pressure departs from the main trend in the pressure data.

Finally, before excluding any outlying pressure point, the point should be compared with the trend for the well for which the pressure was measured. The pressure may be different because the well is part of a different pool, in a partially isolated part of the reservoir, or located near an injector.

Example: In a particular oil field, five wells were completed in the Nisku horizon. It was originally believed that the wells were producing from two independent pools. There is now sufficient pressure history to check this assumption. Figure 6.3.1 shows all of the pressure data from the five wells.

The pressure data are a mix of bottom hole build up tests (BH-BU), SG tests, AWSs, acoustic build up tests (AWS-BU), and a DST test. The bottom hole pressure tests all follow a clear trend, as do most of the acoustic build ups tests. The acoustic tests are scattered, but also follow the same pressure trend. There is one outlier, an acoustic build-up test (Figure 6.3.1). There was likely an error in the fluid levels for this test. The pressure point was eliminated from further consideration.

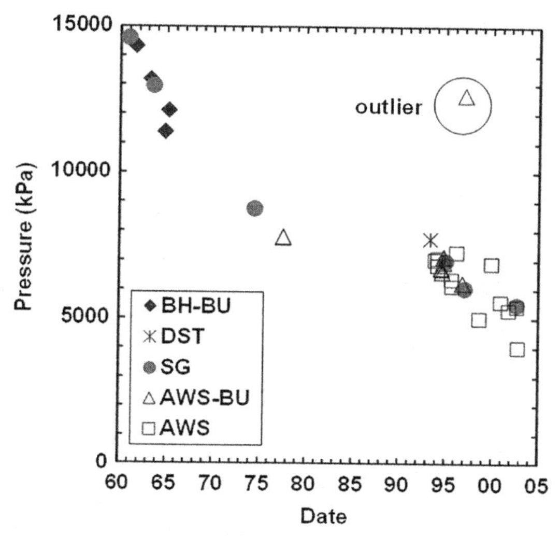

Figure 6.3.1 Pressure data from Nisku pools marked by test type.

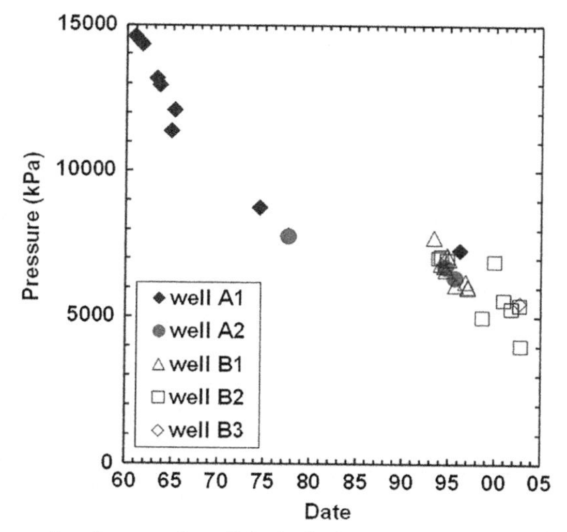

Figure 6.3.2 Pressure data from each well in "A" pool (closed symbols) and "B" pool (open symbols).

The remaining pressure data are plotted per well on Figure 6.3.2. The two pools are here named the "A" pool (solid symbols) and the "B" pool (open symbols). All the wells follow the same pressure trend, and there is sufficient overlap to conclude that all the wells are in pressure communication. Therefore, the two pools are really one reservoir or are in communication through a common aquifer. An assessment of the reservoir structure based on seismic data concluded that all the wells are in a single pool.

6.4 Flow Tests

Flow tests involve flowing a well at a constant rate or a set of constant rates and measuring the bottom hole flowing pressure. The flow rates can be measured in a well site test unit for a new well or at the surface facilities of a tied-in well. Bottom hole pressure can be recorded directly with a tubing-conveyed pressure transducer or can be calculated from sonic measurements. There are three types of flow test: a single point test, a flow after flow test, and a modified isochronal test.

6.4.1 Single Point Test

As the name suggests, the bottom hole pressure is measured at only one flow rate in a single point test. The well is flowed at constant rate until the bottom hole pressure stabilizes. The relationship between flow rate and bottom hole flowing pressure must be extrapolated using an inflow equation such as:

the productivity index (PI):

$$q_o = J(P_{res} - P_{wf}) \tag{6.4.1}$$

the Vogel relationship:

$$q_o = q_{o,\max}\left[1 - 0.2\left(\frac{P_{wf}}{P_{res}}\right) - 0.8\left(\frac{P_{wf}}{P_{res}}\right)^2\right] \tag{6.4.2}$$

the pressure squared inflow equation (P-squared):

$$q_o = C\left(P_{res}^2 - P_{wf}^2\right)^n \tag{6.4.3}$$

in which q_o is oil rate, J is the productivity index, P_{res} is the reservoir pressure, P_{wf} is the bottom hole flowing pressure, C is the productivity coefficient, and n is the productivity exponent. The productivity exponent can have any value between 0.5 and 1.0. The single point test is best used for reservoirs above the bubble point, in which the PI relationship usually applies.

Example single point test: Determine the inflow performance of a well producing from an undersaturated reservoir, based on the following single point measurements:

$P_i = 3300$ psia	$P_{wf} = 2815$ psia	$q_o = 300$ stb/d

The productivity index (J), $q_{o,\max}$ for the Vogel relationship, and productivity coefficient (C) are calculated as follows:

$$J = \frac{q_o}{P_i - P_{wf}} = \frac{300\ \text{stb/d}}{3300 - 2815\ \text{psia}} = 0.619\ \text{stb/d}\cdot\text{psia}$$

$$q_{o,\max} = \frac{q_o}{\left[1 - 0.2\left(\frac{P_{wf}}{P_{res}}\right) - 0.8\left(\frac{P_{wf}}{P_{res}}\right)^2\right]}$$

$$= \frac{300}{1 - 0.2\left(\frac{2815}{2200}\right) - 0.8\left(\frac{2815}{2200}\right)^2} = 1213\ \text{stb/d}$$

$$C = \frac{q_o}{\left(P_{res}^2 - P_{wf}^2\right)^n} = \frac{300}{\left(3300^2 - 2815^2\right)^{1.0}} = 1.01\cdot 10^{-4}\ \text{stb/d}\cdot\text{psia}^2$$

Note that the productivity exponent is assumed to be equal to 1.0 unless known to be otherwise from a multiple point test.

The inflow curves from the PI, Vogel relationship, and P-squared inflow equation are shown in Figure 6.4.1. The PI is the most optimistic equation and the least accurate because it does not account for the effect of evolving solution gas.

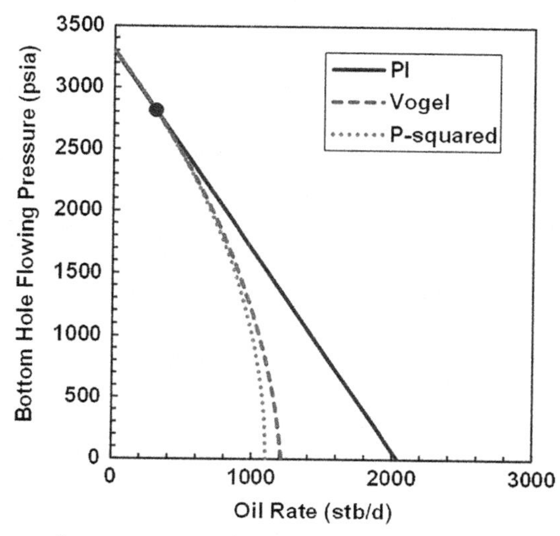

Figure 6.4.1 Inflow performance curves for the example single point test. The symbol is the measured data point.

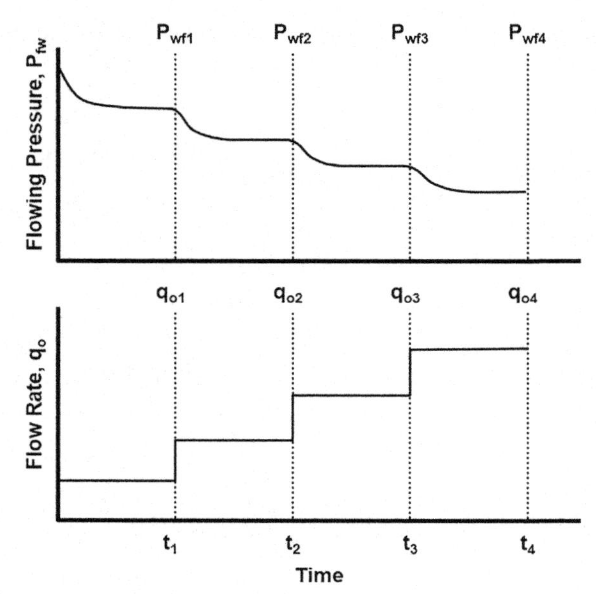

Figure 6.4.2 Schematic of flow after flow test.

6.4.2 *Multiple Point Tests*

A more accurate relationship can be established if multiple rates are tested. In a flow after flow test, the well is flowed at four different rates, each for a time sufficient to reach a stabilized bottom hole flowing pressure, as shown in Figure 6.4.2. For oil reservoirs above the bubble point, the PI is determined from a plot of $(P_i - P_{wf})$ versus flow rate. For oil reservoirs below the bubble point, a pressure-squared deliverability relationship is usually assumed, and the data are usually plotted as $(P_i^2 - P_{wf}^2)$ versus flow rate on a log–log plot. The slope is the productivity exponent, n. Once the exponent is known, the productivity coefficient is calculated from a point on a line of best fit. Note that the initial test pressure, P_i, is taken to be the reservoir pressure. The flow after flow test is the most accurate assessment of the well's inflow performance. However, it may take an impractically long time for the bottom hole flowing pressure to stabilize at each rate.

The modified isochronal test is a compromise designed to establish the inflow performance curve in a practical test time. It is usually used for gas reservoirs and for oil reservoirs below the bubble point. The well is flowed at four different rates, each for a constant but relatively short interval. The well is shut-in after the first three flow rates for the same interval, as shown in Figure 6.4.3. The final flow rate is extended until the bottom hole pressure stabilizes or for some fixed but significantly longer interval. The bottom hole pressure is measured at the end of each shut-in period, P_{ws}, and at the end of each flow period, P_{wf}. A final flowing pressure, $P_{wf,ext}$,

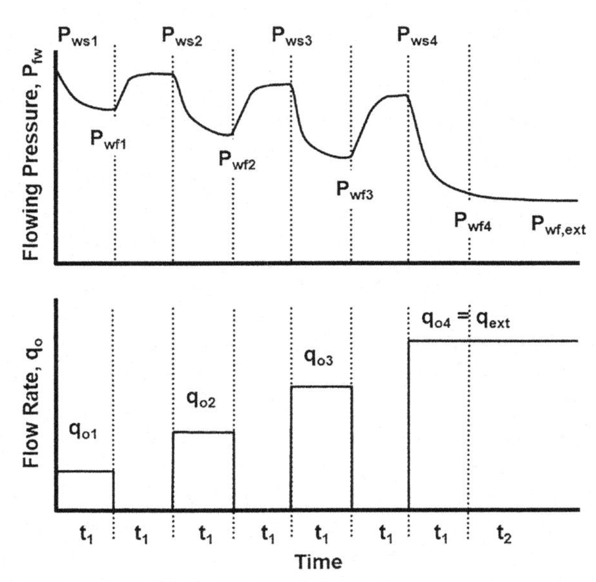

Figure 6.4.3 Schematic of modified isochronal flow test.

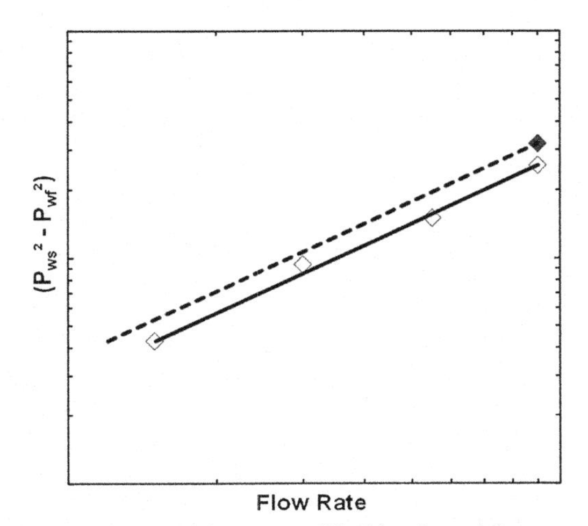

Figure 6.4.4 Oil inflow performance from a modified isochronal flow test.

is recorded at the end of the "extended" flow period. To establish the inflow equation, a pressure squared relationship is assumed. A log–log plot of $(P_{ws}^2 - P_{wf}^2)$ versus flow rate is created, as shown in Figure 6.4.4. The four isochronal (equal time) points form a line with a slope equal to the productivity exponent, n. A line of the same slope is placed through the extended point, and the productivity coefficient is determined.

Example flow after flow test: Determine the P-squared inflow relationship from the following test data:

Flow time (h)	Flow rate (stb/d)	Final bottom hole flowing pressure (psia)	$(P_i^2-P_{wf}^2)$ (psia^2)
Initial	0	3300	–
120	100	3138	1,042,956
120	200	2971	2,063,159
120	300	2800	3,050,000
120	400	2623	4,009,871

Figure 6.4.5 is a plot of $(P_i^2-P_{wf}^2)$ versus flow rate. A line based on Eqn (6.4.3) is fitted to the data using least squares, with C and n as parameters. The productivity exponent, n, is the slope of the best fit line and is 1.0. The productivity coefficient, C, is $9.88 \cdot 10^{-5}$ stb/d psia^2.

Example modified isochronal test: Determine the P-squared inflow relationship from the following test data:

Flow time (h)	Flow rate (stb/d)	Final bottom hole pressure (psia)	$(P_{ws}^2-P_{wf}^2)$ (psia^2)
Initial	0	3300	–
4	100	3164	879,100
4	0	3295	–
4	200	3025	1,706,000
4	0	3288	–
4	300	2884	2,493,000
4	0	3279	–
4	400	2742	3,233,000
116	400	2653	3,713,000

Figure 6.4.6 is a log–log plot of $(P_{ws}^2-P_{wf}^2)$ versus flow rate. Note that P_{sw} is the shut-in pressure prior to each flow period, not the initial pressure. A line based on Eqn (6.4.3) is fitted to the four isochronal data points using the least squares method. The productivity exponent, n, is the slope of the best fit line and is 1.0. The productivity coefficient, C, is calculated from the extended point as was shown in the single point example but using the value of 'n' determined from the isochronal points. Here, C is $1.04 \cdot 10^{-4}$ stb/d$\cdot$psia^2.

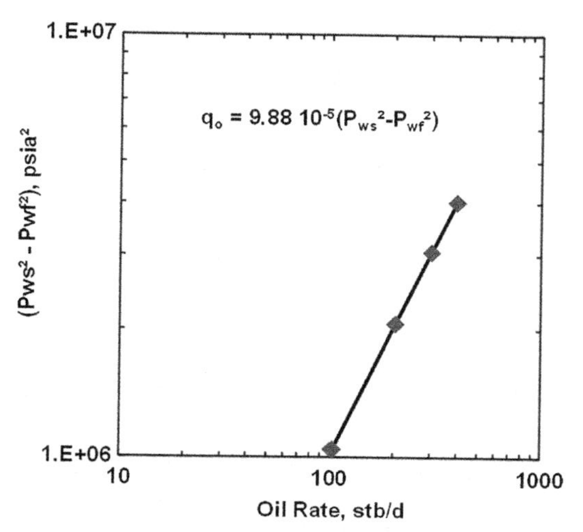

Figure 6.4.5 Analyzed flow after flow test for example problem.

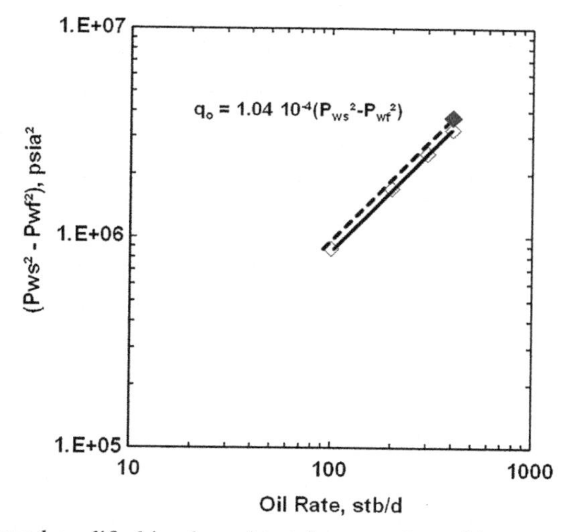

Figure 6.4.6 Analyzed modified isochronal test for example problem.

6.4.3 Estimating Permeability From an Inflow Test

The productivity index inflow equation is derived from Darcy's law. For example, the constant rate, pseudo steady-state radial inflow equation (see Chapter 3) is given by:

$$q = \frac{2\pi kh}{\mu} \frac{P_r - P_{wf}}{\ln\left\{\frac{r_e}{r_w}\right\} - \frac{3}{4} + S} \tag{6.4.4}$$

Equations (6.4.1) and (6.4.4) are equated to obtain a relationship for the productivity index:

$$J = \frac{2\pi kh}{\mu\left[\ln\left\{\frac{r_e}{r_w}\right\} - \frac{3}{4} + S\right]} \tag{6.4.5}$$

The permeability of the reservoir can be determined from an inflow test using Eqn (6.4.5) as follows:

$$k = J\frac{\mu}{2\pi h}\left[\ln\left\{\frac{r_e}{r_w}\right\} - \frac{3}{4} + S\right] \tag{6.4.6}$$

Equation (6.4.6) is restated below in SI and Field units:

SI: $$k = J\frac{\mu_o B_o}{0.0005355h}\left[\ln\left\{0.472\frac{r_e}{r_w}\right\} + S\right]$$

Field: $$k = J\frac{\mu_o B_o}{0.00708h}\left[\ln\left\{0.472\frac{r_e}{r_w}\right\} + S\right]$$

Example: The productivity index of a well was found to be 0.619 stb/d. A build up test indicated that the skin factor was 2.9, and the average permeability was 113 mD. Check the permeability using the productivity index. The well is believed to be draining a quarter section (180 acres). The following data are also available:

$r_w = 0.333$ ft	$B_o = 1.2$ rb/stb
$h = 15$ ft	$\mu_o = 1.5$ cp

It is necessary to estimate the drainage radius from the area of a quarter section, as follows:

$$r_e = \left(\frac{A}{\pi}\right)^{½} = \left(\frac{(180\text{ acres})(43560\text{ ft}^2/\text{acre})}{\pi}\right)^{0.5} = 1580\text{ ft}$$

The permeability is found from Eqn (6.4.3) using Field units:

$$k_o = J\frac{\mu_o B_o}{0.00708h}\left[\ln\left\{0.472\frac{r_e}{r_w}\right\} + S\right]$$

$$= 0.619\frac{(1.5)(1.2)}{0.00708(15)}\left[\ln\left\{0.472\frac{1580}{0.333}\right\} + 2.9\right] = 111\text{ mD}$$

The agreement between the permeability from the inflow test and from the build test is excellent.

6.5 Other Tests—Interference, Pulse, and Tracer Tests

Flow and build-up tests provide information within the drainage area of the tested well. Interference and pulse tests can provide a measure of average properties between well pairs. In an interference test, the production of an active well is shut-in for a fixed period and then restarted. The pressure response at an observation well is measured. An interference test sometimes does not provide enough pressure disturbance to extract much useful information. In a pulse test, the active well is repeatedly shut-in and restarted so that more data are obtained, as shown in Figure 6.5.1. The pressure response in the observation well is then analyzed to determine, for example, the average *kh* between the well pair. Note that multiple observation wells can be monitored and analyzed simultaneously. This technique is useful for identifying flow barriers between wells.

A tracer test involves adding a slug of a water soluble dye or a radioactive compound to a water injection well. The concentration of the dye in the produced water of offset wells is measured and profiles of dye concentration versus time are obtained, as shown in Figure 6.5.2. Short breakthrough times indicate a short flow path or a high

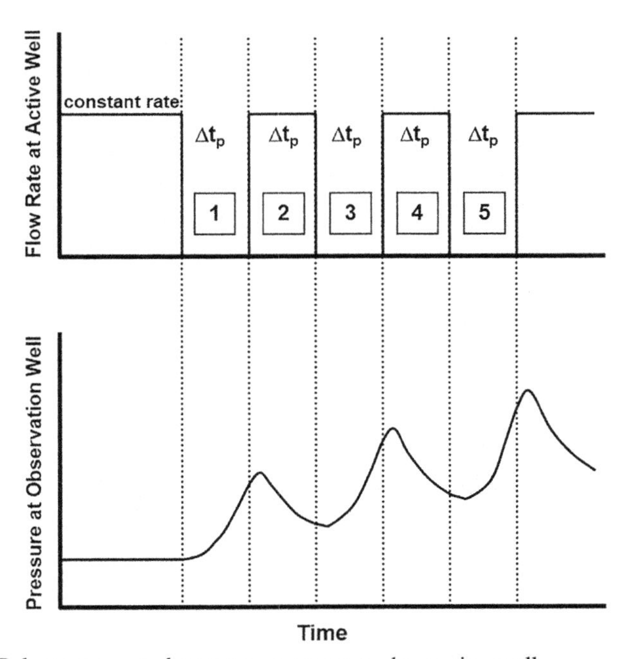

Figure 6.5.1 Pulse test rate and pressure response at observation well.

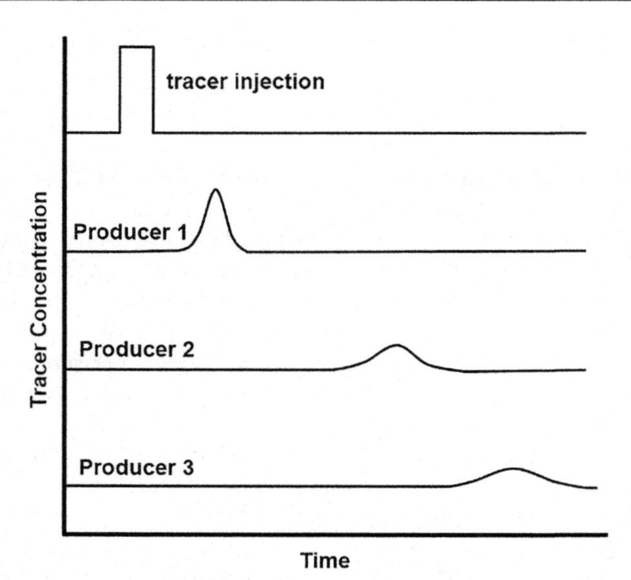

Figure 6.5.2 Tracer test concentration profiles at injector and offset producers.

permeability conduit between the injector and the producer. A compact tracer profile indicates a high permeability conduit such as a connected fracture network. A spread out tracer profile is typical of displacement through a porous matrix where the tracer fronts spread through diffusion and dispersion. Tracer tests are useful for analyzing the direction and connectivity of fracture networks. They are also useful for identifying geological features, such as channels and sheetflood deposits.

Conventional Core Analysis—Rock Properties

7

Chapter Outline

Conventional core analysis includes a lithological description as well as porosity, density, permeability, and residual saturation measurements. Because conventional core data are collected at low pressure, rock compressibility and the effect of overburden pressure are discussed. The measurement of rock compressibility is discussed in this section, although it is not a conventional core measurement.

Core data are required for most reservoir calculations, including volumetrics, material balances, inflow flow calculations, and pressure transient analysis. Some core data are used to aid in log interpretation, for example, to calibrate density logs to determine porosity. Another property of relevance to log interpretation is formation resistivity. The measurement of resistivity is also discussed in this section.

7.1 Core Sampling and Errors

A core is a sample of reservoir rock captured during drilling using a core cutting drill bit, as shown in Figure 7.1.1. An individual core is typically a few inches in diameter and up to 60 ft in length. Consecutive cores can be gathered to sample as much reservoir thickness as desired. In most cases, the core is exposed to the fluid, temperature, and pressure

Practical Reservoir Engineering and Characterization. http://dx.doi.org/10.1016/B978-0-12-801811-8.00007-9

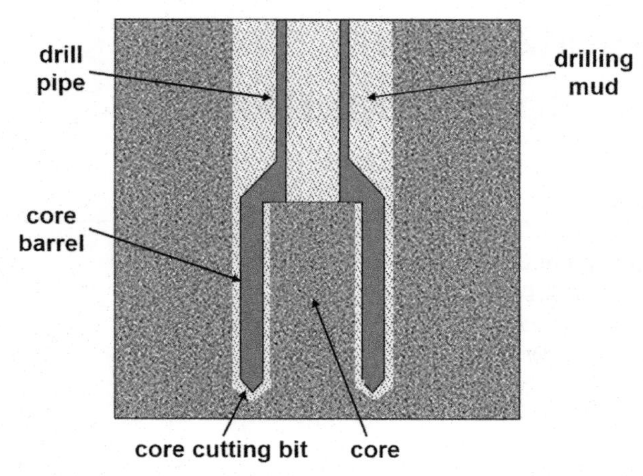

Figure 7.1.1 Schematic of cutting a core.

in the wellbore while it is cut and while it is brought to the surface. Therefore, drilling mud may displace reservoir fluid from the core, and gas may evolve from any oil in the core. It is possible to use a "native state" coring tool to capture a core sample and its reservoir fluids at reservoir conditions. The core is enclosed to maintain reservoir pressure and isolate the core from further drilling fluid invasion while the core is brought to surface. However, this technique is expensive and rarely used.

The core is transported to a laboratory, and a Gamma Ray (GR) log may be run along the length of the core. The log is used to match the core to a well log so that the depths can be correlated. This is particularly helpful when there is less than full recovery of the core. Incomplete recovery happens, for example, when there are shales that break up during the drilling process. In this case, 10 m of formation may have been drilled, but only 8.5 m of core recovered. Depth matching between the well logs and core can be a tricky process, and a match within 30−50 cm may be the best that can be achieved.

A set of cylindrical plugs (small plugs) are cut from the core for further testing. These plugs are typically 3/4 inch in diameter and 1−2 inches in length. Representative whole core sections (full diameter core) may also be used for testing. Leftover edge pieces of core near the core plugs or sections are also used to measure residual saturations.

The core plugs and sections are cleaned to remove oil and water and then dried. The oil and water are removed using solvent extractors, centrifuges, and/or Dean Stark refluxing solvent extractors. The most commonly used solvent is toluene, followed by methanol. The samples are dried in humidity-controlled ovens to preserve the hydration state of hydrated minerals such as clays. A lithological description of the core is taken and the samples can then be tested to measure:

- Density;
- Porosity;
- Permeability;

- Relative permeability;
- Capillary pressure;
- The response to core floods using any choice of displacement fluid.

Residual saturations, density, porosity, and permeability are the most common measurements and are included in a conventional core analysis. Relative permeability and capillary measurements are usually referred to as "special" core analysis methods. Other displacement tests are tailored to the requirements of a particular reservoir and development strategy.

The paradox of core data is that core samples are the only tangible source of data for rock properties and yet core data can be unrepresentative of the reservoir. There are two reasons for the large uncertainties in core data:

1. Core samples may not be representative in two ways. First, a core 2.63 inches in diameter and 20 ft in length has a volume of 0.75 ft^3, while a 20-foot-thick, 1280-acre reservoir has a volume of 1.1×10^9 ft^3. The core samples approximately 1 billionth of the reservoir. Because a single core samples a tiny fraction of the reservoir, core data can only be representative of the reservoir if the whole reservoir (or really continuous layers in the reservoir) has uniform properties. If reservoir properties vary laterally, then cores from many wells may be required to construct an accurate reservoir model. If the lateral variations are random, it may not be possible to accurately portray the reservoir. Instead, statistical distributions of permeability can be used to generate many reservoir characterizations, and a range of possible outcomes can be assessed. Second, the locations for the core plug samples may be biased. For example, core plug locations may avoid silty or shale intervals because, when drilled, the plug may break up so that no permeability measurement would be possible. If plug samples are preferentially taken from the more porous, permeable locations, we will get a biased view of the flow characteristics of the reservoir. Viewing core photographs can help determine whether there was preferential sampling and help identify which rock types were sampled.
2. Core properties may be altered during sampling and preparation: Mud invasion, depressurization, and cleaning can all alter the pore structure and wettability of the rock. Rock flour from core cutting can block pore throats and reduce permeability unless properly cleaned. Hence, the sample analyzed in the laboratory may be significantly different than the same rock in the reservoir. Laboratory-induced changes aside, conventional core samples tend to be more water wet than native state samples, as shown in Figure 7.1.2.

Despite these uncertainties, core data are a valuable tool in reservoir characterization, particularly when supported by other reservoir data. Core data are also a good measure of reservoir porosity because porosity is little affected by alterations to the core during sampling.

7.2 Conventional Core Data

A conventional core analysis includes a lithological description, the residual fluid saturation, the density, the porosity, and the permeability of each core plug. Each type of measurement is discussed below.

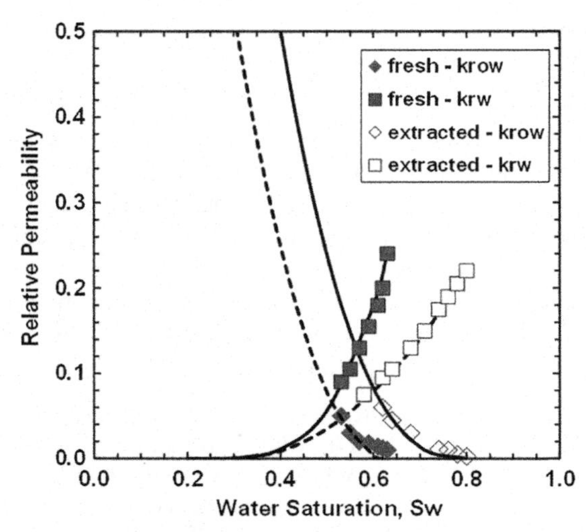

Figure 7.1.2 Comparison of relative permeability of conventional and native state core samples from the same formation.

7.2.1 Lithological Description

The lithological description is obtained from a visual inspection carried out by a geologist or technician. The following factors may be included:

- Main rock type (e.g., sandstone, conglomerate, limestone, dolomite, shale, anhydrite, halite);
- Other mineralology (e.g., pyrite, coal, cementation, fossils);
- Grain or crystal sizes (e.g., coarse, fine, very fine);
- Description of fractures (e.g., horizontal, vertical);
- Description of vugs (e.g., pinpoint, small, medium, large).

The description is reported using shorthand notations, which may vary from one service company to another. A key to the shorthand descriptions for one company is given in Table 7.2.1. Lithological descriptions in core reports are typically brief and may overlook important details; a visit to the core store and/or a review of the core photographs may be very helpful for identifying those details.

The lithological description is used to:

- Understand the depositional environment and geology of a reservoir;
- Interpret porosity and permeability data;
- Aid in well log interpretation.

For example, the presence of vugs will lead to relatively high porosity. The presence of open fractures can lead to high measured permeabilities, while closed (cemented) fractures can reduce the permeability to zero. Grain density is useful for calibrating some well-logging tools, as discussed in Chapter 9.

Extra care must be taken when fractures are observed. Open fractures could be natural fractures that exist in the reservoir or induced fractures that were created when the core was drilled and retrieved. If natural fractures are suspected, an independent test such as a buildup test or a tracer test is recommended.

Table 7.2.1 Example codes for lithological descriptions in conventional core analysis

Code	Description
ACA	Removed for advanced core analysis
anhy	Anhydrite
arc	Argillaceous
AST	Appears similar to
biotur	Bioturbated
bit	Bitumen
bk	Break
c	Coarse
calc	Calcite (calcareous)
carb	Carbonaceous
cbl	Cobble
cem	Cemented
cgl	Conglomerate
cht	Chert
cly	Clay
coal	Coal/coal inclusion
dol	Dolomite
evap	Evaporite
f	Fine
FD	Full diameter analysis including three-directional permeabilities, porosity, and densities
fenst	Fenestral
flor	Fluorescence
foss	Fossil (fossiliferous)
frac	Fracture (undifferentiated)
fri	Friable
glauc	Glauconite (glauconitic)
gml	Granule
gr	Grain
gyp	Gypsum
hfrac	Horizontal fracture
hal	Halite (salt)

Continued

Table 7.2.1 Example codes for lithological descriptions in conventional core analysis—cont'd

Code	Description
IFD	Inner full diameter, (a full-diameter sample is drilled from the bulk portion of the core in the vertical direction for permeability and porosity measurements)
I	Intercrystalline
incl	Inclusions
intbd	Interbedded
lam	Laminae (laminated)
ls	Limestone
lv	Large vug
m	Medium
mi	Mud invaded
mic	Microcrystalline
mold	Moldic
mv	Medium vug
NA	Not analyzed by request
NP	No permeability measurement possible due to poor sample quality
NR	Not received
ool	Oolitic
OB	Overburden sample (permeability and porosity measured at net overburden stress)
P	Preserved for future studies
pbl	Pebble
PFD	Preliminary full diameter sample
PSP	Preliminary small plug sample
PSA	Particle size analysis
ppv	Pinpoint vug
pyr	Pyrite (pyritic)
pyrbit	Pyrobitumen
ru	Rubble
SA	Sieve analysis
sdy	Sandy
SEM	Scanning electron microscope analysis
sh	Shale

Table 7.2.1 Example codes for lithological descriptions in conventional core analysis—cont'd

Code	Description
shy	Moderately shaly (20–40%)
sid	Siderite
sltst	Siltstone
slty	Silty
SPT	Small plug used for tracer analysis
SP	Small plug (sample drilled from core in maximum horizontal direction and parallel to bedding plane when possible) permeability, porosity, and grain density are measured
srt	Sorted
ss	Sandstone
ssdy	Slightly sandy (<20%)
sshy	Slightly shaly (<20%)
sty	Stylolite (ic)
sulf	Sulphur
sv	Small vug
TEG	Thermal extraction chromatography to determine oil richness
TS	Thin section
uncons	Unconsolidated
vc	Very coarse
vfrac	Vertical fracture
vf	Very fine
VIS	Viscosity of oil measured
VOB	Vertical overburden sample (vertical permeability measured at net overburden stress)
vshy	Very shaly (>40%)
VSP	Vertical small plug drilled from whole core to measure vertical permeability and occasionally porosity
vug	Vuggy
ws	Water sand
xln	Crystalline
XRD	X-ray diffraction
*	Perm unavailable due to broken core

7.2.2 Residual Saturations

The residual saturations of oil and water are the volume fractions of the pore space in the core sample occupied by oil and water, respectively. The remainder of the core sample is occupied by gas. To determine residual saturations, an edge piece of core is ground up and heated up to 300 °C, and the volumes of vaporized water and oil are determined. The residual saturations are calculated from these volumes and the pore volume of the core plug. If the core was recovered conventionally, the saturations will be strongly affected by drilling mud invasion and gas evolution, as shown in Figure 7.2.1.

If an oil-based mud was used, the water saturation may be close to the initial water saturation of the rock. If the oil-based mud has displaced free water, the residual water saturation provides an estimate of the irreducible water saturation. If a water-based mud was used, then the residual saturations are usually meaningless. The water in the core is a mixture of formation water and drilling mud. The oil saturation is strongly affected by the evolution of solution gas, and therefore an estimate of the residual oil saturation is not usually possible. An upper limit to the initial water saturation may be obtained.

7.2.3 Porosity

As discussed in Chapter 2, porosity is the ratio of the pore volume (PV) to the bulk volume (BV). Therefore, it is necessary to measure the BV and PV of a core plug to determine its porosity.

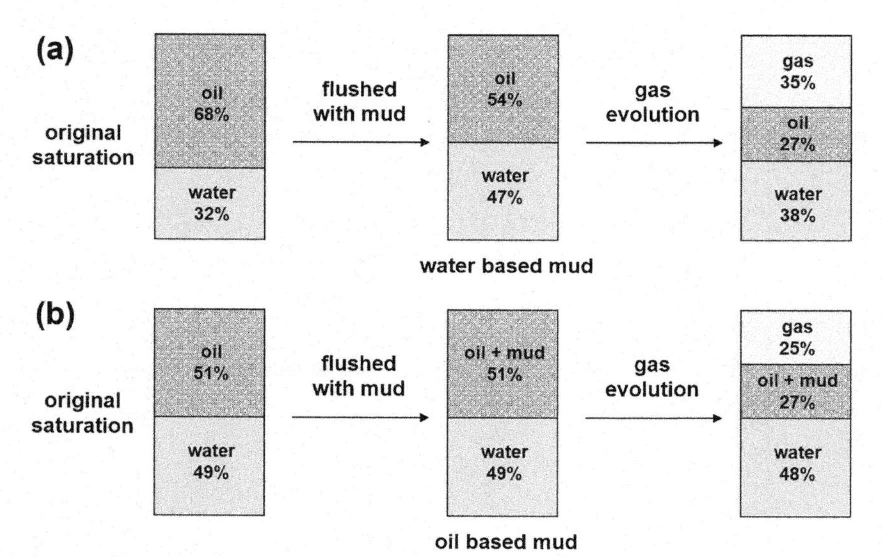

Figure 7.2.1 Typical residual saturations in a recovered core sample. Adapted from Bass (1987).

The BV can be measured from a sample's geometry or in a displacement test. Geometrical evaluation is used when the plug is a right circular cylinder and there are no surface irregularities. For a displacement measurement, the core plug is immersed in a fluid and the volume of displaced fluid is equal to the volume of the core plug. To measure the BV, it is necessary that none of the fluid penetrates the pore space of the core plug. There are several approaches to prevent fluid penetration:

1. Use mercury as the immersion fluid. The surface tension and wetting behavior of mercury prevent it from entering the pore space.
2. Presaturate the core plug with the immersion fluid. As long as the immersion fluid does not enter or leave the pore space, only the BV of the sample will displace fluid.
3. Coat the core plug with a thin layer of paraffin or some other insoluble coating.

The PV is measured based on the introduction or extraction of fluid to or from the pore space. An example of each type of method is given below:

- Extraction method: The core is filled with a gas at atmospheric pressure. The sample is placed in a vacuum chamber and the gas is allowed to expand into the vacuum. The volume of the gas (equal to PV) is determined from the change in pressure using Boyle's law. Usually, helium is used because it does not adsorb significantly on reservoir rock at room temperature.
- Introduction method: The saturation method involves saturating a clean dry core sample with a fluid of known density. The volume of fluid in the core (PV) is determined from the density of the fluid and the mass difference between the dry and the saturated core sample.

There are many other methods, but the general principles are the same.

Once the BV and PV are known, the porosity is calculated. Because only the connected PV is measured, the porosity determined from a core analysis is the effective porosity. The porosity measurement is usually accurate to ± 0.5 porosity percentage units. Anomalously high porosity measurements may be encountered in vuggy rock, and larger core plugs are recommended to obtain a more representative porosity.

7.2.4 Density

The density of the core is determined from the mass of the clean core and the bulk and pore volume measurements. The bulk density is given by:

$$\rho_{bulk} = \frac{m_{core}}{BV_{core}} \tag{7.2.1}$$

and the grain density is given by:

$$\rho_{grain} = \frac{m_{core}}{BV_{core} - PV_{core}} \tag{7.2.2}$$

in which m_{core} is the mass, BV_{core} is the bulk volume, and PV_{core} is the pore volume of the core sample.

Table 7.2.2 **Densities of come common rock types**

Rock type	Density (kg/m^3)
Anhydrite	2977
Limestone	2710
Dolomite	2870
Sandstone	2650
Shale	2200 to 2650

The grain density is the average density of the rock within the core. If the core consists predominantly of one rock type, then the grain density can provide confirmation of the lithology. The average grain density is also used for calibrating density-porosity well logs. The densities of some common rock types are given in Table 7.2.2.

7.2.5 Absolute Permeability

Absolute permeability is determined by flowing a fluid through a core plug and measuring the pressure drop at different flow rates, as shown in Figure 7.2.2. Permeability can also be measured on a complete full diameter section of the core. In this case, the core section is wrapped in a rubber sleeve, and screens are applied to opposing surfaces on the perimeter of the core. Fluid is flowed through the screens

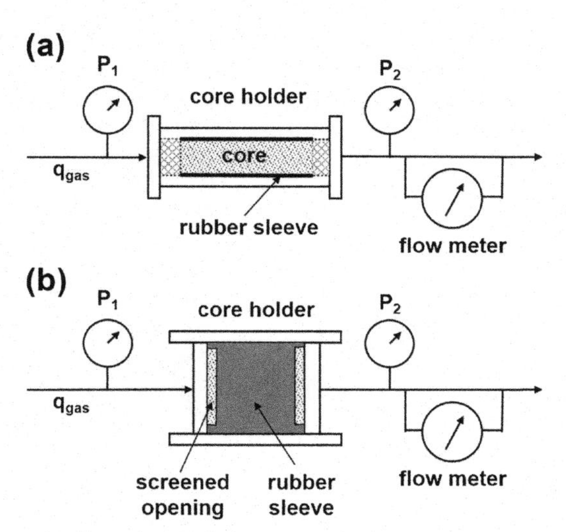

Figure 7.2.2 Schematic of permeability testing apparatus: (a) to measure the permeability of a small plug or the vertical permeability of a full diameter core, flow is directed through the axis of the core section; (b) to measure the horizontal permeability of a full diameter core, flow is directed through the side of the core.

and through a diameter of the core, as shown in Figure 7.2.2(b). The core can be rotated within this set up to measure permeabilities in different directions. Fluid can also be flowed vertically through the core section to measure vertical permeability using the configuration shown in Figure 7.2.2(a).

As discussed in Chapter 2, reservoir permeability may be different in different directions due to depositional effects such as layered deposits or oriented sand grains. Therefore, with whole core samples, permeability is measured in three directions. Two perpendicular measurements are made in the horizontal direction (horizontal in the reservoir), and one measurement is taken in the vertical direction. The measured permeabilities are defined as follows:

- k_{max}—the larger of the two permeabilities measured in the horizontal direction;
- k_{90}—the lesser of the horizontal permeabilities;
- k_v—the vertical permeability.

When core plugs are used for permeability measurements, only one direction of permeability is measured. This is usually in the horizontal direction. If a vertical permeability measurement is also desired, a separate, vertically oriented plug is cut from the core in the vicinity of a horizontal plug. Having the plugs near to each other facilitates comparisons during the data analysis (Figure 7.2.3).

As was discussed in Chapter 3, the flow rate of a liquid or a low-pressure gas through a core section is related to the permeability of the core and the pressure drop as follows:

liquid:

$$\frac{q_{liq}}{A} = \frac{k}{\mu_{liq}} \frac{(P_{in} - P_{out})}{L} \tag{7.2.3}$$

gas below 2000 kPa:

$$\frac{q_g \mu_g P_{atm}}{A} = k \frac{\left(P_{in}^2 - P_{out}^2\right)}{2L} \tag{7.2.4}$$

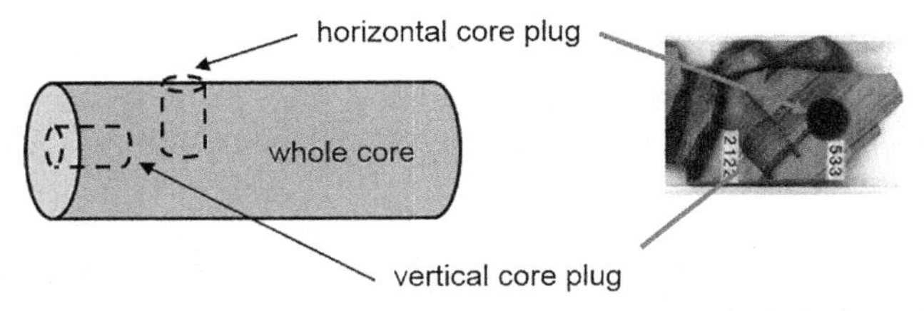

Figure 7.2.3 Schematic positions of core plugs taken to measure both horizontal and vertical permeability (left) and core photograph showing where plugs have been taken in an actual example (right). The horizontal plug is oriented parallel to the laminations, while the vertical plug is taken across the laminations.

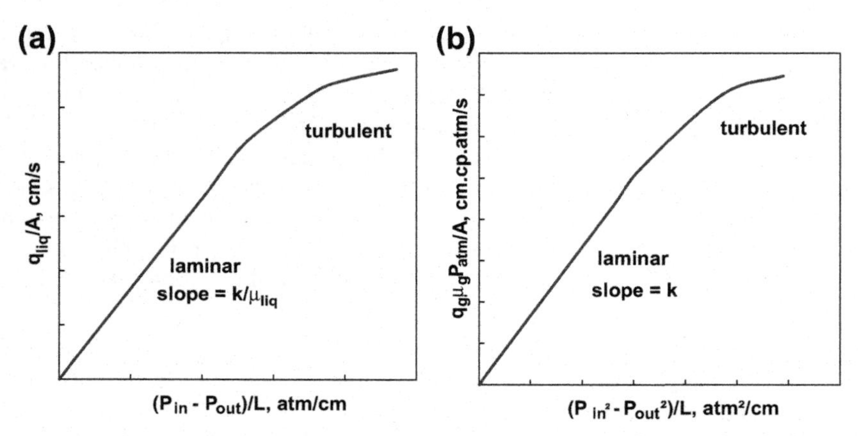

Figure 7.2.4 Flow rate versus pressure plots for permeability determination: (a) liquid flow; (b) gas flow.

in which k is the absolute permeability in D, q is the volumetric flow rate in cm^3/s at atmospheric pressure, A is the cross-sectional area of the core section or plug in cm^2, μ is the viscosity in cp, P is pressure in atm, and L is the length of the core section or plug in cm.

Typically, air or nitrogen is used to determine absolute permeability in a clean, dry core plug. Pressure drops are measured for several flow rates, and a plot of $q_g\mu_g P_{atm}/A$ versus $(P_{in}^2 - P_{out}^2)/2L$ is created, as shown in Figure 7.2.4(b). Note that the gas viscosity is determined at the average pressure of each measured pressure drop. The slope of the plot is the permeability, k. If a liquid is used instead of air, a plot of q_{liq}/A versus $\Delta P/L$ is created, and the slope is k/μ_{liq}, as shown in Figure 7.2.4(a).

Equations (7.2.3) and (7.2.4) are valid when:

1. The flow is laminar;
2. The no slip condition applies;
3. The flow is single phase;
4. There is no interaction between the rock and fluid.

The third and fourth conditions are usually satisfied when air or nitrogen is the fluid. There are usually no reactions between rock and air or nitrogen. However, liquids, particularly water, can swell or mobilize fines significantly, reducing the effective permeability. Water can also dissolve some core materials such as anhydrites, leading to an increase in permeability.

The first condition must be checked through an examination of the experimental data. If the flow rate is high enough for turbulence to occur, there will be more pressure drop than predicted by Eqns (7.2.3) and (7.2.4). Hence, there will be a downward deviation from the linear trends on the rate versus pressure plots, as shown in Figure 7.2.4(a) and (b). It is recommended to always collect a range of rate and pressure points to ensure that some data fall in the laminar flow regime.

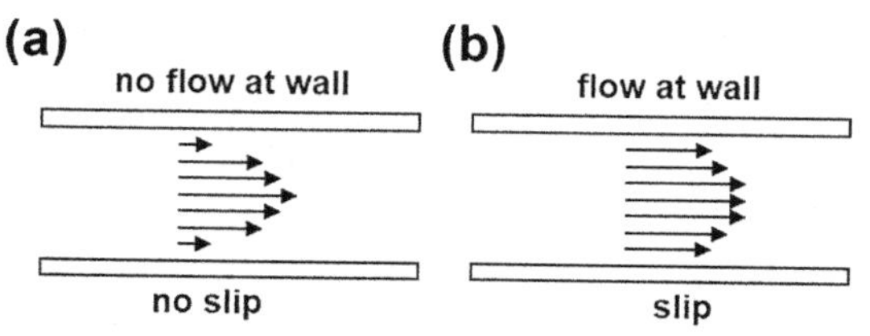

Figure 7.2.5 Laminar velocity profiles with no slip and slip boundary conditions.

The second condition, the no slip boundary condition, means that the velocity of the fluid at the wall, or in this case, the rock surface, is zero. This is usually true for liquid flow. However, in gas flow, the gas can move at the wall; that is, slippage occurs. The difference in flow profiles is illustrated in Figure 7.2.5. The more uniform velocity profile for flow with slip results in less pressure drop at a given flow rate. Therefore, the measured permeability is higher than would be observed for flow with no slip.

Klinkenberg (1941) developed corrections to convert "with slip" (gas) permeability to "no slip" (liquid) permeability. He found a linear relationship between the difference in permeabilities and the inverse of average pressure:

$$k_{liq} = k_{air} - \frac{m}{P_{avg}} \tag{7.2.5}$$

in which m is a constant. The value of m is in the order of 1 mD-atm for most gases. Permeability tests are usually performed within an order of magnitude of atmospheric pressure. Therefore, the Klinkenberg correction is usually in the order of 0.1−1 mD and is only significant for low-permeability rock.

To correct measured gas permeability data, the apparent permeability is calculated using Eqn (7.2.4). The calculated apparent permeabilities are plotted against the reciprocal of the average test pressure, as shown in Figure 7.2.6. The slope of the plot is m, and the intercept at zero reciprocal pressure is the corrected absolute permeability.

Besides core plug and whole core permeability measurements, there is another method for measuring core permeability. The probe permeameter (or minipermeameter) measures the gas permeability on rock having a flat surface (Figure 7.2.7). Once the tip is positioned and pressed against the rock surface, nitrogen gas is injected through the tip into the rock, and the permeability measured. The simplest devices are steady-state systems that measure the gas injection pressure and flow rate and use Eqn (7.2.4). More elaborate devices use unsteady-state permeability measurement by releasing a known volume of gas through the tip and monitoring the pressure drop with time. Steady-state measurements typically have a lower limit of 0.1 mD, while the unsteady-state measurements can measure to 0.01 mD. Both measurement types can be slip-corrected.

The device can be used in the laboratory or in the field, and measurements can be taken at any desired spacing. Probe measurements represent the permeability of

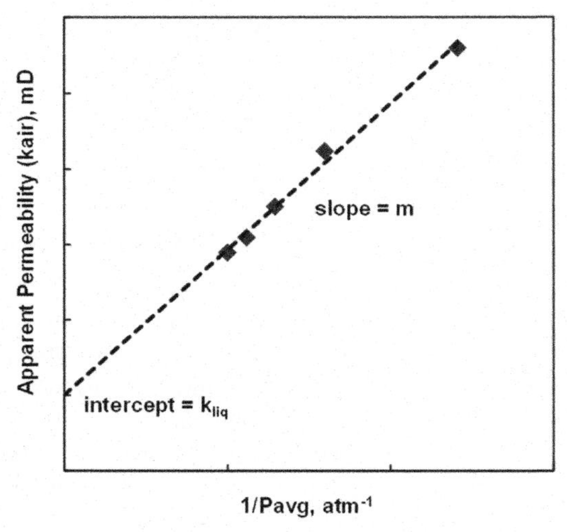

Figure 7.2.6 Klinkenberg correction plot of apparent permeability versus the reciprocal of the average test pressure.

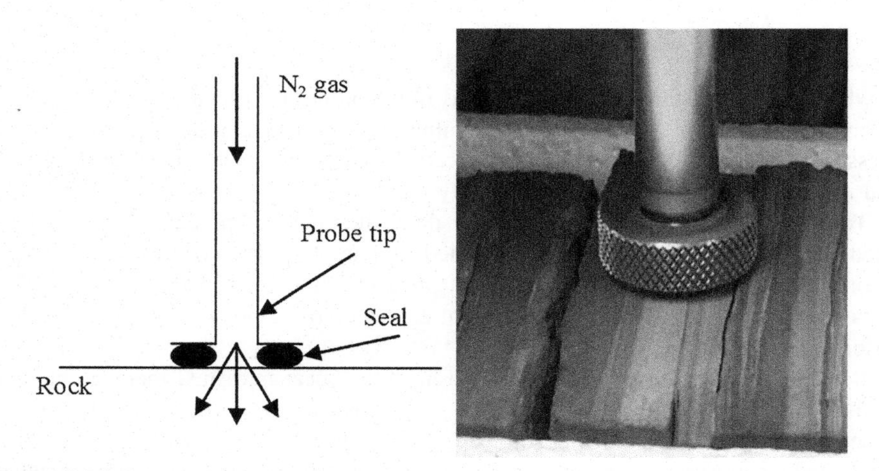

Figure 7.2.7 Schematic of permeameter tip positioned against a rock surface (left). A tip for a core probe permeameter positioned to take a measurement in a laminated rock (right). The tip internal diameter is typically a few millimeters.

unstressed rock and, therefore, may be greater than core plug and whole core measurements made at the same location. For finely layered rock, the probe permeameter helps to assess the fine-scale permeability variations that would otherwise go unsampled using other methods (Figure 7.2.7 right). Knowledge of these variations can assist net pay calculations and depth matching of core and logs.

7.3 Analyzing Conventional Core Data

7.3.1 Core Analysis Report

An excerpt from a core analysis report is given in Table 7.3.1. We will examine the excerpted data first, and then consider the full set of 258 data points from this core. Each sample in the report is an individual full diameter core section. The depth of the core and length of the core are known when the core is recovered. The depth of each core section is determined from its position within the recovered core. Each core sample is chosen to represent a particular section of the core. Therefore, the reported length may not be the length of the core sample, but the length of the core it is assumed to represent. The excerpted data represent 7 m of the formation. Whole core measurements usually are limited to samples between 15 and 40 cm in length, because of the size of the testing apparatus, and every sample in Table 3.5.3 falls in that range. Thus, it is possible that the entire 7.3 m interval was sampled.

The lithology of the example core is predominantly dolomite with some anhydrite. The measured grain densities of 2830–2870 kg/m^3 are very close to that of dolomite (2870 kg/m^3), consistent with the lithological interpretation. Vuggy porosity was observed, including pinpoint (pp/v), small (s/v), and average (vug) sized vugs. If this interval had been sampled by core plugs, there could have been some samples with high porosities (exceeding 40%). The larger volume of the whole core samples reduces the chances of this occurring, but does not avoid it completely. As we will see, the permeability of vuggy whole core samples is highly variable.

A water-based mud was used when the core was cut. Therefore, the residual saturations are not very meaningful. However, the residual water saturations are all less than 20%. The average residual water saturation is determined from a thickness-weighted average as follows:

$$S_{wr,avg} = \frac{\sum h_i S_{wr,i}}{\sum h_i} \tag{7.3.1}$$

and is 7%. The low average residual water saturation even with a water-based mud suggests that the initial water saturation in this reservoir is less than 10%.

The porosity ranges from 2.7% to 11.4%. The wide range in porosity is typical of carbonate deposits that have undergone diagenesis that has both created porosity (vugs) and reduced porosity (compaction and anhydrite cement). This reservoir likely has a random porosity and permeability distribution at least on a scale of a few centimeters. The average porosity can be determined as follows:

$$\phi_{avg} = \frac{\sum h_i \phi_i}{\sum h_i} \tag{7.3.2}$$

The average porosity of the core sample is 5.9%. Note that while the individual porosity measurements are usually accurate, the average may not be an accurate representation of the reservoir. When there is significant variation in the individual

Table 7.3.1 **Excerpt from an example conventional core analysis report**

Conventional full diameter core analysis													
Well name: example well drilled with water-based mud			Pool: Nisku										
			Permeability								Residual saturation (fraction of pore volume)		
Sample number	Top depth (m)	Length (m)	k-max (md)	k-90 (md)	k-vert (md)	Perm. x pay (mD m)	Porosity frac	Porosity pay (m)	Grain density (kg/m^3)	Bulk density (kg/m^3)	Oil	Water	Lithology
SP1	3066.29	0.15	0.79	0.29	0.15	0.12	0.062	0.009	2810	–	0	0.058	dol pp-s/v
SP2	3066.44	0.18	8.43	6.0	1.84	1.52	0.075	0.014	2870	–	0.059	0.029	dol pp-s/v anhy
SP3	3066.62	0.22	1.17	0.91	0.2	0.26	0.038	0.008	2840	–	0	0.052	dol pp-s/v anhy
SP4	3066.84	0.21	1.37	1.21	0.41	0.29	0.048	0.010	2860	–	0.113	0.113	dol pp-s/v anhy
SP5	3067.05	0.15	2.44	1.84	1.93	0.37	0.053	0.008	2850	–	0	0.038	dol pp-s/v anhy
SP6	3067.2	0.15	8.27	6.67	1.28	1.24	0.069	0.010	2870	–	0.063	0.063	dol pp/v anhy
SP7	3067.35	0.22	1.06	0.99	0.95	0.23	0.048	0.011	2850	–	0.104	0.052	dol s/v anhy
SP8	3067.57	0.15	11.2	7.84	0.46	1.68	0.029	0.004	2880	–	0	0.138	dol pp-s/v anhy
SP9	3067.72	0.15	0.5	0.17	4.63	0.08	0.033	0.005	2850	–	0	0.063	dol pp-s/v anhy
SP10	3067.87	0.25	5.12	0.66	0.43	1.28	0.044	0.011	2840	–	0	0.102	dol pp-s/v anhy
SP11	3068.12	0.21	8.65	5.03	3.07	1.82	0.073	0.015	2860	–	0	0.065	dol vug anhy
SP12	3068.33	0.24	114	74.9	24.8	27.36	0.084	0.020	2870	–	0	0.081	dol pp-s/v anhy
SP13	3068.57	0.16	86.2	9.44	25.1	13.79	0.076	0.012	2870	–	0	0.037	dol pp-s/v
SP14	3068.73	0.18	86.2	9.44	25.1	15.52	0.076	0.014	2870	–	0	0.037	dol pp-s/v

SP15	3068.91	0.21	5.36	4.17	51.9	1.13	0.060	0.013	2860	–	0	0.034	dol vug
SP16	3069.12	0.25	7.06	2.6	4.36	1.77	0.051	0.013	2830	–	0	0.072	dol pp-s/v
SP17	3069.37	0.21	20.5	16.2	9.06	4.31	0.092	0.019	2840	–	0	0.032	dol pp-s/v anhy
SP18	3069.58	0.24	7.16	6.35	2.32	1.72	0.045	0.011	2840	–	0.061	0.012	dol pp-s/v anhy
SP19	3069.82	0.19	31.1	29.2	19	5.91	0.094	0.018	2870	–	0.103	0.026	dol pp/v anhy
SP20	3070.01	0.18	0.78	0.65	0.17	0.14	0.041	0.007	2890	–	0	0.068	dol pp/v anhy
SP21	3070.19	0.21	5.37	3.13	2.33	1.13	0.060	0.013	2890	–	0	0.039	dol pp/v anhy
SP22	3070.4	0.25	12.8	2.44	0.77	3.20	0.030	0.008	2890	–	0.151	0.076	dol pp/v anhy
SP23	3070.65	0.27	4.87	2.77	0.98	1.31	0.030	0.008	2870	–	0	0.055	dol pp-s/v anhysh
SP24	3070.92	0.24	31.4	24.1	1.52	7.54	0.043	0.010	2830	–	0	0.097	dol pp-s/v anhy
SP25	3071.16	0.19	2.95	2.03	0.7	0.56	0.029	0.006	2830	–	0	0.087	dol pp/v anhy
SP26	3071.35	0.3	27.8	10.3	0.58	8.34	0.048	0.014	2860	–	0	0.045	dol vug anhy
SP27	3071.65	0.22	6.03	3.53	0.37	1.33	0.027	0.006	2840	–	0	0.076	dol vug
SP28	3071.87	0.27	4.91	3.34	2.7	1.33	0.036	0.010	2840	–	0	0.172	dol pp/v anhy
SP29	3072.14	0.18	23	7.2	9.87	4.14	0.061	0.011	2820	–	0	0.046	dol pp-s/v
SP30	3072.32	0.16	0.15	0.08	0.03	0.02	0.036	0.006	2810	–	0.092	0.037	dol pp-s/v
SP31	3072.48	0.12	39.2	10.4	102	4.70	0.065	0.008	2850	–	0	0.038	dol pp-s/v
SP32	3072.6	0.21	101	94.9	76.2	21.21	0.071	0.015	2830	–	0.095	0.047	dol pp-s/v
SP33	3072.81	0.21	22	13.5	2.72	4.62	0.095	0.020	2840	–	0	0.186	dol pp-s/v anhy
SP34	3073.02	0.19	84.9	54.4	206	16.13	0.114	0.022	2830	–	0.239	0.06	dol pp-s/v
SP35	3073.21	0.39	84.9	54.4	206	33.11	0.114	0.044	2830	–	0.239	0.06	dol pp-s/v
Total		7.310				189.175		0.432					

porosity measurements, a large number of core samples from a number of wells are required to have confidence in the average porosity. Statistical analysis of these results could help define the uncertainty of the average (5.9%) and how many samples are needed to obtain an average that has a predefined level of uncertainty (Jensen et al., 2000).

The maximum horizontal permeability (k_{max}) in Table 3.5.3 ranges by about three orders of magnitude, from 0.1 to 100 mD (Figure 7.3.1); it is common for permeability to vary by several orders of magnitude in a reservoir. A histogram of log(k_{max} or k_{90}) shows a "bell-shaped" curve, similar to a Gaussian or normal distribution. This suggests that the permeability may have a "log normal" distribution. It is common for permeability to be log normally distributed, and one consequence of this is that log(k) has some convenient statistical properties. For this reason, we will be using log(k) instead of k for some of our analyses.

There is only one mode (i.e., peak) for the log(k) histogram (Figure 7.3.1 left), suggesting that the sample permeabilities are statistically similar. Multiple modes might indicate that the data come from two or more rock types with distinctly different permeabilities. One aspect of the dataset is curious: there are sets of repeat permeability values at 3068.57 and 3068.73 m and at 3073.02 and 3073.21 m. To have two samples with all three permeability values identical (to three significant figures) leads one to wonder if samples were confused.

A measure of the "middle" permeability, the average, is determined as follows:

$$k_{H,avg} = \frac{\sum h_i k_{H,i}}{\sum h_i} \tag{7.3.3}$$

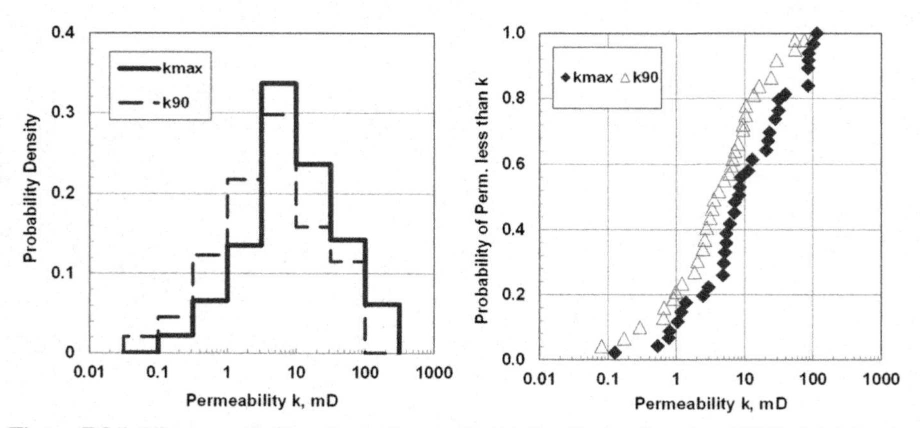

Figure 7.3.1 Histogram (left) and sample cumulative distribution function (CDF, right) for the permeability data in Table 7.3.1. The mode is easier to identify from the histogram, while the median (permeability at 50% probability) is more apparent from the CDF.

and is 26 mD. Another measure of the middle permeability is the median, the value for which about half the values fall below and half are above. In this case, the median is about 8 mD. We often prefer to use the median when we are dealing with a highly variable property, because a few extreme values do not affect the median as much as the average. Both the average and the median are based on a limited number of measurements of a highly variable property, so there is some uncertainty to these values. As was the case with porosity, statistical analysis tools could be used to quantify the uncertainties.

The average orthogonol permeability is 15 mD and the median is approximately 4 mD (Figure 7.3.1 right). The maximum and orthogonal horizontal permeabilities differ by about a factor of two using either the average or the median. This difference could be an indication of directional permeability, but could also be caused by uncertainties in the value. Note that permeability is high enough that a Klinkenberg correction is not required.

A plot of vertical versus horizontal permeability, Figure 7.3.2, shows that the k_V/k_{max} ratio based on the data of Table 3.5.3 mostly fall between the 0.1 and 1 lines, suggesting most of the samples have only a modest amount of anisotropy. A few samples show $k_V/k_{max} > 1$, and these are from the thinner samples taken; perhaps the vug-to-vug connections were better vertically than horizontally. This behavior is not uncommon for vuggy carbonates, in which chance can play a large role in determining whether the rock is permeable. If the vugs happen to touch one another, a very permeable path is created; if the vugs remain isolated, the rock has small permeability (Lucia, 2007).

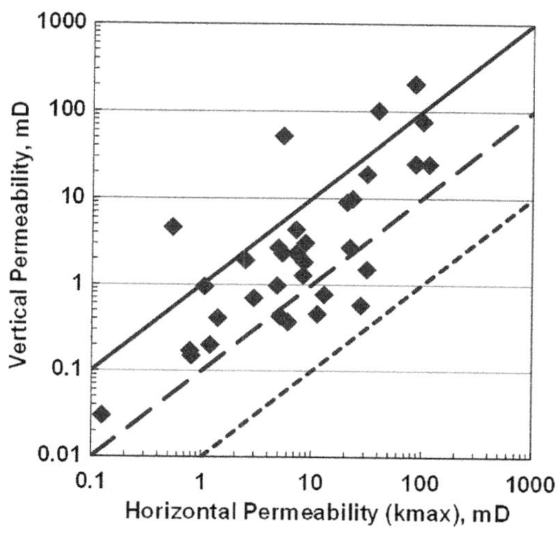

Figure 7.3.2 Cross plot of vertical versus horizontal permeability for excerpted core data. Log–log plot shows lines of $k_V/k_H = 1$ (solid), 0.1 (dashed), and 0.01 (dotted) for comparison with data. Note that some samples show $k_V > k_H$.

7.3.2 Screening Data and the Importance of Representative Sampling

Now, let us consider the full dataset of 258 core samples, all of which were dolomite. Before determining average properties, it is necessary to identify and cull invalid data points that may affect the analysis. For example, a very high permeability may result from a fractured sample or a sample with connected vugs. If we are interested in only evaluating the matrix permeability, the permeability of fractured or vuggy samples could mislead us. Very low vertical permeabilities may result from the presence of rock flour in a poorly cleaned sample. One method to detect questionable data is to examine the probability distribution of the permeabilities and log—log cross plots of k_{90} versus k_{max} and k_V versus k_{max}. The underlying approach is to avoid data exclusion unless a particular reason can be identified. The reason we take this approach is that often we find that the "maverick" or extreme values tell us a lot more about the reservoir than the data that "fall in the middle."

The permeability cumulative distribution functions (Figure 7.3.3) show the permeability is varying by five or six orders of magnitude in each direction. This wide a range suggests that permeability at the whole core sample volume (about 2000 cm^3) is being affected by the presence and connectivity of vugs. There are not large differences (more than 10:1 ratio) between the measurements in any two directions; this indicates that, at the core scale, there is not a strong directionality to the permeability. The vugs may connect vertically almost as easily as they connect horizontally. Five samples have permeabilities exceeding 10 D in one or more directions; these values are unlikely

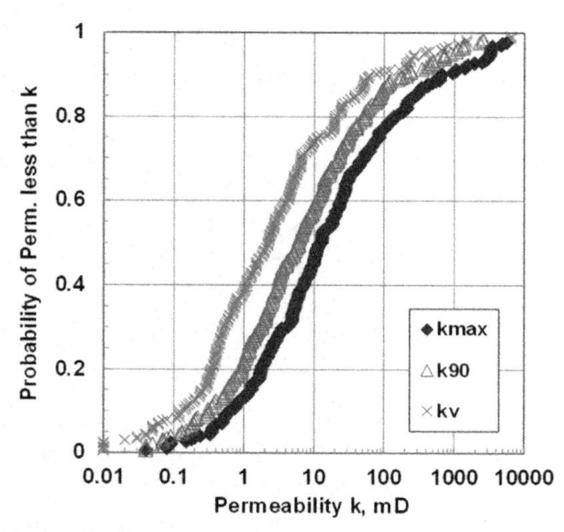

Figure 7.3.3 Cumulative distribution functions for the dataset of 258 measurements. Depths at which the sample interval was identified as too dense to measure or lost core were excluded from the analysis. There is less than a 10:1 variation between the median values of the three permeabilities, but k_V is more variable than the horizontal permeabilities.

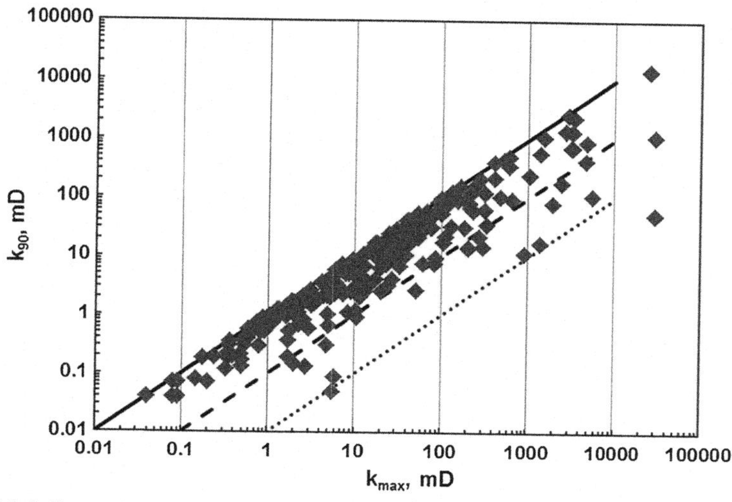

Figure 7.3.4 Cross plot of k_{90} versus k_{max} for the example core. The lines are $k_{90}/k_{max} = 1$ (solid), 0.1 (dashed), and 0.01 (dotted).

to be accurate. The smallest k_V is 0.01 mD, and this probably represents the smallest value that the core laboratory could measure. Some samples were unanalyzed because they were classified as "dense." These may be samples with little or no porosity and/or that had porosity, but were cemented by later diagenesis.

Figure 7.3.4 is a plot of k_{90} versus k_{max} for all of the example core data. There are several samples with permeabilities greater than 10,000 mD. The lithological description of these high permeability samples indicates the presence of vugs. The high measured permeabilities almost certainly result from flow through connected vugs and are not a measure of matrix permeability. k_{90}/k_{max} is greater than 0.1 for more than 90% of the data, and k_{90}/k_{max} is greater than 0.5 for 60% of the data. These results confirm our earlier observation that the anisotropy in the horizontal permeability is very weak.

Figure 7.3.5 is a plot of k_V versus k_{max} for all of the example core data. k_V/k_{max} is more variable than k_{90}/k_{max}; this is common for many rock types because small shale drapes or cemented horizons can disrupt or interrupt the vertical flow paths in the samples. There are several data points with high horizontal permeability and near zero vertical permeability. There is no evidence of laminations or shale breaks in the lithologial description, but these features might have been overlooked. These samples could be affected by rock flour from the core cutting. Alternatively, the sample may have had a vuggy part and a cemented part (e.g., a stylolite), so that horizontal flow was via the vug-dominated portion, while the vertical flow was via both the vuggy and tight, cemented portions. Analysis of the core photos or a visit to the core store could clarify which situation is more likely.

With the exception of the few >10 D samples, we find relatively few measurements are immediately questionable. A few more, however, require reference to the core

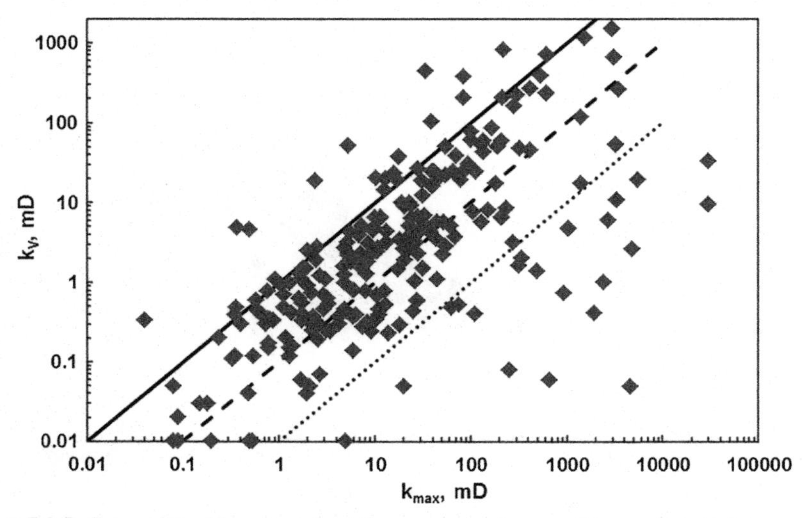

Figure 7.3.5 Cross plot of k_V versus k_{max} for one example core. Line coding is the same as Figure 7.3.4.

photographs or the core store to clarify the reasons for the small k_V/k_{max} values. Other uncertainties regarding these porosity and permeability data include what was not measured. About 58 m of the total formation (103 m) was measured, the remainder was unmeasured because of lost core (6 m), client request (34 m), or the sample was too "dense" (5 m) to merit measurement. Unless we refer to other data, such as visual inspection of the core and well logs, we will only have a partial understanding of the formation properties.

Average core properties for the excerpted and full datasets are compared in Table 7.3.2. Several points merit attention:

1. Core data must be reviewed in conjunction with geological and photographic data to obtain a meaningful analysis. The average maximum permeability is particularly sensitive to very large permeability values; if the >10 D samples are included, $k_{H,max} = 928$ mD, while if these are removed, $k_{H,max} = 262$ mD, a factor of 3.5 difference. Which value is the "right" one? Probably neither value is the right one, and the change in average when five of 258 data points are removed is a strong indicator that neither value is correct. Statistical analysis supports this observation; the 928 mD result is accurate to within ±500 mD for 95% of such datasets. The 262 mD value is accurate to ±110 mD.
2. Sampling a small dataset can lead to significant errors when properties are variable. For example, the average porosity of the full dataset (237 data points) is 19% higher than of the excerpted dataset (38 data points). The average maximum permeability of the full dataset is 30 times higher than of the excerpted dataset. In this case, use of a limited dataset would have led to a 19% underestimate of the OOIP and large underestimates of productivity if the averages were used.
3. Another aspect of the variation between the excerpted and full datasets is the choice of statistic we calculate. The medians show much less change (factor of two) than the averages.

Table 7.3.2 Comparison of average core properties of excerpted and full core data

Property	Excerpt	Full
Number of samples	35	258
ϕ (%), average	5.9	7.2
k_{max}, mD average	25.9	928
k_{max}, mD median	7.7	13
k_{90}, mD average	14.9	243
k_{90}, mD median	3.8	6.0
k_V, mD average	25.7	204
k_V, mD median	2.2	2.2

Thus, if we want to calculate a representative number for the horizontal permeability at this well, we might prefer the median.

4. A suitable value for the vertical permeability needs further evaluation. None of the values in Table 3.5.4 is appropriate if we are seeking a value for flow in the vertical direction. The harmonic average is given by:

$$k_{V,avg} = \frac{\sum h_i}{\sum (h_i/k_{V,i})}$$

$k_{V,avg}$ is 0.2 mD for the full dataset, and 0.6 mD for the excerpted dataset. Both these values, however, are suspect because they do not include the "dense portions of the reservoir. $k_{V,avg}$ could be much smaller than 0.2 mD.

To complete the analysis, we briefly discuss the representativity of these numbers for the formation beyond the wellbore. This is when the geology becomes important, because what we measure at the wellbore could completely change a few meters away. By examining the core and core measurements from several wells, a picture of the continuity of high or low porosity and permeability regions begins to emerge. Ahr (2008) uses the approach of "flow units" to identify zones that are important to fluid flow in carbonate reservoirs. Slatt (2006) describes a similar approach for sandstone reservoirs. The presence and geometries of the flow units will indicate whether we can expect high permeability intervals to extend into the reservoir and, perhaps, connect with other wells (Table 7.3.3).

Finally, recall that the absolute permeabilities from a conventional core report are measured at zero overburden pressure and zero water saturation. Effective permeabilities are usually lower than measured absolute permeabilites, as will be discussed in Chapter 8.

Table 7.3.3 **Average properties based on culled dataset**

Property	Value
ϕ (%)	7.0
k_H, mD	117
k_V, mD	51
k_V/k_H	0.43

7.3.3 Permeability–Porosity Cross Plot

If a sandstone consisted of perfectly sorted uniform spheres, the porosity of the rock would be independent of the size of the spheres. However, the permeability would depend on the size of the spheres, because the pores and pore throats are smaller if the rock spheres are smaller. In this case of perfect sorting, there is no correlation between permeability and porosity. However, in practice, rock grains are neither uniform nor spherical and are not perfectly sorted. Grain sorting affects both porosity and permeability, because small grains can fill the spaces between larger grains and partially block pore throats. When sorting is relatively constant and the grain size changes, permeability will vary. But, when sorting varies, both porosity and permeability are affected, and a correlation between the two will be observed. Typically, both grain size and sorting change in a reservoir, so one will observe a partial correlation of porosity and permeability, Figure 7.3.6. When sorting is the major control, the correlation may be used to estimate permeabilities from porosity data because

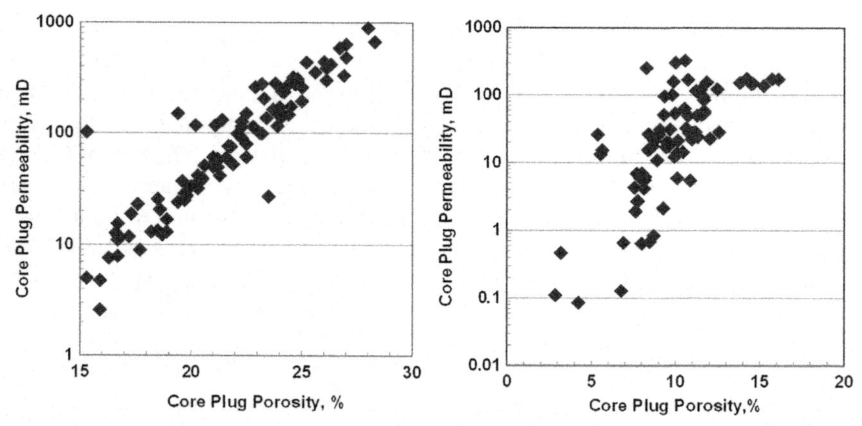

Figure 7.3.6 A uniformly very fine grain size, lower-shoreface sandstone has both permeability and porosity variations controlled by sorting (left). A fluvial sandstone with more variable grain size (silty to medium) shows weak correlation (right).

permeability data are usually more limited than porosity data. The cross plot can also be used to estimate porosity cutoffs at which the reservoir is deemed unproductive. However, when grain size is the major control, permeability estimation using porosity may be unhelpful, and other measurements may be better permeability predictors. A cross plot of permeability and the wireline gamma ray log, for example, may provide a stronger correlation. See Section 9.2.2 and Figure 9.2.4.

In carbonate rocks, the controls on porosity and permeability may be different than in sandstones. Sandstones are primarily created by grains and the porosity is between the grains or intergranular. Carbonate porosity can take many forms besides intergranular, including intercrystalline, moldic, and vug porosity. Carbonate rocks usually have several porosity types present. Molds and vugs increase porosity, but if they are isolated, the permeability will change little. On the other hand, molds and vugs can enhance permeability considerably if they are in contact with each other, but there have to be sufficient so that there is a continuous path through the vugs and molds. Therefore, porosity—permeability cross plots for carbonates tend to show much more scatter than such plots for sandstones, reflecting the weaker relationship between porosity and permeability.

Whatever the porosity—permeability relationship, it is possible to calculate a "best-fit" line through the data using least squares regression. The permeability—porosity cross plot for the dataset for the example core is shown in Figure 7.3.7. The core samples are all from one rock type and follow the same trend. The data were fitted with a polynomial function of the form:

$$\log\{k\} = a + b\phi + c\phi^2 \tag{7.3.4}$$

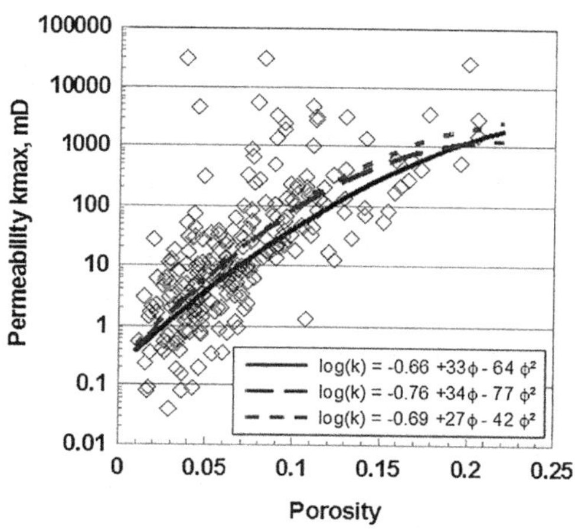

Figure 7.3.7 Three different quadratic lines that could characterize the porosity—permeability relationship. Line 1 (solid) is for all 258 data; line 2 (long dash) is for all data except the >10 D points, and line 3 (short dash) is where a number of extreme data have been excluded. The coefficient values have been reported to only two figures to reflect the uncertainty in the meaningfulness of any of the lines.

in which a, b, and c are constants. Sometimes a linear fit is used, but it is recommended to use a quadratic equation so that the curvature at high porosity is captured. The quadratic equation will usually avoid predicting unrealistically high permeability at high porosities.

The purpose of the line is to predict permeability using porosity measurements, which are usually available using well log measurements. Unfortunately, well log porosity measurements typically have a resolution of 50 or 60 cm, which is not the same resolution as the data used to define the porosity−permeability line. This problem can be eased by taking running averages of the core porosity and permeability before calculating the "best fit" line (Figure 7.3.8).

Note that even for a single rock type, there is considerable scatter in the correlation and any of several lines might be considered as "fitting the data," Figure 7.3.7. For example, using line 3, the predicted permeability at the average porosity of 7.0% is 10.4 mD, while the measured permeabilities range from approximately 1−100 mD. Averaged data are better (Figure 7.3.8), but the scatter at 7% porosity still is significant. For these data, the permeability−porosity correlation may be only useful for order of magnitude estimates of permeability.

To state the obvious, the properties of different rock types are usually different. When a reservoir has more than one rock type present, the core data should be marked according to the rock type on a permeability−porosity cross plot. Usually, the data will cluster according to the rock type, as shown in Figure 7.3.9. In this case, average properties should be determined for each rock type rather than the entire screened dataset. In some cases, different permeability−porosity correlations can be observed for the same rock type if the depositional environment was different.

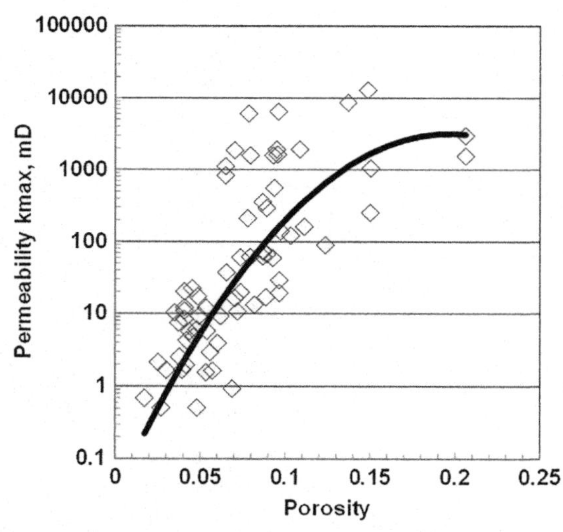

Figure 7.3.8 Porosity−permeability correlation obtained by taking 1 m running averages of the data and the least-squares best-fit line. The scatter is reduced, and the line could be more appropriate for permeability predictions using well log porosities.

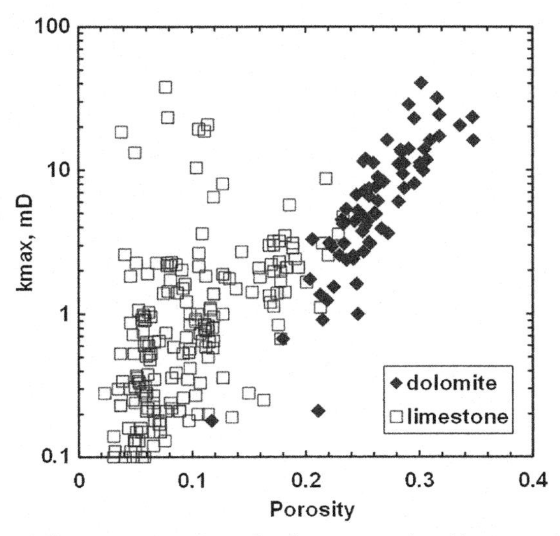

Figure 7.3.9 Permeability—porosity cross plot for a reservoir with two rock types.

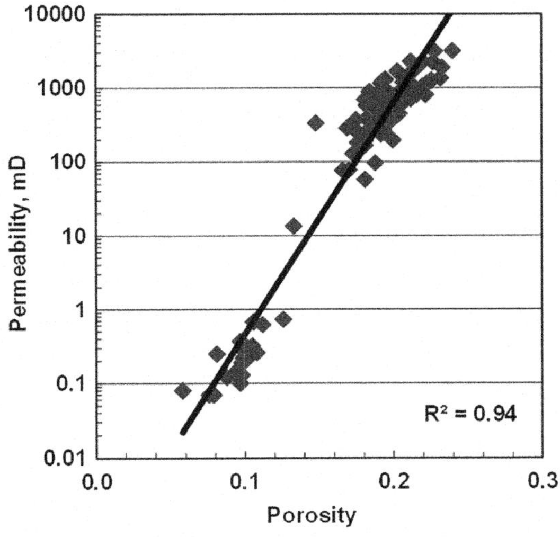

Figure 7.3.10 A large R^2 suggests a good fit of the line to all the data from this formation with two rock types.

The coefficient of determination (R^2) is often used as an indicator of the appropriateness of the line for porosity—permeability data. However, R^2 should be used with caution. It is not a substitute for examining the cross-plot with the line to visually assure an appropriate fit. For example, despite a large R^2, the line in Figure 7.3.10 does not do a good job. It does not reflect the porosity—permeability trend in the more porous and permeable rock. Also, the data suggest there are only three samples with porosities between 0.12 and 0.17, and the permeabilities vary from 0.7 to 350 mD, while the line shows a continuous increase of permeability with porosity. Each cluster should have its own line, but note that the R^2 will decrease for each of the lines. Thus, a more geologically consistent set of trend lines may have smaller R^2 values than a trend line that has a large R^2 but captures the porosity–permeability trend poorly.

7.3.4 Permeability and Porosity Cutoffs and Determination of Net Pay

One use of core data is to determine how much of the formation is capable of producing hydrocarbons. Some definitions used in productivity estimation are:

- Gross pay: the total thickness of the formation;
- Net pay: the thickness of the productive formation;
- Net-to-gross ratio (NGR): the ratio of net pay to gross pay thickness.

The problem is how to define "productive" formation. One approach is to determine the minimum permeability at which the reservoir is capable of flowing hydrocarbons. Only rock at or greater than this permeability cutoff is considered to be net pay. If capillary pressure tests or displacement tests are performed on a number of core samples covering a range of permeabilities, then the cutoff permeability can be estimated experimentally (Chapter 8). Otherwise, a common rule of thumb for oil reservoirs is to use a permeability cutoff of 0.5–1 mD. Note that average reservoir properties should be determined based only on data points that exceed the cutoff.

In practice, most available rock data are porosities, not permeabilities. The permeability–porosity cross plot is used to determine the porosity cut-off that corresponds to the permeability cutoff. For example, if a permeability cutoff of 0.5 mD is used, then the porosity cutoff based on Figure 7.3.7 is 1.5% using any three of the lines. If a permeability cross plot is not available, cutoffs can be obtained from analogous pools with core data or from rules of thumb. Some rules of thumb for porosity cutoffs are provided in Table 7.3.4. Porosity cutoffs are used in log interpretation, as discussed in Chapter 9.

Applying Cutoffs to Conventional Core Data

Let us return to the previous conventional core analysis and determine the average properties of the cored interval using a cutoff of 0.5 mD. The gross pay of the

Table 7.3.4 **Rules of thumb for porosity cutoffs for oil reservoirs**

Rock type	Porosity cutoff
Sandstone	5–8%
Carbonate	1–3%

Table 7.3.5 Average rock properties of net and gross pay intervals of example core dataset

	Excerpted dataset		Screened dataset	
Property	**Gross**	**Net**	**Gross**	**Net**
Gross pay, m	7.31	7.31	53.49	53.49
Net pay, m	7.31	7.15	53.49	49.39
NGR	1.00	0.98	1.00	0.92
Porosity, %	5.9	6.0	7.0	7.3
k_H, mD	20.4	20.8	117	127
k_V, mD	25.7	26.3	50.9	55.1
k_V/k_H	1.26	1.26	0.43	0.43

excerpted cored interval, Table 7.3.5, is 7.31 m. All but one (SP 30) of the maximum permeabilities exceed the cutoff of 0.5 mD. The net pay using a 0.5 mD cutoff is 7.15 m, and the NGR is 0.98. The average core properties of the net and gross pay intervals are compared in Table 7.3.5. The average properties for the screened dataset from the entire core are also summarized in Table 7.3.5. As expected, the average porosities and permeabilities of the net pay interval are slightly higher than for the gross pay interval because some low permeability data points have been removed from the average. Note that reservoir calculations are usually based on the properties of the net pay interval.

7.3.5 Gamma Ray Log and Depth Correction

When interpreting reservoir data, it is necessary to compare and reconcile core and well log data. To do so, it is necessary to match the core depth to the well log depth. Then, a core porosity at a certain depth can be better compared directly with a log porosity at that same depth. These comparisons are used to check the calibration of porosity logs and to confirm geological interpretations, such as layering or lateral porosity/permeability variations.

The depth correlation can be achieved using Gamma Ray logs. A Gamma Ray (GR) log is taken of the core and of the open hole wellbore. Gamma Ray logs are discussed in Chapter 8. At this point, it is sufficient to recognize that the profile of the GR measurement is expected to be the same in the core and the log. Therefore, the profiles are aligned as shown in Figure 7.3.11. If a core GR has not been taken, then depth matching between core porosity and well log porosity or core permeability and well log porosity may be effective. Then the core depths are corrected to match the log depths. The comparison of core and log porosity is discussed in Chapter 9.

Depth correlation using the GR with probe permeability measurements can be useful. The example of Figure 7.3.12 suggests that adjustments to the core depth can be

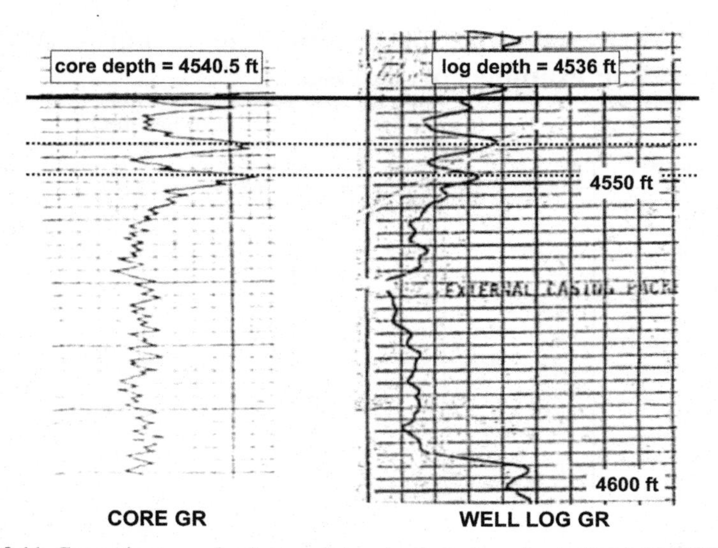

Figure 7.3.11 Correcting core depth to well log depth for Jumpbush 16-14-20-20W4M. Here, the core depth must be corrected upward by 4.5 ft.

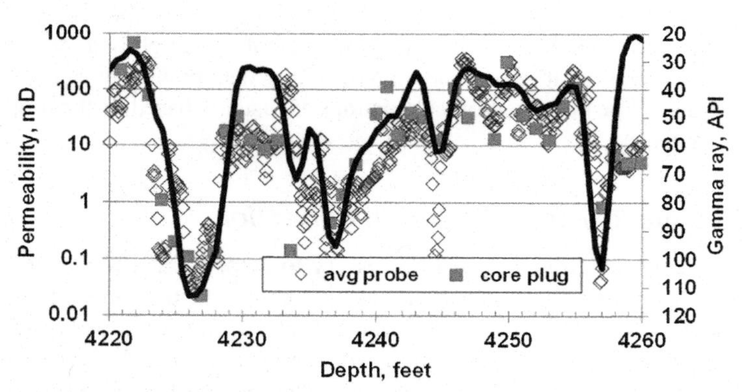

Figure 7.3.12 Permeability correlates well with the wireline gamma ray for this fluvial sandstone. (The GR scale is reversed to correlate with permeability.) Core plug values are each foot, and the probe data are every 1/10th foot and averaged over three measurements.

done with either the core plug or probe data over the interval 4220–4238 ft, at which the silts and shales are sufficiently thick to be sampled with the core plugs. Depth shifts can change, however, at different locations of the core because there may be gaps caused by less than 100% core recovery. Thus, a match over one interval does not guarantee a depth match over the entire cored interval. Below 4238 ft, thinner events (e.g., 4244 and 4257 ft), are clearly defined with the probe data and confirm the core depths still correspond with the log depths.

7.4 Rock Compressibility and the Effect of Overburden Pressure

7.4.1 Rock Compressibility

The isothermal rock compressibility is a measure of the change in rock volume or pore volume with a change in pressure at constant temperature and is defined as follows:

$$c_f = -\left(\frac{dPV_f}{PV_f dP}\right)_T = \left(\frac{dV_f}{V_f dP}\right)_T \tag{7.4.1}$$

in which PV is pore volume, V is total volume, and P is pressure. Rock compressibility is required for material balance calculations, reservoir simulation, and well test analysis.

Rock compressibility is not measured in a conventional core analysis, but can be determined using hydrostatic or triaxial tests. In these tests, the core section is sealed in a rubber sleeve and placed lengthwise in a piston. Pressure is imposed on the rock matrix using the piston. In a hydrostatic test, the pressure is the same in all directions. In a triaxial test, the pressure of the fluid in the pore space is controlled by compression or expansion of the sleeve in the radial direction. Hence, the pressure can be different in the radial (horizontal) direction than in the axial (vertical) direction. The change in core volume is then measured as a function of the fluid and rock pressures. Hydrostatic and triaxial tests give different results as shown in Figure 7.4.1.

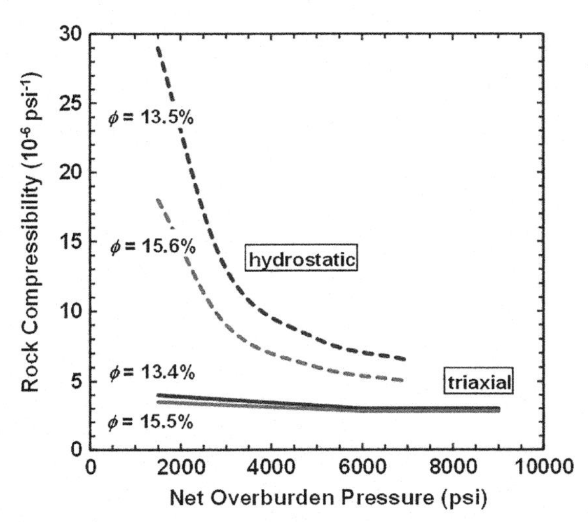

Figure 7.4.1 Effect of net overburden pressure on compressibility in hydrostatic and triaxial tests.
Adapted from Lachance and Anderson (1983).

Table 7.4.1 Values of the conversion constant, *K*, between triaxial and hydrostatic tests from Yale et al. (1993)

Rock type	*K*
Consolidated sandstone	0.45
Friable sandstone	0.60
Unconsolidated sandstone	0.75
Carbonates	0.55

The consensus in the literature is that triaxial tests best represent the formation at reservoir conditions (Cronquist, 2001). However, most compressibility measurements are from hydrostatic tests. An approximate adjustment to convert hydrostatic compressibility to triaxial compressibility was developed by Yale et al. (1993) and is given by:

$$(c_f)_T = K(c_f)_H \tag{7.4.2}$$

in which $(c_f)_T$ and $(c_f)_H$ are the triaxial and hydrostatic rock compressibility, respectively, and K is a rock-type dependent constant with the values listed in Table 7.4.1.

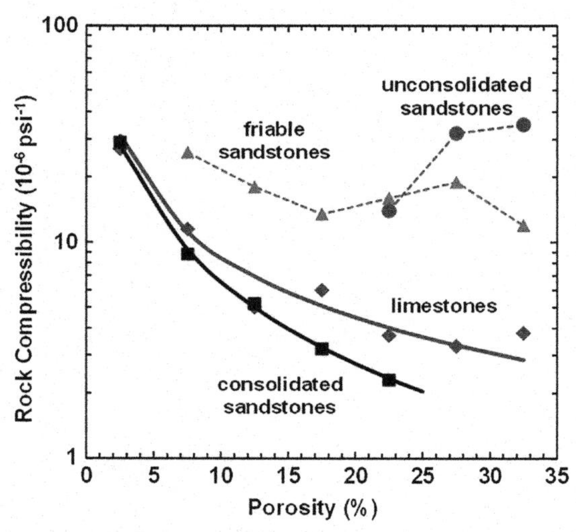

Figure 7.4.2 Average compressibility of different rock types versus porosity. Error bars are omitted for clarity; scatter is approximately ±100%. Data points are from Newman. Solid lines are the Newman correlations. Dotted lines are added for clarity.
Adapted from Newman (1973).

Rock or formation compressibility appears to depend on the type of rock, the porosity, and on consolidation of the rock, as shown in Figure 7.4.2. The types of consolidation considered in the original study were defined as follows:

- Consolidated — no crumbling under handling, edges cannot be broken by hand;
- Friable — no crumbling under handling, edges can be broken by hand;
- Unconsolidated — crumbles under own weight unless frozen.

In general, sandstones at a depth of less than several hundred meters may be friable or unconsolidated. Limestones and dolomites are almost always consolidated. While the extent of consolidation certainly affects compressibility, the effects of pressure and porosity are less certain. The averaged compressibility data of Figure 7.4.2 were based in 79 samples tested with the hydrostatic technique. Triaxial data indicate that compressibility is nearly independent of porosity and much less dependent on pressure, as indicated in Figure 7.4.1.

The data of Figure 7.4.2 were fitted by Newman to obtain the following correlations for isothermal compressibility under hydrostatic pressure:

For consolidate sandstones ($0.02 < \phi < 0.23$):

$$c_f = \frac{97.3 \cdot 10^{-6}}{[1 + 55.9\phi]^{1.429}} \tag{7.4.3}$$

For limestones ($0.02 < \phi < 0.33$):

$$c_f = \frac{0.856}{[1 + 2.48 \cdot 10^6 \phi]^{0.930}} \tag{7.4.4}$$

in which c_f is in units of psi^{-1}, and ϕ is the fractional porosity.

Note that the compressibilities of friable and unconsolidated sandstones were not found to follow any clear trend. Another correlation commonly used in the petroleum industry is the Hall (1953) correlation:

$$c_f = \frac{1.782}{\phi^{0.438}} \tag{7.4.5}$$

Unfortunately, measured rock compressibilities are scattered when compared with the correlations. For example, only average properties are shown in Figure 7.4.2. The scatter in the data within each rock type was approximately $\pm 100\%$. Therefore, it is recommended to measure the rock compressibility of the formation of interest and only use the correlations for order of magnitude estimates.

Example:

The data collected in a hydrostatic test are listed in Table 7.4.2. Determine the reservoir compressibility at a porosity of 7%.

Table 7.4.2 Hydrostatic pressure test results for a Viking sandstone

	Measured porosity		
Sample	**0 psig**	**200 psig**	**3600 psig**
SP1	0.022	0.022	0.020
SP2	0.031	0.031	0.029
SP3	0.038	0.038	0.036
SP4	0.072	0.072	0.069

The approximate compressibility was determined as shown below for sample SP4:

$$c_f = -\left(\frac{dPV_f}{PV_f dP}\right)_T \approx -\frac{1}{\phi}\frac{\Delta\phi}{\Delta P} = -\frac{2}{(0.069+0.072)}\frac{(0.069-0.072)}{(3600-200)}$$
$$= 12.5 \cdot 10^{-6}\text{psi}^{-1}$$

in which the average porosity for the two pressures is used in the denominator. The experimental compressibility at approximately 7% porosity is approximately 13×10^{-6} psi^{-1}.

The results for each sample are compared with correlations in Figure 7.4.3. The Newman sandstone correlation is based on hydrostatic test data and is in good agreement with the example hydrostatic test core data. Even so, this level of agreement is unusually good. The Hall correlation tends to underpredict hydrostatic test data and does so here as well.

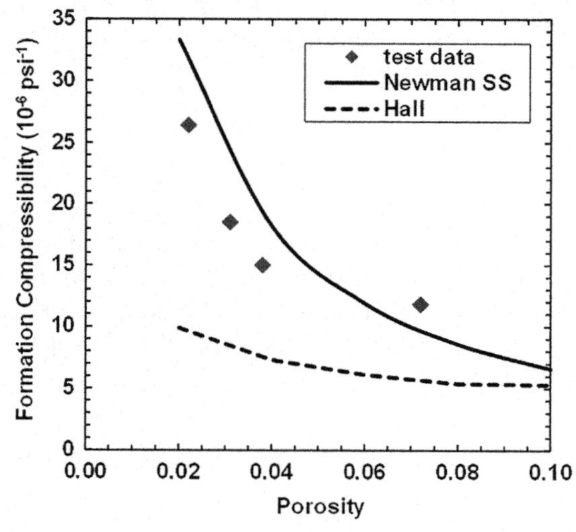

Figure 7.4.3 Comparison of measured rock compressibility of example core with correlations.

7.4.2 The Effect of Overburden Pressure on Porosity and Permeability

Overburden pressure is the pressure on the rock from the weight of the rock and earth above the formation. When the overburden pressure exceeds the fluid pressure in the pore space, the formation is compacted. The porosity, permeability, and compressibility are reduced. To illustrate the effect of net overburden pressure, consider an unconsolidated sandstone. At low net overburden pressures, the formation is loosely packed, and there is a lot of space for sand grains to realign under pressure. In general, the pore throats are relatively large. Hence, the compressibility, porosity, and permeability are high. As the net overburden pressure increases, the sand grains are forced into close contact, and there is less space for realignment. Hence, the porosity and compressibility decrease. The compaction of the sand grains also reduces the size of the pore throats. Therefore, the permeability is also reduced.

Effect on Porosity

In a conventional core test, the measurements are conducted at near atmospheric pressure and with no overburden pressure. Therefore, it is sometimes necessary to correct conventional core test data to account for the effect of overburden pressure. The effect of net overburden pressure on compressibility was discussed in Section 7.4.1. The overall effect of net overburden pressure on the porosity for typical consolidated, friable, and unconsolidated rock is shown in Figure 7.4.4.

Usually, the effect of net overburden pressure on porosity is negligible for consolidated rock, but can be significant for friable and unconsolidated rock. It is recommended to conduct compressibility tests whenever the reservoir rock is expected to be friable or unconsolidated. The effect of overburden on porosity can be determined directly from the compressibility test data.

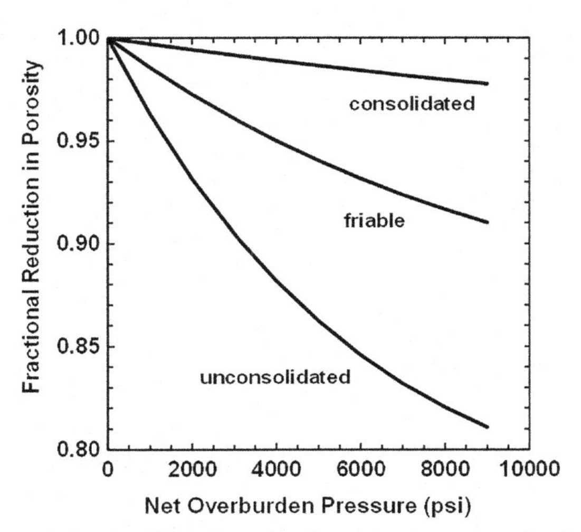

Figure 7.4.4 The approximate effect of overburden pressure on porosity for unconsolidated, friable, and consolidated rock.
Adapted from The Fundamentals of Core Analysis (1973).

Example:

Consider the data in Table 7.4.2. The porosity reduction of sample SP4 at a net overburden pressure of 3600 psi is:

$$\frac{\Delta\phi}{\phi} = \frac{7.2 - 6.9}{7.2} = 0.042$$

This porosity reduction is consistent with the friable sandstone curve on Figure 7.4.4. A compressibility of 19×10^{-6} psi^{-1} was determined previously for this sample. This compressibility is consistent with a rock in the consolidated to friable range on Figure 7.4.4. Even for this friable sample, the porosity reduction with overburden pressure is less than 5%.

Effect on Permeability

The effect of overburden pressure on permeability is greatest for unconsolidated rock and for low permeability rock. The overall effect of net overburden pressure on the permeability for typical consolidated, friable, and unconsolidated rock is shown in Figure 7.4.5. The effect of overburden pressure on some low permeability rock is shown in Figure 7.4.6. Permeability can be reduced in very low permeability or unconsolidated rock by a factor of four. It is recommended to conduct permeability tests at overburden pressure in these cases.

Overburden pressure can reduce the permeability from 10% to 20% even for permeable, consolidated rock. Therefore, conventional absolute permeability measurements

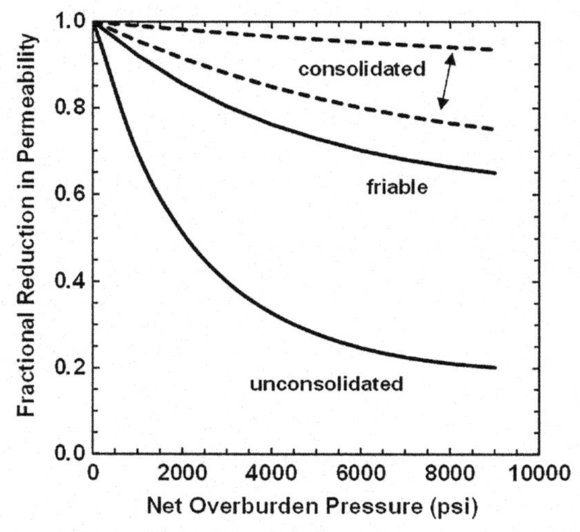

Figure 7.4.5 The approximate effect of overburden pressure on permeability for unconsolidated, friable, and consolidated rock.
Adapted from The Fundamentals of Core Analysis (1973).

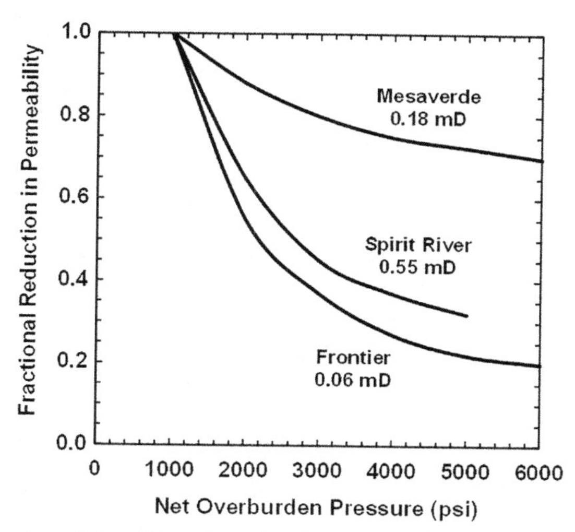

Figure 7.4.6 Examples of the effect of overburden pressure on permeability for very low permeability rock.
Adapted from Jones and Owens (1980).

are almost always optimistic unless the core has been plugged with fines or mud. As will be discussed in Chapter 8, the presence of connate water can also reduce the permeability to oil. Therefore, effective permeability is typically 30% less than the absolute permeability to air measured with zero net overburden pressure. It is always best to confirm average reservoir permeability from an inflow calculation or preferably from a pressure transient test.

7.5 Formation Resistivity

As will be discussed in Chapter 9, the water saturation of a formation can be estimated from a resistivity log. This log measures the total resistivity of the reservoir rock and the fluids in the pore space of the rock. The water saturation is determined as follows:

$$S_w = \left(\frac{R_o}{R_T}\right)^{\frac{1}{n}} \tag{7.5.1}$$

in which S_w is the water saturation, R_o is the formation resistivity when $S_w = 1$, R_T is the total resisitivity, and n is the saturation exponent. Hence, to determine the water saturation, it is necessary to know the values of R_o, R_T, and n. The value of n can be determined experimentally, but is usually assumed to equal 2.

The formation resistivity is related to the porosity and the shape and connectivity of the pore structure, that is, its *tortuosity*. Archie found an empirical relationship between formation resistivity and porosity for water-saturated rock:

$$\frac{R_o}{R_w} = K\phi^{-m} \tag{7.5.2}$$

in which K and m are constants that must be determined for each rock type. The constant m is termed the cementation factor. The ratio of R_o/R_w is often termed the formation factor. Equations (7.5.1) and (7.5.2) are combined to obtain:

$$S_w = \left(K\frac{R_w}{\phi^m R_T}\right)^{\frac{1}{n}} \tag{7.5.3}$$

The values of K and m must be determined so that water saturations can be calculated from log data.

Determination of *K* and *m*

Resistivity is measured on brine-saturated core plugs simply by passing a current through the core plug and measuring the voltage drop. Because the water saturation is 100%, Eqn (7.5.3) can be rearranged to obtain:

$$\log R_o = (\log K + \log R_w) - m \log \phi \tag{7.5.4}$$

or

$$\log\left\{\frac{R_o}{R_w}\right\} = \log K - m \log \phi \tag{7.5.5}$$

A log–log plot of total resistivity or formation factor versus porosity is expected to be linear with a slope of $-m$. The value of K can be determined from the constant term in Eqn (7.5.5). In the limit, as $\phi \rightarrow 1$, $R_o \rightarrow R_w$, so that K should equal 1. In practice, we find that K is near to 1.

Figure 7.5.1 shows representative log–log plots of formation factor versus porosity as a function of cementation (consolidation). The slope increases as the extent of consolidation increases. Some rules of thumb for K and m when core data are unavailable are:

Sandstones (Humble's formula):

$$\frac{R_o}{R_w} = 0.62\phi^{-2.15} \tag{7.5.6}$$

Carbonates:

$$\frac{R_o}{R_w} = 1.0\phi^{-2.0} \tag{7.5.7}$$

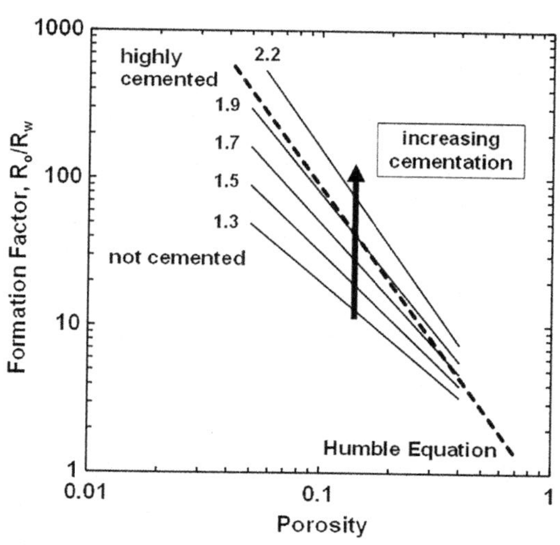

Figure 7.5.1 The effect of porosity and cementation on the formation factor ($K = 1$ in all cases except for Humble equation).

In fractured rocks $m = 1$. For carbonates, m is affected by the porosity type and proportion of that type. For example, in rocks with predominantly intergranular porosity, m has a value near 2. For vuggy rocks, m can be as large as 4.

Determination of *n*

To measure the value of n, the resistivity of the core is measured at different brine saturations. Equation (7.5.1) is rearranged as follows:

$$\log R_T = \log R_o - n \log S_w \tag{7.5.8}$$

Because R_o is constant for a given core plug, a log–log plot of total resistivity versus water saturation is expected to be linear with a slope equal to $-n$.

Example:

Tables 7.5.1 and 7.5.2 list the resistivity data available from a core analysis. Based on this data, determine the values of n, K, and m.

First, a log–log plot of R_T/R_o versus S_w is created as shown in Figure 7.5.2. The data are fit according to Eqn (7.5.8), and n is found from the slope to be 2.20. Then, a log–log plot of R_o/R_w versus ϕ is created, as shown in Figure 7.5.3. The data are fit with Eqn (7.5.2), with the constraint that K is less than or equal to unity. In this case, m and K were found to be 1.76 and 1.0, respectively. The cementation factor, $m = 1.76$, is consistent with a moderately cemented sandstone. The equation relating water saturation to rock properties for this core is then:

$$S_w = \left(1.0 \frac{R_w}{\phi^{1.8} R_T}\right)^{1/2.20}$$

Table 7.5.1 Total resistivity versus saturation data from example core

SP1		SP2	
S_w	R_T/R_o	S_w	R_T/R_o
1.00	1.00	1.000	1.00
0.775	1.81	0.573	3.55
0.708	2.60	0.528	4.84
0.629	3.87	0.470	7.51
0.575	4.25	0.440	7.82
0.558	4.64	0.430	8.38
0.535	5.06	0.419	9.51

Table 7.5.2 Total resistivity versus porosity data from example core at 3600 psia net overburden pressure

Sample	ϕ	R_o/R_w
SP1	0.02	852
SP2	0.029	662
SP3	0.036	243
SP4	0.069	152

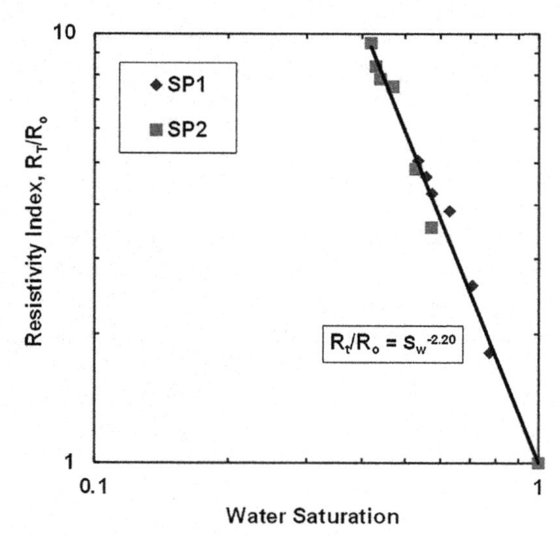

Figure 7.5.2 R_T/R_o versus S_w for the example core.

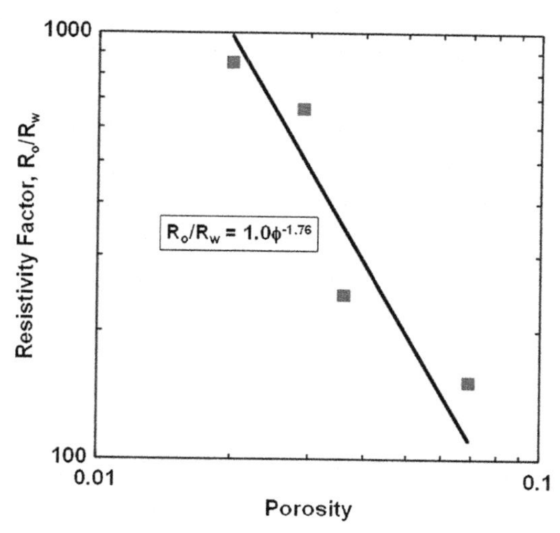

Figure 7.5.3 R_o/R_w versus ϕ for the example core.

Note that this correlation was based on a small dataset. There can be considerable variation in constants fitted with log—log relationships. Therefore, a large dataset is required to build confidence in the correlation. If data are limited, it is best to use Eqn (7.5.6) or (7.5.7). Another factor to consider is that Eqn. (7.5.1) assumes that only the water of resistivity R_w is conducting electricity. If other conductive materials are present, such as clays and pyrite, then Eqn. (7.5.1) requires modification.

Special Core Analysis—Rock–Fluid Interactions

Chapter Outline

Special core analysis includes relative permeability and capillary pressure data. Relative permeability is required to predict flow in the reservoir and to model reservoir displacement processes. In the first part of this chapter, the measurement of relative permeabilities is reviewed, analysis of relative permeability data is discussed, correlations are presented, and empirical guidelines are provided. Capillary pressure is required to determine the thickness of the water–oil transition zone and to perform some displacement calculations. In the latter part of this chapter, the measurement and analysis of capillary pressure are discussed.

Practical Reservoir Engineering and Characterization. http://dx.doi.org/10.1016/B978-0-12-801811-8.00008-0

8.1 Relative Permeability

Relative permeability is the ratio of the permeability to a given fluid in the presence of other fluids to the absolute permeability (see Chapter 2). Some symbols and definitions used in describing relative permeability data are listed below:

S_{wirr} = irreducible water saturation (the water saturation that cannot be displaced by oil);
S_{wi} = initial water saturation (assumed to equal S_{wirr} unless otherwise stated);
S_{orw} = residual oil saturation to water (the oil saturation that cannot be displaced by water);
S_{org} = residual oil saturation to gas (the oil saturation that cannot be displaced by gas);
S_{gc} = critical gas saturation (the gas saturation at which gas begins to flow);
S_{gt} = trapped gas saturation (the gas saturation which cannot be displaced by oil);
k_{row} = relative permeability to oil in an oil–water system;
k_{rog} = relative permeability to oil in an oil–gas system;
k_{rw} = relative permeability to water;
k_{rg} = relative permeability to gas;
$k_{rowe} = k_{row}$ at S_{wi};
$k_{roge} = k_{rog}$ at S_{wi} and $S_g = 0$;
$k_{rwe} = k_{rw}$ at S_{orw};
$k_{rge} = k_{rg}$ at S_{org}.

Relative permeability depends not only on the overall fluid saturations, but also on the geometry of the pore space and the geometry of the fluids distributed within the pore space. These factors depend in turn on:

- Wettability;
- Pore structure;
- Ratio of capillary to viscous forces;
- Interfacial tension;
- Saturation history.

Each factor is reviewed below.

Wettability

In a water-wet rock, water tends to occupy the small pores and to contact most of the rock (Bobek at al., 1958; Raza et al., 1968). The connate water saturations in water-wet rock are typically greater than 20%. At low water saturations, the oil is usually connected throughout the pore structure and has a high relative permeability (typically between 0.6 and 0.9 at S_{wi} relative to the absolute permeability). At high water saturations, the oil exists as disconnected droplets occupying the centers of pores. The residual oil partially blocks the flow of water so that water saturations are relatively low (typically between 0.05 and 0.30 at S_{orw}).

The situation is reversed in oil-wet rock. The oil occupies the smaller pores and is in contact with the rock and the relative permeabilities are a mirror image of those in a water-wet rock. At high water saturations, the water forms a well-connected network with high relative permeabilities (typically greater than 0.5 at S_{orw}). The relative permeability to oil may be high at S_{wi}, but typically decreases sharply as the water saturation

Table 8.1.1 Criteria for assessing rock wettability, Craig (1971)

Criteria	Water-wet	Intermediate	Oil-wet
S_{wi}	>0.20	0.15–0.20	<0.15
S_w at which $k_{row} = k_{rw}$	>0.50	~0.50	<0.50
k_{rw} at S_{orw}	<0.30	0.30–0.50	>0.50

increases and oil flow is blocked by water in the center of the pores. Irreducible water saturations in oil-wet rock are typically less than 15%. Criteria for rock wettability are summarized in Table 8.1.1. Note that gas is always a nonwetting phase.

In general, carbonates tend to be oil–wet, whereas sandstones are equally likely to be oil-wet as water-wet (Anderson, 1986). The oil compostion is also a critical determing factor for wettability.

Pore Structure

As a rule, lower permeability sandstones have more strongly water-wet characteristics with higher irreducible water saturations. There is typically a power law relationship between permeability and irreducible water saturation of the form:

$$k = aS_{wi}^{b} \tag{8.1.1}$$

in which a and b are constants. Figure 8.1.1 shows some permeability–irreducible water saturation correlations. Irreducible water saturations up to 90% were observed in very low permeability rock. Typically, water saturations in productive reservoirs are less than 40%.

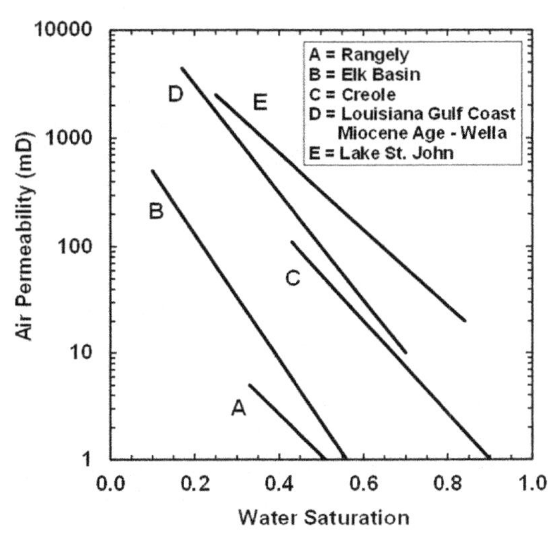

Figure 8.1.1 Relationships between irreducible water saturation and permeability for various formations. Adapted from Welge and Bruce (1947).

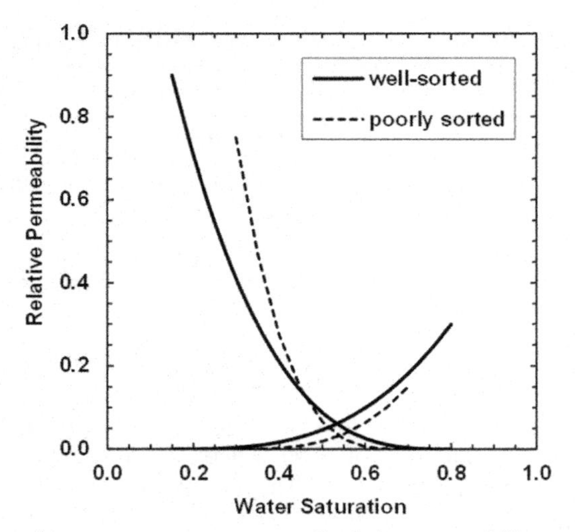

Figure 8.1.2 Effect of pore structure on water–oil relative permeability curves.

The effect of permeability in sandstones is usually related to the degree of sorting and consolidation of the formation. A well-sorted unconsolidated sandstone has relatively large pore throats with a more narrow distribution of diameters. The surface area is smaller in well-sorted rock, and therefore the irreducible water saturation is lower. The large pore throats and good connectivity also result in a low residual oil saturation and a high relative permeability to water at the residual oil saturation. As a sandstone formation becomes more consolidated, S_{wi} and S_{orw} increase and k_{rw} at S_{orw} decreases as shown in Figure 8.1.2. Typical relative permeability endpoints for consolidated water-wet sandstones are:

S_{wi}	30–40%
S_{orw}	25–40%
k_{rw} at S_{orw}	0.05–0.25

When diagenesis occurs, pore structure no longer depends on the original sorting and consolidation. In this case, the pore structure of the reservoir in question must be assessed through a microscopic examination. The important factors are the pore/pore throat diameter ratio and the connectivity of the pore structure. Recall that diagenesis is common in carbonates, and also occurs in sandstones.

Vuggy reservoirs tend to have high residual oil saturations and high trapped gas saturations, because the ratio of the pore diameter to the pore throat size is high. The small pore throats "snap-off" the oil phase as water or gas saturation increases, and the large void (vugs) hold a high residual saturation. Similarly, gas flow is snapped off at the pore throats while a high saturation of gas remains trapped in the vugs. In addition, the pore structure is often poorly connected in vuggy reservoirs resulting in poor sweep and high residual saturations.

Ratio of Capillary to Viscous Forces

At the microscopic level, the distribution of fluid in the reservoir is determined from a combination of capillary (rock–fluid interaction) and viscous (flow) forces. When capillary forces dominate, the fluid saturations depend solely on capillary pressure, as discussed in Section 2.3.2. In this case, finite irreducible water saturations and residual oil saturations are observed. As viscous forces begin to dominate, fluid is able to flow against the capillary pressure gradient, and the immobile saturations decrease. When viscous forces are completely dominant, the immobile saturations equal zero.

The balance of viscous and capillary forces is typically described using the capillary number. The capillary number is a dimensionless ratio of viscous over capillary forces and is defined as:

$$N_c = \frac{v\mu}{\sigma} \tag{8.1.2}$$

in which v is the fluid velocity, μ is the viscosity, and σ is the interfacial tension. It has been shown that capillary forces dominate when N_c is less than approximately 10^{-2}, as shown in Figure 8.1.3.

For most immiscible displacement processes in reservoir rock, N_c is approximately 10^{-6} to 10^{-5}. Hence, immobile saturations are observed, and conventional capillary pressure and relative permeability data apply. In miscible floods, the interfacial tension is very low, and the capillary number exceeds 10^{-2}. At this point, viscous forces are able to mobilize previously immobile fluid. For this reason, miscible floods are able to displace more oil than immiscible floods.

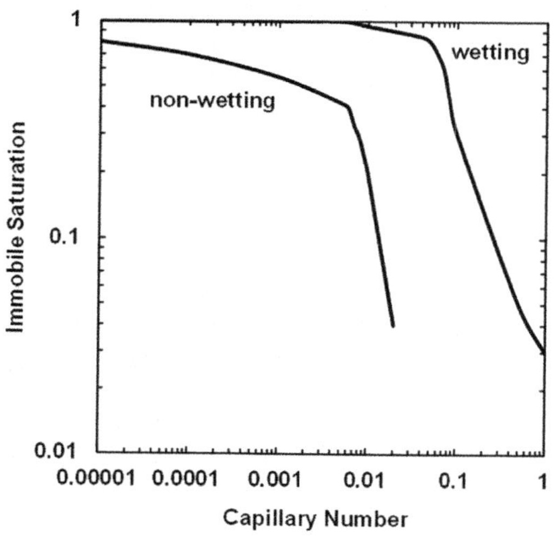

Figure 8.1.3 Effect of capillary number on immobile saturation. Immobile saturation ratio is the ratio of the immobile saturation at a given capillary number to the immobile saturation when viscous forces are zero ($N_c = 0$). Adapted from Willhite (1986).

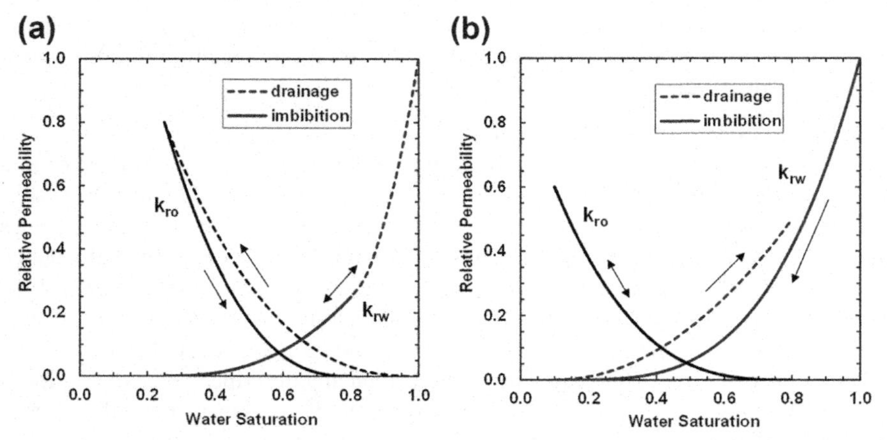

Figure 8.1.4 Schematic of drainage and imbibition hysteresis in water—oil relative permeability data: (a) water-wet rock; (b) oil-wet rock.

Saturation History

The geometry of the fluids depends on the wettability of the rock and the reservoir history. For example, when oil displaces water from a water-wet rock, the oil will form continuous flow channels through the centerlines of the pores. If the same rock is then swept by water, some of the oil is bypassed and trapped. Therefore, the relative permeability to oil (the nonwetting phase) will be reduced at a given saturation. The relative permeability of the wetting phase (water in this example) is usually unaffected. Note that the two types of displacement are defined as follows:

- Drainage: displacement by the nonwetting fluid;
- Imbibition: displacement by the wetting fluid.

Typical oil—water drainage and imbibition curves are shown in Figure 8.1.4. In water-wet rock, drainage is equivalent to oil migration into the originally water-filled reservoir. Imbibition is equivalent to displacement of oil from a waterflood or rising aquifer. In oil-wet rock, the opposite is true. Gas is a nonwetting phase, and therefore gas displacement is always a drainage process and is usually assumed to be independent of wettability. Nonetheless, if oil invades a gas zone, there is a hysteresis in the relative permeability curves as shown in Figure 8.1.5.

Uncertainty in Relative Permeability Data

Because relative permeability is history dependent, there is even more uncertainty in applying laboratory results than for absolute permeability. Not only can the core plugs be altered during retrieval, as discussed in Section 7.1, but the experimental saturation history may not match the reservoir saturation history. Some care must be taken to match the appropriate data to a particular reservoir application.

Usually relative permeabilities are measured for two phases: oil and water or oil and gas. In reservoirs below the bubble point or undergoing gas injection, there may be three phases present simultaneously. Three-phase relative permeabilities are almost always estimated from correlations.

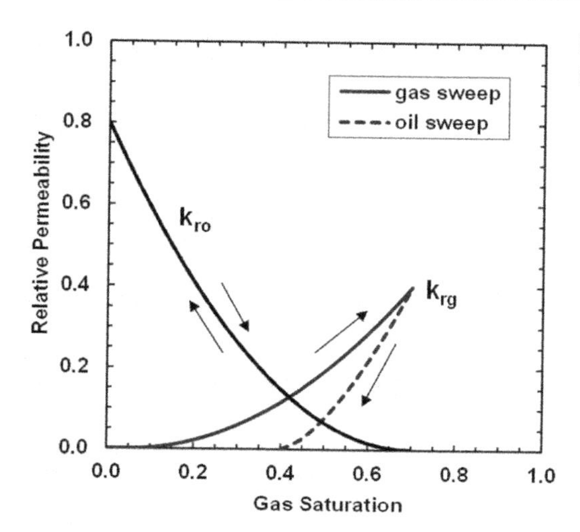

Figure 8.1.5 Schematic of hysteresis in gas-oil relative permeability data.

Given all the uncertainties in relative permeability data, there is sometimes so little confidence in the laboratory data that relative permeability curves are back-calculated from production history. However, a large number of assumptions are required to perform the back-calculation and the resulting relative permeabilities are usually no more reliable than the laboratory data. Fortunately, there are some well-established empirical guidelines that can be used to check experimental data or to tune correlations to a particular reservoir type.

8.2 Measurement of Relative Permeability

Relative permeabilities are usually measured using laboratory fluids rather than reservoir fluids; that is:

- A brine consisting of water and NaCl;
- A clean oil such as a refined mineral oil;
- An inert gas such as nitrogen.

Because relative permeability arises from rock fluid interactions and the test fluids are not identical to reservoir fluids, another degree of uncertainty is added to relative permeability data.

Relative permeabilies are measured on small core plugs using a similar apparatus, as is used to measure absolute permeabilities. The differences are that two fluids are flowed through the core plug instead of one, and separators are added after the core holder so that the fluids can be recycled through the core plug, as shown in Figure 8.2.1. The flow rate of each fluid is measured, and the pressure drop across the core plug is recorded. It is also necessary to determine the fluid saturations in the core plug. This can be accomplished through a material balance on the fluid passed through the core

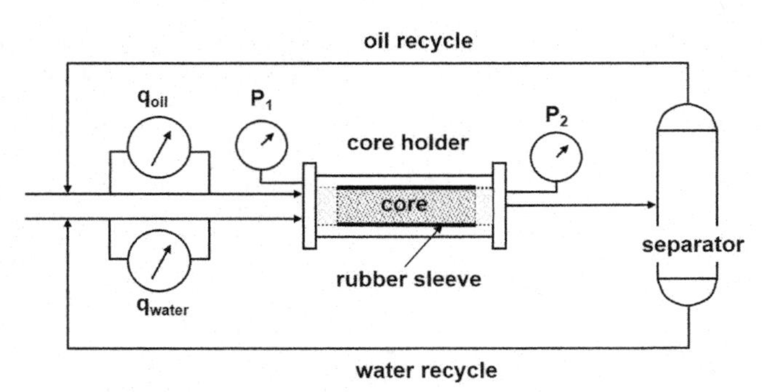

Figure 8.2.1 Simplified schematic of a relative permeability test apparatus.

plug, using tracers, or from the density of the fluids and the change in mass of the core plug after it is saturated.

There are two main procedures for determining relative permeabilities: the steady-state method and the unsteady-state method.

8.2.1 Steady-State Method

The steady-state method involves flowing each of the two fluids at a constant (but not necessarily identical) rate until a steady-state is achieved. At first, the flow rate of each phase at the core exit changes as the saturation within the core plug changes. Eventually, the saturation reaches a steady state, and each phase enters and exits the core at a constant flow rate. It is assumed that the pressure drop across the core plug is the same for both fluids. In this case, the relative permeability is determined from Darcy's law for linear flow as follows:

Liquid:

$$k_{r,liq} = \frac{1}{k}\frac{q_{liq}\mu_{liq}L}{A(P_{in} - P_{out})} \tag{8.2.1}$$

Gas below 2000 kPa:

$$k_{r,g} = \frac{2}{k}\frac{q_g\mu_g P_{atm}L}{A\left(P_{in}^2 - P_{out}^2\right)} \tag{8.2.2}$$

in which k is the absolute permeability in D, q is the volumetric flow rate in cm^3/s at atmospheric pressure, A is the cross-sectional area of the core section or plug in cm^2, μ is the viscosity in cp, P is pressure in atm, and L is the length of the core section or plug in cm. Note that the absolute permeability of the core plug must also be measured.

After a steady-state condition is achieved, the saturation in the core plug is determined. The set of measurements at the given ratio of flow rates gives the relative permeability of each phase for that saturation. To obtain relative permeabilities at another saturation, the ratio of the flow rates of the two fluids is changed, and the experiment is repeated.

To obtain drainage and imbibition curves, see Figure 8.1.4, the flow rates are adjusted in a step-wise fashion. For example in a water—oil relative permeability test, the core is initially saturated with brine and the permeability at 100% water saturation is determined. This is the absolute permeability as long as the water does not alter the core through clay swelling or fines migration. Then, water is flowed through the core at a relatively high flow rate, and oil is flowed at a relatively low flow rate. The relative permeabilities are then measured at what will be a high steady-state water saturation. The oil flow rate is incrementally increased, while the water flow rate is incrementally decreased, so that measurements are obtained at successively lower steady-state water saturations. Eventually, only oil is flowed through the core, and the relative permeability of oil is measured at the irreducible water saturation. The relative permeabilities at successively lower water saturations form the drainage curve. To obtain the imbibition curve, the process is reversed and relative permeabilities are measured at successively higher water saturations. Eventually, only water is flowed through the core plug, and the relative permeability to water at the residual oil saturation is obtained.

The advantage of the steady state method, is that the calculation of relative permeability is straightforward and requires few assumptions. The disadvantage is that considerable time is required to reach steady-state and, therefore, the test is time-consuming.

8.2.2 Unsteady-State Method

The unsteady state involves displacing one fluid within the core by another under a controlled flow regime, such as a constant flow rate of the displacement fluid or a flow rate that maintains a constant pressure. If it is assumed that the displacement is uniform across each cross-section of the core plug (perfect linear displacement), then the saturation profile and flow rates can be described by the Buckley—Leverett approach (Buckley and Leverett, 1942). If capillary pressure effects are neglected, the Buckley—Leverett equations become

$$\phi\frac{\partial S_w}{\partial t}+\frac{\partial}{\partial L}\left\{\frac{q_w+q_{nw}}{(1+M)A}\right\}=0 \tag{8.2.3}$$

in which L is length, A is cross-sectional area, q_w and q_{wn} are the flow rates of the wetting and nonwetting phases, respectively, at the exit, and M is the mobility ratio given by:

$$M=\frac{k_{nw}\mu_w}{k_w\mu_{nw}} \tag{8.2.4}$$

The relative permeability can be back-calculated by fitting the flow rate data. The advantage of the unsteady state method is that the test is quicker. However, the test is often conducted at an unfavorable mobility ratio to spread out the transient period between 100% production of the displaced fluid and 100% production of the displacement fluid. The Buckley—Leverett analysis is only strictly valid for favorable mobility ratios. Overall, the unsteady state analysis requires more assumptions, and there is correspondingly more uncertainty in the relative permeability data. Therefore, it is preferable to use the steady-state method to obtain relatively permeability data.

8.3 Analyzing Relative Permeability Data

8.3.1 Relative Permeability Reports and Plots

A typical relative permeability report includes a sample description, a fluid description, and tabulated relative permeability measurements. A sample description is a summary of the conventional core analysis results for the small core plug used in the relative permeability test. An example set of sample descriptions is given in Table 8.3.1. The fluid description includes the type of fluid (such as mineral oil, humid air, brine) and the fluid viscosity. If a synthetic fluid was prepared, a composition is also provided as shown in Table 8.3.2.

The relative permeability measurements usually include tabulated water or gas saturations, relative permeabilities to each fluid, fractional flow rates of each fluid, and the calculated relative permeability ratio. The relative permeabilities are usually reported relative to the oil permeability at the irreducible water saturation. Therefore, the measured permeability at the irreducible water saturation is also reported. Note that the absolute permeability to air and the porosity of a given sample are known from the sample description. The type of relative permeability test, the overburden pressure applied during the test, and the temperature of the test should also be reported. An example report of a water–oil relative permeability test is given in Table 8.3.3.

Usually, the relative permeability data are plotted on Cartesian and semi-log coordinates, as shown in Figures 8.3.1 and 8.3.2, respectively. The Cartesian plot is the most typical presentation and is the best plot for assessing the curvature in the relative permeability plot. The semi-log plot is preferred when examining the endpoints at low relative permeabilities. Relative permeability plots are discussed in more detail in Sections 8.4–8.6.

The relative permeability ratio is also plotted on semi-log coordinates, as shown in Figure 8.3.3. The relative permeability ratio is used for displacement calculations such as the Tarner–Tracy method for solution gas drive reservoirs. Fractional flow curves, Figure 8.3.4, are also used for displacement calculations, such as the Buckley–Leverett method for waterfloods.

8.3.2 Normalizing, Fitting, and Denormalizing Relative Permeability Data

A reservoir often exhibits a broad range of relative permeabilities, because the rock properties are not uniform, and the relative permeability varies as the permeability and porosity vary. Therefore, as we found for conventional core analysis, a large number of samples may be required to determine a representative relative permeability. The relative permeability data must also be screened for test errors and nonrepresentative samples.

Table 8.3.1 Sample description from a relative permeability test report

SAMPLE DESCRIPTION

Location: Example Well | Formation: Boundary Lake | Field: Progress

Sample Number	Depth (m)	Permeability to Air (mD)	Porosity Fraction	Grain Density (kg/m³)	Lithological Description
7	1781.46	1.14	0.155	2790	dol, gry-brn, f xln, calc, lam
8	1781.53	114	0.305	2830	dol, gry-brn, f-m xln, pp por, lam
10	1782.17	51.1	0.175	2870	dol, gry-brn, f xln, pp vugs, lam, anhy
11	1782.30	28.2	0.237	2820	dol, gry-brn, f-m xln, pp por, lam

Table 8.3.2 Synthetic brine composition for relative permeability test

Constituent	Concentration (mg/L)
Sodium chloride ($NaCl$)	63,525
Potassium chloride (KCl)	1073
Calcium chloride ($CaCl_2 \cdot 2H_2O$)	15,861
Magnesium chloride ($MgCl_2 \cdot 6H_2O$)	11,777
Sodium bicarbonate ($NaHCO_3$)	379
Sodium sulfate (Na_2SO_4)	3161

One procedure to determine representive relative permeability curves is as follows:

1. Normalize the data so that the saturations and relative permeabilities are scaled from zero to one.
2. Eliminate nonrepresentative samples.
3. Determine representative saturation and relative permeability endpoints.
4. Fit the normalized data with modified Corey equations.
5. Denormalize the fitted data to the representative endpoints.

Each step is outlined below and illustrated through an example at the end of this section.

Step 1: Normalize the Data

The saturation is normalized as follows:

Water saturation:

$$S_{wN} = \frac{S_w - S_{wi}}{1 - S_{wi} - S_{orw}} \tag{8.3.1}$$

Table 8.3.3 Relative permeability test data

WATER OIL RELATIVE PERMEABILITY

Location:	Example Well	Formation:	Boundary Lake
		Field:	Progress
Sample No:	8	Sample State:	Restored
Depth (m)	1781.53	Test Method:	Steady State
kair (mD)	114	Temperature (°C):	22.0
porosity	0.305	Pore Pressure (kPa)	0
		Overburden P. (kPa)	23526
Brine Used:	Synthetic	Flow Method	Constant Rate
Oil Used:	Mineral Oil		
Brine Viscosity (cp):	0.994	Permeability to oil	
Oil Viscosity (cp):	1.656	at S_{wi} (mD)	59.34

Water Saturation Fraction	Water-Oil Relative Permeability Ratio	Relative Permeability to Water Fraction *	Relative Permeability to Oil Fraction *	Fractional Flow of Water **	Fractional Flow of Oil **
0.163	0.00	0.000	1.000	0.0000	1.0000
0.252	0.06	0.032	0.520	0.0540	0.9463
0.290	0.12	0.048	0.392	0.1030	0.8973
0.345	0.30	0.082	0.273	0.2190	0.7810
0.390	0.61	0.118	0.195	0.3610	0.6394
0.501	2.97	0.243	0.082	0.7360	0.2642
0.555	6.06	0.315	0.052	0.8500	0.1499
0.605	12.10	0.374	0.031	0.9190	0.0811
0.851	---	0.840	0.000	1.0000	0.0000

* Relative to oil permeability at irreducible water saturation

** Calculated at room conditions, assuming a reservoir oil/water viscosity ratio of 0.936

Gas saturation:

$$S_{gN1} = \frac{S_g}{1 - S_{wi} - S_{org}} \tag{8.3.2}$$

$$S_{gN2} = \frac{S_g - S_{gc}}{1 - S_{gc} - S_{wi} - S_{org}} \tag{8.3.3}$$

The first normalized gas saturation, S_{gN1}, is used for the oil relative permeability curve in a gas-oil system. The second normalized gas saturation, S_{gN2}, is used for the gas relative permeability curve. If the critical gas saturation is zero, only S_{gN1} is required.

The relative permeabilities are normalized as follows:

$$k_{riN} = \frac{k_{ri}}{k_{rie}} \tag{8.3.4}$$

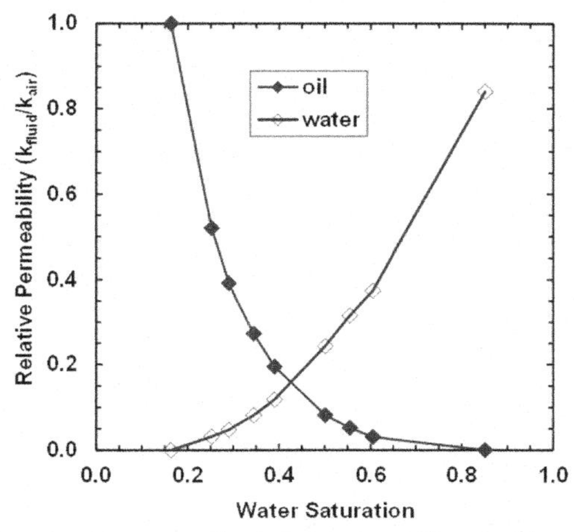

Figure 8.3.1 Cartesian plot of water–oil relative permeability data from example test.

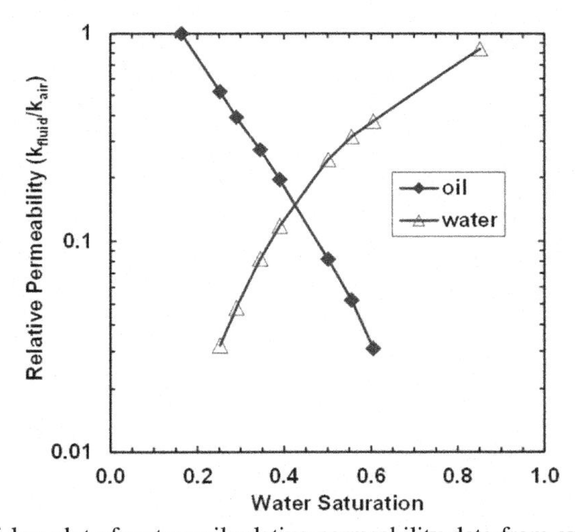

Figure 8.3.2 Semi-log plot of water–oil relative permeability data from example test.

in which i represents the fluid (oil, gas, or water), and k_{rie} is the endpoint relative permeability for that fluid.

Step 2: Eliminate Non-Representative Samples

The normalized data are plotted on a single graph so that the sets of relative permeability can be compared. In many cases, most relative permeability curves follow a similar path, and outliers can be identified easily. Outliers can also be identified from the saturation and permeability endpoints. Sometimes most of the endpoints fall within a narrow range, and a few outliers stand out from the rest. Outliers may result from a poor test or from natural variations in rock properties. In either case, if a representative analysis is sought, the outliers can be eliminated.

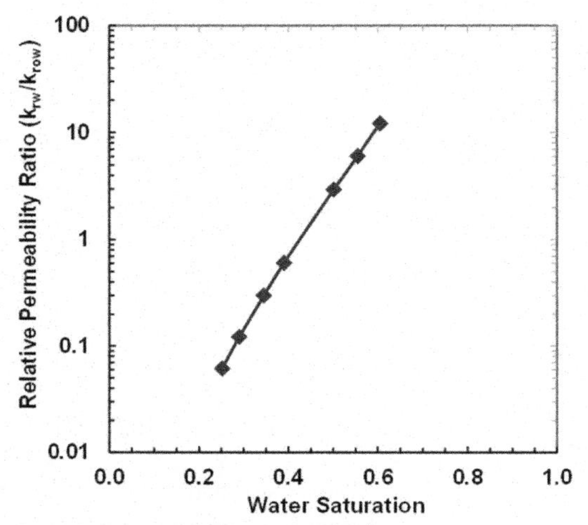

Figure 8.3.3 Semi-log plot of relative permeability ratio from example test.

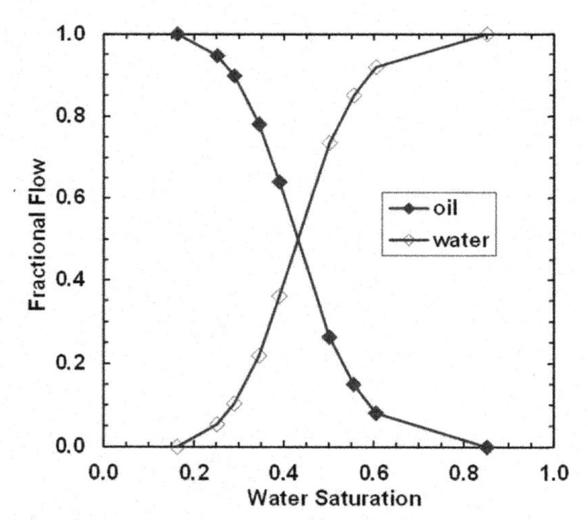

Figure 8.3.4 Cartesian plot of fractional flows from example test.

Step 3: Determine Representative Endpoints

The saturation and relative permeability endpoints can be determined from an average of the endpoints of the culled experimental data. Wherever possible, these average values should be checked against other data, such as:

- Capillary pressures (S_{wi}, S_{orw}, and S_{org});
- Well log data (S_{wi});
- Honarpour correlations (k_{rowe}, k_{roge}, k_{rwe});
- Analog reservoir data (all);
- Ultimate oil recovery ($1 - S_{wi} - S_{orw}$ or $1 - S_{wi} - S_{org}$);

- Solution gas drive production profile (S_{gc}, S_{org}, k_{rg}/k_{rog});
- Waterflood performance (S_{wi}, S_{orw}, k_{rw}/k_{row}).

Reconciling data from different sources is discussed in Chapter 10. Guidelines for selecting endpoints when data are limited are given in Section 8.6.

If relative permeability data are to be used for reservoir simulation, the constraints of the given simulation program must be considered. Most commercial simulators require that:

$$S_{orw} = S_{org}$$

$$k_{orwe} = k_{orge}$$

Note that these constraints are not satisfied if the Honarpour correlations (Section 8.4.4) are used to determine the endpoints.

Step 4: Curve Fit Normalized Data

The most common approach to curve fit relative permeability data is to adapt the Corey relative permeability correlations (see Section 8.4.1) as follows:

$$k_{rowN} = (1 - S_{wN})^a \tag{8.3.5}$$

$$k_{rwN} = (S_{wN})^b \tag{8.3.6}$$

$$k_{rogN} = \left(1 - S_{gN1}\right)^c \tag{8.3.7}$$

If the critical gas saturation is zero, the relative permeability to gas is fit as follows:

$$k_{rgN} = \left(S_{gN1}\right)^d \tag{8.3.8}$$

If the critical gas saturation is greater than zero, the gas relative permeability can be fit as follows:

$$k_{rgN} = \left(S_{gN2}\right)^d \tag{8.3.9}$$

The exponents a, b, c, and d are the fit parameters and are usually determined with a least squares regression.

Step 5: Denormalize Curve Fits

To denormalize the curve fits, the endpoint saturations and relative permeabilities that were determined in step three are required. The normalized saturations input into the curve fit equations are denormalized as follows:

$$S_w = S_{wi} + S_{wN}(1 - S_{wi} - S_{orw}) \tag{8.3.10}$$

$$S_{gN1} = \frac{S_g}{1 - S_{wi} - S_{org}} \tag{8.3.11}$$

$$S_{gN2} = \frac{S_g - S_{gc}}{1 - S_{gc} - S_{wi} - S_{org}} \tag{8.3.12}$$

The denormalized relative permeabilities are simply given by:

$$k_{ri} = k_{rie} \cdot k_{riN} \tag{8.3.13}$$

in which i denotes the phase (oil, gas, water).

Example: Water–oil relative permeability

Relative permeability data from four core plugs are summarized in Table 8.3.4. Determine representative water–oil relative permeability curves for these data.

In the following discussion, the second data point of sample 7 ($S_w = 0.323$, $k_{rw} = 0.009$, $k_{row} = 0.5891$) is chosen to illustrate the calculations.

Step 1: Normalize

Notice that the relative permeabilities are all scaled to the relative permeability to oil at S_{wi}. The oil relative permeabilities are already normalized. The water saturation and water relative permeability are normalized as follows:

$$S_{wN} = \frac{S_w - S_{wi}}{1 - S_{wi} - S_{orw}} = \frac{0.323 - 0.253}{0.819 - 0.253} = 0.124$$

$$k_{rwN} = \frac{k_{rw}}{k_{rwe}} = \frac{0.009}{0.955} = 0.0094$$

The normalized relative permeability curves are plotted in Figure 8.3.5.

Step 2: Eliminate Outliers

The relative permeability curves from three of the four core samples have similar curvature. The relative oil permeability of sample 10 clearly does not follow the same trend. Sample 10 has similar rock properties to the other samples, although the permeability/porosity and oil/air permeability ratios are somewhat higher than the other samples. However, the shape of the oil relative permeability curve for this sample does not follow the typical exponential form for relative permeability data. It is likely that there were some testing errors with this sample. Therefore, sample 10 was eliminated from consideration.

Step 3: Determine Endpoints

The average endpoints for the remaining three samples are:

$$S_{wi} = (0.253 + 0.163 + 0.190)/3 = 0.202$$

$$S_{orw} = (0.181 + 0.149 + 0.202)/3 = 0.177$$

Table 8.3.4 Example water–oil relative permeability data from a Boundary Lake dolomite (relative permeabilities scaled to k_{row} at S_{wi})

Sample 7			Sample 8			Sample 10			Sample 11		
$\phi = 0.155$, $k_{air} = 1.14$ mD, k_o at $S_{wi} = 0.177$ mD			$\phi = 0.305$, $k_{air} = 114$ mD, k_o at $S_{wi} = 59.3$ mD			$\phi = 0.175$, $k_{air} = 51.1$ mD, k_o at $S_{wi} = 29.3$ mD			$\phi = 0.236$, $k_{air} = 27.1$ mD, k_o at $S_{wi} = 9.08$ mD		
S_w	k_{rw}	k_{row}	S_w	k_{rw}	k_{row}	S_w	k_{rw}	k_{row}	S_w	k_{rw}	k_{row}
0.253	0.000	1.0000	0.163	0.000	1.0000	0.097	0.000	1.0000	0.190	0.000	1.0000
0.323	0.009	0.5891	0.252	0.032	0.5203	0.236	0.011	0.7422	0.257	0.008	0.5180
0.347	0.014	0.4801	0.290	0.048	0.3915	0.310	0.018	0.6183	0.291	0.012	0.3906
0.377	0.023	0.3780	0.345	0.082	0.2725	0.375	0.029	0.4949	0.321	0.018	0.2961
0.41	0.033	0.2784	0.390	0.118	0.1950	0.438	0.044	0.3636	0.349	0.026	0.2196
0.499	0.079	0.1308	0.501	0.243	0.0817	0.544	0.100	0.1666	0.412	0.069	0.1152
0.593	0.176	0.0580	0.555	0.315	0.0520	0.655	0.209	0.0699	0.483	0.178	0.0598
0.63	0.241	0.0402	0.605	0.374	0.0309	0.714	0.272	0.0456	0.527	0.227	0.0380
0.673	0.335	0.0278	0.851	0.840	0.0000	0.775	0.346	0.0290	0.572	0.287	0.0241
0.819	0.955	0.0000	–	–	–	0.871	0.496	0.0000	0.798	0.524	0.0000

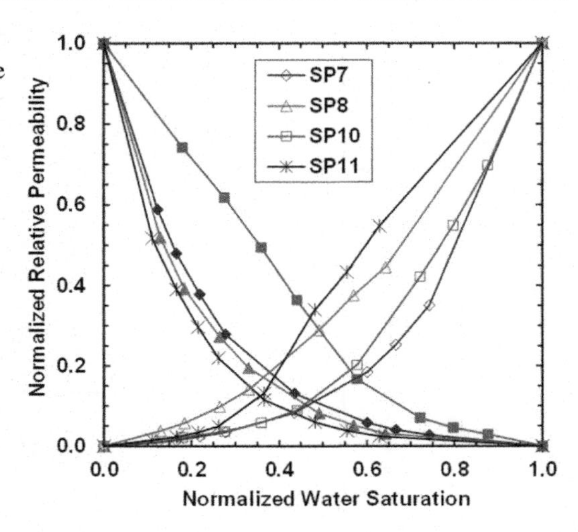

Figure 8.3.5 Normalized water–oil relative permeabilities for the example special core analysis.

$$k_{rwe} = (0.955 + 0.840 + 0.524)/3 = 0.773$$

$$k_{rowe} = 1.000$$

in which $k_{rowe} = 1.0$ by definition, because the relative permeability data are scaled to k_{row} at S_{wi}. In this example, we will not compare the endpoints with other data sources.

Step 4: Curve Fit

The modified Corey equations were used to fit the data using a least squares regression. The best-fit values of a and b are 4.4 and 2.1, respectively. These values fall within the typical range observed for Corey exponents (see Section 8.4.1). The fitted relative permeability curves are shown on Figure 8.3.6.

Step 5: Denormalize

The original oil relative permeability was already normalized and does not require denormalization. The water saturation and water relative permeability are denormalized as follows:

$$S_w = 0.202 + S_{wN}(1 - 0.202 - 0.177)$$

The denormalized relative permeabilities are simply given by:

$$k_{rw} = k_{rwe} \cdot k_{rwN} = 0.773 k_{rwN}$$

The denormalized relative permeabilities are given in Table 8.3.5 for an arbitrary set of normalized saturations and are plotted in Figure 8.3.7. The relative permeabilities are compared with the criteria for oil-wet rock (see Chapter 2) in Table 8.3.6. The rock appears to be oil-wet based on two of three criteria, with a relative permeability cross-over at a water saturation of 0.45, an S_{wi} of 20%, and a k_{rwe} of 0.77.

Recall that the relative permeabilities in Table 8.3.4 were scaled to the oil permeability at the irreducible water saturation, not to the absolute permeability. In fact, the

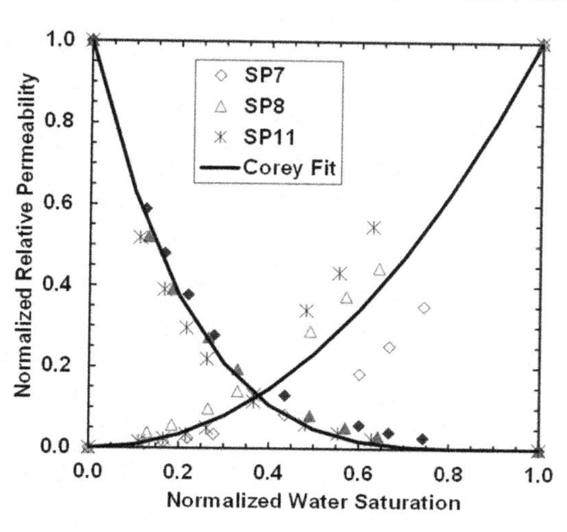

Figure 8.3.6 Curve fitting of normalized water–oil relative permeabilities for the example special core analysis.

oil permeability at S_{wi} is approximately 34% of the measured air permeability based on samples 7, 8, and 11. When characterizing this reservoir, the representative relative permeability curves in Figure 8.3.7 can be used without further modification as long as the absolute permeabilities are reduced to 34% of their nominal value. Alternatively,

Table 8.3.5 Curve-fitted normalized relative permeabilities and denormalized relative permeabilities for example special core analysis

Normalized			Denormalized				
S_w	k_{rw}	k_{row}	S_w	k_{rw}	k_{row}	k_{rw}	k_{row}
0.0	0.0000	1.0000	0.202	0.0000	1.0000	0.0000	0.3400
0.1	0.0079	0.6290	0.264	0.0061	0.6290	0.0021	0.2139
0.2	0.0341	0.3746	0.326	0.0262	0.3746	0.0089	0.1274
0.3	0.0798	0.2082	0.388	0.0614	0.2082	0.0209	0.0708
0.4	0.1460	0.1056	0.450	0.1124	0.1056	0.0382	0.0359
0.5	0.2333	0.0474	0.512	0.1796	0.0474	0.0611	0.0161
0.6	0.3421	0.0177	0.574	0.2634	0.0177	0.0896	0.0060
0.7	0.4728	0.0050	0.636	0.3641	0.0050	0.1238	0.0017
0.8	0.6259	0.0008	0.699	0.4819	0.0008	0.1639	0.0003
0.9	0.8015	0.0000	0.761	0.6172	0.0000	0.2098	0.00001
1.0	1.0000	0.0000	0.823	0.7700	0.0000	0.2618	0.0000

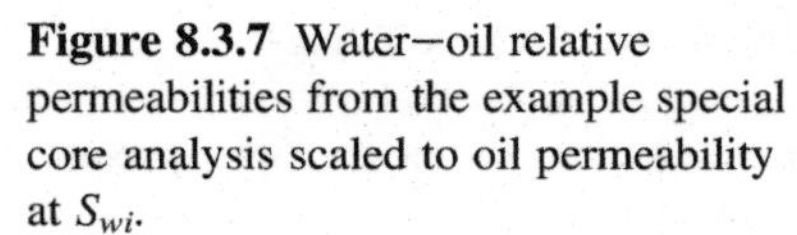

Figure 8.3.7 Water–oil relative permeabilities from the example special core analysis scaled to oil permeability at S_{wi}.

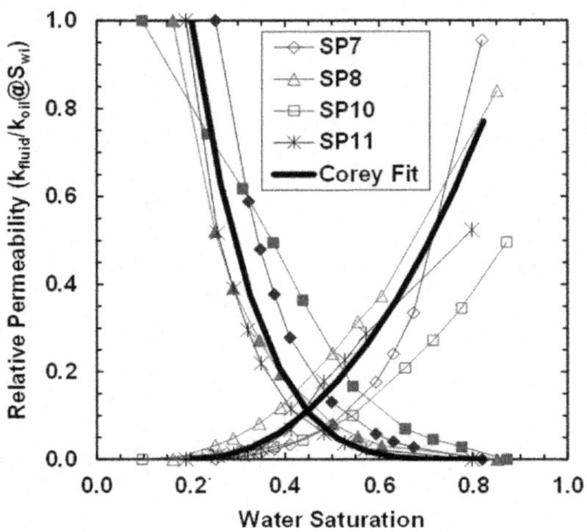

Table 8.3.6 Comparison of relative permeability with oil-wet criteria

Criterion	Example data	Wettability
$S_{wi} < 0.15$	0.20	Intermediate
$k_{row} = k_{rw}$ at $S_w < 0.5$	0.45	Oil-wet
k_{rw} at $S_{orw} > 0.5$	0.77	Oil-wet

the relative permeabilities can be denormalized so that they are scaled to the absolute permeability as shown in Figure 8.3.8.

8.4 Two-Phase Relative Permeability Correlations

Some of the most commonly used correlations for water–oil and gas-oil relative permeability are presented below. Their applicability is summarized in Table 8.4.1. Overall, the Corey correlation is recommended for curve fitting data and the Honarpour et al. correlations are recommended if relative permeability data are not available. Guidelines for determining relative permeability endpoints are discussed in Section 8.6.

8.4.1 Corey Water–Oil and Gas-Oil Relative Permeability Correlations

Corey et al. (1956) developed the following curve-fitting equations for water–oil systems and an explicit correlation for gas-oil systems.

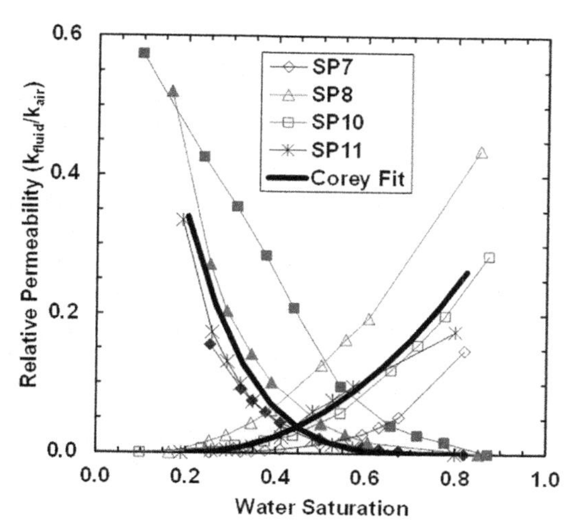

Figure 8.3.8 Water–oil relative permeabilities from the example special core analysis scaled to absolute permeability.

Water–Oil Systems

The Corey fit equations for water–oil systems are given by:

$$k_{row} = k_{rowe}(1 - S_{wD})^a \tag{8.4.1}$$

$$k_{rw} = k_{rwe}(S_{wD})^b \tag{8.4.2}$$

Table 8.4.1 Applicability of two-phase relative permeability correlations

Correlation	Fluids	Drainage/ imbibition	Wettability	Rock type
Corey	W/O	Water imbibition	All[a]	All[a]
	G/O	Gas drainage		
Pirson	W/O	Both	All	N/A
Burdine	G/O	Drainage	All	N/A
Hornapour et al.	W/O	Water imbibition	All	Sandstones/ conglomerates
	G/O	Gas drainage		Carbonates

[a]Accounted for through curve-fit parameters.

in which k_{rowe} (k_{row} at S_{wi}) and k_{rwe} (k_{rw} at S_{orw}) are the relative permeability endpoints, a and b are constants, and S_{wD} is the dimensionless water saturation defined as follows:

$$S_{wD} = \frac{S_w - S_{wi}}{1 - S_{wi} - S_{orw}} \tag{8.4.3}$$

Recall that S_{wi} is the irreducible water saturation, and S_{orw} is the residual oil saturation to water. The relative permeability endpoints and endpoint saturations must be known or estimated to use the Corey water–oil relative permeability correlations. Some guidelines for establishing endpoints are provided in Section 8.6. The exponents a and b depend on the rock type, consolidation, and degree of sorting. They are generally greater than unity and less than or equal to four. Occasionally, exponents up to six are required to fit data.

Gas-Oil Systems

The Corey correlations for gas-oil systems are given by:

$$k_{rog} = k_{roge}\left(1 - S_{gD}\right)^4 \tag{8.4.4}$$

$$k_{rg} = k_{rge}\left(S_{gD}\right)^2\left(S_{gcD}\right)^2 \tag{8.4.5}$$

in which k_{roge} (k_{rog} at $S_g = 0$) and k_{rge} (k_{rg} at $S_g = 1 - S_{wi} - S_{org}$) are the relative permeability endpoints, and the dimensionless saturations are defined as follows:

$$S_{gD} = \frac{S_g}{1 - S_{wi} - S_{org}} \tag{8.4.6}$$

$$S_{gcD} = \frac{S_g - S_{gc}}{1 - S_{gc} - S_{wi} - S_{org}} \tag{8.4.7}$$

The Corey gas-oil correlation is applicable to moderately consolidated rock with intergranular porosity. The correlations are not valid for stratified, fractured, or extensive consolidated reservoir or when solution channels are present.

8.4.2 Pirson Water–Oil Relative Permeability

Pirson in 1958 derived the following water–oil relative permeability correlations for imbibition and drainage of water-wet and oil-wet rock.

Imbibition of water-wet rock:

$$k_{ro} = (1 - S_{wD})^2 \tag{8.4.8}$$

$$k_{rw} = \left(S_w^*\right)^{0.5} S_w^4 \tag{8.4.9}$$

Drainage of water-wet rock:

$$k_{ro} = \left[1 - \left(S_w^*\right)\right]\left[1 - \left(S_w^*\right)^{0.25}\left(S_w\right)^{0.5}\right]^{0.5} \tag{8.4.10}$$

$$k_{rw} = \left(S_w^*\right)^{0.25} S_w^3 \tag{8.4.11}$$

Imbibition of oil-wet rock:

$$k_{ro} = \left(S_o^*\right)^{0.5} S_o^3 \tag{8.4.12}$$

$$k_{rw} = \left(S_{wD}\right)^2 \tag{8.4.13}$$

Drainage of oil-wet rock:

$$k_{ro} = \left(S_o^*\right)^{0.5} S_o^3 \tag{8.4.14}$$

$$k_{rw} = \left[1 - \left(S_o^*\right)\right]\left[1 - \left(S_o^*\right)^{0.25}\left(S_o^*\right)^{0.5}\right]^2 \tag{8.4.15}$$

in which the saturations are given by:

$$S_{wD} = \frac{S_w - S_{wi}}{1 - S_{wi} - S_{orw}} \tag{8.4.16}$$

$$S_w^* = \frac{S_w - S_{wi}}{1 - S_{wi}} \tag{8.4.17}$$

$$S_o^* = \frac{S_o - S_{orw}}{1 - S_{orw}} \tag{8.4.18}$$

The only required inputs for the Pirson correlations are the irreducible water saturation and the residual saturation to oil. The Pirson correlations do not account for the effects of rock type or degrees of consolidation. The correlations are shown in Figure 8.4.1.

8.4.3 Brooks-Corey Gas-Oil Relative Permeability

Brooks and Corey (1964) used the relationship between capillary pressure and relative permeability to derive the following correlations for drainage:

$$k_{rwet} = \left(S_{wet}^*\right)^{(2+3\lambda)/\lambda} \tag{8.4.19}$$

$$k_{rnw} = \left(1 - S_{wet}^*\right)^2\left[1 - \left(S_{wet}^*\right)^{(2+\lambda)/\lambda}\right] \tag{8.4.20}$$

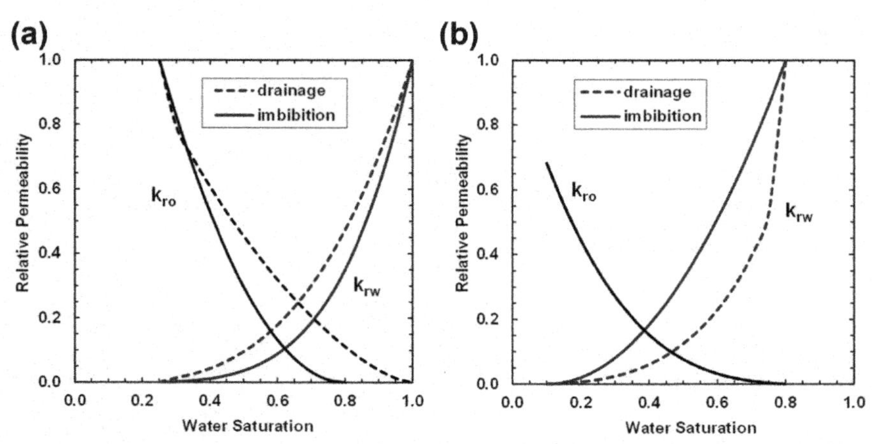

Figure 8.4.1 Pirson correlations for water–oil relative permeability in (a) water-wet rock and (b) oil-wet rock ($S_{wi} = 0.25$, $S_{orw} = 0.20$).

in which k_{rwet} and k_{rnw} are the relative permeabilities of the wetting and nonwetting phases, respectively, and λ is the pore size distribution index. The dimensionless wetting phase saturation is given by:

$$S^*_{wet} = \frac{S_{wet} - S_{wetr}}{1 - S_{wetr}} \tag{8.4.21}$$

in which S_{wet} is the saturation of the wetting phase, and S_{wetr} is the residual saturation of the wetting phase.

Required inputs for the Burdine correlation are the residual saturation of the wetting phase and the pore distribution index. Typical values of the pore distribution index for different rock types are listed in Table 8.4.2. If capillary pressure data are

Table 8.4.2 Typical values of pore distribution index for different rock types

Range of pore sizes	Rock type	λ
Very wide	–	0.5
Wide	Cemented sandstone Oolitic limestone	2
Medium	Poorly sorted unconsolidated sandstone	4
Uniform	Well-sorted unconsolidated sandstone	∞

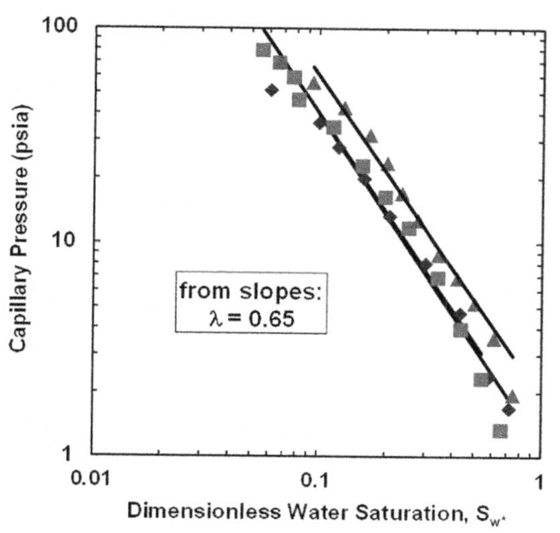

Figure 8.4.2 Determination of pore size distribution parameter (λ) from capillary pressure data (symbols are for three different core samples).

available, the pore distribution index can be determined from a log–log plot of capillary pressure versus S^*_{wet}. Brooks and Corey (1964) showed that S^*_{wet} is often related to capillary pressure as follows:

$$\log P_c = \log P_e - \frac{1}{\lambda}\log\{S^*_{wet}\} \tag{8.4.22}$$

in which P_c is the capillary pressure, and P_e is the capillary entrance pressure. The slope of a log–log plot of S_{wet} versus P_c is equal to $1/\lambda$, as shown in Figure 8.4.2.

The Brooks-Corey correlations are shown for a water-wet, water–oil system in Figure 8.4.3. For a gas-oil system, the liquid is treated as the wetting phase and gas as a nonwetting phase to obtain the correlations shown in Figure 8.4.4. Brooks-Corey correlation is useful in fields that only have capillary pressure data but no relative permeability data. The Brooks-Corey correlations are little used because capillary pressure data are required for best results and the procedure is then more complicated than with other correlations. Also, the correlation is usually only appropriate for gas-oil systems because it applies only to drainage and it neglects the irreducible water saturation for water–oil systems in oil-wet rock. However, the correlation does not account for the critical gas saturation in gas-oil systems.

8.4.4 Honarpour et al. Water–Oil and Gas-Oil Relative Permeability

Honarpour (Honarpour et al., 1982; Honarpour, 1986) developed correlations for water displacement of oil for water–oil systems and drainage for gas-oil systems. Different correlations were developed for different rock types and wettabilities. The

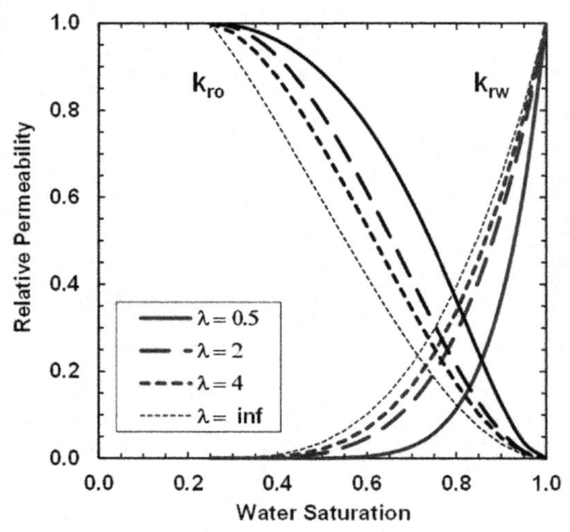

Figure 8.4.3 Brooks-Corey correlation for water–oil relative permeability in a water-wet rock ($S_{wr} = S_{wi} = 0.25$).

Honarpour et al. correlations are based on several hundred sets of relative permeability data from reservoirs throughout the world. Note the significant figures on the correlations below are reproduced as reported by the authors; however, the accuracy of the correlations (or any other relative permeability correlation) is no greater than two significant figures. The correlations are listed below according to rock type, fluid system, and wettability.

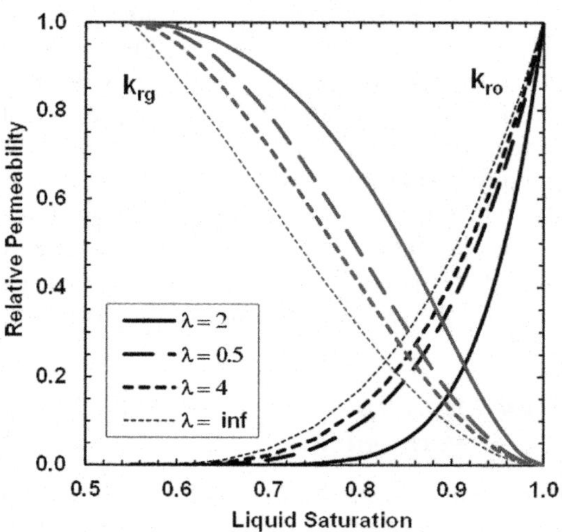

Figure 8.4.4 Brooks-Corey correlation for gas-oil relative permeability ($S_{liqr} = S_{wi} + S_{org} = 0.55$).

Sandstone and Conglomerate; Water–Oil Systems
Relative permeability to oil—any wettability:

$$k_{row} = 0.76067\left[\frac{\left(\frac{1-S_w}{1-S_{wi}}\right)-S_{orw}}{1-S_{orw}}\right]^{1.8}\left(\frac{1-S_w-S_{orw}}{1-S_{wi}-S_{orw}}\right)^{2.0} + 2.6318\phi(1-S_{orw})(1-S_w-S_{orw}) \quad (8.4.23)$$

Relative permeability to water—water-wet:

$$k_{rw} = 0.035388\left(\frac{S_w-S_{wi}}{1-S_{wi}-S_{orw}}\right) - 0.010874\left(\frac{S_w-S_{orw}}{1-S_{wi}-S_{orw}}\right)^{2.9} + 0.56556(S_w)^{3.6}(S_w-S_{wi}) \quad (8.4.24)$$

Relative permeability to water—intermediate or oil-wet:

$$k_{rw} = 1.5814\left(\frac{S_w-S_{wi}}{1-S_{wi}}\right)^{1.91} - 0.58617\left(\frac{S_w-S_{orw}}{1-S_{wi}-S_{orw}}\right)(S_w-S_{wi}) + 1.2484\phi(1-S_{wi})(S_w-S_{wi}) \quad (8.4.25)$$

in which ϕ is the fractional porosity. An example of the Honarpour et al. correlations for water–oil relative permeabilities in sandstone and conglomerates is shown in Figure 8.4.5. Note that when S_w is less than S_{orw}, Eqn (8.4.25) will give a negative result. In this case, the absolute value of $(S_w - S_{orw})$ should be used to obtain a positive k_{rw}.
Sandstone and Conglomerate; Gas-Oil Systems; Any Wettability
Relative permeability to oil:

$$k_{rog} = 0.98372\left(\frac{1-S_g-S_{wi}}{1-S_{wi}}\right)^4\left(\frac{1-S_g-S_{wi}-S_{org}}{1-S_{wi}-S_{org}}\right)^2 \quad (8.4.26)$$

Relative permeability to gas:

$$k_{rg} = 1.1072\left(\frac{S_g-S_{gc}}{1-S_{wi}}\right)^2 k_{rge} + 2.7794\left(\frac{S_{org}(S_g-S_{gc})}{1-S_{wi}}\right)k_{rge} \quad (8.4.27)$$

in which k_{rge} is the gas relative permeability endpoint at S_{org}. Note that the endpoint predicted by Eqn (8.4.27) is usually less than the input endpoint. An example of the Honarpour gas-oil relative permeability correlations for sandstones and conglomerates is shown in Figure 8.4.6.

Figure 8.4.5 Honarpour et al. correlations for water–oil relative permeabilities in sandstones and conglomerates ($S_{wi} = 0.25$, $S_{orw} = 0.30$, $\phi = 0.15$).

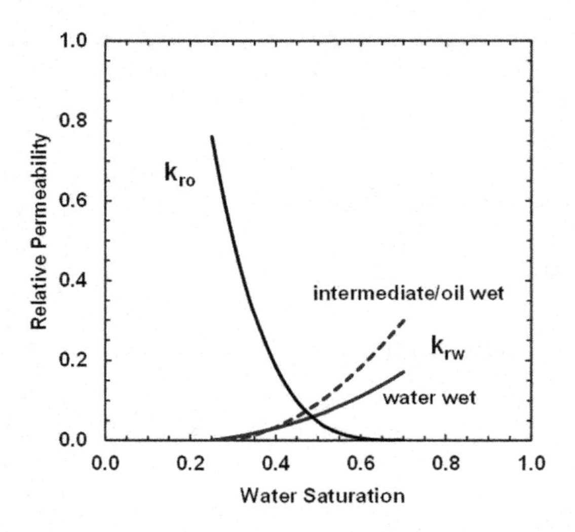

Limestones and Dolomites; Water–Oil Systems
Relative permeability to oil—any wettability:

$$k_{row} = 1.2624\left(\frac{1 - S_w - S_{orw}}{1 - S_{orw}}\right)\left(\frac{1 - S_w - S_{orw}}{1 - S_{wi} - S_{orw}}\right)^{2.0} \quad (8.4.28)$$

Relative permeability to water—water-wet:

$$k_{rw} = 0.0020525\left(\frac{S_w - S_{wi}}{\phi^{2.15}}\right) - 0.051371(S_w - S_{wi})\left(\frac{1}{k_{abs}}\right)^{0.43} \quad (8.4.29)$$

Figure 8.4.6 Honarpour et al. correlations for gas-oil relative permeabilities in sandstones and conglomerates ($S_{wi} = 0.25$, $S_{orw} = 0.30$, $S_{gc} = 0.05$, $k_{rge} = 0.40$).

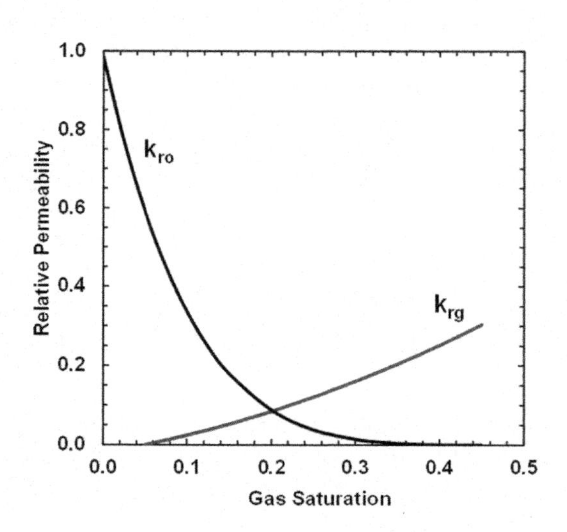

Relative permeability to water—intermediate or oil-wet:

$$k_{rw} = 0.29986\left(\frac{S_w - S_{wi}}{1 - S_{wi}}\right) - 0.32797\left(\frac{S_w - S_{orw}}{1 - S_{wi} - S_{orw}}\right)^2 (S_w - S_{wi}) + 0.413259\left(\frac{S_w - S_{wi}}{1 - S_{wi} - S_{orw}}\right)^4 \quad (8.4.30)$$

in which k_{abs} is the absolute permeability (usually to air) in mD. An example of the Honapour et al. water–oil relative permeability correlations for limestones and dolomites is shown in Figure 8.4.7.

Limestone and Dolomite; Gas–Oil Systems; Any Wettability

Relative permeability to oil:

$$k_{rog} = 0.93752\left(\frac{1 - S_g - S_{wi}}{1 - S_{wi}}\right)^4 \left(\frac{1 - S_g - S_{wi} - S_{org}}{1 - S_{wi} - S_{org}}\right)^2 \quad (8.4.31)$$

Relative permeability to gas:

$$k_{rg} = 1.8655\left(\frac{(S_g - S_{gc})S_g}{1 - S_{wi}}\right)k_{rge} + 8.0053\left(\frac{(S_{org})^2(S_g - S_{gc})}{1 - S_{wi}}\right) - 0.02589(S_g - S_{gc})\left(\frac{1 - S_{wi} - S_{org} - S_{gc}}{1 - S_{wi}}\right)^2\left(\frac{S_{org} + S_{gc}}{1 - S_{wi}}\right)^2\left(\frac{k_{abs}}{\phi}\right)^{0.5} \quad (8.4.32)$$

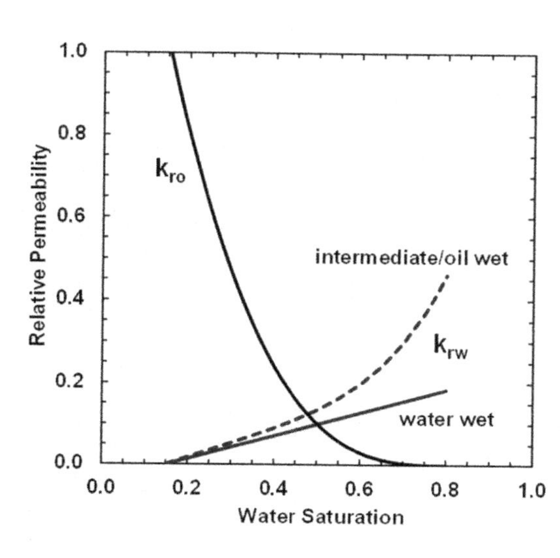

Figure 8.4.7 Honarpour et al. correlations for water–oil relative permeabilities in limestones and dolomites ($S_{wi} = 0.15$, $S_{orw} = 0.20$, $\phi = 0.10$, $k_{abs} = 100$ mD).

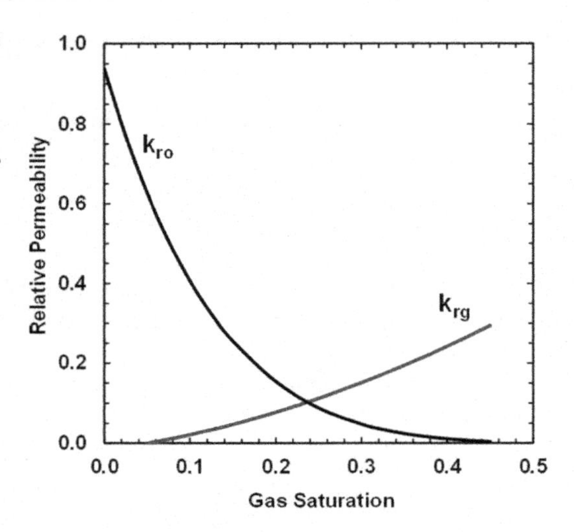

Figure 8.4.8 Honarpour et al. (1982) correlations for gas-oil relative permeabilities in limestones and dolomites ($S_{wi} = 0.15$, $S_{orw} = 0.20$, $S_{gc} = 0.05$, $\phi = 0.10$, $k_{abs} = 100$ mD, $k_{rge} = 0.40$).

Note that the k_{rge} predicted by Eqn (8.4.32) is usually less than the input k_{rge}. An example of the Honapour et al. water–oil relative permeability correlations for limestones and dolomites is shown in Figure 8.4.8.

The Honarpour et al. correlations are widely used because they are applicable to most reservoir rock types and displacement mechanisms. There are some mathematical quirks, as noted above. These correlations also require a relatively large number of inputs: the initial water saturation (S_{wi}), the residual oil saturation to water (S_{orw}) and to gas (S_{orw}), the porosity (ϕ), the absolute permeability (k_{abs}), and the gas relative permeability endpoint ($k_{rge} = k_{rg}$ at S_{org}).

8.4.5 Accuracy of Correlations

There is considerable variation in relative permeability data from one reservoir to another and sometimes within a single reservoir. Therefore, correlations should only be used when no other alternative is available. For example, in Figure 8.4.9, the Pirson and Honarpour correlations are compared with the measured relative permeabilities of the oil-wet dolomite example from Section 8.4.4. Neither correlation accurately represents this particular example.

8.5 Three-Phase Relative Permeability Correlations

Two-phase relative permeability data are sufficient for many reservoir applications, such as solution gas drive, gas cap drive, strong water drive, and waterfloods in undersaturated reservoirs. However, if a reservoir is under combination drive, three phases can flow simultaneously. In this case, three-phase relative permeabilities are required.

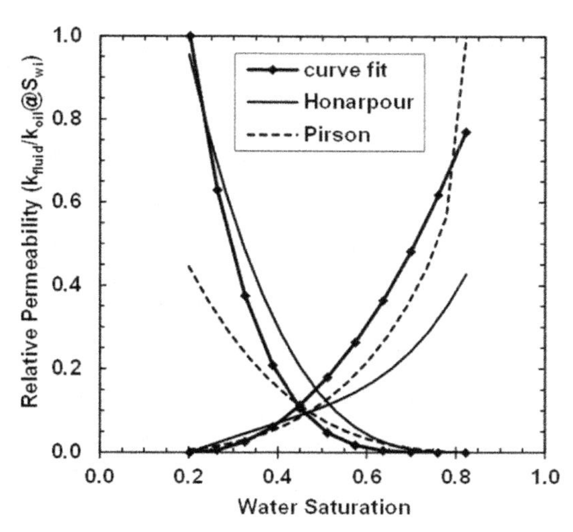

Figure 8.4.9 Comparison of Honarpour and Pirson correlations with curve-fitted relative permeability from an example core analysis.

Three-phase permeabilities are rarely measured. Instead, they are estimated using averaging techniques or correlations.

For all the methods discussed here, the water and gas relative permeabilities are determined from the respective water and gas saturations in the same way as for two-phase relative permeabilities:

$$k^*_{rw} = k_{rw} \text{ at } S_w \tag{8.5.1}$$

$$k^*_{rg} = k_{rg} \text{ at } S_g \tag{8.5.2}$$

in which k_{ri} and k^*_{ri} are the two- and three-phase permeabilities of phase i, respectively. There are several methods for determining the three-phase relative oil permeability.

Averaging Method

The three-phase relative permeability to oil, k^*_{ro}, is given by:

$$k^*_{ro} = \left(\frac{S_g}{S_g + S_w - S_{wi}}\right) k_{rog} \text{ at } S_o + \left(\frac{S_w - S_{wi}}{S_g + S_w - S_{wi}}\right) k_{row} \text{ at } S_o \tag{8.5.3}$$

in which k_{rog} and k_{row} are the two-phase oil relative permeabilities in a gas-oil and a water−oil system, respectively. The two-phase relative permeabilities are determined at the oil saturation in the three-phase system, that is, $S_o = 1 - S_g - S_w$.

Stone's Method I

To use Stone's Method I (Stone, 1973), the oil relative permeability endpoints must be identical in the water−oil and gas-oil systems; that is:

$$S_{orw} = S_{org} \tag{8.5.4}$$

$$k_{rowe} = k_{roge} \tag{8.5.5}$$

The three-phase relative permeability to oil is given by:

$$k^*_{ro} = S_{oD}\left(\frac{k_{row} \text{ at } S_w}{1 - S_{wD}}\right)\left(\frac{k_{rog} \text{ at } S_g}{1 - S_{gD}}\right) \tag{8.5.6}$$

in which:

$$S_{oD} = \frac{S_o - S_{or}}{1 - S_{wi} - S_{or}} \tag{8.5.7}$$

$$S_{wD} = \frac{S_w - S_{wi}}{1 - S_{wi} - S_{or}} \tag{8.5.8}$$

$$S_{gD} = \frac{S_g}{1 - S_{wi} - S_{or}} \tag{8.5.9}$$

Stone's Method II

To use Stone's Method II (Stone, 1973), $k_{rowe} = k_{roge} = k_{roe}$. The three-phase relative permeability to oil is given by:

$$k^*_{ro} = k_{roe}\left[\left(\frac{k_{row}}{k_{roe}} + k_{rw}\right)\left(\frac{k_{rog}}{k_{roe}} + k_{rg}\right) - k_{rw} - k_{rg}\right] \tag{8.5.10}$$

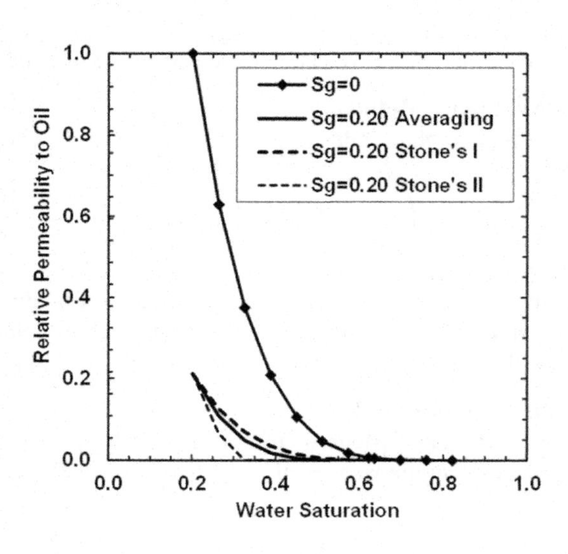

Figure 8.5.1 Comparison of three-phase oil permeabilities determined from example core analysis.

in which k_{row} and k_{rw} are evaluated at S_w, and k_{rog} and k_{rg} are evaluated at S_g. Note that Stone's Method II can give negative values. In this case, the relative permeability to oil is set to zero.

Figure 8.5.1 shows the three-phase oil relative permeabilities determined at a 20% gas saturation for the example from Section 8.4. The averaging method and Stone's I Method give similar results, while the Stone's Method II provides a conservative estimate of the three-phase oil permeability. It is recommended to use the averaging method or Stone's I Method unless there is evidence of low three-phase oil relative permeability.

8.6 Guidelines for Determining Endpoints

In many cases, relative permeability data are not available for a reservoir, or, if available, it is not certain whether the data are valid or representative. While every reservoir is unique, there is usually a range of expected endpoint values for a particular reservoir type. The best approach is to use data from analogous pools. Otherwise, some guidelines for assessing or choosing relative permeability endpoints are provided below. The ratio of oil permeability in the presence of connate water to the absolute permeability is also discussed.

8.6.1 Water–Oil Systems

Irreducible Water Saturation (S_{wi})

There are usually sufficient reservoir data to determine the irreducible water saturation including:

- Well logs;
- Relative permeability data;
- Capillary pressure data.

Open hole logs are almost always available. However, it is best to correlate the logs to core data because there is a great deal of uncertainty in calculating water saturation from well logs, particularly in shaly formations, as discussed in Chapter 9.

Irreducible water saturations can vary from approximately 5% to almost 100%. If the rock wettability is known from offset pool data, some limits can be set on the saturation. The irreducible water saturation is less than 15% in oil-wet rock and greater than 20% in water-wet rock. Productive water-wet sandstones usually have irreducible water saturations between 20% and 40%. In general, the irreducible water saturation increases as permeability decreases, as was shown in Figure 8.1.1.

In some fields, there is a hyperbolic relationship between the irreducible water saturation and porosity. In the absence of core data, we can estimate S_{wi} from analogous field data using a plot of porosity versus irreducible water saturation (Buckles plot), Figure 8.6.1, or air permeability versus irreducible water saturation, as was shown in Figure 8.1.1.

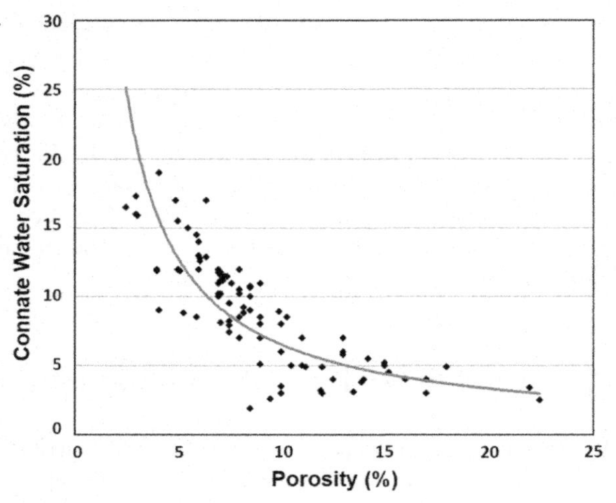

Figure 8.6.1 Buckles plot of oil-based core data from two neighboring D3 reefs; the line is $S_{wi} = 641/\phi$.
Adapted from Buckles (1965).

Note that often there are inconsistencies between the field-derived S_{wi} measurements and laboratory data for S_{wi}, such as capillary pressure test and relative permeability (Keelan, 1972; Welge and Bruce, 1947). Laboratory data are believed to be less reliable due to the changes to wettability during core handling, cleaning, and laboratory work. Also, core can dry out if left exposed before analysis. However, there may be large uncertainties associated with field values determined from induction logs because the water resistivities are uncertain or because of other uncertainties in the log interpretation (Chapter 9).

Residual Oil Saturation to Water

For some waterflood or water drive systems, we must sometimes work with analogous field data. Felsenthal (1979) collected residual oil saturation data on a number of sandstone and carbonate cores from North America and Libya, as shown in Table 8.6.1. The average residual oil saturation for both rock types was approximately 27%;

Table 8.6.1 Residual oil saturation to water from special core data (Felsenthal, 1979)

Statistic	Sandstone	Carbonate
Mean (% PV)	27.7	26.2
Median (% PV)	26.6	25.2
Standard deviation (% PV)	8.8	8.8
Number of core samples	316	108
Number of reservoirs	75	20

however, the standard deviation of 8.8% is high. The high deviation occurs because the residual saturation depends on wettability and pore structure, which vary from reservoir to reservoir. In general, sandstone reservoirs usually have residual saturations between 20% and 40%. Carbonate reservoirs can have more extreme residual saturations.

Felsenthal's data were essentially derived from reservoirs with intercrystalline porosity. His data showed a poor correlation between S_{orw} and porosity or permeability. This lack of correlation is probably because some samples have a high porosity, but may also have high pore size/pore throat ratios, giving high S_{orw}. Also, some partially fractured carbonates may have good permeability through fractures, but high trapped oil/gas saturations because of high tortuosity and poor connectivity.

It is not usually possible to accurately predict the residual saturation of a given reservoir without special core data or an extensive production history. However, it is usually possible to predict whether the residual saturation will be higher or lower than the average. Reservoirs with a high contrast between pore size and pore throat diameter will have higher residual saturations. In general, the residual saturation is expected to increase as permeability decreases. If diagenesis has not occurred, the residual saturation is expected to increase as consolidation increases. Vuggy reservoirs tend to have high residual oil saturations.

Stoian and Telford (1966) developed the following correlations for S_{orw} based on the major oil pools of Alberta:

Sandstones:

$$S_{orw} = 0.565 - 0.51S_{wi} - 0.69\phi \tag{8.6.1}$$

Carbonates:

$$S_{orw} = 0.45 - 0.41S_{wi} - 0.25\phi \tag{8.6.2}$$

The correlation coefficients (R^2) were low (0.81 for sandstones and 0.21 for carbonates). Therefore, the correlations provide a rough estimate at best. The average residual saturations of the Alberta pools used in the study were 33.5% for sandstones and 35.3% for carbonates.

Ratio of Oil Permeability at S_{wi} to Absolute Permeability

The ratio of k_o at S_{wi} to k_{abs} is typically from 0.7 to 0.9 for moderate- to high-permeability rock. The ratio decreases as the permeability decreases and the connate water saturation increases. Land (1968) developed the following correlation based on water saturation:

$$k_{ro} \text{ at } S_{wi} = 1.08 - 1.11S_{wi} - 0.73S_{wi}^2 \tag{8.6.3}$$

The correlation is intended for sandstone reservoirs, but was based on a small number of samples and should be used with caution.

Table 8.6.2 Effect of permeability on the water relative permeability endpoint for sandstones and carbonates (Felsenthal 1979)

	Sandstones			Carbonates		
Statistic	**1–10 mD**	**10–100 mD**	**100–2000 mD**	**1–10 mD**	**10–100 mD**	**100–2000 mD**
Mean	0.065	0.133	0.256	0.211	0.357	0.492
Median	0.033	0.095	0.210	0.179	0.303	0.428
# samples	30	213	143	33	45	24

Relative Permeability to Water at S_{orw}

The water relative permeability endpoint increases as permeability increases, as shown in Table 8.6.2. Note that the endpoints in Table 8.6.2 are scaled to the oil permeability at S_{wi}. The mean endpoints for sandstones range from 0.06 to 0.26, while the mean endpoints for carbonates range from 0.21 to 0.49. The difference in the two ranges can be attributed to differences in wettability, because sandstones are usually water-wet while carbonates are usually intermediate to oil-wet. Recall that k_{rw} at S_{orw} is less than 0.3 for water-wet rock and greater than 0.5 for oil-wet rock.

The following correlation was developed for US Gulf Coast reservoirs:

$$k_{rw} = k_{row \text{ at } Sorw} \frac{(\log k_{air})^2 + 6}{40} \tag{8.6.4}$$

The correlation applies to rock with predominantly intergranular porosity; that is, sandstones and some carbonates.

8.6.2 Gas-Oil Systems

Critical Gas Saturation

Critical gas saturations from core data are measured with an external gas drive. Gas is injected into the core, displacing oil. In reservoirs under solution gas drive, the gas flow is an internal drive. The critical gas saturation under internal drive is often much less than under external drive. Therefore, critical gas saturations from special core data should usually be adjusted to lower values.

Critical gas saturation depends on both interfacial and viscous forces, that is, on the interfacial tension and the rate of pressure depletion. For moderate- to high-permeability sandstones, the critical gas saturation is expected to be less than 15% and is often less than 5% for conventional crude oils.

Trapped Gas Saturation

In general, the trapped gas saturation depends on the initial gas saturation, the pore structure, and the wettability. However, there are no specific guidelines for trapped gas saturation. In general, the trapped gas saturation increases as the initial gas saturation increases, as shown in Figure 8.6.2. Experimental data such as shown schematically in Figure 8.6.2 can be fitted with a Land (1968) curve of the form:

$$\frac{1}{S_{gt}^*} - \frac{1}{S_{gi}^*} = C \tag{8.6.5}$$

in which

$$S_{gt}^* = \frac{S_{gt}}{1 - S_{org}} \tag{8.6.6}$$

$$S_{gi}^* = \frac{S_{gi}}{1 - S_{org}} \tag{8.6.7}$$

S_{gi} is the initial gas saturation, and C is a constant that must be determined from experimental data. Note that the initial gas saturation in the reservoir itself depends on the production history of the reservoir.

The trapped gas saturation increases as the ratio of pore size to pore throat size increases. Therefore, the trapped gas saturation tends to be higher in low porosity, low permeability reservoirs, and in more vuggy reservoirs. A typical relationship of trapped gas saturation to porosity for sandstones is shown in Figure 8.6.3.

As a rule of thumb, the trapped gas saturation in carbonates is approximately double that of sandstones. Carbonates are usually intermediate to oil-wet, while sandstones are usually water-wet. Carbonates often have a greater ratio of pore size to pore throat size. Both factors increase the trapped gas saturation.

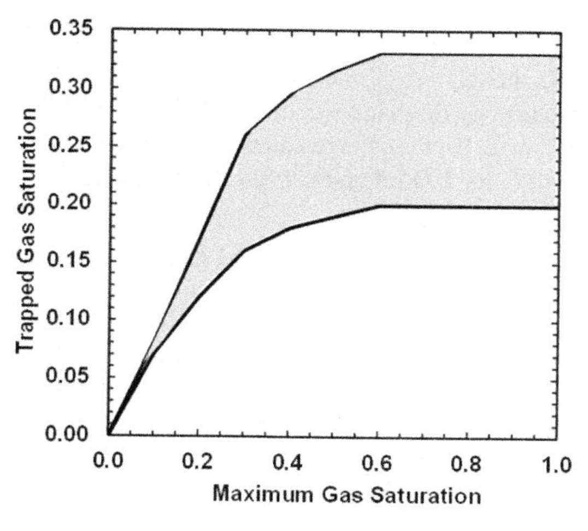

Figure 8.6.2 Effect of maximum gas saturation on trapped gas saturation in sandstones.
Adapted from Keushnig (1976).

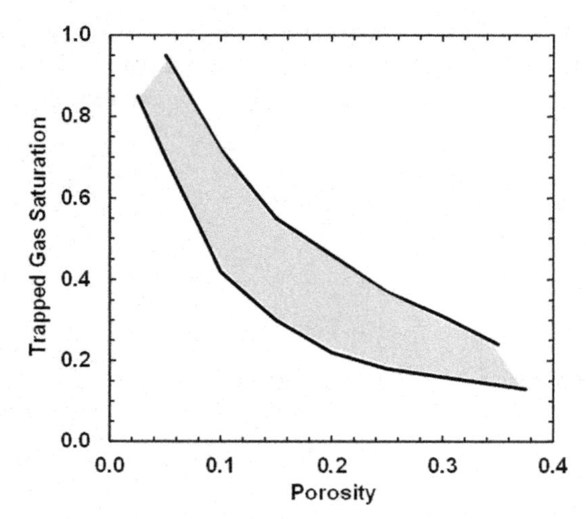

Figure 8.6.3 Effect of porosity on trapped gas saturation for sandstones. Maximum gas saturation was 1.0 in all cases (Jerauld, 1997).

Trapped gas saturation is often measured on air saturated cores with zero water saturation. The effect of the water saturation can be incorporated using an adaptation of the Land equation:

$$S_{gt} = \frac{S_{gi}}{1 + \left(\frac{1}{S^*_{gtmax}} - 1\right)\left(S_{gi}\right)^{1/\left(1 - S^*_{gtmax}\right)}} \tag{8.6.8}$$

in which S^*_{gtmax} is the value of S^*_{gt} at $S^*_{gi} = 1$.

Residual Oil Saturation to Gas and Gas Relative Permeability at S_{org}

There are few published guidelines for gas permeability endpoints. As an approximation, the residual oil saturation to gas is usually similar to the residual oil saturation to water. Because gas is a nonwetting phase, the relative permeability to gas at S_{org} is usually greater than 0.5. However, in stable gravity drainage situations, the residual saturation to gas may be markedly lower ($S_{org} < 0.10$) than the residual saturation to water, even in immiscible situations (Da Sie and Guo, 1990; Lange, 1998; Bennett and Geoghegan, 1992).

8.7 Capillary Pressure

In reservoir applications, capillary pressure is the force-per-area arising from rock-fluid interactions within a porous medium (see Chapter 2). Capillary pressure is formally defined as:

$$P_c = \frac{2\sigma \cos\theta}{r} \tag{8.7.1}$$

in which σ is the interfacial tension between the two interacting fluids, θ is the contact angle between the denser fluid and the solid surface, and r is the radius of a capillary.

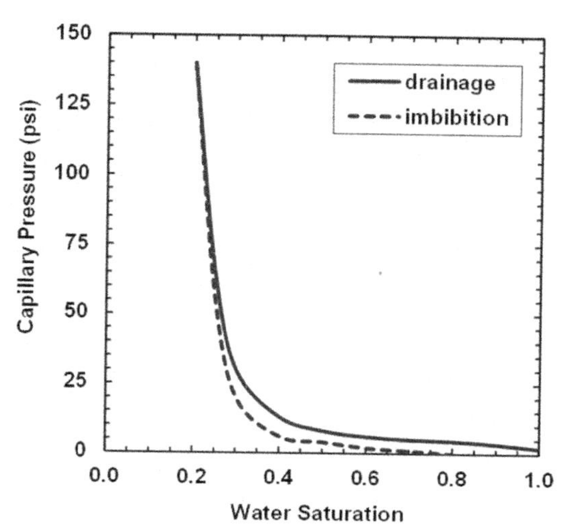

Figure 8.7.1 Water–oil drainage and imbibition capillary pressure curves.

A reservoir can be considered as a complex network of capillaries. However, the network is too complicated to model analytically. Instead, capillary pressure is measured as a function of water saturation, as shown in Figure 8.7.1. Capillary pressure follows a J-shaped curve versus water saturation. The position of the curve depends on the wettability, permeability, and porosity of the rock, as well as the interfacial tension. For a given rock of fixed wettability, the capillary curves can sometimes be grouped according to a dimensionless semi- *empirical* function given by:

$$J(S_w) = \frac{P_c}{\sigma}\left(\frac{k}{\phi}\right)^{0.5} \tag{8.7.2}$$

which $J(S_w)$ is simply called the J-function or Leverett J-function (Leverett, 1940). Because the J function is an empirical fit to data, it is recommended to plot different rock samples using different symbols and then compare the data to the derived average curve (see Section 8.9.2)

As was discussed for relative permeability, capillary pressure measurement depends on which fluid is the displacing fluid. If the wetting fluid is displaced, the process is termed drainage, and if the nonwetting fluid is displaced, the process is termed imbibition. Capillary pressures for both types of displacement are shown in Figure 8.7.1.

8.8 Measurement of Capillary Pressure

Capillary pressures are usually measured on small core plugs. All the concerns regarding representative sampling and sample preparation outlined in Chapter 7 apply to capillary pressure data as well. There are four commonly used methods for measuring capillary pressure:

- Displacement through a porous diaphragm;
- Mercury injection;
- Centrifuge;
- Dynamic.

Each method is discussed below.

Displacement Through a Porous Diaphragm

With this method, the core sample is saturated with one fluid that is to be displaced, for example, water. The saturated sample is placed on a porous plate in a sealed chamber. The plate is constructed of a material that will allow the passage of the displaced fluid, but not of the displacing fluid. Typical porous plate materials are fritted glass, porcelain, and cellophane. The core sample is surrounded with a displacement fluid, such as oil, and pressure is increased in increments. At each pressure increment, the system is allowed to equilibrate and the amount of displaced fluid is measured. The remaining saturation of the displaced fluid in the core sample is determined at each pressure increment. This method is usually used to measure drainage curves, but can be modified to measure imbibition. The advantage of this method is that reservoir fluids can be used. The disadvantages are that several hours may be required to achieve equilibrium at each pressure and that the range of pressures that can be tested is restricted by the diaphragm. At higher pressures, some of the displacing fluid may pass through the diaphragm.

Mercury Injection

With this method, a dry (air-saturated) core plug is placed in a mercury chamber, and pressure is applied incrementally to displace mercury into the core. Mercury is almost always the nonwetting phase. The amount of mercury displaced into the core is measured, and the saturation of the core is determined at each pressure increment. The advantages of this method are that each measurement can be obtained in a few minutes and a greater range of pressure can be investigated. The disadvantages are that the interfacial tension and wettability are different than the reservoir system and the sample is permanently altered and cannot be used for further testing.

Mercury—air capillary pressure data can be converted to water—air capillary pressure data as follows:

$$P_{cw} = \frac{\sigma_w \cos \theta_w}{\sigma_m \cos \theta_m} P_{cm} \tag{8.8.1}$$

in which P_{cw} and P_{cm} are the capillary pressure in air—water and air-mercury systems, respectively, θ_w and θ_m are the contact angles, and σ_w and σ_m are the surface tensions of water and mercury, respectively.

Centrifuge Method

The centrifuge method is an adaptation of the diaphragm method. The core sample is centrifuged to accelerate the displacement process. The effective pressure in the center of the core sample is determined from the centrifugal force. The saturation of the core is determined at each rotational speed—in effect, at each pressure. The advantage of the centrifuge method is that the test can be completed within a few hours. The disadvantage is the centrifugal force varies across the length of the sample, and therefore the effective pressure is not uniform.

Dynamic Method

With this method, two fluids are flowed through the core. The saturation of the core depends on the ratio of the flow rates. The pressure of each fluid is measured through a pair of wetted disks. Each disc allows hydraulic communication of only one of the two fluids. Therefore, the pressure of each fluid can be measured. The difference in pressure is the capillary pressure. The flow rates are altered so that a full set of capillary pressure versus saturation can be obtained.

The porous diaphragm is considered to be the most accurate test. Both the centrifuge method and dynamic method have been shown to give similar results. The mercury injection test is the least accurate, because the differences in wettability must be accounted for.

Laboratory tests are usually performed with representative fluids, on cleaned core samples at room temperature and relatively low pressures. If the interfacial tension of the reservoir fluids and the wettability of the formation is known at reservoir conditions, the laboratory data can be corrected to reservoir conditions as follows:

$$P_{cR} = \frac{\sigma_R \cos\theta_R}{\sigma_L \cos\theta_L} P_{cL} \tag{8.8.2}$$

in which the subscripts R and L denote the reservoir and the laboratory, respectively.

8.9 Analyzing Capillary Pressure Data

8.9.1 Capillary Pressure Reports and Plots

A typical relative permeability report includes a sample description, a fluid description, and tabulated capillary pressure measurements. A sample description is a summary of the conventional core analysis results for the small core plug used in the capillary pressure test. An example set of sample descriptions in Table 8.3.1. The fluid description may include the type of fluid (such as mercury, mineral oil, air, brine), the fluid interfacial tension, the density or density gradient of the fluid, and the contact angle between the rock and that fluid.

The capillary pressure measurements usually include tabulated saturations of the wetting fluid and capillary pressures. In some cases, the pore throat size is calculated from Eqn (8.7.1) as follows:

$$r = \frac{2\sigma\cos\theta}{P_c} \tag{8.9.1}$$

An example report of an air–mercury capillary pressure test is given in Table 8.9.1. In this example, the equivalent air–water capillary pressure is also calculated based on Eqn (8.8.1).

Capillary pressure data are usually plotted on Cartesian and semi-log coordinates, as shown in Figures 8.9.1 and 8.9.2, respectively. The Cartesian plot is the most typical presentation and is useful when comparing irreducible saturations or when determining the height of a transition zone. The semi-log plot is preferred when examining low capillary pressures at relatively high wetting phase saturations.

Table 8.9.1 Air–mercury capillary pressure test data

AIR MERCURY CAPILLARY PRESSURE

Location:	Example	Formation:	Example
		Field:	Example
Sample No:	6		
Depth (ft)	4819.4	Temperature (°C):	22.0
kair (mD)	35.0		
porosity	0.12	Mercury IFT (dynes/cm)	480
		Mercury contact angle (deg)	140
Sample State:	Restored	Water/air IFT (dynes/cm)	72
		Water/air contact angle (deg)	0

Wetting Phase Saturation (fraction)	Measured Air-Mercury Capillary Pressure (psi)	Calculated Pore Throat Radius (microns)	Calculated Air-Water Capillary Pressure (psi)
1.000	36.4	2.93	7.1
0.973	45.4	2.35	8.9
0.930	56.6	1.88	11.1
0.870	71.1	1.50	13.9
0.815	88.5	1.21	17.3
0.768	110.6	0.964	21.7
0.727	138.8	0.768	27.2
0.692	172.5	0.618	33.8
0.659	216.3	0.493	42.4
0.626	271.1	0.393	53.1
0.597	340.0	0.314	66.6
0.570	424.7	0.251	83.2
0.546	526.6	0.203	103.1
0.521	663.1	0.161	129.8
0.500	829.5	0.129	162.4
0.476	1026.1	0.104	200.9
0.449	1295.2	0.082	253.6
0.414	1618.0	0.066	316.8
0.385	2020.6	0.053	395.7
0.363	2528.0	0.042	495.0
0.346	3149.3	0.034	616.7
0.330	3952.9	0.027	774.0
0.315	4940.2	0.022	967.3
0.303	6180.7	0.017	1210.2
0.293	7694.1	0.014	1506.6
0.285	9544.7	0.011	1869.0
0.280	11910	0.009	2332.1
0.276	15433	0.007	3022.0
0.272	18735	0.006	3668.5
0.270	23429	0.005	4587.7
0.268	29436	0.004	5763.9
0.268	34898	0.003	6833.4
0.268	39848	0.003	7802.7
0.268	44895	0.002	8790.9
0.268	49867	0.002	9764.5

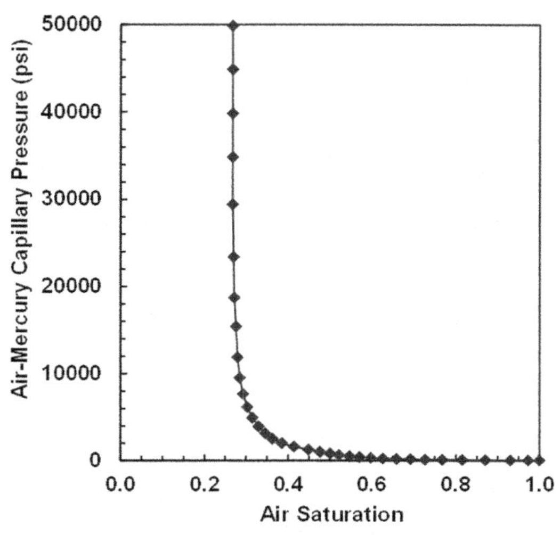

Figure 8.9.1 Cartesian plot of air—mercury capillary pressure data from example test.

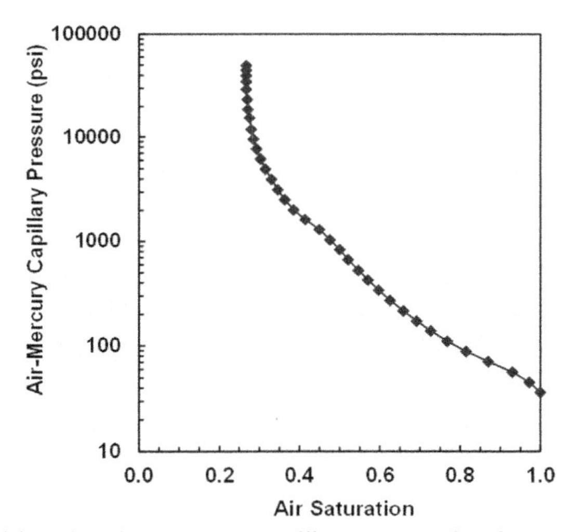

Figure 8.9.2 Semi-log plot of air—mercury capillary pressure data from example test.

The pore throat diameters are plotted on semi-log coordinates as a cumulative volume frequency distribution, in which the cumulative volume frequency is the wetting phase saturation, as shown in Figure 8.9.3. The pore size distribution is not used in conventional reservoir calculations, but does provide a physical description of the reservoir that can aid in interpretation or in the design of model reservoir studies.

8.9.2 Averaging Capillary Pressure Data

As was discussed with relative permeability, a reservoir often exhibits a broad range of rock—fluid interactions because the rock properties are not uniform. Therefore, a large

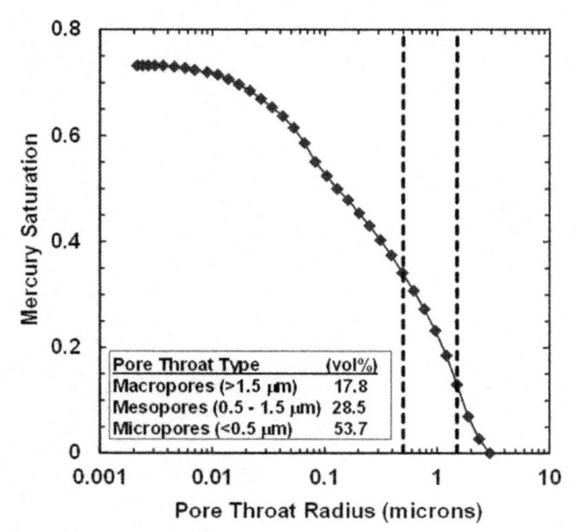

Figure 8.9.3 Pore throat radius distribution from example test.

number of samples may be required to determine a representative capillary pressure curve. The capillary pressure data must also be screened for test errors and nonrepresentative samples.

There are two approaches to averaging relative permeability data: the J-function method and the permeability plot method. Each method is outlined below and illustrated on an example at the end of the section.

J-Function Method

This method uses the J-function as a normalizing tool. The normalized data are averaged and then denormalized to reservoir conditions. One procedure to implement the J-function method is as follows:

1. Plot J-function versus normalized wetting phase saturation for each cap press curve

The J-function is calculated for each sample using Eqn (8.7.2):

$$J(S_w) = \frac{P_c}{\sigma}\left(\frac{k}{\phi}\right)^{1/2}$$

The normalized wetting phase saturation is determined as follows:

$$S_{wetN} = \frac{S_{wet} - S_{weti}}{1 - S_{weti}} \tag{8.9.2}$$

in which S_{weti} is the irreducible wetting phase saturation.

2. Eliminate outliers

The normalized data are plotted on a single semi-log graph so that the sets of J-functions can be compared. In many cases, most of the relative permeability curves follow a similar path, and outliers can easily be identified. Outliers may result from a poor test or from natural variations in rock properties. In either case, if a representative analysis is sought, the outliers can be eliminated or used to define a curve for a second rock type.

3. Calculate average J-function
There is no generalized curve-fitting equation for capillary pressure data. Therefore, the retained J-functions are simply averaged to obtain a table of averaged data.

4. Denormalize to average reservoir properties
The J-functions are converted back to capillary pressure using the average permeability and porosity of the reservoir. The wetting phase saturation is denormalized using either the average irreducible saturation of the retained data, an irreducible saturation determined from a permeability-saturation plot (see permeability plot method), or an irreducible saturation for the reservoir based on reconciled datasets such as capillary pressure data, relative permeability data, and well log data.

Permeability Plot Method
The permeability plot method assumes that there is a linear relationship between log permeability and wetting phase saturation (see Figure 8.1.1). A best-fit line is obtained for permeability versus wetting phase saturation for a set of capillary pressures (Amyx et al., 1960). A capillary pressure curve is then constructed from the best-fit solutions at the average reservoir permeability. One procedure to implement the permeability plot method is as follows:

1. Determine the wetting phase saturations of each sample for a given set of capillary pressure values
An arbitrary set of capillary pressure values is selected, from which an average capillary pressure curve will be constructed. The selected pressures should cover the full range of capillary pressures with a sufficient number of points to capture the shape of the curve.

For each sample, the wetting phase saturation is determined through interpolation for each of the arbitrary capillary pressure values. In this way, a set of wetting phase saturations is obtained at each of the chosen capillary pressures.

2. Construct a table of permeability, wetting phase saturation, and capillary pressure
Since each of the wetting phase saturations is determined for a particular sample, a table can be constructed of capillary pressure, wetting phase saturation, and sample permeability, as shown in Table 8.9.2. Each of the columns in Table 8.9.2 provides a set of permeability versus wetting phase saturation data at a fixed capillary pressure.

3. Plot permeability versus wetting phase saturation at each capillary pressure and eliminate outliers
Each of the permeability versus wetting phase saturation datasets from Table 8.9.2 are plotted on semi-log coordinates. Each dataset is expected to follow a straight line. Data points that deviate from a straight line may be outliers. To confirm the outliers, plot the capillary pressures or J-functions on semi-log coordinates, and proceed as described in the J-function method. Eliminate outliers or decide if a second curve is needed for another rock type.

Table 8.9.2 Schematic of construction of permeability versus wetting phase saturation dataset

		Wetting phase saturation at given capillary pressure			
Sample	**Permeability**	P_{c1}	P_{c2}	...	P_{cm}
1	k_1	$S_{wet,11}$	$S_{wet,12}$	...	$S_{wet,1m}$
2	k_2	$S_{wet,21}$	$S_{wet,22}$	...	$S_{wet,2m}$
...	...	...	...	...	...
n	k_n	$S_{wet,n1}$	$S_{wet,n2}$	...	$S_{wet,nm}$

4. Curve-fit the permeability versus wetting phase saturation datasets
Curve-fit permeability versus wetting phase saturation with an equation of the following form:

$$\log k = a + bS_{wet} \tag{8.9.3}$$

in which a and b are the curve-fit coefficients.

5. Construct an average capillary pressure curve at the average reservoir permeability
At each of the chosen capillary pressures, determine the wetting phase saturation at the average reservoir permeability from the curve-fit equations. Combine the results to obtain the average capillary pressure and wetting phase saturation data.

Example: Air–mercury capillary pressures
Capillary pressure data from five core plugs are summarized in Table 8.9.3 and Figure 8.9.4. Determine a representative capillary pressure curve.

In the following discussion, the fifth data point of sample 4 ($S_{air} = 0.587$, $P_c = 71.1$, $k_{air} = 25.8$ mD, $\phi = 0.179$) is chosen to illustrate the calculations.

J-Function Method
The J-functions for each capillary pressure dataset are determined as follows:

$$J(S_w) = \frac{P_c}{\sigma}\left(\frac{k}{\phi}\right)^{0.5} = \frac{71.1\ \text{psi}}{480\ \text{dyne/cm}}\left(\frac{25.8\ \text{mD}}{0.179}\right)^{0.5} = 1.778\frac{\text{psi}\cdot\text{mD}^{0.5}}{\text{dyne/cm}}$$

in which the surface tension of mercury is 480 dyne/cm.

Air is the wetting phase in an air-mercury system. The air saturations are normalized as follows:

$$S_{airN} = \frac{S_{air} - S_{airi}}{1 - S_{airi}} = \frac{0.587 - 0.126}{1 - 0.126} = 0.527$$

Three of the J-functions (samples 4, 20, and 24) follow a similar trend. Sample 6 is clearly an outlier, perhaps from a different rock type. Sample 17 has a kink in the curve, which could be a result of sample alteration or an indication of a bimodal

Table 8.9.3 Example air–mercury capillary pressure data from a Hana sandstone

Sample 4		Sample 6		Sample 17		Sample 20		Sample 24	
$k = 25.8$ mD $\phi = 0.179$		$k = 35$ mD $\phi = 0.120$		$k = 113$ mD $\phi = 0.161$		$k = 1.91$ mD $\phi = 0.153$		$k = 947$ mD $\phi = 0.223$	
S_{air}	P_c (psi)	S_{air}	P_c (psi)	S_{air}	P_c (psi)	S_{air}	P_c (psi)	S_{air}	P_c (psi)
1.000	29.30	1.000	36.4	1.000	6.1	1.000	72.8	1.000	1.51
0.947	36.40	0.973	45.4	0.975	7.5	0.991	90.4	0.999	2.47
0.780	45.50	0.930	56.6	0.940	9.5	0.892	112.4	0.984	3.03
0.659	56.60	0.870	71.1	0.892	11.9	0.706	140.1	0.976	3.90
0.587	71.10	0.815	88.5	0.835	14.9	0.607	173.6	0.963	4.92
0.535	88.50	0.768	110.6	0.777	18.5	0.534	217.3	0.929	6.03
0.491	110.60	0.727	138.8	0.723	23.3	0.476	271.0	0.728	7.48
0.451	138.60	0.692	172.5	0.716	29.1	0.429	340.4	0.549	9.51
0.412	172.30	0.659	216.3	0.704	36.4	0.389	423.1	0.463	11.8
0.374	216.00	0.626	271.1	0.687	45.4	0.356	529.1	0.404	14.9
0.340	269.70	0.597	340.0	0.668	56.8	0.328	666.6	0.359	18.5
0.310	339.20	0.570	424.7	0.648	70.8	0.306	828.4	0.322	23.3
0.284	421.90	0.546	526.6	0.622	88.7	0.286	1033	0.303	29
0.262	527.90	0.521	663.1	0.601	111.0	0.269	1297	0.284	36
0.243	665.30	0.500	829.5	0.582	138.7	0.255	1618	0.266	46
0.227	827.20	0.476	1026.1	0.562	173.1	0.241	2020	0.252	57

Continued

Table 8.9.3 Example air–mercury capillary pressure data from a Hana sandstone—cont'd

Sample 4		Sample 6		Sample 17		Sample 20		Sample 24	
k = 25.8 mD, ϕ = 0.179		k = 35 mD, ϕ = 0.120		k = 113 mD, ϕ = 0.161		k = 1.91 mD, ϕ = 0.153		k = 947 mD, ϕ = 0.223	
S_{air}	P_c (psi)	S_{air}	P_c (psi)	S_{air}	P_c (psi)	S_{air}	P_c (psi)	S_{air}	P_c (psi)
0.213	1031	0.449	1295.2	0.545	216.2	0.229	2517	0.239	71
0.202	1296	0.414	1618.0	0.527	271.1	0.215	3141	0.225	90
0.191	1617	0.385	2020.6	0.511	337.5	0.202	3943	0.212	111
0.181	2018	0.363	2528.0	0.495	423.5	0.190	4929	0.199	139
0.174	2516	0.346	3149.3	0.477	526.6	0.175	6218	0.186	173
0.165	3139	0.330	3952.9	0.462	664.9	0.165	7698	0.175	216
0.158	3942	0.315	4940.2	0.446	825.5	0.156	9568	0.165	273
0.152	4928	0.303	6180.7	0.432	1032.4	0.149	11,941	0.154	339
0.145	6217	0.293	7694.1	0.417	1293.1	0.141	14,945	0.146	423
0.140	7697	0.285	9544.7	0.405	1617.6	0.138	18,752	0.138	531
0.136	9567	0.280	11,910	0.391	2010.2	0.134	23,420	0.131	659
0.129	11,940	0.276	15,433	0.379	2529.5	0.134	29,371	0.124	823
0.128	14,944	0.272	18,735	0.368	3146	0.134	34,846	0.118	1037
0.126	18,751	0.270	23,429	0.355	3935	0.134	39,849	0.113	1285
0.126	23,419	0.268	29,436	0.346	4939	0.134	44,921	0.109	1609
0.126	29,370	0.268	34,898	0.336	6179	0.134	49,833	0.105	2013

0.126	34,845	0.268	39,848	0.328	7663	0.134	54,818	0.102	2518
0.126	39,848	0.268	44,895	0.323	9543	0.134	58,820	0.098	3147
0.126	44,920	0.268	49,867	0.317	11,982			0.095	3933
0.126	49,832	0.268	54,826	0.314	14,943			0.092	4926
0.126	54,817	0.268	58,778	0.312	18,766			0.091	6180
0.126	58,819			0.310	23,404			0.089	7663
				0.310	29,361			0.087	9542
				0.310	34,992			0.086	11,952
				0.310	39,978			0.086	14,925
				0.310	44,919			0.086	18,721
				0.310	49,750			0.086	23,386
				0.310	54,766			0.086	29,359
				0.310	58,774			0.086	34,901
								0.086	39,896
								0.086	44,868
								0.086	49,749
								0.086	54,825
								0.086	58,812

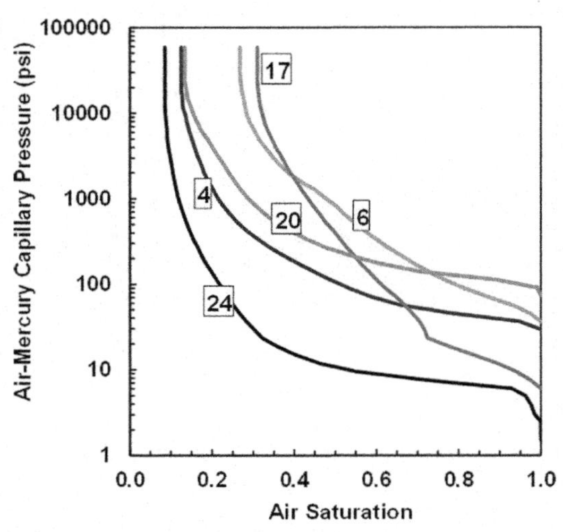

Figure 8.9.4 Capillary pressure curves for the samples in Table 8.9.3.

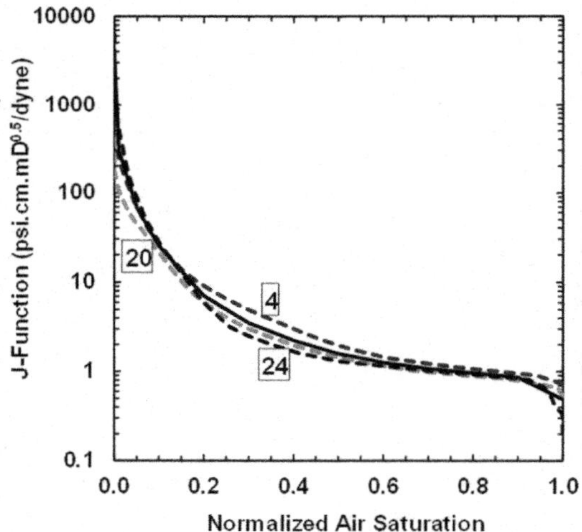

Figure 8.9.5 Average normalized J-function (solid line) compared with J-functions calculated from measured data (dotted lines) for the example set of capillary pressure data.

pore distribution. In any case, it is also an outlier. Therefore, samples 4, 20, and 24 were used to create a representative capillary pressure curve.

An average J-function was calculated for samples 4, 20, and 24, as shown in Figure 8.9.5. The average J-function was denormalized using the following average reservoir properties:

$k_{air} = 70$ mD (based on core and log data)
$\phi = 0.148$ (based on core and log data)
$S_{air} = 0.115$ (average of capillary pressure endpoints)

Note that the average value of S_{air} (0.115) is close to the value of 0.110 determined later at 70 mD from a permeability plot. The denormalized average capillary pressure curve is compared with the measured capillary pressure data in Figure 8.9.6.

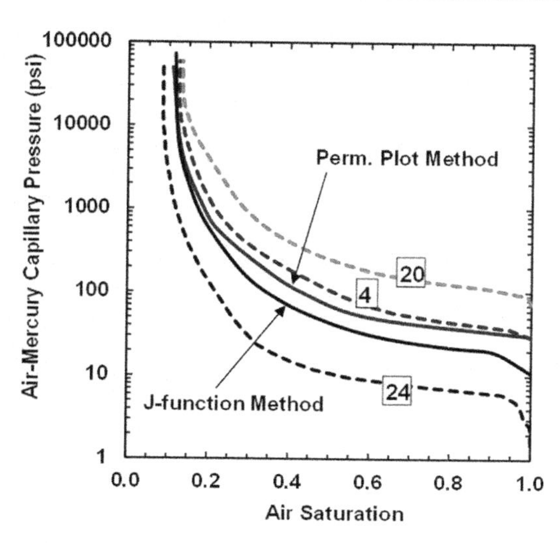

Figure 8.9.6 Average capillary pressure curves (solid line) compared with measured data (dotted lines) for the example set of capillary pressure data.

Permeability Plot Method

Because the outliers were culled using the J-function method, only the remaining samples (4, 20, and 24) are considered here. Capillary pressures of 30, 50, 100, 200, 500, 1000, 5000, and 50,000 psi were selected to construct an average curve. The interpolated air saturations at each capillary pressure are summarized in Table 8.9.4.

A plot of permeability versus air saturation was constructed from each row of data in Table 8.9.4 and is shown in Figure 8.9.7. A line was fit through each set of permeability and air saturation data. Then a representative capillary pressure curve was calculated at the average reservoir permeability of 70 mD. The best-fit parameters

Table 8.9.4 **Interpolated air saturations for selected capillary pressures**

Capillary pressure (psi)	Air saturation		
	Sample 20 1.91 mD	**Sample 4 25.8 mD**	**Sample 24 947 mD**
30	1.000	1.000	0.300
50	1.000	0.731	0.261
100	0.948	0.512	0.219
200	0.553	0.384	0.178
500	0.362	0.266	0.14
1000	0.288	0.214	0.119
5000	0.183	0.149	0.093
50,000	0.134	0.126	0.086

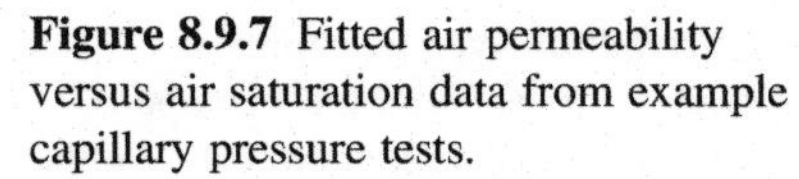

Figure 8.9.7 Fitted air permeability versus air saturation data from example capillary pressure tests.

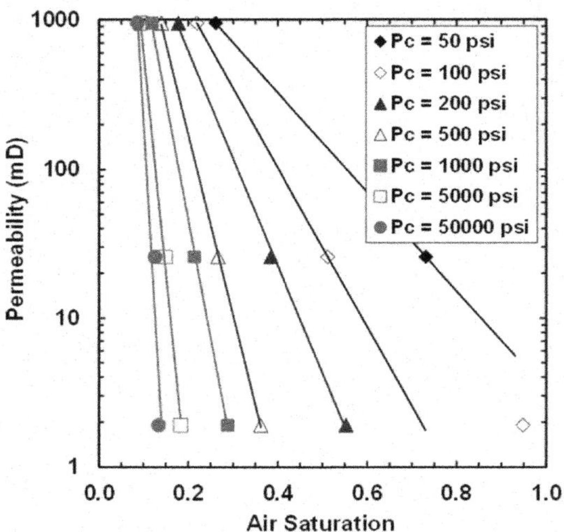

and calculation of the average curve are given in Table 8.9.5. The average capillary pressure curve is compared with the measured data in Figure 8.9.6.

The two averaging methods give similar results. In this case, to determine the most representative average capillary pressure curve, it was necessary to match the predicted water–oil transition zone to the actual reservoir transition zone (see Section 8.9.4). Note that air–mercury capillary pressure data must be transformed to the appropriate fluid system for the reservoir in question: that is, water–oil, gas-oil, or gas–water.

Table 8.9.5 Permeability versus air saturation fitting parameters ($\log k = a + b \cdot S_{air}$) and calculated average capillary pressures at 70 mD

Capillary pressure (psi)	*a*	*b*	S_{air} calculated at 70 mD
30	–	–	1.000
50	3.845	−3.329	0.601
100	4.146	−5.340	0.431
200	4.233	−7.202	0.332
500	4.668	−12.155	0.232
1000	4.863	−15.973	0.189
5000	5.771	−29.748	0.132
50,000	7.399	−50.657	0.110

8.9.3 Conversion of Capillary Pressure Data to Different Fluid System

If the interfacial tensions and contact angles for the fluid systems are known, capillary pressure can be adjusted from one fluid system to another using Eqn (8.8.1):

$$P_{cw} = \frac{\sigma_W \cos \theta_W}{\sigma_m \cos \theta_m} P_{cm}$$

Example: Convert the average air–mercury capillary pressure curve of the example problem to air–water capillary pressure. The surface tension of mercury is 480 dyne/cm, and the surface tension of water is 72 dyne/cm. The contact angle of water in an air–water system is zero. The mercury contact angle was reported to be 140°. Hence, the contact angle of the wetting phase, air, is 180 − 140 = 40°. The capillary pressure correction becomes:

$$P_{cw} = \frac{72 \cos 0}{480 \cos 40} P_{cm} = 0.196 P_{cm}$$

The calculated air–water capillary pressure curve is shown in Figure 8.9.8.

It is much more challenging to obtain a capillary pressure curve for a water–oil system at reservoir conditions because the water–oil interfacial tension and the contact angle are not known. The interfacial tension of paraffinic components is typically in the order of 40–50 mN/m (dyne/cm). The interfacial tension of aromatic components

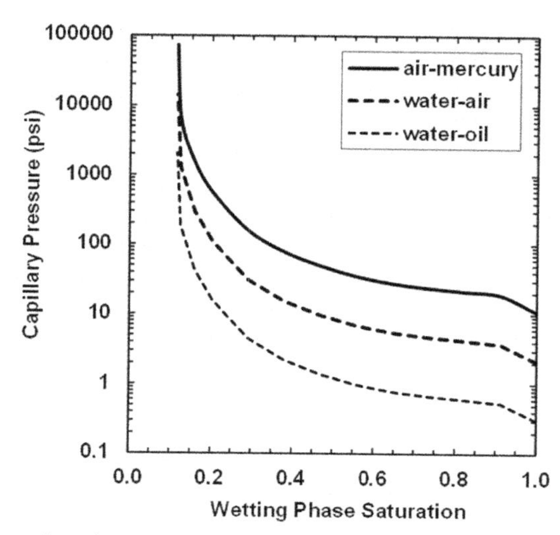

Figure 8.9.8 Water–air and water–oil capillary pressure curves obtained from air–mercury capillary pressure data. An interfacial tension of 40 dyne/cm and a contact angle of 75° was used to calculate the oil–water capillary pressure.

is in the order of 30–40 mN/m. However, interfacial tension is strongly affected by the presence of surface active components such as acids or bases and the water chemistry. If measured data are unavailable and there is no reason to suspect significant surface active components, a reasonable approximation is 40 mN/m. As was discussed in Section 8.9.1, the contact angle of a reservoir can vary between 0° and 180°. If relative permeability data are available, the approximate wettability of the rock at laboratory conditions can be estimated. However, the best approach is use the contact angle as a fitting parameter when fitting a calculated transition zone to the observed transition zone in a reservoir.

8.9.4 Calculation of a Water–Oil Transition Zone and Initial Water Saturation

In Section 2.3.2, it was shown that the height of water above the WOC in water-wet rock is given by:

SI units:

$$h = \frac{1000P_c}{\Delta\rho(g/g_c)} \tag{8.9.4}$$

in which h is in meters, P_c is in kPa, ρ is in kg/m^3, g is in m/s^2, and $g_c = 1$ kg m/N s^2.

Field units:

$$h = \frac{144P_c}{\Delta\rho(g/g_c)} \tag{8.9.5}$$

in which h is in ft, P_c is in psia, ρ is in lb$_m$/ft^3, g is in ft/s^2, and $g_c = 32$ lb$_m$ ft/lb$_f$ s^2.

Example: Determine the saturation profile of the water–oil transition zone based on the air–mercury capillary pressure data of the example problem. The oil and water densities are 53.8 and 63.3 lb$_m$/ft^3, respectively.

As was discussed in Section 8.9.3, the air–mercury capillary pressure data must first be converted to the water–oil system. The capillary pressure was tuned to the actual transition zone using a water–oil interfacial tension of 40 dyne/cm and a fluid contact angle of 75°. It was found the average curve from the J-function method provided the best match to the actual transition zone. An example calculation is given below.

If the air–mercury capillary pressure is 19.0 psia, then the corresponding water–oil capillary pressure is determined as follows:

$$P_{wo} = \frac{\sigma_{wo} \cos \theta_R}{\sigma_{H_g} \cos \theta_{H_g}} P_{cm} = \frac{40 \cos 75}{480 \cos 40} 19.0 = 0.54 \text{ psia}$$

Note that, in this case, the water–oil capillary pressure is less than 3% of the air-mercury capillary pressure.

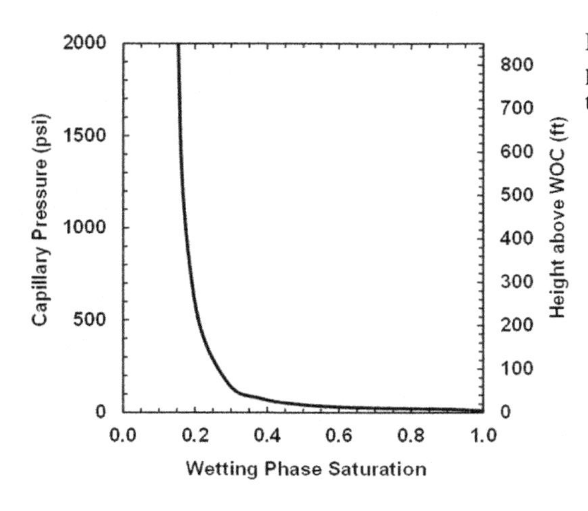

Figure 8.9.9 Water–oil capillary pressure curve and calculated height of transition zone for example problem.

The height of the transition zone for a capillary pressure of 0.54 psia is found as follows:

$$h = \frac{144P_c}{\Delta\rho(g/g_c)} = \frac{144(0.54\text{ psia})}{(63.3 - 53.8\text{ lb}_\text{m}/\text{ft}^3)\left(\frac{32\text{ ft/s}^2}{32\text{ lb}_\text{m}\cdot\text{ft/lb}_\text{f}\cdot\text{s}^2}\right)} = 8.1\text{ ft}$$

The average water–oil capillary pressure curve and the corresponding height above the water–oil contact are plotted in Figure 8.9.9. Note that the transition zone is thick and the irreducible water saturation is reached only several hundred feet above the free water level. The reservoir itself is less than 180 ft thick. Therefore, the transition zone curve is replotted to a maximum of 180 ft in Figure 8.9.10. The effective connate water saturation for this reservoir is approximately 25%, significantly greater than the irreducible saturation of 12% achieved with the high-pressure air–mercury capillary pressure data. A water–air or water–oil capillary pressure test could not reach such high pressures and would likely report an irreducible water saturation in the order of 20%. It is recommended to always check the thickness of a transition zone and the effective water saturation when capillary pressure data are available.

8.9.5 Determination of Permeability Cutoffs

The permeability at which a given reservoir rock will not flow oil can sometimes be assessed from the permeability versus water saturation plots shown in Figures 8.1.1 and 8.9.7. Consider the permeability versus irreducible water saturation data of Figure 8.9.11. The line for each reservoir rock can be extrapolated to an irreducible water saturation of unity, and the extrapolated permeability can be calculated. For example, the extrapolated permeability for lines A and B on Figure 8.9.11 are 0.4 and 6.0 mD, respectively. At $S_{wi} = 1.0$, the pore space is filled with immovable water,

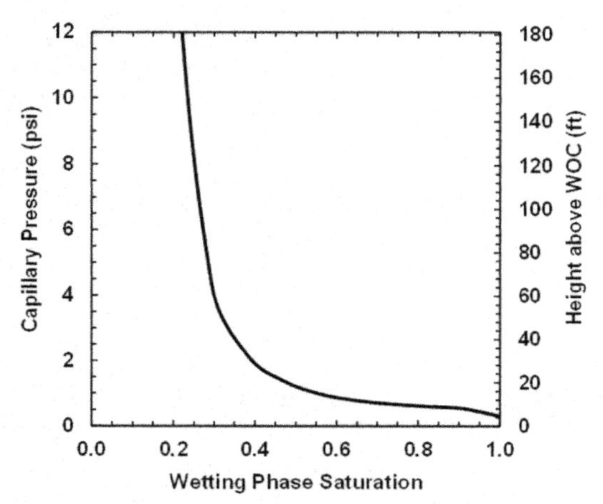

Figure 8.9.10 Water–oil capillary pressure curve and calculated height of transition zone in first 180 ft above WOC.

and there is zero permeability to any fluid. Hence, the extrapolated permeability at S_{wi} is the cutoff permeability below which fluid flow is not possible.

The permeability versus saturation curves should be generated using capillary pressures that match the expected pressure gradients in the reservoir, typically less than 500 psi. Also, note that the cutoff for oil flow will be higher than this absolute permeability cutoff, and the cutoff for economic oil rates will be still higher. Nonetheless, the extrapolated permeability versus water saturation plot can provide an order of magnitude estimate of the permeability cutoff.

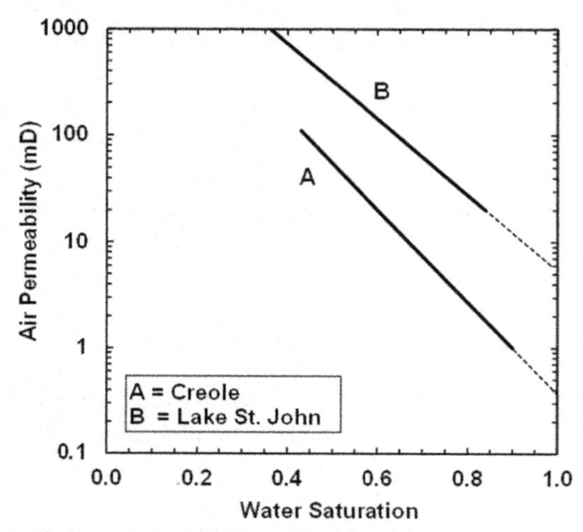

Figure 8.9.11 Extrapolation of permeability versus irreducible water saturation for two formations.
Data from Welge and Bruce (1947).

In many cases, the extrapolated permeability cutoff is low (less than 0.01 to 0.1 mD). In this case, the cutoff is not applicable, and an economic flow rate calculation based on Darcy's Law will provide a more useful cutoff. Also note, the rule-of-thumb cutoff of 1 mD is too conservative for new completion techniques such as horizontal wells and horizontal wells with induced multifractures. There is more discussion on cutoffs in Chapter 9.

Openhole Well Logs—Log Interpretation Basics

Chapter Outline

Conventional openhole well logs include measurements of hole size, gamma ray emissions, various indirect measurements of porosity (compressional wave acoustic, density, and neutron logs), spontaneous potential, and formation resistivity. From these measurements, the thickness, porosity, and water saturation of any give formation interval can be estimated. The location of gas-oil and water–oil contacts can also be determined. Openhole well log data are used for most reservoir calculations and particularly for the creation of reservoir maps, the determination of volumetric reserves, and predicting lateral continuity of reservoir bodies.

There are also so-called "non-conventional" well log measurements, including shear-wave acoustic, magnetic resonance, and acoustic and resistivity microimaging devices. These are more sophisticated measurements than the conventional logs listed above and are beyond the scope of this book.

In this section, conventional openhole logging and factors that affect the accuracy of the log data are discussed. The major types of conventional wireline logging tools are reviewed, and basic log interpretation for vertical wells is introduced. Detailed log interpretation is not covered here, and the interested reader is referred to publications from the major well servicing companies (Dresser Atlas, 1995; Schlumberger, 1987; Welex(Haliburton), 1987; Baker Hughes, 1995).

Practical Reservoir Engineering and Characterization. http://dx.doi.org/10.1016/B978-0-12-801811-8.00009-2

9.1 Openhole Logging

Openhole logs are measured in the wellbore either during drilling (logging-while-drilling) or after drilling and before the casing is set (wireline logging). For wireline logging, a logging tool string is lowered on a wireline to the bottom of the well, and data are collected and transmitted through the wireline as the logging string is pulled up the wellbore, as shown in Figure 9.1.1, or recorded in memory and played back at surface. Each measurement is correlated to depth based on the length of the wireline. The length of the wireline is measured from the Kelly Bushing (KB) on the rig floor. The elevation of the KB is recorded on the log header, so that log depths can be converted into elevations.

Figure 9.1.1 also illustrates some of the down hole factors that can affect the accuracy of the log data including hole sloughing, the presence of mudcake, and invasion of the formation by drilling mud. Each factor is discussed below.

When a well is drilled, the sides of the wellbore can be enlarged due to shale sloughing in which the formation is water sensitive or spalling due to geomechanical failure. After the cuttings are washed away, a small cavern remains. Some logs (microresistivity and porosity logs) may require that part of the logging tool maintains contact with the hole wall. These logs may be unreliable when hole sloughing occurs and this contact cannot be maintained. A similar problem can arise when the wellbore experinces brittle fracture due to stress in the formation. If the stress in the earth is stronger in one direction than another, the wellbore shape may

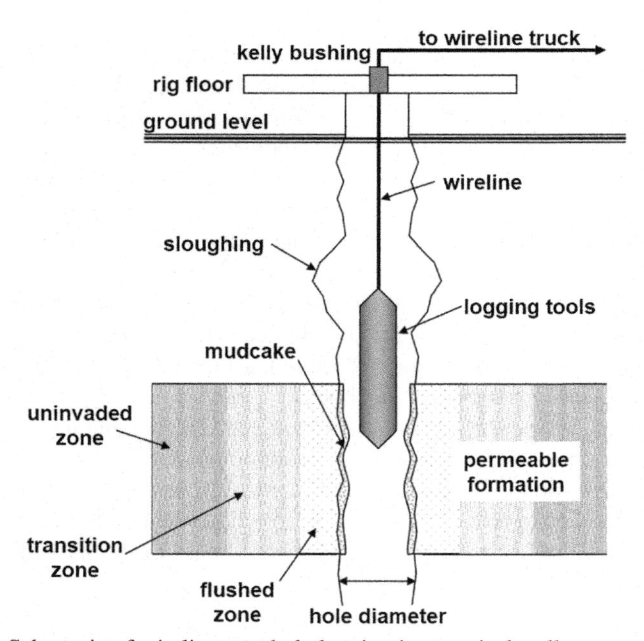

Figure 9.1.1 Schematic of wireline openhole logging in a vertical well.

become more elliptical. Most logging tools will maintain contact with the formation in an elliptical hole, but occasionally the contact is lost, and the data become unreliable.

Most wells are drilled overbalanced, that is, with a mud with sufficient density that the bottom hole pressure exceeds the formation pressure When a permeable formation is drilled, the fluid in the drilling mud flows into or invades the formation. However, the solids in the drilling mud are too large to penetrate the formation. The formation then acts as a filter passing through the fluid (filtrate) and stopping the solids (the filter cake). In some cases, a layer of solids remains on the surface of the wellbore. This layer is termed "mudcake." The presence of a mudcake affects openhole log readings. Many logging tools are designed to minimize the effects of mudcake when it is present, but there still may be some effects.

When mud invasion occurs, the formation fluid is displaced by the drilling fluid. Near the wellbore, the formation fluid may be displaced. This region is termed the "flushed" zone. Next there is a region of partial displacement, the "transition" zone. Eventually, the original reservoir fluid is encountered in the "uninvaded" zone. The presence of drilling mud changes the resistivity of the formation and therefore affects resistivity logs. This effect must be accounted for to obtain accurate water saturations.

Other factors that can affect the accuracy of log data are mineralogy and formation thickness. The effect of mineralogy will be discussed for each log later on. Formation thickness is significant, because logging tools measure average properties over a certain length, typically 0.6–1 m for post-1990s logging equipment, but longer for older tools. If the bed is 1 m or less, then the measurement will be influenced by the beds above and below the formation and may require adjustment. Logging companies provide "correction charts" to help compensate for shoulder bed effects, but these are only approximate. Today, forward modeling and inversion are used to combine logs with different vertical resolution and to correct for invasion and shoulder bed effects.

9.2 Types of Openhole Log

The openhole logs considered in this section and their applications are listed in Table 9.2.1. The basis for each measurement, the presentation of the log data, and potential sources of error are discussed. More detailed log interpretation is presented in Section 9.3.

9.2.1 *Caliper Log*

A caliper log records the hole diameter measured using spring loaded caliper arms. Figure 9.2.1 shows a variety of two, three, four, and six arm calipers. In general, the more arms, the more accurate the measurement of the shape and cross-sectional area of the wellbore. Caliper logs are usually run simultaneously with an acoustic or a neutron-density log. Typically, an average hole diameter is calculated and recorded.

Table 9.2.1 Types and application of openhole logs

Log	Application
Caliper	Hole size and condition
Gamma ray	Lithology
Spontaneous potential	Lithology, formation water resistivity
Acoustic	Porosity, lithology/mineralogy
Density	Porosity, mineralogy, gas-oil and gas-water contacts
Neutron	Porosity, mineralogy, gas-oil and gas-water contacts
Resistivity	Water saturation, water-hydrocarbon contact

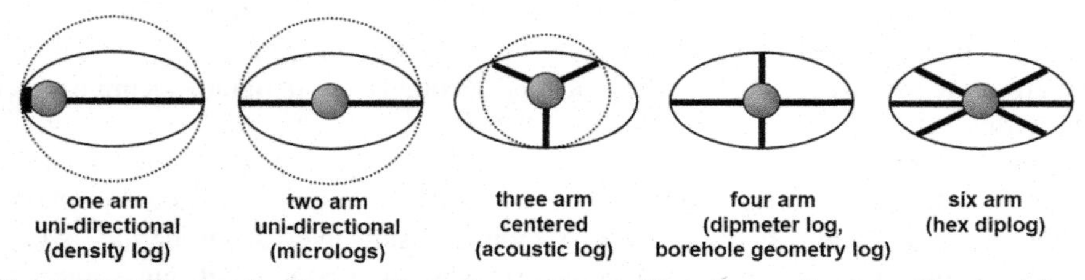

Figure 9.2.1 Types of caliper configurations and typical logging applications, illustrated for an elliptical wellbore. The dotted lines indicate the measured hole size and shape.

An example caliper log is shown in Figure 9.2.2. The bit size is usually shown on the log as well. A comparison of the measured hole diameter with the bit size is used to identify enlargement, hole roughness, and mudcake (under gauge), as shown in Figure 9.2.2. The caliper log is also used to determine the volume of the wellbore for cementing calculations.

9.2.2 Gamma Ray Log

The Gamma Ray (GR) logging tool measures the natural radioactivity from potassium, thorium, and uranium isotopes in the earth surrounding the wellbore. The GR detector is a scintillation counter that is 10–30 cm long; the resolution of a GR log is thus about 30–50 cm. The counter calibration can be traced back to one of the American Petroleum Institute (API) test facilities in Houston and the radioactivity is expressed in API units, which range from almost zero in anyhydrites to more than 200 in some shales and sylvite.

The measured GR response is a function of the natural radioactivity of the formation, the density of the formation, the type of mud, the hole size, and the position of the tool in the wellbore. In modern logs, the effects of the mud type, hole size, and tool

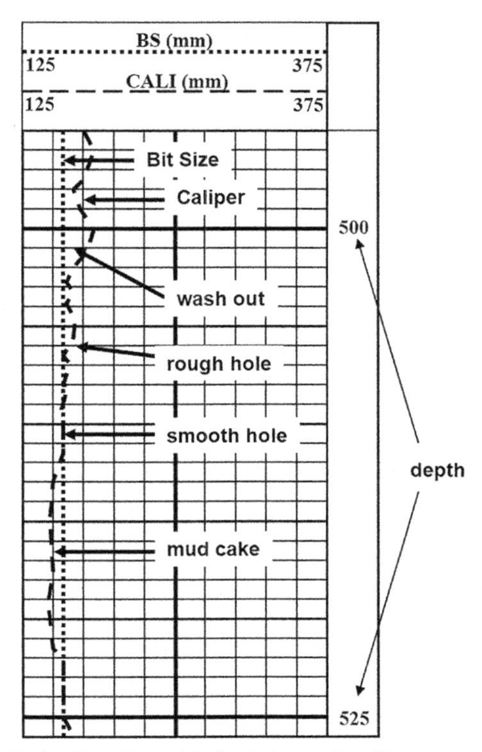

Figure 9.2.2 Hypothetical caliper log with depth intervals of 1 m.

location are generally small, except when there is potassium (increased GR readings) or barite (attenuated GR readings) in the mud. If the log is not compensated, adjustments can be made using the appropriate borehole correction charts produced by logging companies. The GR reading is less reliable in washed out zones, because the hole dimensions and tool location are not well defined.

After adjustment, the GR reading responds primarily to the natural radioactivity of the surrounding formation. Heavy radioactive elements tend to concentrate in shales and clays. Also, K^{40} is an element in some clay minerals. Therefore, the radioactivity of shales and shaly sands are higher than clean sands and carbonates. Typical GR readings for different rock types are listed in Table 9.2.2. A hypothetical GR log is shown in Figure 9.2.3.

Application: The GR log is used to help identify lithology. In practice, the log best serves to identify shale zones. The GR log is also used to determine the shale content of shaly sand formations. To estimate the shale volume, a version of the lever arm rule is used based on the following ratio:

$$X = \frac{GR_{form} - GR_{clean}}{GR_{sh} - GR_{clean}} \tag{9.2.1}$$

Table 9.2.2 Typical Gamma Ray log response in different rock types

Rock type	GR reading (API units)
Shale	80–140
Sandstone	15–30
Limestone	10–20
Dolomite	8–15
Anhydrite	~15
Salt	5–10
Gypsum	5–10
Coal and lignite	5–10

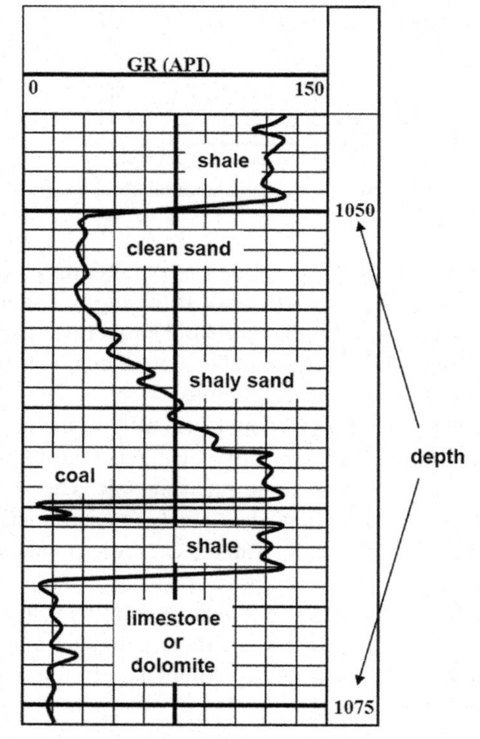

Figure 9.2.3 Hypothetical GR log with depth intervals of 1 m.

in which GR_{form}, GR_{clean}, and GR_{sh} are the measured GR responses for the formation of interest, a clean sand, and a shale zone, respectively. The clean sand and shale zone readings should be obtained from intervals as close and of similar age to the formation of interest as possible. Eqn (9.2.1) inherently assumes that GR_{sh}, taken from a nearby shale, represents the same type of shale/clay in the formation. This is not always true. For example, the clays in the formation may have developed in place after the formation was deposited, and so have no relation to the clays present in the shale. This issue may need to be discussed with the field geologist.

Some empirical equations for the shale content are given by:

$$V_{sh} = X \tag{9.2.2a}$$

$$V_{sh} = 0.33[2(2^{2X} - 1.0] \quad \text{(Larianov, 1969; older rocks)} \tag{9.2.2b}$$

$$V_{sh} = 0.33[2(2^{2X} - 1.0] \quad \text{(Larianov, 1969; Tertiary rocks)} \tag{9.2.2c}$$

$$V_{sh} = \frac{X}{A - BX} \quad \text{(Stieber, 1970)} \tag{9.2.2d}$$

$$V_{sh} = 1.7 - \left[3.38 - (X + 0.7)^2\right]^{0.5} \quad \text{(Clavier et al., 1977)} \tag{9.2.2e}$$

in which V_{sh} is the volume fraction of the shale in the formation, A and B are constants (original values are 3.0 and 2.0, respectively; variations are 2.0 and 1.0 or 4.0 and 3.0).

The GR log may also be useful for permeability prediction in sandstones. Permeability is affected by grain size, with smaller grains reducing permeability. With finer grain size, clays and heavy minerals may also be present. These minerals tend to be radioactive, so that the GR log correlates with permeability (Figure 9.2.4). The GR-permeability correlation may be weak for several reasons, including changes in the clay mineral composition, changes in the amounts of nonradioactive cements, and changes in sorting. Also, permeability variations may occur at the cm-scale, while the GR log is responding to dm-scale changes in radioactivity.

The GR log has a relatively high resolution and can also be used in cased holes. It is therefore useful for locating specific intervals in the wellbore for workover operations. It is also used to align suites of logs and to match cored intervals to openhole logs.

9.2.3 Spontaneous Potential Log

The spontaneous potential (SP) is the electrical potential between an electrode placed at the borehole wall and a fixed electrode at surface. The SP log records the change in the potential as the mobile electrode is moved up or down the wellbore. The electrical potential is the sum of the potentials arising from contacts between different formations (membrane potential), contact between different fluids (fluid junction potential), and the motion of fluid through the formation (electrokinetic potential).

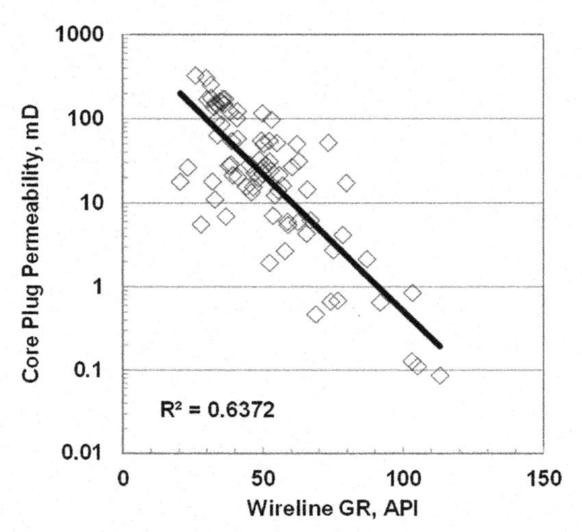

Figure 9.2.4 GR correlation with permeability for a fluvial sandstone. The porosity−permeability correlation for this formation is shown in Figure 7.3.6, (right).

Membrane potential: The SP response is primarily determined by shales. Shales consist of layers of clays that are permeable to Na+, but impermeable to Cl−. The positive ions move from the more concentrated fluid (usually a more saline formation water) to the less concentrated fluid (usually a less saline mud). The movement of the positive ions creates a positive current across the shale, as shown in Figure 9.2.5a. The potential created across the shale is termed the membrane potential, E_m.

Fluid junction potential: A second potential is created between the mud filtrate in the invaded zone that is in direct contact with the formation water. The Cl− ions are more mobile than Na+ ions, and therefore a negative current is generated from the more concentrated solution to the less concentrated solution, as shown in Figure 9.2.5a. The potential associated with this current is termed the fluid junction potential, E_{fj}. It is approximately 20% of the magnitude of the membrane potential and acts in the same direction.

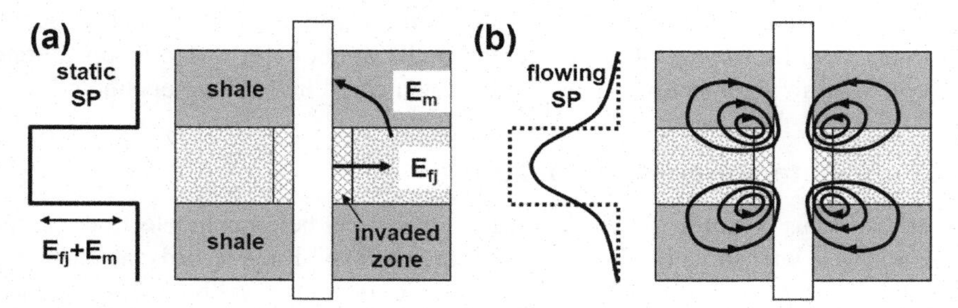

Figure 9.2.5 Static potential (a) and measured flowing potential (b) for a fresh mud and a saline formation. The arrows indicate shape and direction of potential.
Adapted from Doll (1949).

Electrokinetic potential: The flow of an electrolyte through a permeable medium can also generate a current and a potential termed the electrokinetic potential or streaming potential. Electrokinetic potentials can occur in the mudcake and in the formation. The net potential is usually negligible, but can be significant if there is a large pressure drop and flow through the mudcake or in very low permeability formations (less than a few millidarcies).

The static spontaneous potential (SSP) is the sum of all the potentials at the wellbore if no current is flowing. The SSP is the potential required to determine formation water resistivity. However, the SP is measured while current is flowing. The creation of electrical flow paths in the formation alters the SP, and therefore, the flowing potential differs from the static potential, as shown in Figure 9.2.5b. In general, the SP is only measured in thick formations and corrections are required for thin beds.

A hypothetical SP log is shown in Figure 9.2.6. Note that the operator sets the scale and the sensitivity so that all the signal is within the track. The potential in mV is plotted versus depth. A baseline potential is established in the shale zones and the shale baseline and is the reference for interpreting SP changes. When a permeable zone is

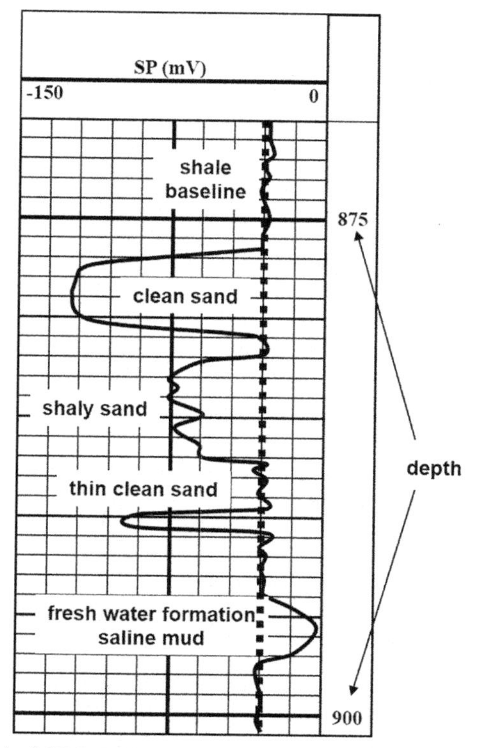

Figure 9.2.6 Hypothetical SP log with depth intervals of 1 m. All responses are for fresh mud and saline formation water unless otherwise indicated. The scale values of −150 and 0 mV indicate the SP curve is displayed with a sensitivity of 15 mV per division.

encountered, the SP deviates from the baseline to more negative potentials if the mud filtrate is less saline ("fresher") than the formation water. If the mud filtrate is more saline, then the SP deviates from the baseline to more positive potentials. If the mud filtrate salinity is the same as the formation water, then there is no SP deflection.

The shape and amplitude of the SP response depends on:

- The thickness (h) and true (uninvaded) resistivity (R_t) of the formation;
- The diameter (d_I) and resistivity (R_{xo}) in the invaded zone;
- The resistivity of the adjacent shale (R_{sh});
- The resistivity of the mud (R_m) and the diameter of the wellbore (d_w)

In general, the best results are obtained when the formation is more than 2 ft thick, the formation resistivity is the same as the mud resistivity, and there is little invasion.

Some other factors to consider when evaluating SP logs are baseline shift, high resistivity formations, and invasion effects.

Baseline shift: A baseline shift occurs when formations containing water of different salinity are separated by a thin shale zone that does not act as a perfect cationic membrane. The baseline shifts across the imperfect membrane leading to potential misidentification of the lower formation, as shown in Figure 9.2.7(a).

High-resistivity formations: When the formations have high resistivity, the SP currents are largely confined to the borehole, and there is a linear change in SP response across the formations as long as the borehole has constant diameter. As Figure 9.2.7(b) shows, it can be difficult to identify shale boundaries from the SP curve in high-resistivity formations.

Invasion effects: Figure 9.2.7(c) illustrates some effects of invasion. Deep invasion of a lower density mud filtrate can occur at the top of a formation or below a shale streak. The SP response is reduced where there is deep invasion. At the bottom of the formation, there may be no invasion. In this case, the fluid junction potential disappears, and then there may be a potential across the mudcake that reduces the SP response.

Application: The SP log is used in the same manner as the GR log to detect shales and potentially permeable formations, Figure 9.2.6. The SP log can also be used to determine the resistivity of the formation water if the SSP can be determined from the log response. In practice, this is usually only possible for thick, clean (non-shaly) formations in which the SP response is close to the SSP.

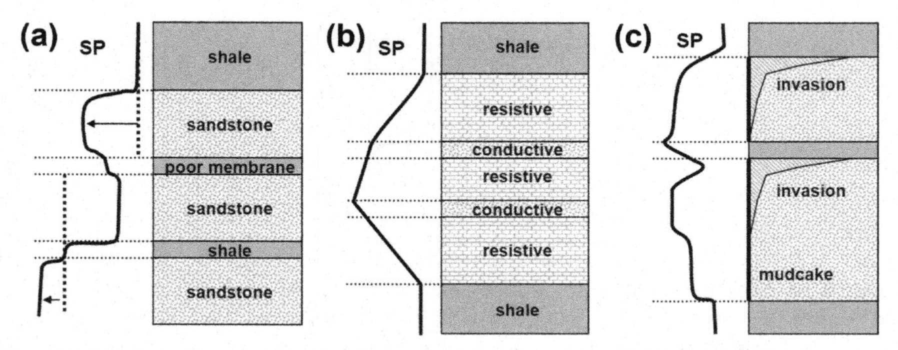

Figure 9.2.7 Some factors affecting SP logs: (a) baseline shift, (b) high-resistivity formations, (c) invasion effects.

From electrochemistry, the SSP is related to the activity of the formation water, a_w, and the mud filtrate, a_{mf}, as follows:

$$SSP = -K_c \log\left\{\frac{a_w}{a_{mf}}\right\} = -K_c \log\left\{\frac{R_{weq}}{R_{mfeq}}\right\} \tag{9.2.3}$$

in which K_c is a temperature dependent constant given by:

$$K_c = 65 + 0.24T_f \tag{9.2.4}$$

in which T_f is the formation temperature in °C. For relatively dilute pure NaCl solutions, the resistivity is directly proportional to the activity, and the SSP can be related to equivalent resistivities. The equivalent formation water resistivity is then given by:

$$R_{weq} = R_{mfeq} 10^{-SSP/K_c} \tag{9.2.5}$$

in which R_{weq} and R_{mfeq} are the equivalent resistivities of the formation water, and mud filtrate at the formation temperature, respectively.

The equivalent resistivity of the mud filtrate is determined from a measured resistivity of the mud filtrate, $R^{\circ}{}_{mf}$. This value is usually reported in the header of the SP log at a specified temperature, T_{ref}. The resistivity can be corrected to formation temperature as follows:

$$R_{mf} = R^{o}_{mf}\left(\frac{T_{ref} + 21.5}{T_f + 21.5}\right) \tag{9.2.6}$$

in which R_{mf} is the mud resistivity at the formation temperature, T_f (in °C).

For NaCl and lime-based muds, the equivalent mud filtrate resistivity is approximately:

for R_{mf} @ 24 °C > 0.1 Ωm: $R_{mfeq} = 0.85\ R_{mf}$ @ T_f
for R_{mf} @ 24 °C < 0.1 Ωm: R_{mfeq} from solid lines on Figure 9.2.8

For gypsum-based muds, use Figure 9.2.8.

Once R_{mfeq} is determined, R_{weq} is calculated from Eqn (9.2.5). Then, the actual water resistivity, R_w, is determined using Figure 9.2.8. If R_{weq} is less than 0.1 Ωm, the solid lines are used; otherwise, the dashed lines are used.

The SP is not as important to modern log interpretation as it was in the 1970s and earlier. Old logging suites used the SP for bed boundary delineation and R_w evaluation. Sometimes, the SP was also used to predict permeability because of the effect clay minerals may have on both reducing the SP and reducing permeability. These correlations, however, were only partially successful because the clays affect only the fluid junction component, which is a small part of the SSP.

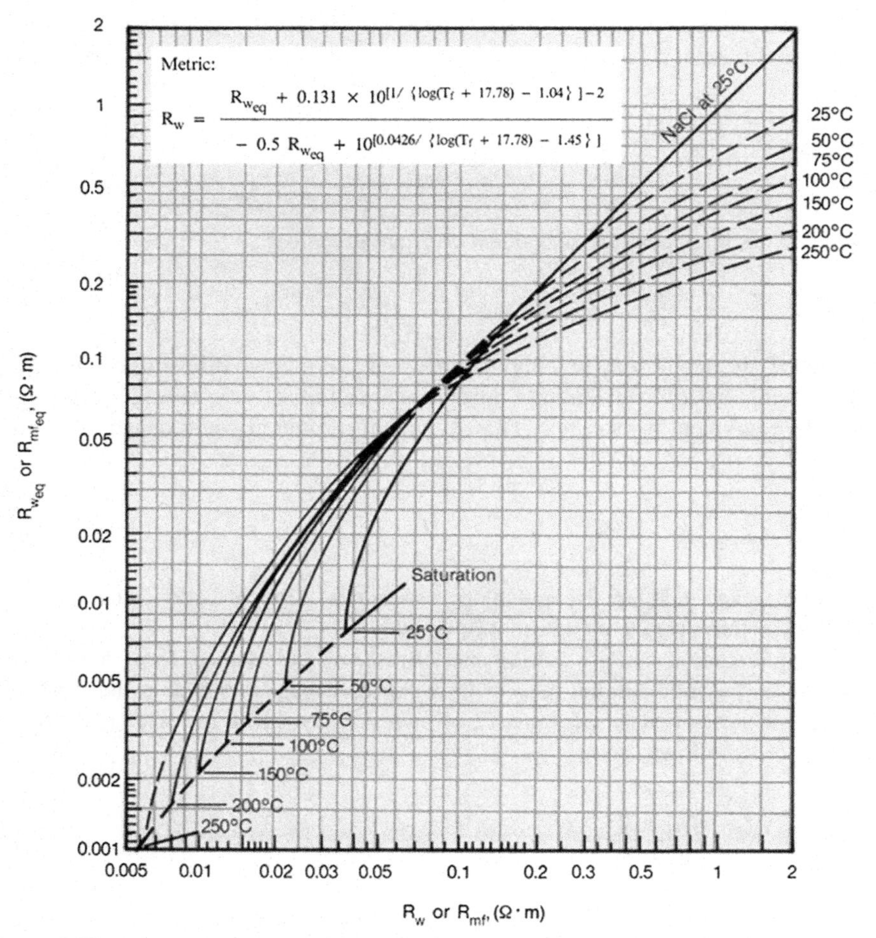

Figure 9.2.8 Relationship between formation water resistivity and mud filtrate resistivity (Baker Hughes, 1995).

9.2.4 Acoustic Log

The acoustic logging tool measures the travel or transit time of sound waves through formation. In the simplest tool configurations, there are two transmitters and a series of two or four receivers as shown in Figure 9.2.9. The transmitters are pulsed alternately. The transit times from the two transmitters are averaged to compensate for errors from changing hole size and tool position. The distance between transmitters and detectors (and therefore the resolution of the log) ranges from 3 to 5 ft for a conventional porosity measurement. Spacings of 10–12 ft can be used to measure deeper into the formation. The vertical resolution is about 2 ft for older logs, while newer tools provide resolutions in the $^1/_2$ to 1 ft range.

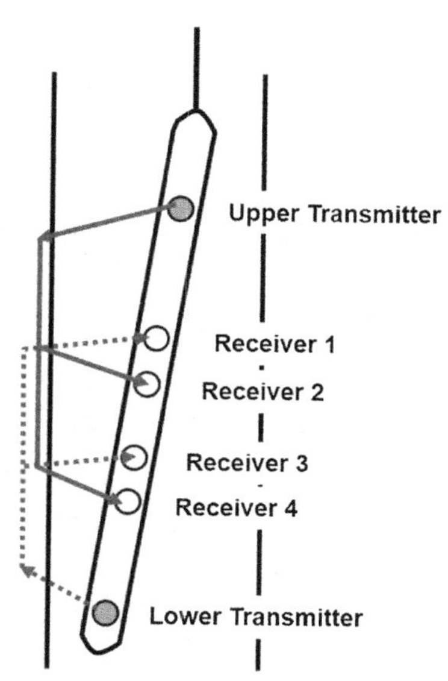

Figure 9.2.9 Schematic of a compensated acoustic logging tool.

The sound emitted from the transmitter travels through the fluid in the wellbore and strikes the wellbore wall, generating compression waves and shear waves through the formation, surface waves along the borehole wall, and guided waves within the fluid column in the wellbore. The waves are refracted, reflected, and converted as they encounter the borehole wall roughness, formation boundaries, fractures, and other discontinuities. Therefore, many waves are recorded at the detectors, including compressional, shear, mud, and Stoneley waves.

Compressional wave: The compressional wave travels through the fluid as a fluid pressure wave, through the formation at the formation compressional wave velocity, and back through the fluid as a fluid pressure wave. It is the first arrival at the detector and is weak.

Shear wave: The shear wave travels through the fluid as a fluid pressure wave, through the formation at the formation shear wave velocity, and back through the fluid as a fluid pressure wave. It is stronger than the compressional wave.

Mud wave: The mud wave travels directly through the fluid in the wellbore at the compressional wave velocity of the borehole fluid.

Stoneley wave: The Stoneley wave is a large amplitude wave that travels at a velocity below that of the compressional wave velocity of the wellbore fluid. The Stonely wave depends on the frequency of the acoustic source, the fluid compressional wave velocity, the formation shear wave velocity, the hole diameter, and the densities of the formation and fluid.

Most acoustic tools are designed to detect compressional waves because they are the first to arrive, and their transit time is related to formation transit times and porosity. A hypothetical acoustic log is shown in Figure 9.2.10. More complex tools are used to obtain a complete full-wave image, but these tools are beyond the scope of this book.

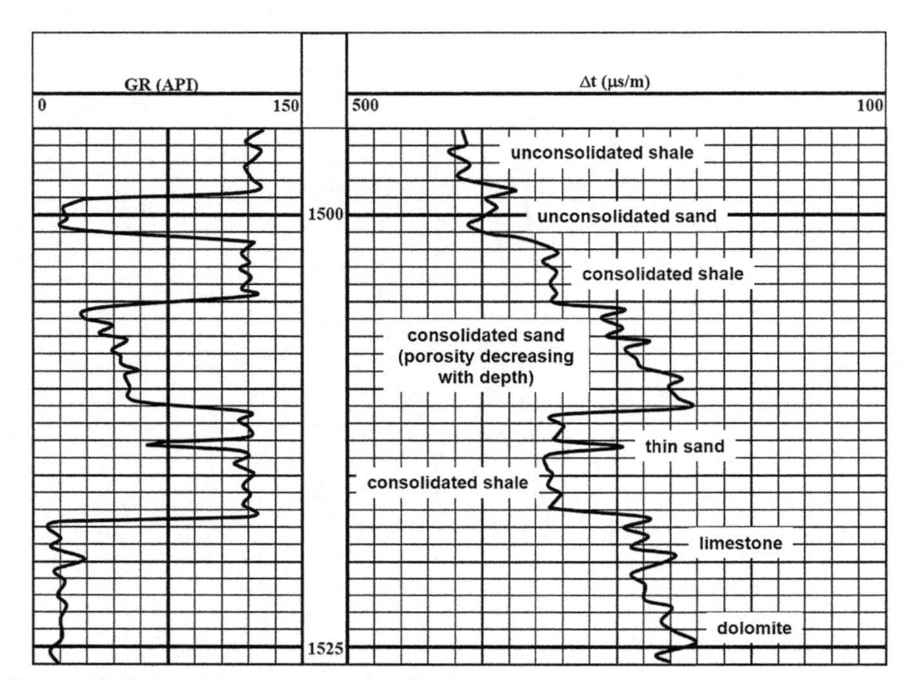

Figure 9.2.10 Hypothetical acoustic log (on right) including GR trace (on left) and with depth intervals of 1 m. Δt is the compressional wave travel time.

Note that an acoustic log with 3–5 ft spacing between transmitters and receivers detects waves that have only penetrated a short distance (typically inches) into the formation. Therefore, they read the flushed zone. While acoustic logs are compensated for hole size and tool position, the readings can be inaccurate in rough hole. Newer tools have larger transmitter-receiver spacings (e.g., 8 and 10 ft) so the tool reads from deeper into the formation, reducing near-wellbore effects.

Application: The acoustic log is often used to estimate porosity. The speed of the compressional wave depends on the density of the medium through which it travels. Therefore, the travel time depends on the type of rock matrix, the density of the fluid phase, and the porosity of the rock. The porosity may be determined from measured travel times using the Wyllie time-average model (Wyllie et al., 1956):

$$\phi_S = \frac{t - t_{ma}}{t_f - t_{ma}} \frac{1}{B_p} \tag{9.2.7}$$

in which ϕ_S is the acoustic porosity, t, t_f, and t_{ma} are the measured travel times (μs/m or μs/ft) of the formation, fluid, and solid matrix, respectively, and B_p is an empirical correction factor used in unconsolidated sandstones. Because the acoustic log reads

Table 9.2.3 Acoustic travel times for typically encountered fluids and solid matrices

Matrix	Travel time range (μs/m)	Typical travel time (μs/m)	Fluid	Travel time (μs/m)
Sandstone	167−182	167 or 182	Fresh water mud	620
Limestone	156−143	156	Salt water mud	607
Dolomite	143−126	143		
Anhydrite	164	164		
Salt	219	220		
Shale	328	328		

the flushed zone, the fluid in the rock through which the wave travels is mud filtrate. The travel times for typical fluids and solid matrices are provided in Table 9.2.3.

In unconsolidated sandstones, the correction factor B_p is applied in Eqn (9.2.8). One indication of an unconsolidated zone is that the transit time in the adjacent shale beds exceeds 100 μs/ft. In this case, the correction factor can be determined as follows:

$$B_p = t_{sh}/100 \tag{9.2.8}$$

A more accurate correction can be determined from an independent measurement of porosity from core data or a neutron-density log. In this case, the correction factor is simply the ratio of the known porosity to the acoustic porosity at an interval at which both were measured.

Experience has shown that acoustic log porosity predictions may systematically deviate from the above relationships between the transit time and porosity. An alternative is given by Raymer et al. (1980):

$$\phi_S = -\alpha - \left[\alpha^2 + \frac{t_{ma}}{t} - 1\right]^{1/2} \quad \text{where } \alpha = \frac{t_{ma}}{2t_f} - 1 \tag{9.2.9}$$

It is recommended to use the transit time−porosity relationship (theoretical or empirical) that best matches independent porosity data. If no independent data are available, it is recommended to use the Eqn (9.2.9). Equations (9.2.7) and (9.2.9) are compared in Figure 9.2.11.

In general, the acoustic response is independent of the hydrocarbon saturations in consolidated sandstones ($<$ 25% porosity). In some higher porosity sandstones with low water saturation, the transit times may be higher than in water-saturated zones. Shale laminae can also increase the measured transit time in proportion to their volume

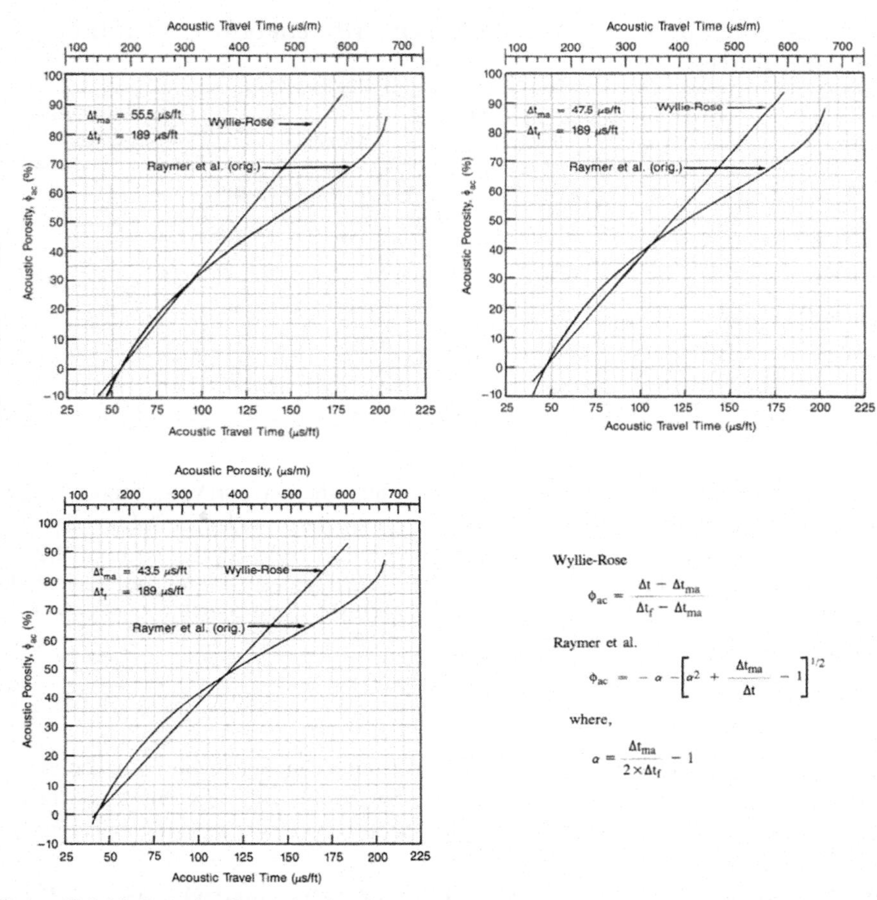

Figure 9.2.11 Porosity determination from acoustic log transit times (Baker Hughes, 1995).

fraction in the formation. In unconsolidated sandstones, unflushed gas can cause high transit times and overestimated porosity.

In carbonates, the acoustic porosity depends mainly on the primary intergranular or intercrystalline porosity and only little on the secondary vuggy or fracture porosity. Therefore, if the total porosity is known from another measurement (core or neutron-density porosity), the amount of secondary porosity can be estimated as follows:

$$\phi_2 = \phi_T - \phi_S \tag{9.2.10}$$

in which ϕ_2 and ϕ_T are the secondary and total porosity, respectively.

As any linear algebra instructor will tell you, you cannot determine more unknowns than you have equations. Thus, the acoustic log (one equation, Eqn (9.2.7)) cannot be used in isolation to determine lithology and porosity (two unknowns assuming

$B_p = 1$). Figure 9.2.10 showed typical acoustic log responses in different rock types. It is virtually impossible to distinguish a low-porosity sandstone from a carbonate or an unconsolidated sandstone from an unconsolidated shale based solely on the measured transit times. An acoustic log is usually coupled with a GR log to help distinguish shale beds. Independent data from drilling cutting returns, core data, other openhole logs, or offset well data are required to determine formation rock types, and precise porosities as is discussed in Section 9.3. An engineer must integrate the data from various logs to be able to estimate porosity, lithology, and other well log-derived properties.

9.2.5 Density Log

The density logging tool measures the amount of GRs scattered by the formation. The GRs are emitted from a radioactive source that is held against the borehole wall, as shown in Figure 9.2.12. In the basic tool, there are two detectors, a short-spaced and a long-spaced detector. To minimize wellbore effects, the source and the detectors are located on a skid, pressed against the borehole wall.

GRs are scattered when they collide with electrons (Compton scattering). The amount of scattering is proportional to the electron density of the medium. The electron density is proportion to the bulk density of the medium. Hence, the GR count at the detector is proportional to the bulk density of the formation. Density logging tools are calibrated in the laboratory using fresh water filled limestones to determine the

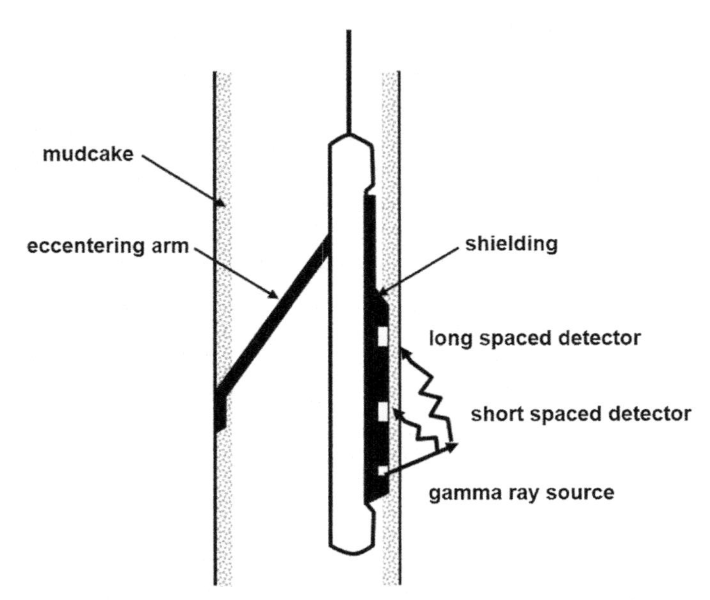

Figure 9.2.12 Schematic of a compensated density logging tool. The GRs from the source are scattered back toward the detectors. Modern tools may have more than two detectors. Adapted from Wahl et al. (1964).

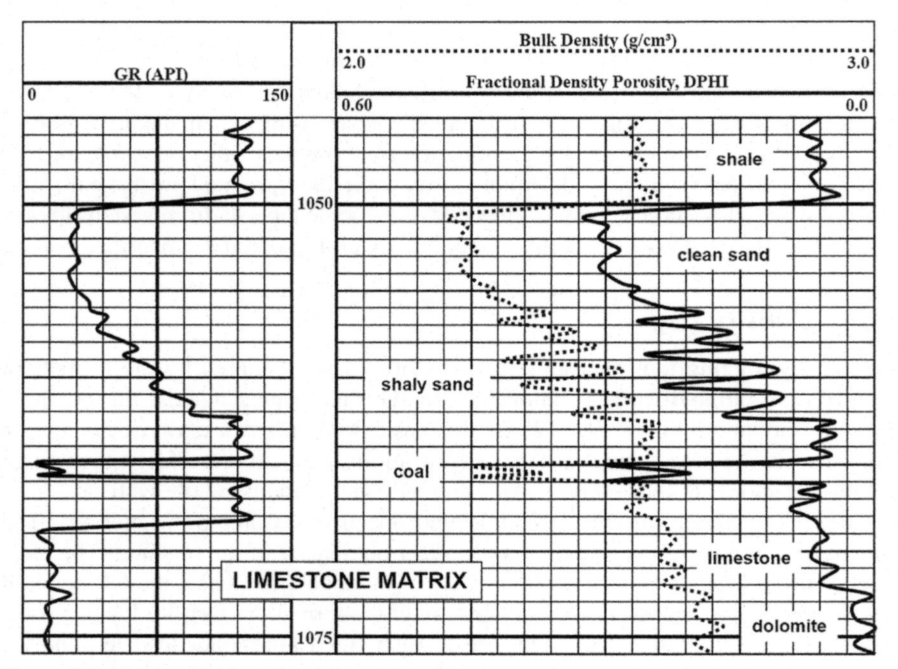

Figure 9.2.13 Hypothetical density and density porosity log (on right) including GR trace (on left) and with depth intervals of 1 m.

proportionality constant. The calibration is used to convert the GR counts to density, which is plotted versus depth, as shown in Figure 9.2.13.

The scattering of the GRs may be affected by mudcake. However, the difference in readings between the short- and long-spaced detectors correlates to the effect of the mudcake and is used to calculate a compensated log reading. After compensation, the log resolution is comparable to that of an acoustic log (approximately 2 ft). The density tool is shallow reading and usually only detects the flushed zone. While the log is compensated for mudcake, the readings can be unreliable in rough hole.

Application: The density log is used to determine porosity. The bulk density is related to the density of the formation rock, the density of the fluid in the formation, and the porosity of the formation. The porosity is determined from the measured density as follows:

$$\phi_D = \frac{\rho_{ma} - \rho_b}{\rho_{ma} - \rho_f} \tag{9.2.11}$$

in which ϕ_D is the density porosity, and ρ_b, ρ_f, and ρ_{ma} are the measured, fluid, and matrix densities (g/cm^3 or kg/m^3), respectively. Because the density log reads the flushed zone, the fluid through which the GRs travel is mud filtrate. The densities for typical fluids and solid matrices are provided in Table 9.2.4.

Table 9.2.4 Densities for typically encountered fluids and solid matrices

	Actual density (g/cm^3)	Measured density (g/cm^3)
Matrix		
Quartz	2.654	2.648
Limestone	2.710	2.710
Dolomite	2.870	2.870
Anhydrite	2.960	2.957
Halite (NaCl)	2.165	2.032
Sylvite (KCl)	1.984	1.863
Gypsum	2.320	2.351
Anthracite coal	1.4–1.8	1.3–1.8
Bituminous coal	1.2–1.5	1.2–1.5
Shale	2.2–2.65	2.2–2.65
Fluid		
Fresh water	1.000	1.000
Salt water (200,000 ppm)	1.146	1.135
Oil	0.85	0.85

As with the acoustic log, it is assumed that the log reads a flushed zone filled with mud filtrate. If there is residual gas in the flushed zone, the measured density will be lower than in a water-filled zone. The porosity calculated based on the filtrate density will therefore be high.

In most modern logs, the density is converted to density porosity and the porosity is plotted as well as or instead of the density, Figure 9.2.13. Porosity logs are typically scaled from 0% to 60% or from −15% to 45% porosity. A matrix density must be assumed to convert density to porosity. Typically, a limestone density (2.710 g/cm^3) is used and a note is made on the log, such as "limestone matrix." Some logs are based on a sandstone matrix density of 2.65 g/cm^3. A log based on a given matrix density will show the correct measured porosity value only in that type of matrix filled with fresh water. The porosity must be corrected for all other rock types, as is discussed in Section 9.3.

9.2.6 Neutron Log

The neutron logging tool measures the amount of neutrons scattered (post-1960s tools) or GRs emitted (older tools) from the formation after it is exposed to a neutron source. The neutrons are emitted from a radioactive source that is held against the borehole

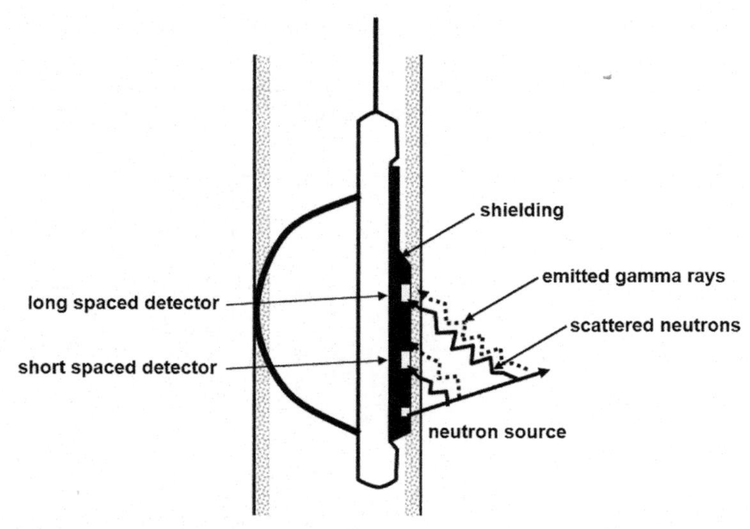

Figure 9.2.14 Schematic of a compensated neutron logging tool with an eccentering spring that helps to keep the tool pressed against the borehole wall. Newer tools may have more than two detectors.
Adapted from Wahl et al. (1964).

wall, as shown in Figure 9.2.14. A typical tool will have a pair of detectors (short spaced and long spaced) that are sensitive to thermal neutrons or epithermal (just above thermal velocity) neutrons.

Neutrons lose energy when they collide with an atomic nucleus, and they lose the most energy when the nucleus has similar mass as the neutron, that is, a hydrogen nucleus. After approximately 20 collisions with hydrogen nuclei, the neutron is slowed to thermal velocity, and it can be scattered or captured by elements in the formation. When it is captured, a gamma ray is emitted. The more hydrogen in the formation, the less distance the neutrons travel. Hence, the count of neutrons or emitted gamma rays is related to the amount of hydrogen in the formation. Neutron logging tools are calibrated in the laboratory using fresh water-filled limestones of known porosity. The calibration is used to convert the neutron or GR counts to neutron porosity, which is plotted versus depth, as shown in Figure 9.2.15.

Neutron logs are compensated in approximately the same manner as density logs based on the difference in readings between the short- and long-spaced detectors. The log resolution is somewhat less than that of a density log (approximately 3 ft). The neutron tool is deep reading (10–14 in), so rough hole effects are less than for the density log. The neutron capture rate is strongly effected by elements with a high thermal neutron capture cross-section, such as boron, chlorine, and other rare earth elements. These elements are often present in shales, and therefore the porosity can be overestimated in shaly formations.

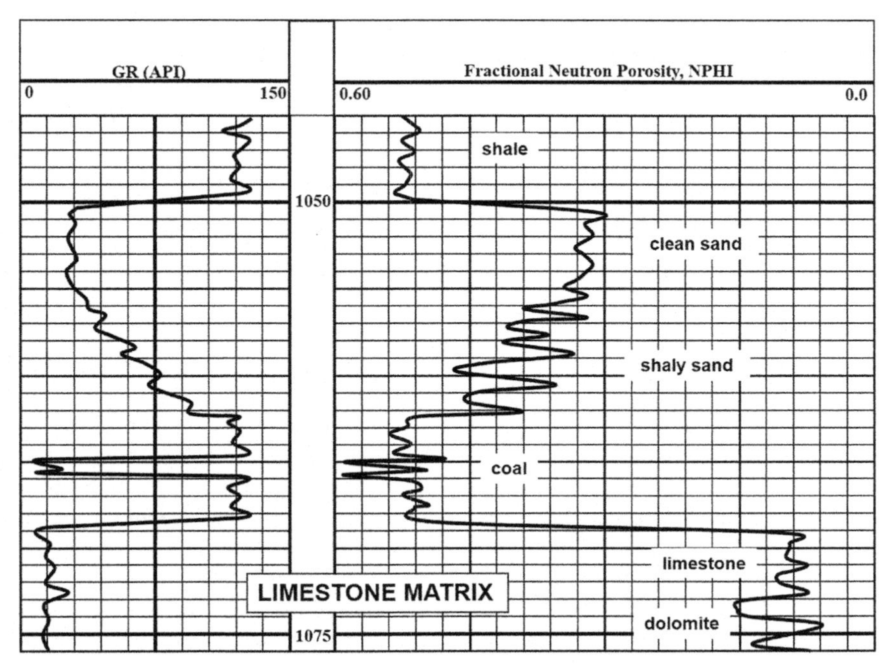

Figure 9.2.15 Hypothetical neutron porosity log (on right) including GR trace (on left) and with depth intervals of 1 m. The neutron log has an elevated response in the shale because of clay-bound water.

Application: The neutron log is calibrated to a porosity scale (0 to 60% or −15 to 45%) and therefore no calculations are required to obtain the porosity. The standard conditions for the calibration are:

- 7 7/8th inch borehole diameter;
- Fresh water in borehole and formation;
- No mudcake or standoff;
- Too eccentered in hole;
- 75 °F temperature;
- Atmospheric pressure.

When there are departures from these conditions, corrections are required although most corrections are small. As with density logs, the neutron log is usually calibrated to a limestone matrix but can be calibrated to a sandstone matrix. A log based on a given matrix density will show the correct measured porosity value only in that type of matrix. The porosity must be corrected for all other rock types, as is discussed in Section 9.3.

The neutron log is a deeper reading log than the density log and is often affected by the hydrocarbon saturation of the formation. The hydrogen content of oil is similar to that of water, and its effect is small. However, the hydrogen content of gas is significantly lower, leading to lower neutron counts and an underestimated porosity. Shales tend to absorb neutrons and therefore have high apparent neutron porosities. The neutron response in shales can change with different neutron tools. Newer tools, which respond to epithermal neutrons, tend to show a lower porosity than older, thermal-neutron tools.

9.2.7 Resistivity Log

There are main two types of resistivity logs: electrode logs and induction logs. Each are discussed below.

Electrode Logs: The electrode logging tool measures the voltage drop between electrodes held adjacent to the formation as a current is passed through the formation, as shown schematically in Figure 9.2.16. In a "normal" tool configuration, the measuring electrode is connected to a galvanometer at the surface. Typically, the spacing between the current electrode and the measuring electrode is 16 in (short-normal) or 64 in (long-normal). In a "lateral" tool, the voltage drop is measured between a pair of electrodes. The spacing between the current electrode and the midpoint of the pair of measuring electrodes is 18 ft 8 in. Note that the greater the spacing, the greater the depth of investigation.

The electrode logs of the 1940s and 1950s all suffered from significant borehole and shoulder bed effects. These logs have been replaced by focused-current devices called

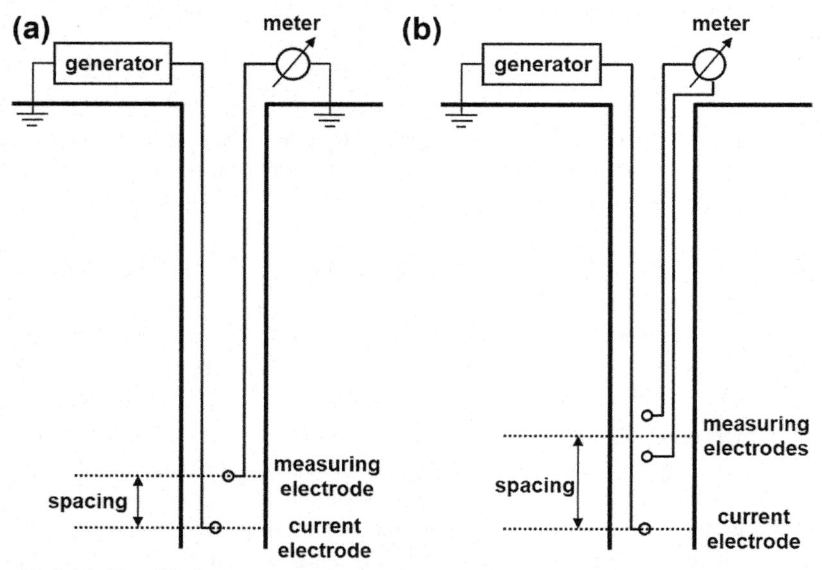

Figure 9.2.16 Simplified schematic of electrode arrangements in old-style electrode logs: (a) normal configuration, (b) lateral configuration.

guard logs or laterologs. Vertical resolution of modern tools is typically 1–2 ft. Modern electrode logs usually have a deep resistivity measurement, corresponding to readings which represent the resistivity of the uninvaded zone. So-called shallow measurements are usually not shallow enough to measure only the flushed zone; they read some combination of the flushed, transition, and uninvaded zones (Figure 9.1.1).

In order to measure the flushed zone resistivity, a very shallow (1 or 2 in) reading device is needed but one that is not affected by the borehole mud. To achieve this, the electrodes for the "microresistivity" logs are placed on a rubber pad, which is pressed against the borehole wall. The term "micro" in the name of the log, e.g., microlaterolog and micro spherically focused log, denotes this is a very shallow measurement. The vertical resolution of such logs is about 2–6 in.

Induction logs: The induction logging tool measures a voltage induced by an alternating magnetic field so there is no need for electrodes. The field is generated by an alternating current inside an insulated transmitter coil. The magnetic field induces alternating current loops in the formation around the wellbore. The circulating currents in turn induce an alternating current in a receiver coil as shown in Figure 9.2.17. The nominal spacing of induction tool transmitter-receiver placements ranges from 16 to 40 in.

Electrode logs are recommended for wells drilled with highly conductive drilling muds, such as salt muds. Induction logs are recommended for wells drilled with moderately conductive or nonconductive muds, such as air, fresh water, or oil-based muds.

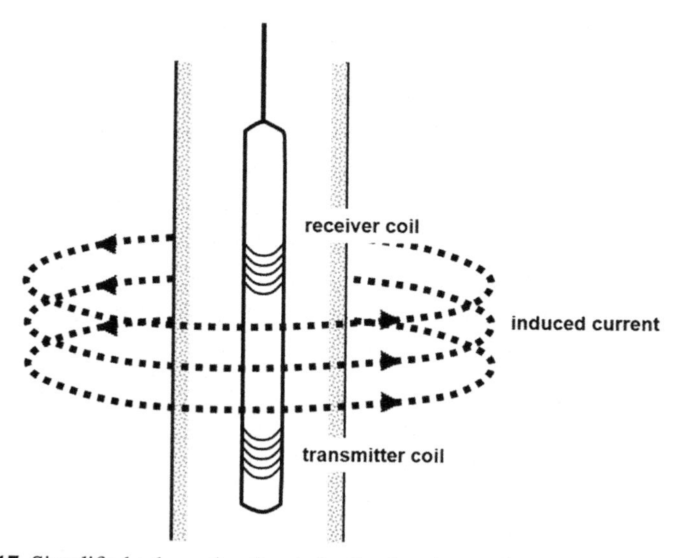

Figure 9.2.17 Simplified schematic of an induction logging tool.

Older induction logs suffered from poor vertical resolution (5–10 ft). The resolution of a modern induction tool is 2–4 ft. Modern induction logs include several sets of coils with focused currents used to minimize the effects of the borehole and surrounding formations. Most modern resistivity log suites include some combination of measurements having different depths of investigation:

- Shallow (R_{msfl}): approximately 1–2 in into the formation;
- Medium (R_{ILm}, R_{sfl}, R_{LLs}): approximately 6–12 in into formation;
- Deep (R_{ILd}, R_{LLd}): more than 12 in into formation.

An example dual induction log (ILm and ILd) is shown in Figure 9.2.18.

Application: Resistivity logs are used to determine water saturations. Both electrode and induction logs measure the resistivity of the formation, which depends primarily on the porosity, the water saturation, and the salinity of the water in the formation. The relationship between water saturation and formation resistivity is estimated with Archie's equation:

$$S_w^2 = \frac{KR_w}{\phi^m R_t} \tag{9.2.12}$$

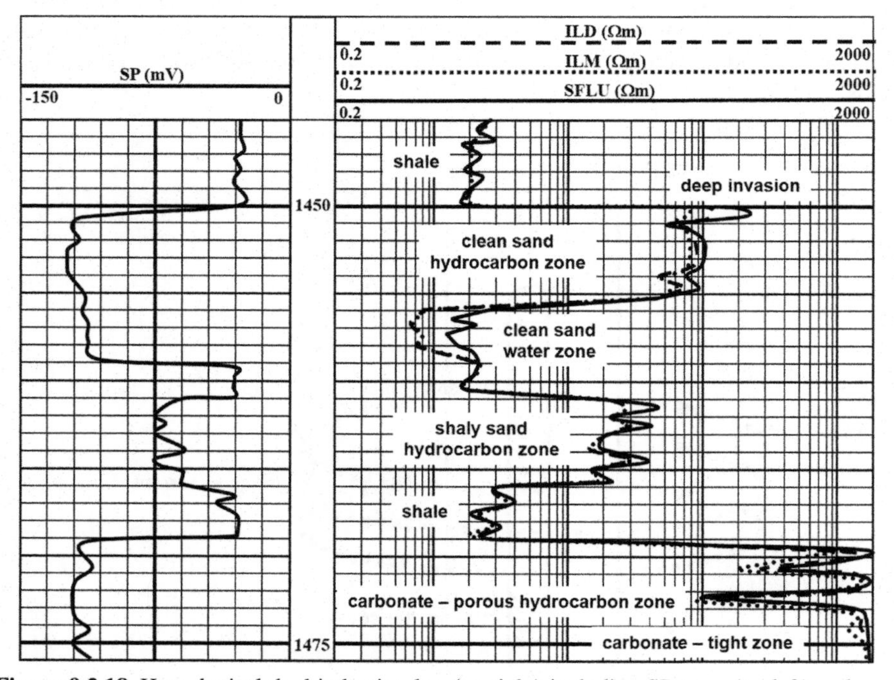

Figure 9.2.18 Hypothetical dual induction log (on right) including SP trace (on left) and with depth intervals of 1 m. The spherically focused log (SFLU) is an electrode-based measurement and designed to be run in combination with the medium and deep induction. The SFLU has a shallower depth of investigation and better boundary definition than the inductions (e.g., 1,450 m).

in which R_t is the measured resistivity (Ωm), R_w is the formation water resistivity (Ωm), ϕ is porosity, and K and m are constants for a given formation. The determination of K and m is discussed in Section 7.5. Rule of thumb values are:

sandstone: $K = 0.62 \quad m = 2.15$

carbonate: $K = 1.0 \quad m = 2.0$

There are four methods to estimate R_w:

1. Measure resistivity and porosity of water zone near formation of interest and calculate R_w assuming $S_w = 100\%$ and known values of K and m.
2. Obtain R_w from formation water samples. The resistivity must be corrected to formation temperature using Eqn (9.2.6).
3. Refer to a water catalogue (e.g., Canadian Well Logging Society, 1987) and look up R_w for the formation of interest in the same geographical area. The resistivity must be corrected to formation temperature using Eqn (9.2.6).
4. Calculate R_w from SP response as was discussed in Section 9.2.3.

In practice, the resistivity measurement is affected by invasion when the resisitivity of the mud differs from that of the formation water. The shallow and medium measurements can be used to assess mud invasion. If the invasion is too deep, it may not be possible to determine an accurate water saturation. The determination of water saturations from resistivity logs is discussed in more detail in Section 9.3.

9.3 Basic Log Interpretation

Basic log interpretation for the reservoir engineer is introduced in this section, including determination of lithology, porosity, water saturation, fluid contacts, and net pay. Analysis of shaly sands is also reviewed. The methods are illustrated through several examples. Detailed log interpretation (including bore hole corrections, thin bed effects, and other log corrections) is not covered here. It is recommended to refer detailed log interpretation to a petrophysicist.

9.3.1 *Analysis of a Partly Shaly Sandstone with a Water–Oil Contact*

First Pass Analysis

Figure 9.3.1 is a typical suite of modern logs (SP/Dual Induction and GR/Caliper/Neutron-Density) obtained for a Dina formation. The caliper log is nearly constant showing no evidence of sloughing and little indication of mudcake. Therefore, accurate log readings should be obtainable. In this first pass, we treat the Dina interval as a clean sand zone. Afterward, in a second pass, the shale content is accounted for.

Lithology: The GR has a shale response of approximately 110 API (right dashed line in Figure 9.3.1). The top of the Dina formation appears as a decrease in the GR reading at 846 m kB. The low resistivity values in the low porosity regions and the high porosities

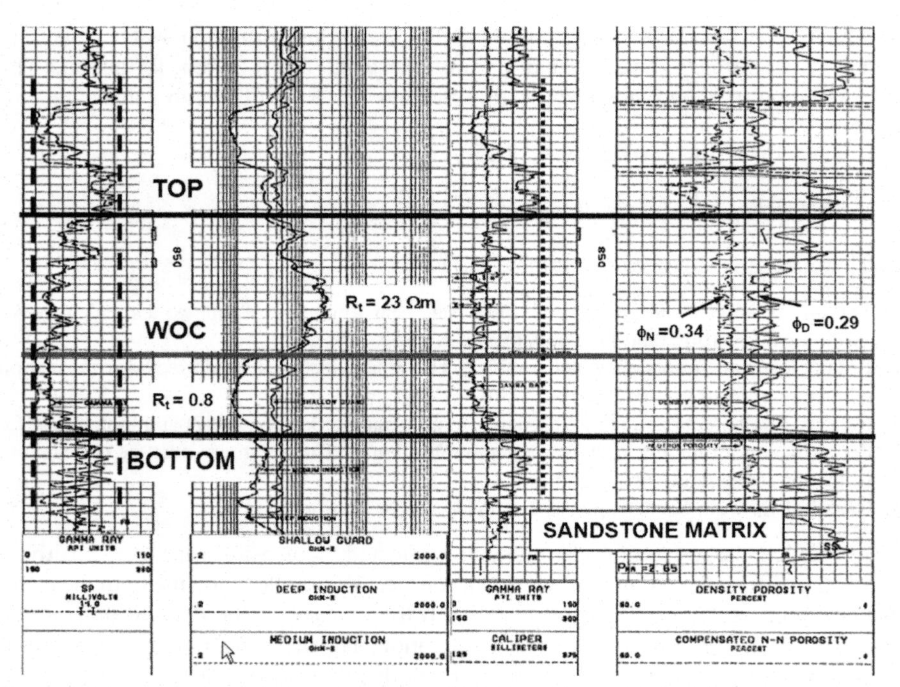

Figure 9.3.1 Set of openhole logs from Dina C2C 1A-05-11-37-5W4 with the customary presentation of dashed curves for the neutron porosity and deep resistivity and solid lines for the density and shallowest resistivity.

in the clean GR regions are indicative of a sandstone formation. There appear to be shale breaks in the top of the formation as the GR reading spikes between 25 and 75 API. The lower half of the formation is cleaner until the lower bounding shale is reached at a depth of 872 mkB. The KB elevation is 706.0 m. Therefore, the formation top and bottom elevations are −140 and −166 m (140 and 166 m subsea), respectively.

Porosity: The porosity is high throughout the Dina formation. Consider the interval at 855.5 mKB (denoted with arrows on Figure 9.3.1). The density porosity is 0.29, and the neutron porosity is 0.34. There are several formulas for estimating the average porosity including:

$$\phi = \phi_D \tag{9.3.1}$$

$$\phi = 0.67\phi_D + 0.33\phi_N \tag{9.3.2}$$

$$\phi = 0.5\phi_D + 0.5\phi_N \tag{9.3.3}$$

The corresponding average porosity estimates are 0.29, 0.305, and 0.315. The best choice of averaging formula can be determined from a comparison of the average log porosity with the core porosity for the same interval (see Section 9.3.4). A better average can also be obtained if the acoustic log is available as will discussed in Section 9.3.4.

Fluid contacts: The resistivity decreases from 10 to 23 Ωm above 862 mKB to 0.8 to 2 Ωm below 863 mKB. The higher resistivities indicate a hydrocarbon-filled zone, and the lower resistivities indicate a water zone. Therefore, the water–oil contact (WOC) is at approximately 862.5 mKB (−156.5 m elevation).

As will be discussed in Example 2, a gas-oil contact is indicated by a closer approach or a crossover of the neutron and density porosities. There is no evidence of a gas-oil contact on the logs of Figure 9.3.1.

Water saturation: The deep measured resistivity is 23 Ωm for the 31% porosity interval analyzed previously. The parameters K, m, and R_w are required to determine a water saturation. In this case, let us use the rule-of-thumb values for sandstone formation parameters: $K = 0.625$, $m = 2.15$. Now let us determine R_w using the four methods listed in Section 9.2.7.

Method 1: *R_w from water zone*

The measured resistivity in the cleanest sand (lowest GR) within the water zone is approximately 0.85 Ωm at a depth of 868 mKB. The average porosity at this depth is 31%. If we assume that the water saturation is 100%, then the water resistivity is:

$$R_w = \frac{\phi^m R_t}{K} = \frac{(0.31)^{2.15} 0.85\ \Omega\text{m}}{0.625} = 0.11\ \Omega\text{m}$$

Method 2: *R_w from water analysis*

The water resistivity from a water analysis is 0.125 Ωm at 25 °C. The formation temperature is 28 °C. Therefore, the water resistivity at formation temperature is:

$$R_w = R_w^o \left(\frac{T_{ref} + 21.5}{T_f + 21.5} \right) = 0.125\ \Omega\text{m} \left(\frac{25 + 21.5}{28 + 21.5} \right) = 0.12\ \Omega\text{m}$$

Method 3: *R_w from water catalogue*

The water catalogue provides a sampling of water analyses from formations from a similar geological horizon. It is useful for establishing a range of possible water resistivities, but is less specific and almost always less accurate than using a water analysis from the reservoir of interest. In this case, water analysis from the Lower Cretaceous horizon range from 0.09 to 0.18 Ωm at 25 °C in the Provost area. The corresponding range of R_ws at formation temperature is 0.085–0.17 Ωm.

Method 4: *R_w from SP response*

To determine R_w from the SP response, the following data are required: the static SP response, the mud resistivity, and the formation temperature. The shale line and the clean sand line are marked with dashed lines on Figure 9.3.1. The SSP is approximately 100 mV (each grid is 15 mV). The formation temperature is 28 °C. The mud resistivity is 3 Ωm at 37 °C. The mud resistivity at 28 °C is 2.8 Ωm, and the equivalent mud resistivity is 85% of this value or 2.40 Ωm.

The constant K_c given by:

$$K_c = 65 + 0.24T_f = 65 + 0.24(28) = 71.72 \text{ mV}$$

and the equivalent water resistivity is then:

$$R_{weq} = R_{mfeq}10^{-SSP/K_c} = (2.40 \ \Omega\text{m})10^{-100/71.72} = 0.096 \ \Omega\text{m}$$

From Figure 9.2.8, the actual water resistivity is approximately 0.12 Ωm.

The estimated water resistivities range from 0.11 to 0.12 Ωm. The calculated water saturation for the 31% porosity and 23 Ωm resistivity interval based on a water resistivity of 0.11 Ωm is then:

$$S_w = \left(\frac{KR_w}{\phi^m R_T}\right)^{0.5} = \left(\frac{0.625(0.11 \ \Omega\text{m})}{(0.31)^{2.15}(23 \ \Omega\text{m})}\right)^{0.5} = 0.19$$

Because the four methods evaluate R_w differently, the good agreement between the results gives confidence to several assumptions in the above analysis.

- The porosity, K, and m values chosen are appropriate for this formation interval.
- The surface measured value of R_{mf} is reasonably accurate.
- The shale and clean base lines are appropriate.

Second-pass Analysis

In a properly calibrated log based on a sandstone matrix, the neutron and density porosity are expected to be identical in the absence of borehole effects or unusual mineralogy. In Figure 9.3.1, the neutron porosity consistently reads approximately 5% higher porosity than the density porosity. This positive gap is a possible indication of a shaly sand because shale leads to high neutron porosity readings.

The GR log indicated the presence of shale intervals. The GR reading in the "clean" sand is 25–30 API, greater than a typical clean sand reading for the area of 15 API. It appears that the formation is slightly shaly. Using an average value of 28 API for the formation, 15 API for clean sand, and 110 API for shale, the shale parameter, X, is:

$$X = \frac{GR_{form} - GR_{clean}}{GR_{sh} - GR_{clean}} = \frac{28 - 15}{110 - 15} = 0.13$$

and the shale volume is:

$$V_{sh} = 1.7 - \left[3.38 - (X + 0.7)^2\right]^{0.5} = 1.7 - \left[3.38 - (0.13 + 0.7)^2\right]^{0.5} = 0.06$$

We can then correct the porosity to account for the presence of the shale based on the shale volume and the total porosity, ϕ_T. For this calculation, we take the total porosity to be the average of the neutron and density porosity; that is 31.5%, for the interval indicated on Figure 9.3.1. The effective porosity is given by:

$$\phi_{eff} = (1 - V_{sh})\phi_T = (1 - 0.060)0.315 = 0.296$$

The effective porosity is the porosity that should be used in volumetric calculations because the shale volume is accounted for. Note that the effective porosity is similar to the density porosity of 0.29. The density porosity is less affected by the shale and provides a good estimate for the effective porosity. For this reason, many log analysts use the density porosity alone for oil and water zones in sandstone formations.

The resistivity response in a shaly sand is influenced by the shale content and can be approximated with the following expression:

$$\frac{1}{R_t} = \frac{\phi_{eff}^m}{KR_w(1 - V_{sh})}S_w^2 + \frac{V_{sh}}{R_{sh}}S_w$$

in which R_{sh} is the measured resistivity in an adjacent shale zone (2.5 Ωm).

When the appropriate values for the interval of interest are substituted into the equation, a quadratic equation for S_w is obtained as follows:

$$\frac{\phi_{eff}^m}{KR_w(1 - V_{sh})}S_w^2 + \frac{V_{sh}}{R_{sh}}S_w - \frac{1}{R_t} = \frac{0.296^{2.15}}{0.625(0.11)(1 - 0.060)}S_w^2 + \frac{0.060}{2.5}S_w - \frac{1}{23}$$

$$= 1.129S_w^2 + 0.0240S_w - 0.0435$$

The solution here is $S_w = 18.5\%$. This value is nearly the same as the clean sand estimate of 19% because V_{sh} is rather small. If capillary pressure data are available, these values could be compared to the irreducible water saturation. In general, the shale corrections for water saturation become significant when the shale volume exceeds 20%. A shalier formation could require more involved analysis from an experienced petrophysicist.

Gross and Net Pay

Gross pay is the total thickness of the oil or gas zone. Net pay is the thickness of the parts of the oil or gas zone that are deemed to be productive. The net pay must exceed thresholds (cutoffs) of porosity, hydrocarbon saturation, and clean sand content. Cutoffs are discussed in Section 9.4. For a sandstone, typical cutoffs are:

> 7–10% porosity
< 50% water saturation
< 50% shale volume

If possible, the porosity cutoff would be identified from core data using a permeability-porosity cross plot and a permeability cutoff of 0.5–1 mD. Based on the porosity log response in the surrounding shales, a porosity cut-off of 15% (as shown on Figure 9.3.2) may be more appropriate for this formation.

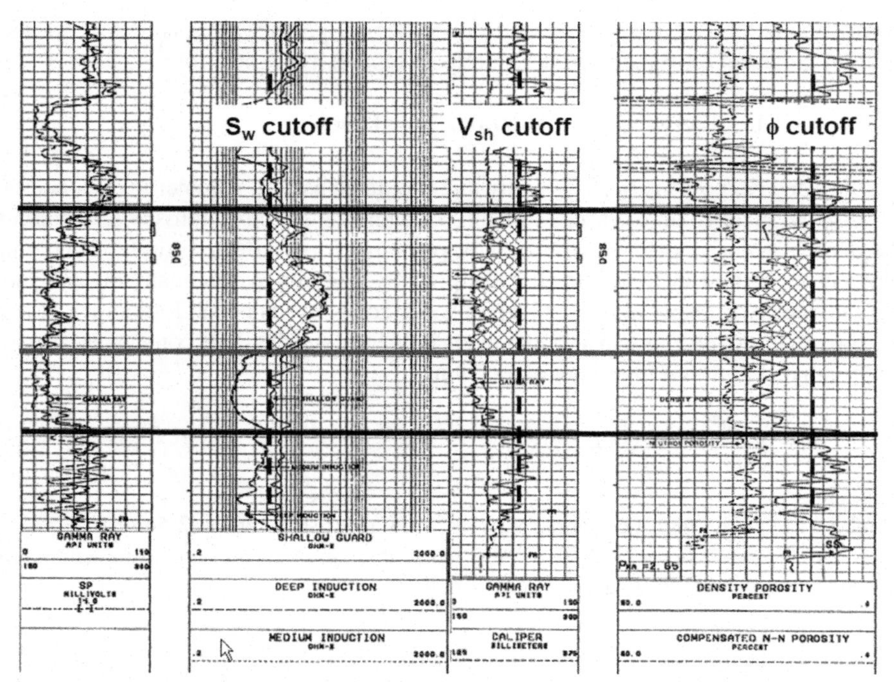

Figure 9.3.2 Cutoffs marked on openhole logs from Dina C2C 1A-05-11-37-5W4.

In this example, a 50% shale volume corresponds to a GR response of 85 API. A 50% water saturation corresponds to a resistivity of 3.4 Ωm. These cutoffs are plotted on Figure 9.3.2. The water saturation cutoff eliminates the water zone below the WOC at 862.5 mKB. The shale cutoff defines the top of the formation at a depth of 846.5 mKB. The gross oil pay therefore lies between 846.5 and 862.5 mKB, with a total thickness of 16.0 m.

The water saturation cutoff does not eliminate any of the pay zone above the WOC. An 8% porosity cutoff also would not eliminate any gross pay. However, the shale cutoff eliminates approximately 1 m of pay at 850 mKB.

9.3.2 Analysis of a Sandstone with a Gas-Oil Contact

Figure 9.3.3 shows SP/Dual Induction and GR/Caliper/Neutron-Density log suite obtained for a Wabiskaw sandstone formation. The caliper log matches the bit size over the interval of interest and therefore, accurate log readings should be obtainable. Based on the GR and neutron-density logs, the top of the porous interval is at a depth of 405 mKB, and the bottom is at a depth of 410 mKB. The resistivity is above 30 Ωm throughout the porous interval, and the density porosity is approximately 30%, suggesting a hydrocarbon-filled zone.

Let us focus on the neutron-density log. In the lower half of the porous interval (407–410 mKB), the density porosity is approximately 30–33%, while the neutron porosity is 36–42%. The higher neutron response suggests that this interval may be shaly. The GR response is higher in the bottom half of the interval, also indicating

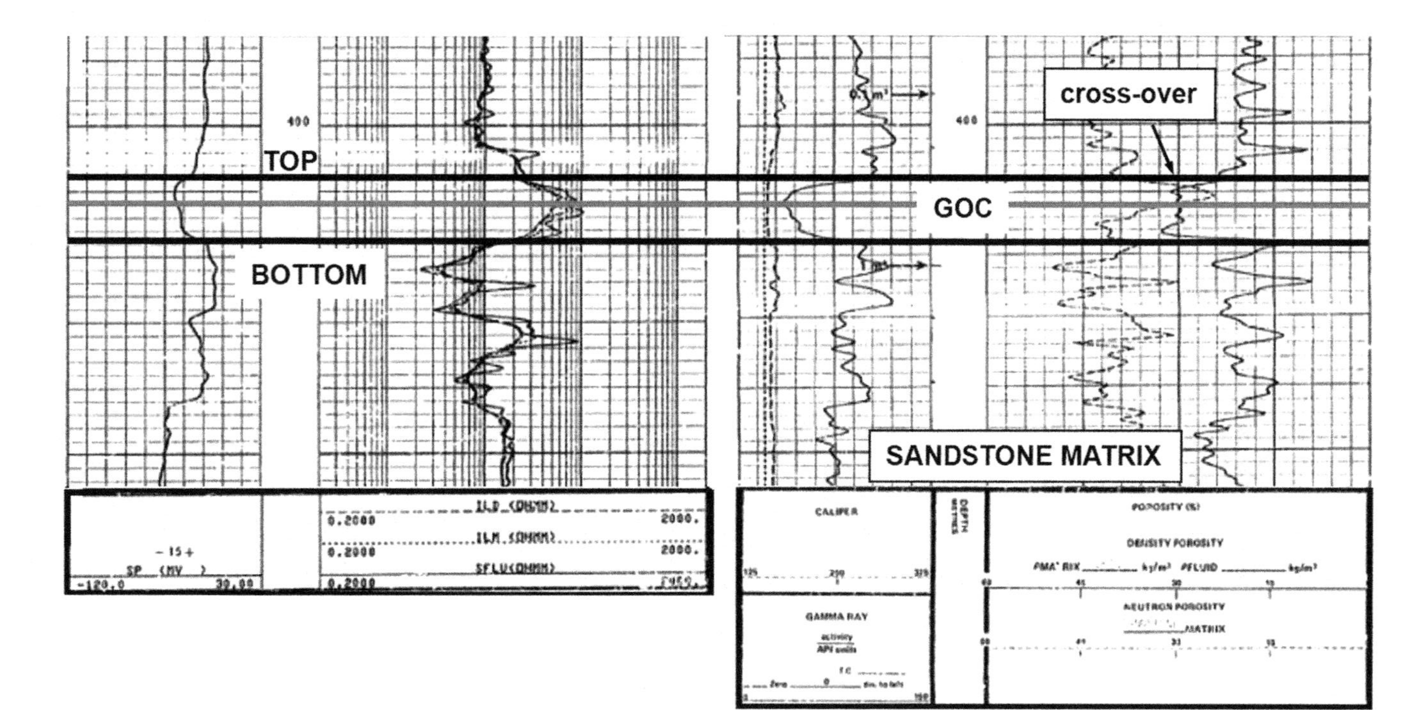

Figure 9.3.3 SP/Dual Induction and GR/Neutron/Density logs from AA14-17-82-21W4.

that some shale may be present. The effective porosity is expected to be approximately equal to the density porosity of 30−33%.

In the interval from 404 to 406 mKB, the neutron porosity decreases to 26%, while the density porosity remains at 30%, and the log responses cross-over. This is a typical response to gas, because gas reduces the neutron porosity and slightly increases the density porosity. The gas-oil contact is at the depth at which the neutron density suddenly decreases, that is, 406 mKB.

The porosity in the gas zone can be estimated with the averaging formulae of Section 9.3.1. The range of possible average porosities is then given by:

$$\phi = \phi_D = 0.30$$

$$\phi = 0.67\phi_D + 0.33\phi_N = 0.29$$

or

$$\phi = 0.5\phi_D + 0.5\phi_N = 0.28$$

The best choice of averaging formula is that which best matches core data.

In this example, the entire porous interval exceeds the standard cutoffs (see Section 9.3.1). Hence, the gross and net pays are identical. The gas pay is 2.0 m (404.0−406.0 mKB) with a porosity of approximately 29% and the oil pay is 3.0 m (406.0−409.0 mKB) with a porosity of 30−33%. The water saturation is determined as described in Section 9.3.1.

9.3.3 Analysis of a Carbonate Sequence with a Water−Oil Contact

Figure 9.3.4 shows SP/Dual Induction and GR/Caliper/Neutron-Density log suite obtained for a Keg River carbonate sequence. The consistently low GR readings and the high-resistivity readings in low-porosity intervals are typical of a carbonate formation. The caliper log is reading lower than the bit size indicating mud cake formation, but no evidence of sloughing. Compensated log readings are expected to be reliable. Production data indicated that there is no gas zone in this well.

Mineralogy and Porosity

Three rock types frequently encountered in carbonate sequences are limestone, dolomite, and anhydrite. All three rock types have low GR responses and cannot be distinguished from each other based on the GR log. Typically two or three types of porosity log are required to determine the lithology and an accurate porosity. Figure 9.3.4 shows that most of the interval of interest has a GR response between 5 and 15 API. Six intervals are marked for lithology and porosity determination. The porosity readings are plotted on a Baker Hughes neutron-density cross plot in Figure 9.3.5. The analysis is summarized below:

1. ($\phi_D = -0.16$, $\phi_N = 0.01$) falls on the anhydrite line (density is 2.96 g/cm^3). The anhydrite provides a caprock seal for this reservoir and the bottom of the anhydrite marks the top of the Keg River formation.

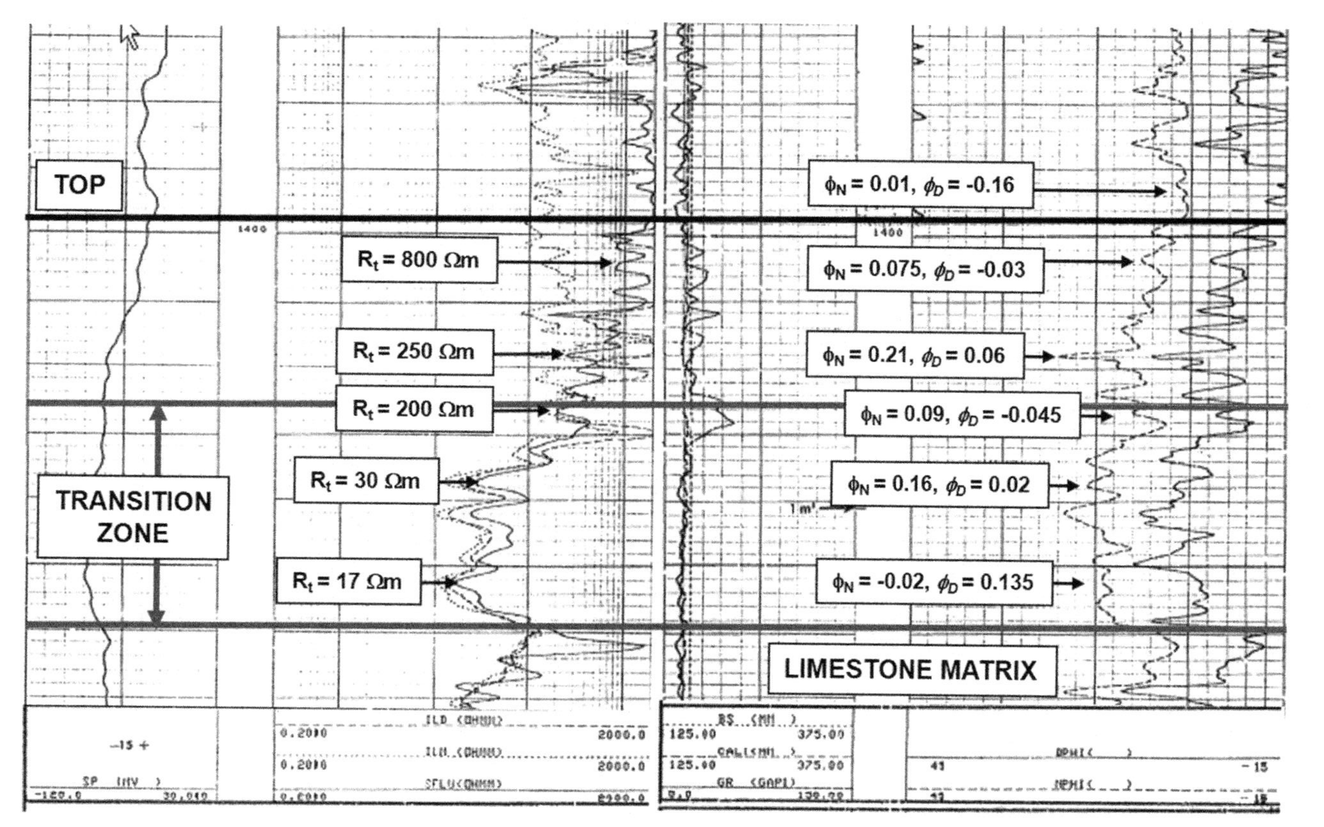

Figure 9.3.4 Set of openhole logs from 11-25-115-4W6.

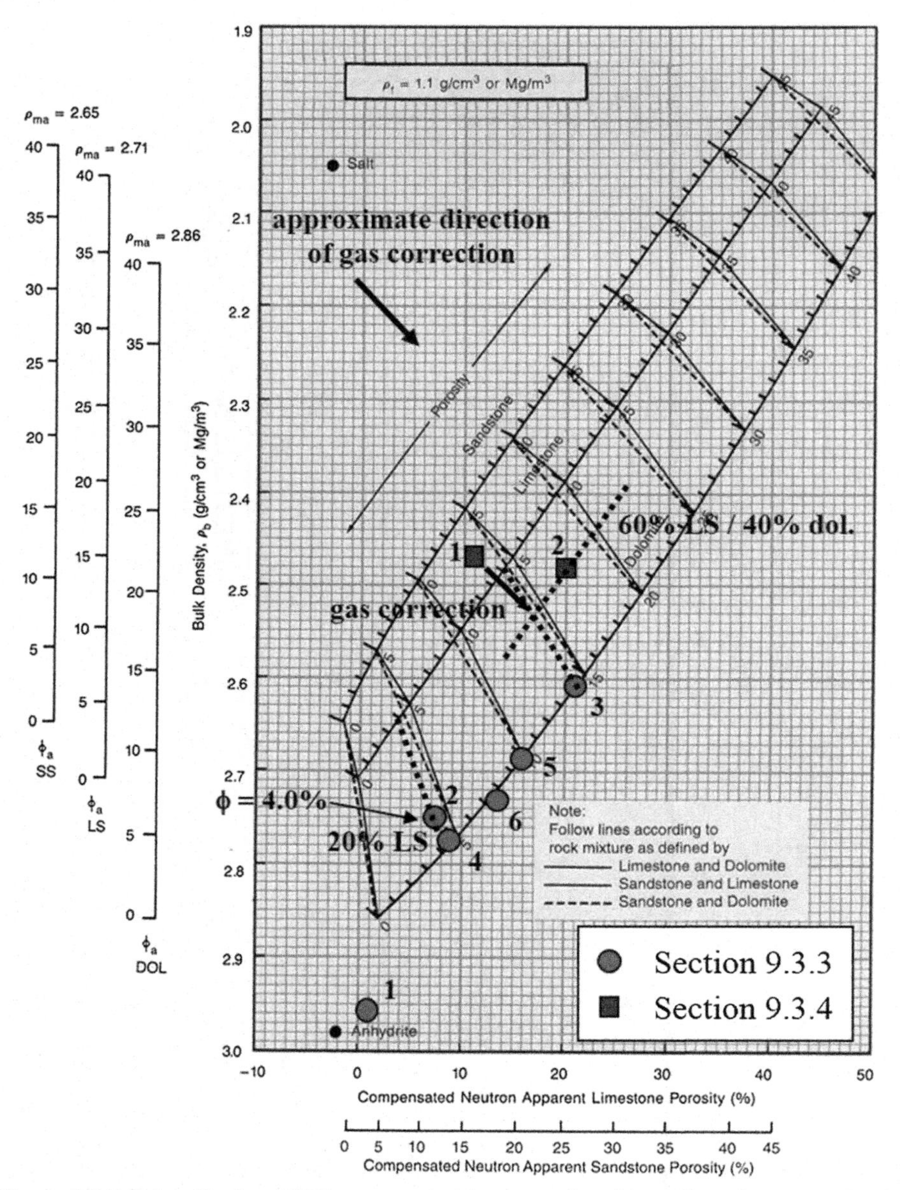

Figure 9.3.5 Baker Hughes (1995) neutron-density cross plot with readings from examples in Section 9.3.3 and Section 9.3.4.

2. ($\phi_D = -0.03$, $\phi_N = 0.075$) falls between the limestone and dolomite line, indicating mixed lithology. The proportion of limestone is determined from a lever arm rule as shown on Figure 9.3.5 and is approximately 20%. The average porosity is approximately 4.0%, as shown on Figure 9.3.5.
3. ($\phi_D = 0.06$, $\phi_N = 0.21$) is approximately 100% dolomite with an average porosity of 14.0%.
4. ($\phi_D = -0.045$, $\phi_N = 0.090$) is approximately 100% dolomite with an average porosity of 4.5%.
5. ($\phi_D = 0.02$, $\phi_N = 0.16$) is approximately 100% dolomite with an average porosity of 9.5%.
6. ($\phi_D = -0.02$, $\phi_N = 0.135$) is approximately 100% dolomite with an average porosity of 7.5%.

There is no evidence of a gas zone on the logs. Recall that if gas is present, the neutron density reads low, and cross-over with the density log is often observed. There is no cross-over in this case. Furthermore, all of the neutron-density porosity readings are on or near the dolomite curve. If a gas correction were applied, the data would fall outside the expected range of log readings. Therefore, it is unlikely that gas is present.

Note that if the formation were primarily limestone, then the effect of gas or dolomite are difficult to distinguish based on neutron-density porosity data alone. Core data or an acoustic log would also be required to assess the reservoir lithology, porosity, and hydrocarbon type.

Water Saturations and Water–Oil Contact

The water resistivity was determined from an underlying water zone and is 0.026 Ωm. The formation resistivities for the five limestone/dolomite intervals are marked on Figure 9.3.4. The water saturations are determined using Archie's Law, as shown below for interval 2:

$$S_w = \left(\frac{KR_w}{\phi^m R_T}\right)^{0.5} = \left(\frac{1.0(0.038\ \Omega\text{m})}{(0.040)^{2.0}(800\ \Omega\text{m})}\right)^{0.5} = 0.17$$

The depths, lithologies, porosities, and water saturations for the six marked intervals are summarized in Table 9.3.1. The value $m = 2$ in the above equation is based on the assumption of intergranular porosity being the dominant porosity type in this formation. If moldic or vuggy porosity is present in significant amounts, m may need to be increased. This would increase the value of S_w.

There appears to be a transition zone from approximately 1413 mKB and 1430 mKB. There is an anhydrite zone with some dolomite streaks from 1430 to 1441 mKB. The porous intervals below this streak (not shown on Figure 9.3.4) have low resistivity and approximately 100% water saturation. The precise location of the WOC cannot be determined because the anhydrite streak intervenes, but it is between 1430 and 1441 mKB.

9.3.4 Analysis of a Carbonate Sequence with a Gas-Oil Contact

Figure 9.3.6 shows SP/Dual Induction, GR/Caliper/Neutron-Density, and GR/Caliper/Acoustic logs obtained for a D3 carbonate sequence. As with the previous example, the high-resistivity readings are consistent with a carbonate formation. A core analysis for

Table 9.3.1 Summary of log interpretation for six Keg River intervals from 11-25-115-4W6

Interval	Depth (mKB)	Lithology	Porosity (%)	Water saturation (%)
1	1396.5	Anhydrite	0	–
2	1402.0	20L/80D[a]	4.0	17
3	1409.0	D	14.0	9
4	1413.5	D	4.5	31
5	1419.0	D	9.5	37
6	1426.5	D	7.5	63

[a]30L/70D = 30% limestone, 70% dolomite.

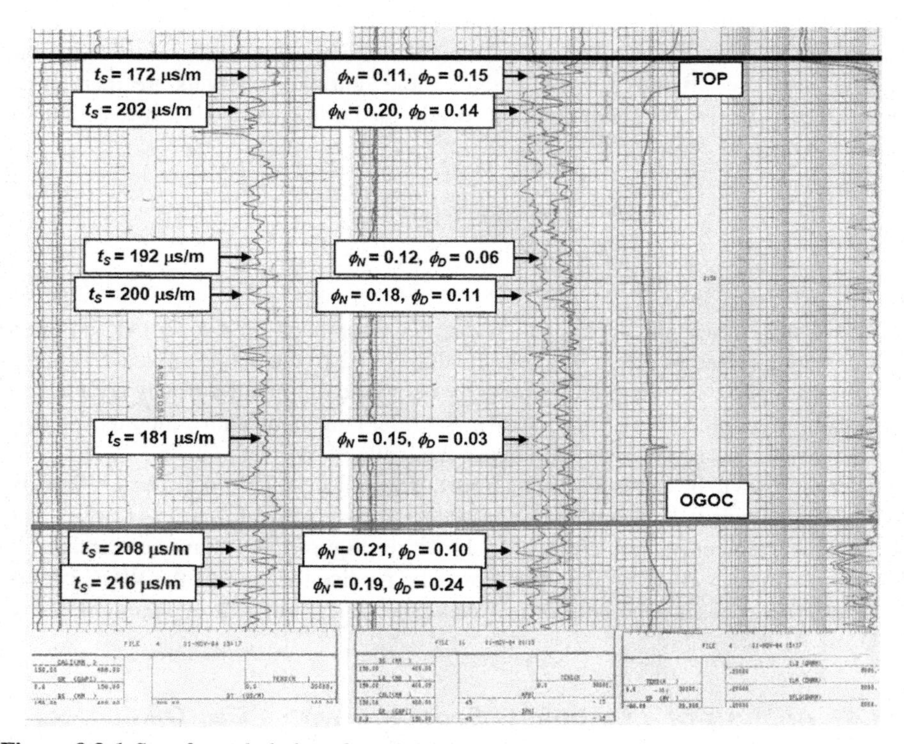

Figure 9.3.6 Set of openhole logs from 2-36-45-1W5 for the interval 2118 - 2199 mKB.

a different interval indicated that the zone was approximately 30−40% limestone and 60−70% dolomite. The original gas-oil contact (OGOC) was established independently and is shown on Figure 9.3.6. The contact has moved downward since the pool was discovered, and the interval below the OGOC on Figure 9.3.6 was gas swept at the time the well was drilled.

The analysis of a carbonate gas zone is complex because both the lithology and the gas effect influence the log readings. Consider for example the top two intervals marked on Figure 9.3.6 at 2121 mKB and 2126 mKB. The top interval shows cross-over on the neutron-density log, while the second interval shows no cross-over. When plotted on a neutron-density cross plot, Figure 9.3.5, the first interval plots above the limestone lithology line and clearly requires a gas correction. Note that the gas correction shifts the plotted point to higher neutron porosities and slightly lower density porosities, as indicated by the gas correction arrow on Figure 9.3.5. The second interval, on the other hand, plots as a 60% limestone, 40% dolomite zone. If there is no gas effect then the lithology is probably correct. However, if there is a mild gas effect, the plotted point must be corrected, and the interval must contain a greater proportion of dolomite than at first indicated. Therefore, it is not possible to reach a conclusive interpretation based on the neutron-density logs alone. In other words, two measurements (density and neutron) cannot uniquely identify three characteristics (porosity, limestone/dolomite proportions, and gas fluid).

As a first pass, the mineralogy and average neutron-density porosity, ϕ_{ND}, were determined from the neutron-density cross plot for the seven marked intervals of Figure 9.3.6. The results are listed in Table 9.3.2. The acoustic porosity was then determined using Figure 9.2.11 and the apparent lithology from the neutron-density log. In most cases, the acoustic porosity and the neutron-density porosity agreed within two porosity units. This agreement is within the error of the log readings and the calculated lithologies. Therefore, it appears that there is little gas effect, and the neutron-density logs do not require correction except for the relatively rare intervals where cross-over occurred. It appears that there is usually little residual gas in the invaded zone of this formation.

Now let us consider the two intervals where cross-over was observed at 2121 mKB and 2192.5 mKB. The average neutron-density porosity was estimated by extrapolating along the gas effect correction line to the expected lithology, as shown in Figure 9.3.5. In both cases, the neutron-density porosity far exceeded the acoustic porosity. One possible explanation is that there is both primary and secondary porosity in these two intervals. The neutron-density porosity is the total porosity, while the acoustic porosity is the primary porosity. The substantial secondary porosity may indicate a vuggy zone. The vugs may be partially isolated from the mud invasion and still contain evolved solution gas.

It is not possible to identify the original or the current gas-oil contact from these logs because the gas effect is inconsistent. When the gas effect is strong and consistent, the GOC can be determined as discussed in Section 9.3.2. Otherwise another method such as production logging must be used.

Table 9.3.2 Summary of log interpretation for six D3 intervals from 02-36-45-1W5. Gas effect is assumed to be negligible unless otherwise stated

Interval	Depth (mKB)	ϕ_D (%)	ϕ_N (%)	Apparent Lithology	ϕ_{ND} (%)	Transit Time (μs/m)	ϕ_S (%)
1	2121.0	15	11	a	~14	172	7
2	2126.0	14	20	60L/40D	17	202	15
3	2147.0	6	12	60L/40D	9.5	192	12
4	2152.0	11	18	55L/45D	14.5	200	15
5	2172.5	3	15	20L/80D	10	181	11
6	2188.0	10	21	35L/65D	16	208	18
7	2192.5	24	19	a	~23	216	20

[a]Gas effect, lithology assumed equal to that of nearest interval without gas effect.

9.4 Comparison of Log and Core Porosity

Core porosity is usually an accurate measure of local porosity because the porosity is measured directly. Also, the core porosity is relatively insensitive to overburden pressure except in unconsolidated rock, and therefore porosities measured in the laboratory are expected to be close to the in situ porosity. Openhole logs provide an indirect measure of an average porosity representing approximately a half-meter interval at the wellbore. Whenever possible, it is desirable to correlate the log porosity to the core porosity. If there is a discrepancy, the log porosity or porosity averaging formula can be corrected to match the core data.

Figure 9.4.1 shows a GR from the core and the GR-Neutron-Density logs from a Mannville formation in the well 16-14-20-20W4. The core was depth corrected to match the log depths, as discussed in Section 7.3.5. The core porosities are listed in Table 9.4.1. The neutron and density porosities at the same depths are also listed in Table 9.4.1.

Because the density log has somewhat better resolution than the neutron log and the lithology is simple, we will use only the density to compare with the core. A cross-plot of density porosity and depth-corrected core porosity (Figure 9.4.2 left) shows a weak correlation. The least-squares regression line has a slope and intercept different than the expected values of 1 and 0, respectively. A simple, three-point running average applied to the core data gives a different result (Figure 9.4.2 right). The difference occurs because the core data represent smaller volumes of the reservoir, so the running average has helped to provide core data at a resolution more consistent with the density log measurements. The regression line suggests that the density porosity is about one porosity unit too low, which is likely caused by having calculated the density porosity

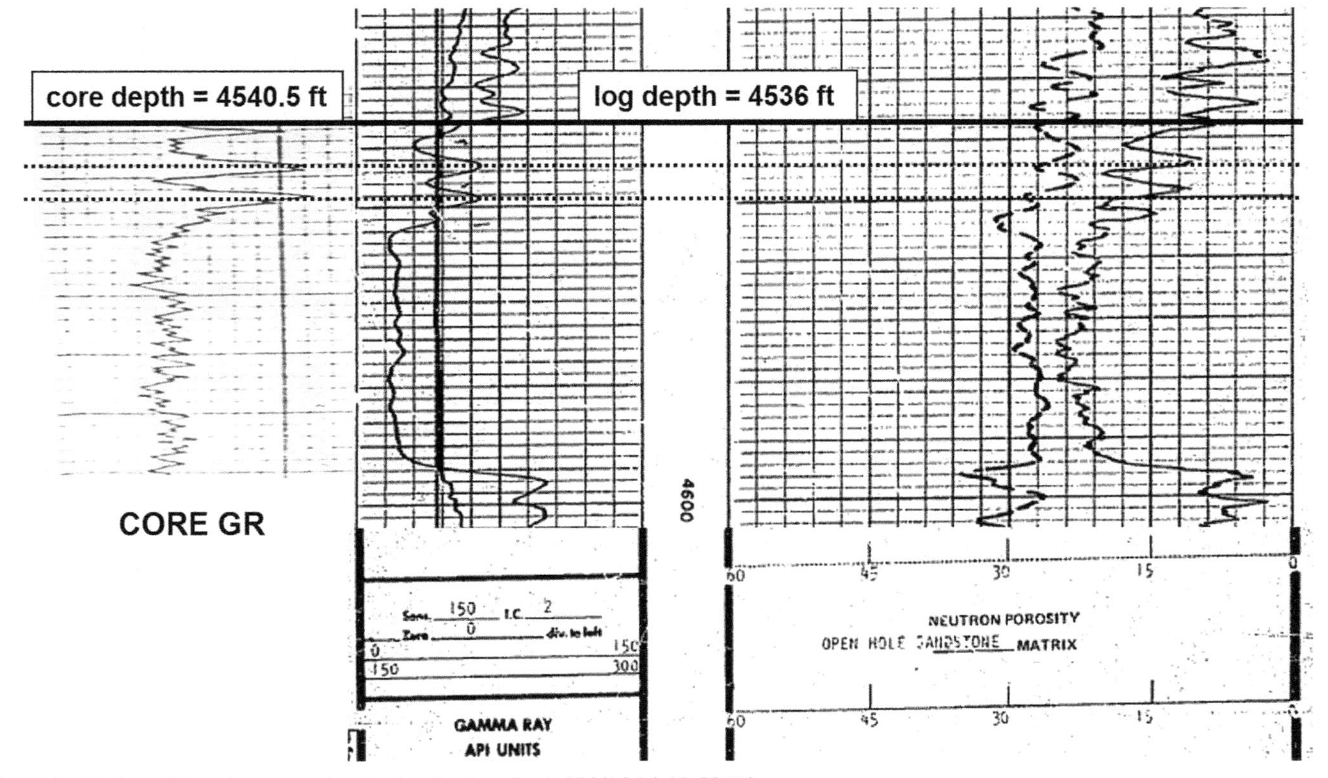

Figure 9.4.1 Core GR and neutron-density log for Jumpbush 100/16-14-20-20W4.

Table 9.4.1 **Comparison of core and log porosities at different depths**

Core depth (ft)	Log depth (ft)	Core porosity	Density porosity	Neutron porosity
4555.45	4550.95	0.209	0.195	0.280
4556.10	4551.60	0.200	0.205	0.270
4557.09	4552.59	shale	–	–
4558.07	4553.57	0.185	0.170	0.310
4558.73	4554.23	0.236	0.215	0.310
4559.38	4554.88	0.236	0.215	0.310
4560.04	4555.54	0.236	0.200	0.305
4561.02	4556.52	0.233	0.200	0.305
4561.68	4557.18	0.231	0.210	0.270
4562.34	4557.84	0.237	0.230	0.265
4562.99	4558.49	0.237	0.225	0.270
4563.65	4559.15	0.232	0.235	0.280
4564.30	4559.80	0.223	0.225	0.285
4564.96	4560.46	0.240	0.210	0.280
4565.62	4561.12	0.232	0.210	0.275
4566.93	4562.43	0.203	0.220	0.290
4567.59	4563.09	0.228	0.210	0.280
4568.24	4563.74	0.240	0.215	0.275
4568.90	4564.40	0.231	0.225	0.275
4569.88	4565.38	0.230	0.250	0.270
4570.21	4565.71	0.230	0.250	0.270
4570.87	4566.37	0.233	0.230	0.270
4571.85	4567.35	0.229	0.225	0.280
4572.83	4568.33	0.221	0.220	0.275
4573.82	4569.32	0.233	0.240	0.270
4574.48	4569.98	0.246	0.220	0.285
4575.13	4570.63	0.241	0.225	0.290
4576.12	4571.62	0.243	0.225	0.285
4576.77	4572.27	0.238	0.230	0.280
4577.76	4573.26	0.238	0.240	0.280

Table 9.4.1 Comparison of core and log porosities at different depths—cont'd

Core depth (ft)	Log depth (ft)	Core porosity	Density porosity	Neutron porosity
4578.74	4574.24	0.263	0.250	0.285
4579.40	4574.90	0.245	0.240	0.295
4580.38	4575.88	0.240	0.250	0.290
4581.04	4576.54	0.241	0.240	0.280
4581.36	4576.86	0.241	0.230	0.275
4582.02	4577.52	0.245	0.230	0.270
4582.68	4578.18	0.269	0.240	0.270
4584.32	4579.82	0.234	0.250	0.265
4585.63	4581.13	0.193	0.220	0.270
4585.96	4581.46	0.229	0.210	0.270
4587.27	4582.77	0.243	0.225	0.270
4587.93	4583.43	0.243	0.230	0.260
4588.58	4584.08	0.241	0.230	0.260
4590.55	4586.05	0.239	0.220	0.280
4592.55	4588.05	shale	–	–
4593.18	4588.68	0.230	0.200	0.280
4594.49	4589.99	0.221	0.220	0.275
4596.46	4591.96	0.229	0.210	0.270
Average	–	**0.233**	**0.223**	**0.279**
SSR[a]	–	–	**0.014**	**0.121**

[a]SSR = sum of square residuals = $\sum(\phi_{avg,i} - \phi_{core,i})^2$.

using $\rho_{ma} = 2.65$ g/cm^3 instead of a slightly denser 2.67 g/cm^3, which is still appropriate for sandstones. Note core grain density data, if available, can be used to select a matrix density appropriate for the formation of interest.

Some caution is required when comparing log and core porosities. In some formations, there is little local variation in porosity, and the porosity of a sample from the core is likely to be similar to the average porosity sampled with the log. In other formations, such as carbonate reefs, there can be considerable variation in porosity on the millimeter scale. Hence the core plug may differ considerably from the average obtained with a log. In this case, a large number of samples are required, and comparisons are best made between average core porosities and average log porosities.

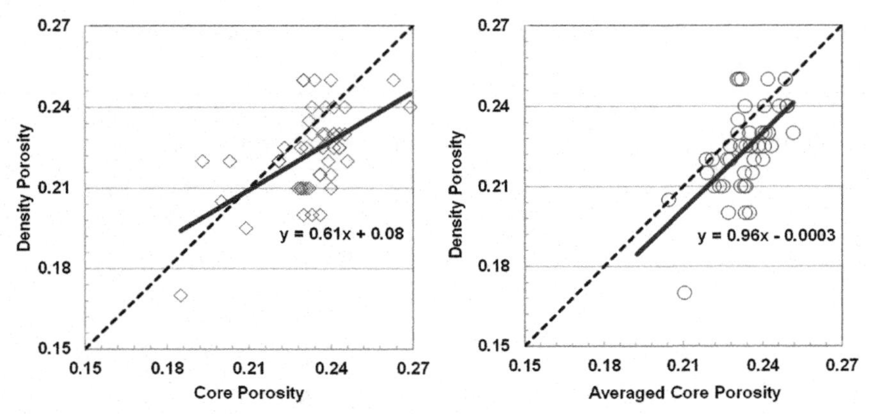

Figure 9.4.2 Cross-plots of core and density log porosity without averaging (left) and with averaging (right) of the core porosity data. The dotted line is the 1:1 line, and the solid line is the least-squares regression line.

9.5 Net Pay Cutoffs

When examining net pay cutoffs, it is important to differentiate among gross rock volume, net sand volume, net reservoir volume, and net pay. For example, the concept of net pay is used in reserves determination, but the concept of net reservoir should be used in flow simulation analysis. Reservoir simulators usually require gross pay with appropriate gross-to-net pay multipliers so that fluid contacts and vertical gradients are modeled correctly. Hence, the appropriate cutoffs depend on the application.

In practice, cutoffs are applied to well logs or cores, and we therefore examine thickness (pay) rather than volume. Figure 9.5.1 shows four categories of rock: gross pay, net sand, net reservoir, and net pay.

Gross pay comprises all rocks within the evaluation interval, that is, the thickness between the top and bottom of the zone(s) of interest. In sandstone reservoirs, the inflection point of the SP curve or the GR curve inflection points are used to identify the boundaries (Cronquist, 2001). In carbonates, porosity cutoffs are used to define the lithological tops and bottoms. Depending on the application, gross pay may be set as the thickness of the reservoir or the hydrocarbon interval. Gross pay can include low permeability intervals, shaly intervals, and water-saturated intervals through which hydrocarbon fluid will not flow.

Net sand is those rocks that might have useful reservoir properties. Net reservoir is those sand intervals that do have useful reservoir properties. Net pay comprises those intervals that do contain movable hydrocarbons (Worthington and Costentino, 2005). Gaynor and Sneider (1993) define net pay as “the hydrocarbon bearing volume of the reservoir that would produce at economic rates using a given production method.” There are two general considerations controlling net pay:

- Does the rock have sufficient permeability to produce oil or gas at economic rates?
- Does the rock contain movable oil volumes (fluid saturation)?

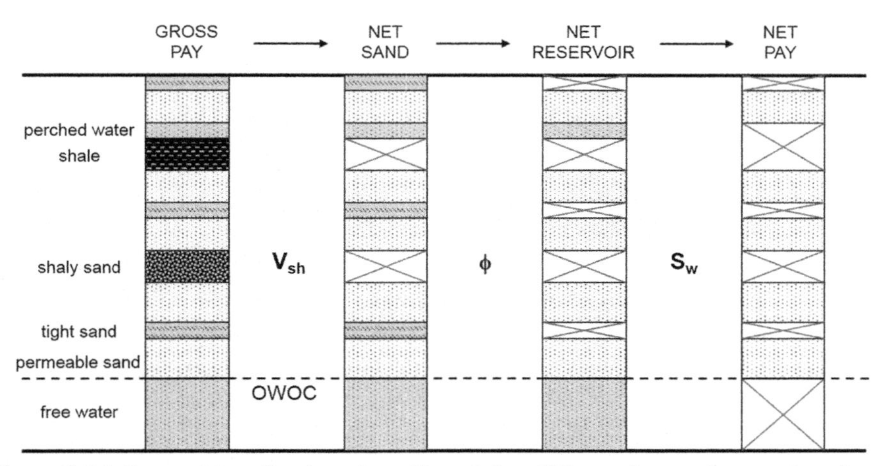

Figure 9.5.1 Sequential application of cutoffs to define different classes of reservoir rock. Adapted from Worthington and Costentino (2005).

Worthington and Costentino (2005) is an excellent source of information on pay cutoff criteria.

There are typically three factors that differentiate gross pay, net sand, net reservoir, and net pay: shale volume (V_{shale}), porosity, and saturation. The first two criteria address whether or not the rock has sufficient permeability. The last criteria addresses whether or not the rock has sufficient hydrocarbons. But what exactly are the values of each factor that differentiate between the categories; that is, what are the cutoffs?

Factors Controlling Net Pay Criteria

A common definition of net pay is the thickness of that part of the hydrocarbon-bearing zone of the reservoir that is capable of producing at economic rates using a given production method. However, economic rate is an imprecise concept. Recall Darcy's law for pseudo steady-state radial inflow:

$$q \propto \frac{k_{eff}}{\mu} \frac{P_r - P_{wf}}{0.472 \ln\left\{\frac{r_e}{r_w}\right\} + S} \tag{9.5.1}$$

Darcy's law illustrates that the rate at which a well flows depends on the effective permeability (and therefore on the oil, gas, and water saturation and the overburden pressure), the reservoir fluid viscosity, the pay thickness, the drainage radius, type of completion, drive mechanism, and the pressure drawdown. In some reservoirs, tight rock can flow into more permeable rock, and thus reservoir heterogeneity and flow path tortuosity become factors. The economics of that rate depend on operating costs and the price of oil and gas. Hence, there are many factors involved in determining net pay including:

- Oil or gas prices;
- Completion or well design (horizontal versus vertical well, hydraulically fractured versus nonstimulated well);

- Near wellbore flow;
- Reservoir heterogeneity and its distribution;
- Total pay thickness;
- Effective vertical thickness and flow path tortousity;
- Drive mechanism;
- Mobility ratio;
- Oil viscosity;
- Lateral continuity of interval for injection processes;
- Amount and quality of data.

Many of those factors are difficult to determine, such as the permeability distribution or the drainage radius of a particular interval. For this reason, there is no general consensus on how to determine net pay but some progress is being made (Worthington and Cosentino, 2005; Jensen and Menke, 2006).

Cutoffs must be adapted to changing conditions. Using a definition of pay that may have been valid at a low oil price is not correct if oil prices increase. Technology also strongly effects the definition of net pay. For example, hydraulically fracturing wells of low-permeability fields and horizontal/multilateral wells have allowed economic development of fields that were once considered below pay cutoff criteria. Therefore net pay cutoffs often need to be fit for purpose. In the authors' experience, rule-of-thumb cutoffs may underestimate tighter oil and gas reservoirs, especially when horizontal wells and multifractured horizontal wells are considered.

Cutoff criteria also need to account for different drive mechanisms (Cobb et al., 1996). For waterfloods, only reasonably permeable and laterally continuous intervals between the injectors and producers should be counted as net pay. Primary production will generally have a lower net pay cutoff criterion compared to waterflood or injection processes.

Often data availability and quality dictate the use of pay cutoff criteria. For example, the probe and core plug data in Figure 9.5.2 give differing net pay proportions. If the cutoff is 1 mD, the core plug permeabilities show 95% net pay, while the probe data show 81% net pay. The probe measurements have better resolution, and because they are nondestructive, they can be made at locations where cutting a plug is not possible. Also, in old fields without core, modern logs, or SCAL data, we must frequently rely on rules of thumb at least initially. In practice, there needs to be an iteration between the dynamic data (production and reservoir pressures) and the cutoffs used.

Rules of Thumb for Net Pay

Determining a pay cutoff is sometimes a difficult decision, because of small sample numbers compared to the size of the reservoir and general lack of data. As Worthington and Cosentino (2005) put it: "Unfortunately, there is no universal definition of net pay nor is there general agreement on how it should be delineated." Nonetheless, there are some rules of thumb for determining cutoffs.

Most rules of thumb for pay cutoffs are built on three considerations. First, does the rock have enough permeability? Second, does the rock have sufficient movable hydrocarbon pore volume to yield oil or gas at economic rates? Third, does that volume of

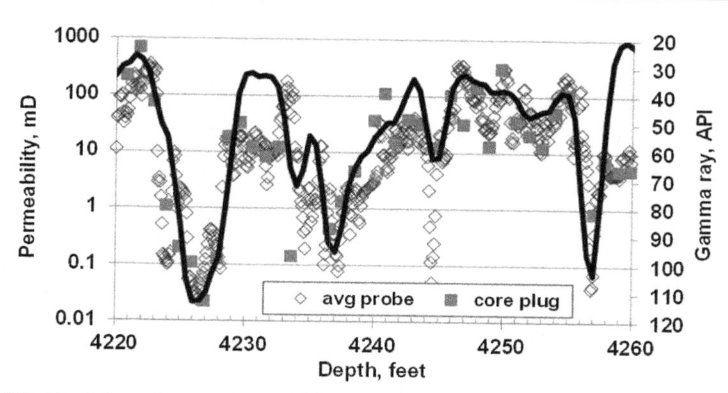

Figure 9.5.2 Fluvial sandstone shows different net pay values, depending on the permeability measurement. Core plug measurements are approximately every one foot, while probe data are taken at 1/10th foot spacing.

rock contain sufficient permeability to flow substantial volumes of hydrocarbon relative to other rocks in the reservoir (relative processing or drainage)?

Most rules of thumb for pay cutoffs are built on an implicit assumption of air permeability, but effective permeability is a more relevant criterion. Yet effective permeability is one of the hardest parameters to determine. Usually, we use log data of porosity, water saturation, and volume of shale (V_{sh}) as proxies for air permeabilities.

Probably the most common cutoffs are permeability cutoffs, with typical values of:

$k_{eff} > 0.5$ to 1.0 mD for oil reservoirs

$k_{eff} > 0.1$ to 0.5 mD for conventional gas reservoirs

$k_{eff} > 0.001$ to 0.000 001 mD for tight gas with horizontal multifrac wells

in which the pay that has a permeability exceeding the cutoff is considered to be net pay (Cronquist, 2001). It is less common, but likely more accurate to include the viscosity as follows:

$k_{eff}/\mu > 1.0$ mD/cp for oil or gas reservoirs

in which μ is the in situ fluid viscosity. A method for determining permeability cutoffs was discussed in Section 7.3.4. The experience of the authors however is that these rules of thumb are usually too pessimistic, especially when horizontal wells, hydraulic fracturing, and horizontal wells with hydraulic fracturing are concerned.

Note that the cutoffs are based on effective permeability and not measured air permeability. The effective permeability includes the effects of overburden pressure and connate water saturation and may be up to 30% lower than measured air permeabilities. Because effective permeabilities are often not known, air permeability is often used as an approximate effective permeability for the reservoir instead. Using air permeability without correction may lead to an overestimation of the net pay.

Usually, cutoffs are applied using openhole logs. Cutoff values for porosity, water saturation, and shale content are required. The porosity cutoff can be determined from a given permeability cutoff using a porosity-permeability cross plot. If a cross plot is not available, the following rules of thumb can be used:

$\phi > 1-3\%$ for gas-bearing carbonates;
$\phi > 2-4\%$ for oil-bearing carbonates;
$\phi > 5-8\%$ for gas-bearing sandstones;
$\phi > 7-10\%$ for oil-bearing sandstones;
$\phi > 26-28\%$ for heavy oil-bearing sandstones.

A high-connate water saturation correlates to a lower effective permeability. Therefore, another rule of thumb is:

$$S_w < 50\%$$

In this case the interval must have a water saturation below the cutoff to count as net pay.

A part of the pay zone that has the same properties as the overlying or underlying sealing rock cannot be considered as net pay. In practice, sandstones with high shale contents also have low effective permeability. Hence, a shale cutoff is often applied as well as follows:

$$V_{sh} < 50\%$$

in which V_{sh} is the volume of shale in the gross pay interval.

Old logs such as the Microlog™ can be very useful in estimating net pay. A common rule of thumb is that zones with less than about 1 mD permeability will not exhibit Microlog™ separation. Vertical intervals with filter cake may indicate permeable zones.

Always remember that the rules of thumb and even the permeability cutoffs are approximations for an economic rate cutoff. It is often useful to perform a sensitivity study to determine whether the volumetric oil or gas-in-place determination is sensitive to the choice of cutoffs. If the initial reserves are sensitive to the cutoffs, it may be necessary to re-evaluate the cutoffs used for a given reservoir if operating practices, oil and gas prices, or operating costs change significantly.

Using Field Production Data to Assist in Determining Net Pay

Often during the initial phases of a field development, drill stem testing is used, and intervals are tested for oil, gas, and water production. Sometimes wells are abandoned after this initial test. This information can be extremely useful in determining net pay cutoffs.

Reserves Approach Versus Modelling Approach

It is critical to realize the fundamental difference between net oil pay and net reservoir. Many reservoirs have highly mobile zones filled with gas and water that strongly affect oil production behavior, yet these intervals may not yield substantial economic volumes of oil. In simulation or analytical models, it is important to account for net reservoir volumes and net pay. For example, it may be necessary to include this net reservoir volume rather than net pay in a material balance calculation. For reserves definition, net pay rather than net reservoir is the critical parameter. People often confuse net pay with net reservoir.

When selecting net reservoir, it is important to use the permeability criteria, and its proxies alone. Therefore, porosity and shale volume may be used to determine net reservoir, but not water saturation because higher water saturation regions may flow into the oil reservoir and alter production forecasts.

Summary

Net pay cutoffs are a function of many parameters. Some of these parameters are not well known (for example flow path tortuosity). Pay cutoff determination is an iterative and fit-for-purpose task.

We recommend using the following plots when data availability permits:

- Porosity versus permeability;
- Porosity versus water saturation;
- Water saturation versus permeability;
- GR versus permeability;
- Cumulative frequency plots of porosity and permeability;
- Cumulative frequency plots of mobile oil ($\phi[S_{oi}-S_{or}]$);
- Initial water saturation versus residual oil saturation.

The trends and scatter of the data in these plots can be used to relate one cutoff to another or to demonstrate that such correlations do not exist.

We recommend the following approach to establish cutoffs:

1. Use rules of thumb and the above plots for an initial guess of net pay. For example, start with a permeability cutoff and use the plots to see if water saturation and porosity cutoffs can be determined from the permeability cutoff.
2. Examine drill stem testing or production test data for abandoned wells at the edges of the pool. Use the permeability, porosity, and water saturation of these wells to refine the cutoffs.
3. Finally, as production and reservoir pressure data become available, iterate the cutoff to match volumetrics with a material balance or flow simulation based on field pressure and production data.

Part Three

Reservoir Characterization

Most reservoir characterization work is done to determine the remaining recoverable oil or to optimize recovery and production rate from that field. Recovery is maximized by optimizing the existing production scheme or by implementing a new improved recovery method. There are a number of interrelated questions that need to be addressed in most reservoir characterization studies:

1. What is the original oil in place (OOIP)?
2. What is the remaining recoverable oil in place (RROIP)?
3. What is the permeability distribution and reservoir architecture?
4. What are the RROIP conditions (pressure, saturation, temperature, permeability)?
5. Which wells communicate strongly and are there bypassed regions?
6. What are the limiting factors in oil recovery?
7. What can I do to improve recovery and or oil rates?
8. Where can I make changes? (Which wells can I convert or drill?)
9. Are the changes proposed economically viable?

For injection processes, two additional questions need to be addressed:

10. What is the displacement efficiency?
11. What is the volumetric sweep?

Although it is important to consider the above questions when addressing reservoir characterization, it is not critical that each question have a 100% accurate answer. Many times only an approximate answer is possible even after substantial production has occurred.

Reservoir characterization involves combining all of the reservoir data discussed in Part 2 to form a one-, two-, or three-dimensional representation of the reservoir. This representation can be a simple conceptual (or mental) model or more advanced analytical, geological, or flow simulation model. The reservoir characterization exercise involves determining four components:

- fluid type (heavy oil, conventional oil, volatile oil, retrograde condensate, gas),
- architecture (size and structure, porosity, permeability, and saturation distributions),
- drive mechanism (expansion, solution gas, gas cap, water, combination), and
- dominant flow direction

Characterizing a reservoir from what are usually limited and sometimes inconsistent data is probably the greatest challenge in reservoir engineering. Figure 10.0.1 presents the major data sources and activities needed to develop

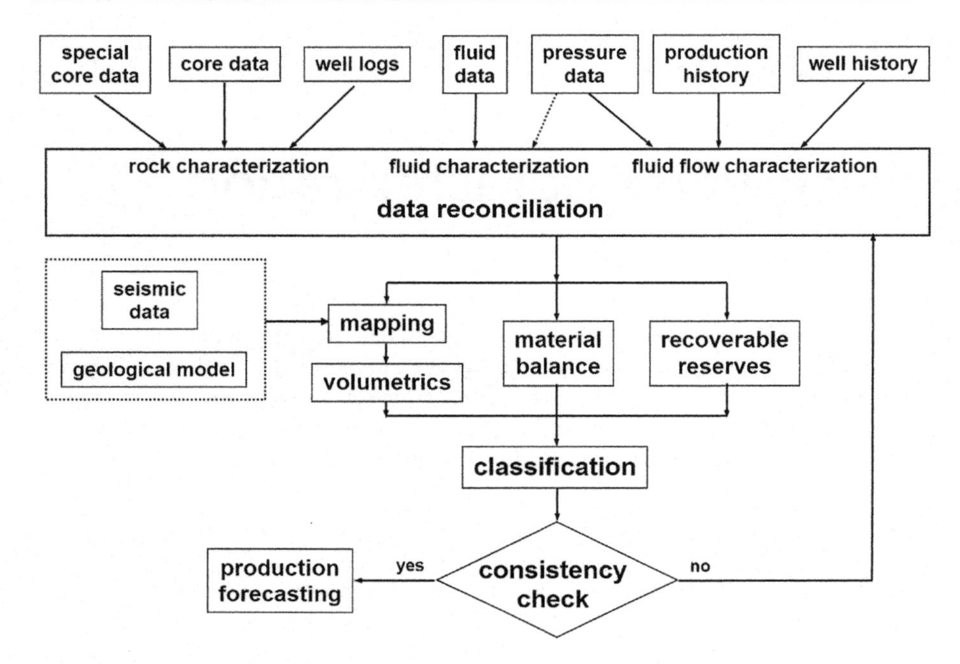

Figure 10.0.1 Components involved in the reservoir characterization process.

a characterization. Note that geological modeling is outside the scope of this book. It involves combining general knowledge of geological depositional environments with diagenesis (changes in rock fabric after burial) with petrophysical and geophysical data. Books such as Slatt (2006) cover geological model development.

The first step in reservoir characterization is to reconcile the data from different sources. For example, the fluid data from a PVT study (bubble-point pressure and solution gas–oil ratio) are reconciled with the produced GOR and pressure behavior (Figure 10.0.1). Sometimes, advanced methods can be used to "fingerprint" the fluids to help identify hydrocarbon sources and assess reservoir compartmentalization. The porosity data (rock model) from core and logs and (if available) seismic data are compared and adjusted as required to prepare a self-consistent data set. Permeability is reconciled between core and log as well as large scale measurements such as productivity index and pressure transient (build up) measurements. Often an understanding of reservoir architecture is critical to reconciling permeability from the reservoir scale to the smaller scale core data.

It is in this step that preliminary rock, fluid, and fluid flow characterizations are prepared. Rock and fluid characterizations include creating an appropriate fluid model, identifying the rock property distributions, and modeling the rock-fluid properties. The fluid flow characterization includes an analysis of the production and injection data as well as reservoir pressure data. The objective of the analysis is to determine continuity, effective permeability, and assist in determining OOIP. In addition, sometimes analysis of fluid flow will give us information on flow geometry, such as coning, linear, radial, or spherical flow. Fluid characterization usually relies on laboratory tests.

The second step in reservoir characterization is to prepare maps of the reservoir and determine the volumetric oil and gas in place. At the same time, the fluid flow characterization is used to determine the likely drive mechanism and expected recoverable reserves. A material balance can also be done in some reservoirs and this is based on the identified drive mechanism. Comparing volumetric OOIP numbers to material-balance OOIP or simulation-derived OOIP can be a very useful exercise.

In the third step, a "mental model" of the reservoir is constructed and a reservoir classification is made based on the volumetrics, fluid flow characterization, and material balance. The reservoir architecture, fluid type, and drive mechanism interpretation is finalized at this point. The main things to reconcile at this stage are (1) the volumetric and material balance hydrocarbon volumes in place, (2) production and pressure history with reservoir architecture, (3) the recovery factor and production profile with the volumes in place, architecture, and drive mechanism. If the analysis is not consistent, then the data and interpretation are revisited and the reservoir characterization team must then judge which data to use and which interpretation is most likely. Often, there are insufficient data to perform the full step-by-step analysis and the reservoir engineer must adjust the approach according to the data available. In some cases, several iterations are required to obtain a satisfactory characterization. If the analysis is self-consistent, then production forecasts can be made.

Chapter 10 focuses on the methods used to develop a reservoir characterization, which include:

- data reconciliation
- mapping
- volumetrics
- analysis of production data
- material balance

The iterative process to combine these methods is the art of reservoir engineering and is discussed in Chapter 11. In particular, the reconciliation of production and pressure data with drive mechanisms and reservoir architecture is examined.

Reservoir Characterization Methods

10

Chapter Outline

Practical Reservoir Engineering and Characterization. http://dx.doi.org/10.1016/B978-0-12-801811-8.00010-9

The five main methods used to develop a reservoir characterization are:

- Data reconciliation;
- Mapping;
- Volumetrics;
- Analysis of production data;
- Material balance.

Data reconciliation is presented in Section 10.1 and includes log and core porosity comparisons, permeability—porosity cross plots, core and pressure transient permeability comparison, evaluation of permeability distributions, transition zones, and assessing the fluid characterization based on production and pressure data. Section 10.2 includes structure maps, pay maps, and cross-sections. Section 10.3 demonstrates how to determine volumes in place from the appropriate reservoir maps. Fluid flow characterization to help determine drive mechanism and reservoir architecture is discussed in Section 10.4. Material balance methods are presented in Section 10.5.

10.1 Data Reconciliation

The data required to characterize a reservoir are: porosity, permeability, water saturation, pay thickness, fluid properties, and initial pressure. Reconciliation of each of these properties is discussed below.

10.1.1 Porosity

Porosity is determined from conventional core and well log data. Typically, core data are more accurate because porosity is measured directly. Exceptions are unconsolidated sandstone and vuggy/fractured reservoirs. In unconsolidated sandstones, the large mechanical disturbances incurred during coring and core retrieval alter the porosity in the core sample, and well log values of porosity are considered to be more accurate than disturbed core samples. In formations with vugs and fractures, it is challenging to obtain a representative core sample because large vugs or a wide fracture and yet small core size may mean that the core sample is not representative.

Although the core porosities may be more accurate, unless every well is cored, core data sample much less of the reservoir than well logs. The recommended approach is to compare core and log porosities at the same depth for each well that has been both cored and logged. The log porosities are adjusted to match the core data either by adjusting the inputs into the log calibration (e.g., matrix density or matrix transit time) or by using an appropriate averaging formula (e.g., $\phi = 0.5\phi_N + 0.5\phi_D$ or $\phi = 0.33\phi_N + 0.67\phi_D$). Porosity reconciliation is discussed in more detail in Chapters 7 and 9.

If there are insufficient data to match core and log data directly, then the overall average core porosity can be compared with overall average log porosity. Some judgment is required to ensure that an appropriate comparison is made. For example, it is not valid to calibrate log data coming from one rock type (e.g., coarse sandstone) with core data coming from a different rock type (e.g., shaly sandstone).

Sometimes, seismic information is available for the reservoir. Seismic attributes such as amplitude and impedance may be compared to average porosity at the wells to establish a relationship between the attributes and porosity. Maps of the attributes can then be used to predict porosity away from the wellbores.

Porosity is one of the more reliable reservoir property measurements. Once an average porosity is determined, it is not recommended to adjust the average porosity by more than 0.01−0.03 (i.e., one to three porosity units). One exception to this rule of thumb is highly heterogeneous reservoirs with limited porosity data. If there are large areas that have not been drilled, the average porosity derived from core, and logs may not be representative. In this case, it may be valid to adjust the average porosity based on reservoir performance, material balance, and seismic information. Porosity distributions are discussed in Section 10.2.

10.1.2 Permeability

Permeability is one of the most critical factors determining reservoir performance, and it is often the most difficult reservoir parameter to determine accurately because it is highly variable. Permeability is also scale dependent, which adds to its complexity. Permeability is determined from conventional core data, permeability−porosity correlations, pressure transient analysis, and production rates. None of these methods provides a direct measurement of in situ permeability and the permeability distribution. Reconciliation and extrapolation of permeability data is probably the most challenging aspect of reservoir characterization. The best results are obtained if all four methods for determining permeability can be used together. The methods are briefly reviewed below.

Conventional core data: As discussed in Chapter 7, permeability in core samples is influenced by sample alteration, the Klinkenberg effect, overburden pressure, and the presence of connate water. Hence, there is a great deal of uncertainty in converting measured air permeability to in situ absolute permeability. Core data also only sample a small fraction of the reservoir. Nonetheless, core data are often the best tool for assessing small-scale, vertical, and lateral permeability variation.

Permeability−porosity correlation: In some cases, a semi-log cross plot of permeability versus porosity provides a good correlation (Section 7.3.3). The correlation can be used to predict permeability based on well log porosity. Average permeability and vertical and lateral permeability variation can also be assessed. Permeability−porosity correlations may have large amounts of scatter, and this method may be only suitable for order of magnitude estimates.

Pressure transient analysis: Pressure transient analysis (PTA) measures the reservoir response within the drainage area of the pressure test (Section 6.2). Hence, the permeability calculated from a drawdown, build-up, or fall-off test can provide a good estimate of the in situ average permeability within the drainage radius. PTA cannot provide a quantitative determination of the local vertical or lateral permeability distribution.

Flow rate equation: Permeability can be back calculated from Darcy's law if the flow rate, reservoir pressure, and bottom hole flowing pressure are known (Section 6.4.3). It is also necessary to correctly identify the boundary conditions. This method can work well in undersaturated reservoirs, but is uncertain when multiphase flow occurs and the local saturation profile and relative permeabilities are unknown. It is not usually possible to identify boundary conditions accurately enough to obtain better than an order of magnitude estimate of the permeability. As with PTA, this method will provide an average in situ permeability and cannot provide a quantitative determination of the local vertical or lateral permeability distribution.

Some factors to consider when comparing permeability data are listed in Table 10.1.1. In general, PTA provides the most representative estimate of the permeability that controls well deliverability because a build-up or fall-off test samples the average properties within the drainage area. This greater sampling area is important when there are lateral variations in permeability and when there are baffling effects due to permeability barriers, as shown in Figure 10.1.1.

If there are sufficient core data to assess the average permeability, the core data can be corrected to match the permeability from PTA. The permeability–porosity

Table 10.1.1 Factors to consider when reconciling permeability data (Haldorsen, 1986)

Factor	Effect
Core permeability not corrected for Klinkenberg effect	k_{core} too high
Core permeability not corrected for overburden pressure	k_{core} too high
Core permeability not corrected for effect of initial water saturation	k_{core} too high
Core permeability not corrected for temperature effect	$k_{core} \neq k_{\text{in situ}}$
Core altered during sample recovery and preparation	$k_{core} \neq k_{\text{in situ}}$
Natural fractures not present in analyzed core sample	$k_{core} < k_{PTA}$
Natural fractures present in analyzed core sample	k_{core} usually too high
Missing unconsolidated core—not included in calculation of average permeability	k_{core} too low
Core averaging techniques give different average	$k_{harm} < k_{geom} < k_{arith}$
PTA performed when multiple phases are flowing	$k_{PTA} < k_{core}$
Incorrect choice of PTA flow model (spherical, hemispherical, radial, linear)	k_{PTA} incorrect
The cored interval and the perforated interval for PTA do not match	$kh_{core} \neq kh_{PTA}$
Bedding, baffling, and tortuosity reduce effective permeability	$k_{PTA} < k_{core}$

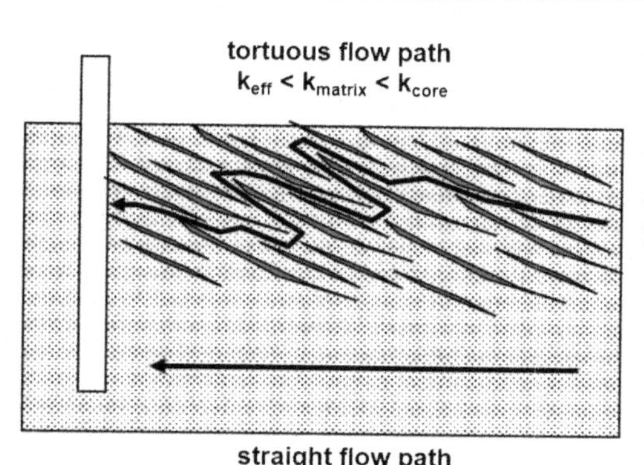

Figure 10.1.1 Effect of tortuosity on effective permeability. In a reservoir with permeability barriers, the fluid flow must take a longer path over the same radial distance and therefore the effective permeability is lower.

correlation should be scaled accordingly. It is recommended to use the permeability that was determined from flow data as a check only. Saleri and Toronyi (1988) recommended the following approach to assigning permeability:

1. Examine the core for bedding, cross bedding, and other heterogeneities.
2. Correct the core permeability data for overburden pressure, Klinkenberg effect, and initial water saturation.
3. Average the core permeability data using arithmetic, harmonic, and geometric averages; obtain separate averages for each zone that is in good hydraulic communication and use an average appropriate to the permeability distribution (Figure 10.1.2).
4. Compare average effective permeability from core data to permeability derived from PTA or productivity tests accounting for location of well and completed interval.
5. Calibrate core data and permeability–porosity data to obtain best estimate of in situ permeability from PTA or productivity tests.
6. Refine through history match of reservoir performance using analytical methods and simulation.

Note that some reservoirs have more than one rock type. Separate calibrations should be used for each rock type.

Averaging Permeability Data

Consider the permeability distributions shown in Figure 10.1.2. Each geometry requires a different expression to determine the average permeability, as outlined below.

For horizontal flow through layered permeability, use the arithmetic average (also called the mean):

$$k_{avg} = k_{arith} = \frac{\sum k_i h_i}{\sum h_i} \tag{10.1.1}$$

For vertical flow through layers, use the harmonic average:

$$k_{avg} = k_{harm} = \frac{\sum h_i}{\sum h_i/k_i} \tag{10.1.2}$$

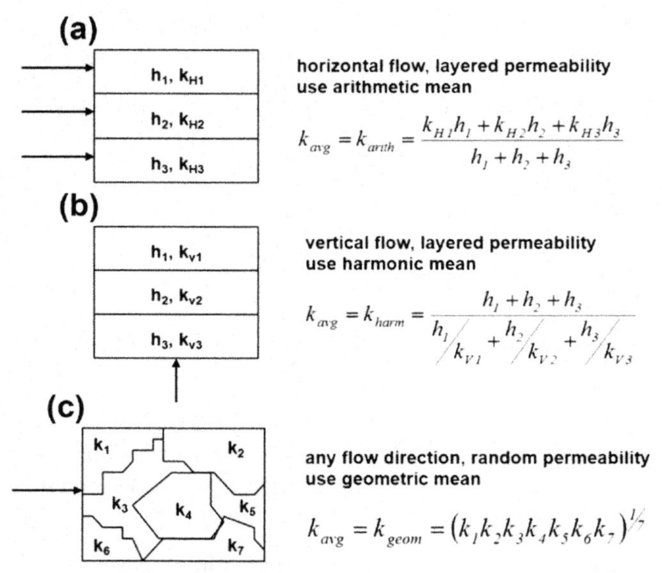

Figure 10.1.2 Averaging permeability for different permeability distributions: (a) layered permeability and horizontal flow; (b) layered permeability and vertical flow; (c) random permeability. Case (c) assumes the regions are roughly equal in size.

For flow in any direction through a disorganized permeability distribution, use the geometric average:

$$k_{avg} = k_{geom} = \sqrt[n]{k_1 k_2 \ldots k_n}. \tag{10.1.3}$$

The above formula for k_{geom} is a simplified version of the full form, which includes the volume V_i of region i and the total system volume V_t:

$$k_{avg} = k_{geom} = k_1^{V_1/V_t} k_2^{V_2/V_t} \ldots k_n^{V_n/V_t}. \tag{10.1.4}$$

For the same set of data, $k_{harm} \leq k_{geom} \leq k_{arith}$, and they are equal only in a homogeneous reservoir. These formulas apply for any scale of heterogeneity. That is, the averages apply equally whether the heterogeneities are in the form of cm-thick laminations, meter-scale beds, or km-size regions.

Vertical-to-Horizontal Permeability Ratio

The k_V/k_H ratio can be determined from a variety of measurements, including core, well test, and tidal influence. The most common source is conventional core data. This value may be representative of the matrix k_V/k_H ratio if there is sufficient core data. However, core data cannot determine the effective vertical permeability when there are large-scale barriers to vertical flow, as shown in Figure 10.1.3. For example, interbedded sandstones can have laterally extensive shale barriers that reduce the effective vertical permeability to near zero.

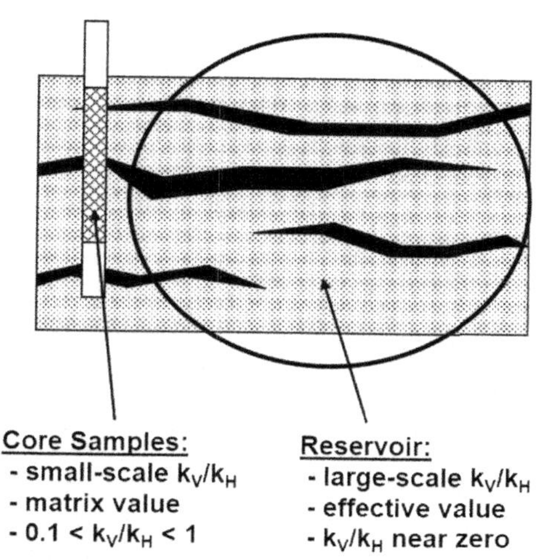

Figure 10.1.3 Differences between matrix k_V/k_H ratio from core data and effective k_V/k_H ratio in reservoir.

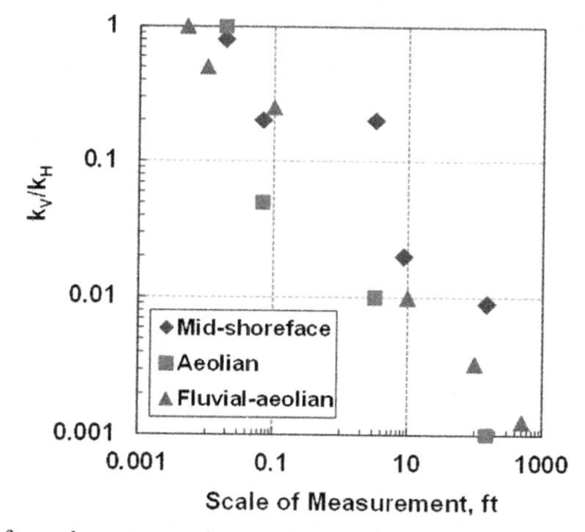

Figure 10.1.4 Data from three reservoirs show the k_V/k_H ratio generally decreases with increasing scale of measurement, but not in a smooth fashion; changes occur according to the geological character of the reservoir. Small-scale measurements are based on core, while the larger estimates come from wireline mini-tester, reservoir simulation, and tidal influence measurements.
Data sources are from Corbett (1993, mid-shoreface) and Cowan and Bradney (1997, fluvial-aeolian).

The reservoir architecture and the measurement scale both influence the k_V/k_H ratio measured (Figure 10.1.4), so there is unlikely to be a unique value for a reservoir that applies at all scales. The appropriate value will depend on the application; small-scale values are needed for horizontal well productivity, while larger-scale values are used for reservoir simulation models. With simulation, the effective vertical permeability can be assessed by history matching reservoir performance. For example, the k_V/k_H ratio can be estimated when tuning a coning model to production data.

For fluvial sandstones interbedded with shales, there are several methods for estimating the effective vertical permeability (Begg and King, 1985, Haldorsen, 1989, Mijnssen et al., 1992). For example, Mijnssen et al. (1992) derived the following expression for the effective k_V/k_H ratio:

$$\frac{k_V}{k_H} = \frac{1 - V_{sh}}{\left(1 + f_s \frac{w}{4}\right)^2} \tag{10.1.5}$$

in which V_{sh} is the volume fraction of shale in the formation, f_s is the number of shale beds per meter of formation thickness, and w is the average thickness of a shale bed.

10.1.3 Permeability and Porosity Distributions

It is also necessary to assess the permeability (and porosity) distribution, a more challenging task because so little of the reservoir is sampled. There are several types of permeability distribution:

- Homogeneous (uniform)
- Heterogeneous
  - Random small scale variations (<10 m)
  - Layered (vertical variation)
  - Pocketed (large-scale lateral variation)
- Fractured

Each type of distribution is discussed below.

A homogeneous reservoir is relatively easy to identify because there will be relatively little porosity variation both vertically and between wells. Typically, porosities will be within approximately three porosity units (0.03) of each other. Permeability may vary within approximately an order of magnitude; some greater variation is likely near pool boundaries. Well performance in a continuous homogenous reservoir under solution gas drive (SGD) will have similar oil rate decline and gas-oil ratio for individual wells. The reason for this consistency is that the pressure distribution and gas saturation tend to be relatively constant, at least in the early stages of development. The low viscosity of the evolved gas acts to equilibrate the pressure gradients within the reservoir. Similarly for water drive or gas cap drives, if the permeability and porosity are homogenous, then the water cut or GOR signatures will be associated with the location of the aquifer and gas cap and standoff distances. Standoff distance is the vertical distance from the water–oil contact (WOC) to the bottom of the perforations or from the gas-oil contact (GOC) to the top of the perforations.

Layered permeability can be assessed using cross-sections (see Section 10.2) in which the porosity profiles of each well are stratigraphically aligned. Laterally extensive layers of distinct porosity can usually be identified. Layered permeability can be observed in core data as well. Typically, porosity and permeability within a layer have similar variation as observed in a homogeneous reservoir.

It is far more challenging to distinguish random small-scale permeability variations from pocketed permeability variations. But it is an important distinction because reservoirs with each type of permeability variation behave very differently. Cross-sections and lithological considerations can help to identify regions of similar permeability. For example, interbedded sandstones and many reefs are pocketed. In general, a reservoir with small-scale variations performs like a homogeneous reservoir. The individual well behavior and a comparison to offset wells will often show whether the reservoir is continuous or discontinuous, but these indicators also depend on drive mechanism. A pocketed or compartementalized reservoir will have regions and wells of different performance in terms of initial production, oil decline rate, and GOR and/or water cut. In many cases, pocketed reservoirs are identified through production data such as water breakthroughs and water cuts. A partially compartmentalized field example is discussed Section 10.4.2.3.

Fractured or faulted reservoirs can sometimes be difficult to identify from core and log data alone. Fractures on cores may be induced by drilling rather than naturally occurring. Naturally occurring fractures can be open or cemented, and their effects on flow are different. Sometimes faults can be identified from repeated log sections. Many faulted reservoirs are also fractured, especially near the fault, Figure 10.1.5. However, the best indication of fractures is from well test analysis and production performance. The pressure transient response of a fractured reservoir can be distinct and can be identified sometimes in a build-up test. Usually, the permeability of a reservoir having open fractures is much higher than the matrix permeability and can be observed in the pressure transient response. Fractures can also be identified from interference or tracer tests. Water and gas breakthrough can be rapid in fractured reservoirs, and there are often clear directional trends in GOR and water cut. Often water production is associated with faults, Figure 10.1.6. The use of production data to guide the interpretation of reservoir architecture is discussed in more detail in Chapter 11.

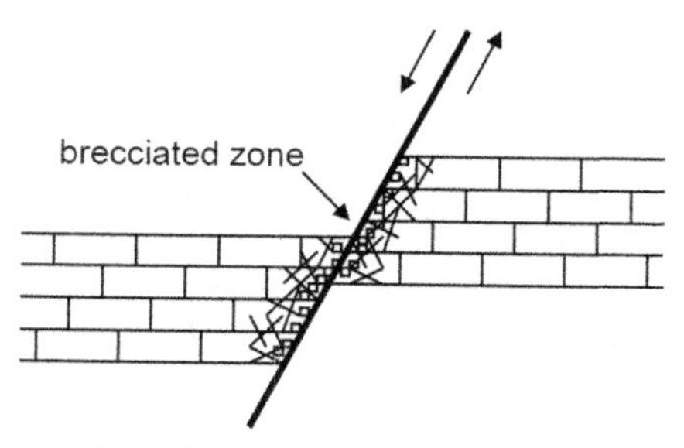

Figure 10.1.5 Fault and associated fracture region. *(Adapted from Heffer et al. (2007).)* Fractures and rubble in brecciated zone created by friction from fault.

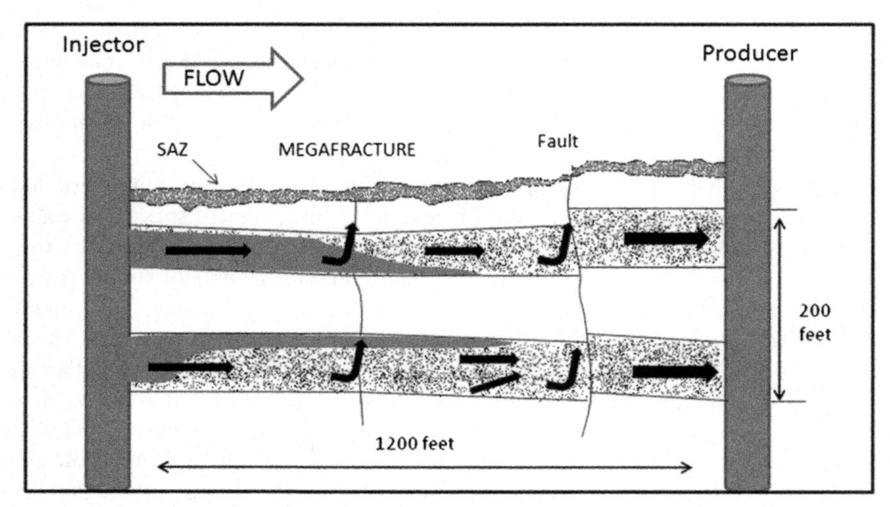

Figure 10.1.6 Movement of water in waterflood and interactions with fractures and a fault. Adapted from Belfield (1988).

10.1.4 Water Saturation

Water saturation is determined from well logs, relative permeability data, and capillary pressure data. If the wettability of the core has not been altered, the relative permeability and capillary pressure data usually give the best measure of the irreducible water saturation. The best results are obtained from cores drilled with an oil-based mud and collected from the oil zone above the transition zone. However, even accurate special core data sample a small fraction of the reservoir. Well log data usually give a better sampling but can be inaccurate because resistivity logs are sensitive to many factors, including invasion, mineralogy, accuracy of water resistivity, and thin bed effects. If the special core data are believed to be representative, water saturations from logs should be scaled to match the special core data.

Sometimes there can be large differences in initial water saturation versus water saturations from special core tests. The difference can be due to changes in wettability of rock from core cleaning and removal of asphaltenes from the surface of the rock. Another explanation for this phenomena is that the rock is at conditions in which $S_{wi} < S_{wir}$ due to hydrocarbon migration during geological times (Bennion et al., 1994). In these cases, the log data should be honored.

If there is a transition zone, the height and profile of the transition zone are identified from the well logs. The capillary pressure data are tuned, as described in Section 8.9.4, to match the well log profile.

It is helpful to consider the porosity distribution when assessing water saturation data. Water saturation increases as permeability decreases. Because porosity often correlates with permeability, higher water saturations are often observed in lower porosity regions. Therefore, the distribution of water saturations is expected to be a mirror image of the distribution of porosities.

In some cases, no reliable water saturation data are available. In this case, water saturations can be estimated from offset reservoirs from the same geological horizon (same lithology) and with similar porosity and permeability. The guidelines presented in Section 8.6 can be used to limit the range of possible water saturations.

10.1.5 Pay Thickness and Fluid Contacts

Pay thickness is determined from well logs and core data. Often there are insufficient core data to determine pay thickness independently. Instead, the core data provide a check on the log interpretation. Net pay thickness depends on the choice of porosity, water saturation, and shale content cutoffs, as discussed in Chapter 9.5. These cutoffs are usually inexact, and there can be considerable subjectivity in the determination of net pay. If the original oil in place (*OOIP*) or original gas in place (OGIP) determined from volumetrics is not consistent with reservoir performance or material balance, it is recommended to consider the sensitivity of the net pay determination to the cutoff values (see Section 10.3).

The distribution and continuity of net pay is best assessed using cross-sections and reservoir performance. The approach is the same as was discussed for porosity distributions.

Fluid contacts are determined from well logs and are sometimes inferred from production data. In transition zones, the WOC can be defined as the water saturation at which the relative permeability to oil is zero, as shown in Figure 10.1.7. The elevation of an initial fluid contact is expected to be consistent across the reservoir to within 2 m. In some cases, the wells do not penetrate the fluid contacts, but the presence of a contact becomes apparent through production performance, for example, high GORs near the top of the structure or high water cuts low in the structure. In other cases, a material

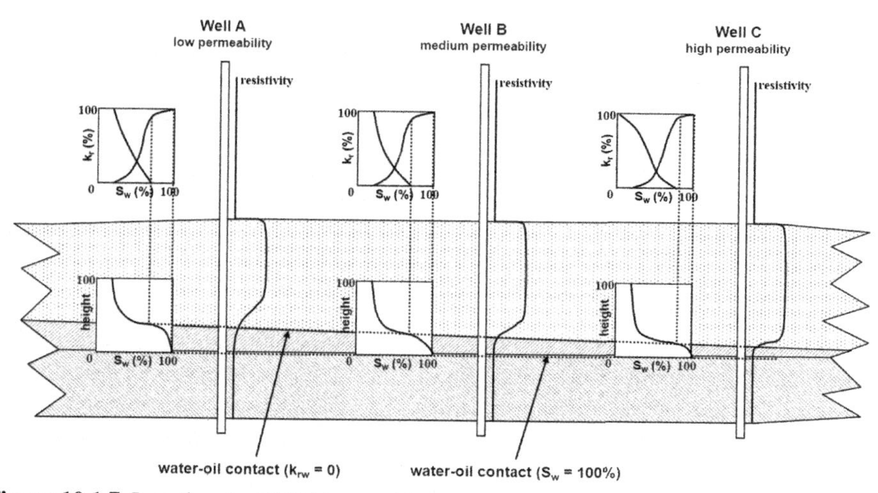

Figure 10.1.7 Locating the WOC in transition zones.
Adapted from Knutsen (1954).

balance indicates that an oil zone is obtaining pressure support from another source, either a gas cap or an aquifer. The appropriate fluid contact must be estimated from the reservoir structure.

10.1.6 Fluid Properties, Initial Pressure, and Initial GOR

Fluid properties are obtained from a fluid study or from correlations. The main checks on fluid properties are on the bubble point pressure and the solution gas-oil ratio (R_S). If a reservoir has a gas cap, then the bubble point pressure must equal the initial pressure. In thinner reservoirs, the producing gas-oil ratio (GOR) is expected to increase continuously as the reservoir is produced. If the reservoir is undersaturated, the initial pressure must exceed the bubble point pressure. As an undersaturated reservoir is produced, the pool GOR is initially constant and is expected to increase soon after the bubble point pressure is reached. The bubble point pressure can sometimes be identified from the GOR trend and pressure data, as discussed in Section 10.4.2.2.

The initial producing GOR is expected to equal the solution gas-oil ratio at the bubble point. The only exception is when coning occurs in a gas cap reservoir. In this case, examine the GORs from wells low in the structure that may not be con-ing gas.

If the bubble point pressure is inconsistent with pressure and production data, then check the pressure data. If the pressure data are valid, then adjust the fluid study or correlations, as described in Chapter 5. If the initial solution gas-oil ratio does not match the initial GOR, then check the production accounting for the reservoir. If the gas rates were measured and reported correctly, then adjust the fluid data.

10.2 Reservoir Mapping

There are two categories of reservoir maps: geological maps and production maps. Geological maps include structure, gross pay, net pay, porosity-pay, and hydrocarbon porosity-pay maps as well as structural and stratigraphic cross sections. Production maps include permeability pay, isobaric maps, and bubble maps. Each geological map type and its application are discussed below. Production maps are discussed in Section 10.4.

10.2.1 Geological Contour Maps

The geological mapping of a reservoir is an expression of the geologist and reservoir engineer's understanding of the reservoir. At this step, the reservoir architecture is interpreted and put on paper. Almost always, the geologist prepares reservoir maps, bringing in his understanding of the regional geology and depositional environment. Unfortunately, engineers sometimes lack the background knowledge to properly map reservoirs, particularly when there is little well control and a large degree of geological interpretation. However, engineers often rely on geological maps to help interpret pool performance. Also, the mapping may be tuned based on the engineer's understanding of the reservoir performance. Therefore, geological mapping is briefly reviewed below.

The data used to prepare geological maps include:

- Seismic structure;
- Formation elevations;
- Fluid contact elevations;
- Pay thicknesses;
- Porosities;
- Water saturations.

To illustrate how the data are combined to prepare maps, we will use a simple hypothetical dataset, Table 10.2.1. The well locations are identified on the area map in Figure 10.2.1.

Structure Maps

A structure map is an aerial contour map showing lines of constant elevation for either the top or the bottom of a reservoir. Structure maps are prepared from seismic data and formation elevations from well logs. Seismic data are subject to interpretation, but provide regional trends in structure. Elevations from well logs are a direct measurement and are more accurate than elevations derived from seismic data. Typically, the first structure map of a reservoir is prepared from seismic data before any wells are drilled. The mapping is improved as wells are drilled and more exact elevations are obtained.

Seismic techniques and interpretation are in the realm of geophysics and are outside the scope of this book. In brief, a series of small explosions are set off at ground level on a line or grid over the area of interest. The acoustic waves travel through the layers of rock underground. The time required for the signal to travel through a particular layer of rock depends on the rock density. Part of the signal reflects off the rock layers and returns to the surface. These return signals are recorded. The pattern of the return signals depends on the structure of all the layers of rock the acoustic signals passed through on their journey down and back to the surface. A three-dimensional representation of the underground structure is reconstructed from the pattern of returned signals. An example of a structural cross-section derived from seismic data is provided in Figure 10.2.2. Part of the interpretive process involves converting the vertical axis, which is the time measured for the waves to travel from the source to the reflector and back to the receiver, into depth. The cross section shows a domed or anitclinal structure that can act as a trap. Seismic reconstructions are three-dimensional, so structure maps can be prepared for any formation within the area examined.

It is more straightforward to prepare a structure map from well data. One approach is to mark the well elevations on a well location map, as shown for the example dataset in Figure 10.2.3. Points of constant elevation are interpolated between wells, as shown in Figure 10.2.3(a). Then the points are linked to form a contour line, as shown in Figure 10.2.3(b). When there is insufficient well control, the elevations are extrapolated based on the prevalent slope of the structure. The shape of the surrounding contour lines also provides a guide. The completed contour map is shown in Figure 10.2.3(c). In this

Table 10.2.1 Hypothetical dataset for example mapping

Well location	Top structure elevation (m)	Pay thickness[a] (m)	Porosity (%)	Water saturation (%)
14-01	−1524	1	7	50
06-02	−1524	0	–	100
14-02	−1491	29	13	11
16-02	−1510	10	12	13
06-03	−1512	8	8	15
08-03	−1513	7	10	12
14-03	−1482	38	11	12
16-03	−1469	51	9	13
08-04	−1522	1	7	50
14-04	−1523	0	–	100
16-04	−1513	7	12	10
08-09	−1523	0	–	100
06-10	−1513	7	12	11
08-10	−1479	41	11	10
16-10	−1523	1	8	55
06-11	−1463	57	12	11
08-11	−1471	49	13	11
14-11	−1511	9	8	16
16-11	−1503	17	11	10
06-12	−1493	27	10	11
08-12	−1512	8	9	14
14-12	−1509	11	9	14
16-12	−1523	2	7	35
06-13	−1523	0	–	100
08-14	−1525	0	–	100

[a]Gross pay = net pay for this case.

case, the reservoir is dome-shaped or anticlinal, with a top structure at −1463 m and with approximately 60 m of structural relief. Some judgment is needed to locate the contours. For example, the location of the outer contour in sections 13 and 14 is near the dry holes, which maximizes the areal extent of the reservoir, but could also be positioned closer to the productive wells in the northern parts of sections 11 and 12. This

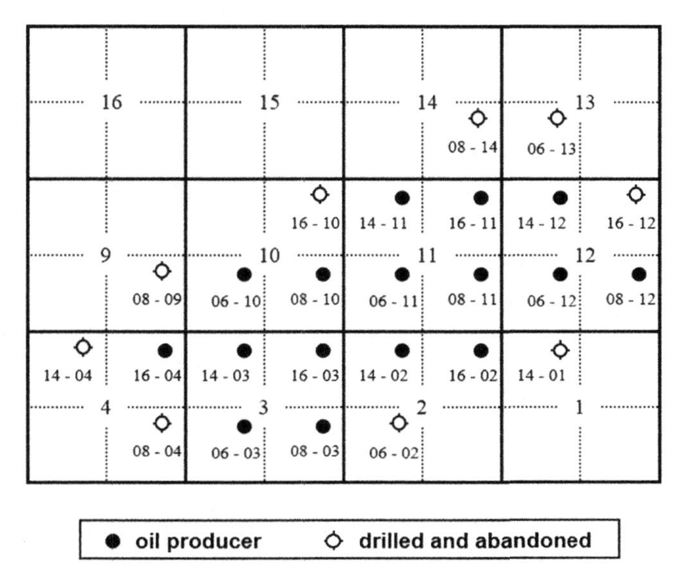

Figure 10.2.1 Well location map for example mapping dataset. Grid lines are section lines and are one mile apart.

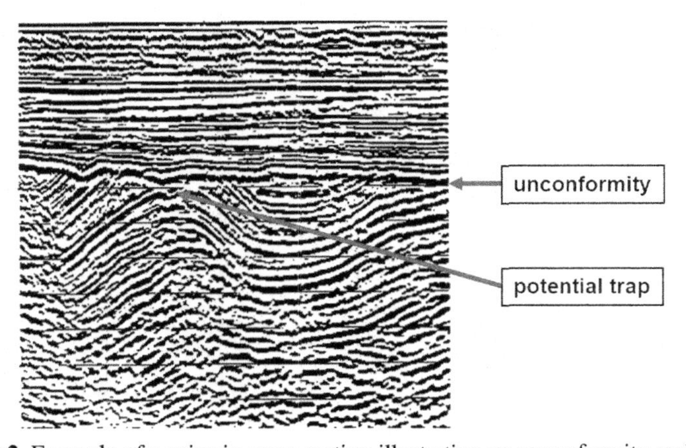

Figure 10.2.2 Example of a seismic cross section illustrating an unconformity and a potential structure trap. The horizontal axis is location, and the vertical axis is time.
From http://pangea.stanford.edu/ ~ sklemp/bering_chukchi/reflection.html

uncertainty in the area may need to be included during the volumetrics calculations to help define the uncertainty of the predicted reservoir volume.

Some care is required when interpreting the depths on well data. Well log depth measurements are often referenced to the Kelly bushing or drill floor, which can change in elevation as the rig moves from location to location. Also, several drilling rigs with different heights may have drilled the wells in one area. Therefore, formation

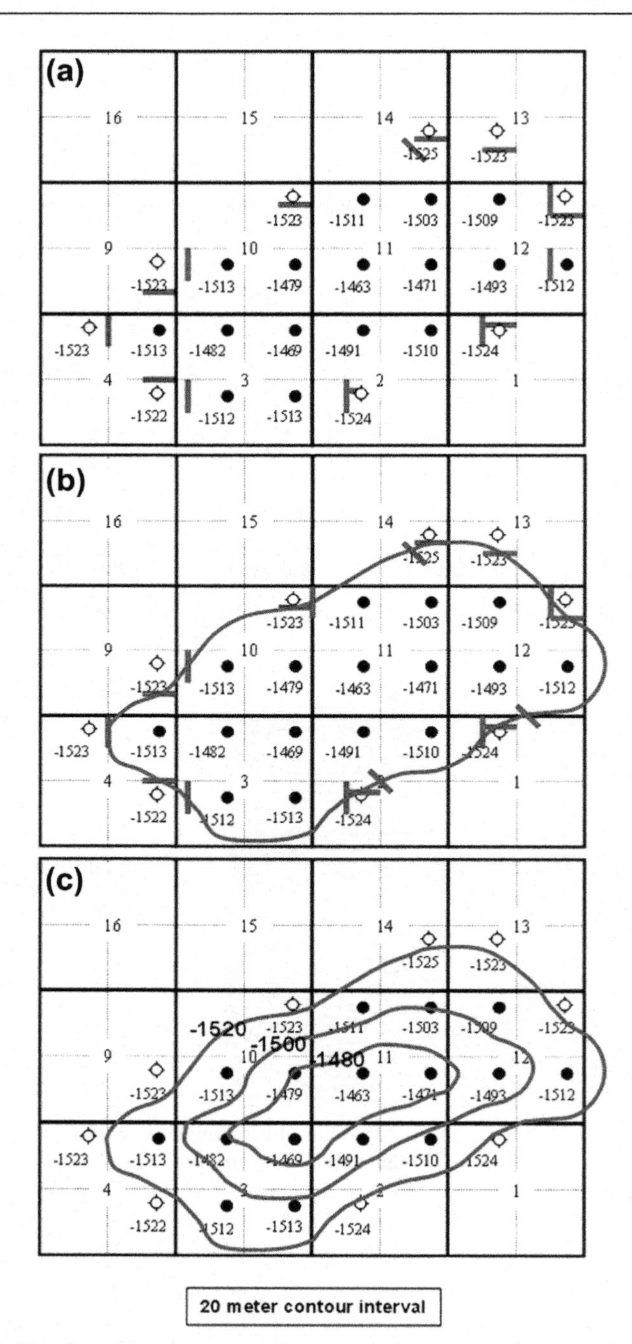

Figure 10.2.3 Creation of a structure map from example elevation data: (a) identifying location of specific elevation between wells; (b) drawing contour line at that elevation; (c) completed map. Grid lines are section lines and are one mile apart. The red lines (Dark gray in print versions) indicate approximate locations of the reservoir boundaries based on the positions of dry and productive wells at the periphery.

top elevations need to be corrected to a common reference, such as mean sea level. This sounds simple to do, but care is required to be sure it is done correctly, otherwise many later results may be wrong and have to be redone.

The same approach should be used for all geological contour maps. It is now common to input the well information into commercial software, which automatically carry out the interpolations and extrapolations and generate contour maps. As mentioned previously, the interpretation may be adjusted based on the understanding of the local depositional environment and the reservoir performance.

Isopach (Pay Thickness) Maps

There are several types of isopach maps including:

- Gross pay;
- Net pay;
- Porosity-pay;
- Net hydrocarbon porosity-pay thickness;
- Permeability-pay.

All are contour maps of a given form of formation thickness.

Gross pay: the total thickness of the formation including nonproductive rock. Gross pay maps are useful for assessing the depositional environment and are often used in conjunction with net-to-gross ratio (NGR) maps as inputs for reservoir simulations of formations with variable quality rock.

Net pay: the thickness of the formation that is deemed productive, that is, exceeds specified permeability, porosity, water saturation, and shale cutoffs. Net pay maps are used for volumetrics when porosity and water saturation are uniform across the reservoir. Net pay maps are used as input to reservoir simulations when the NGR is uniform. A net pay map of the example dataset is given in Figure 10.2.4. The reservoir has a dome structure with a maximum net pay thickness of 57 m.

Porosity-pay: the product of the porosity and the net pay. A porosity-pay map is a contour map of the "thickness" of the reservoir pore volume. Porosity-pay maps are used for volumetrics when the porosity varies across the reservoir. In the example dataset, porosity varies between 7% and 13%. This variation is likely beyond the normal variation expected from log interpretation of thick intervals (greater than 3–5 m). Hence, a porosity-pay map is recommended for volumetric calculations. A porosity-pay map of the example dataset is given in Figure 10.2.5.

Net hydrocarbon porosity-pay: the product of porosity, net pay, and oil saturation, $\phi h(1-S_w)$. A hydrocarbon porosity-pay map is a contour map of the "thickness" of the reservoir oil volume. Hydrocarbon porosity-pay maps are used for volumetrics when the porosity and water saturation vary across the reservoir. In the example dataset, the water saturation is 100% below an elevation of −1523 m and varies from 10% to 15% above an elevation of −1522 m. In this case, there appears to be a WOC at approximately 1523 m depth with a 1–2 m transition zone. The variation in water saturation is well within typical variations in well logs. Hence, there appears to be a nearly uniform water saturation of approximately 12% above the transition zone. In this case, it is unnecessary to use a net hydrocarbon porosity-pay map. Instead, it is recommended to use an average water saturation.

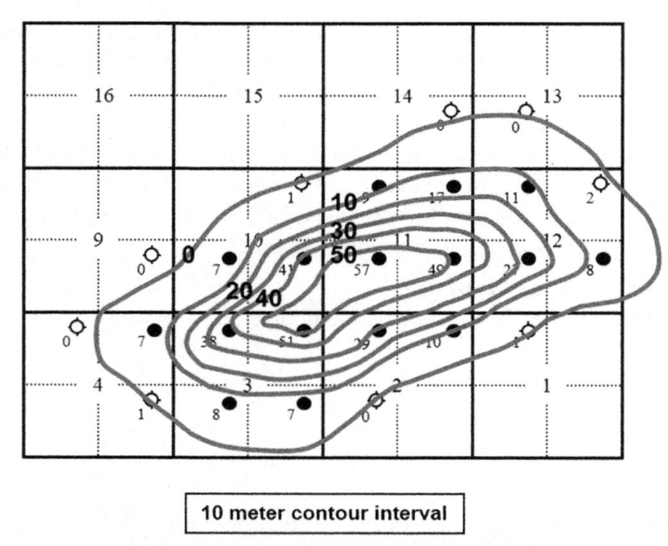

Figure 10.2.4 Net pay map from example thickness data. Grid lines are section lines and are one mile apart.

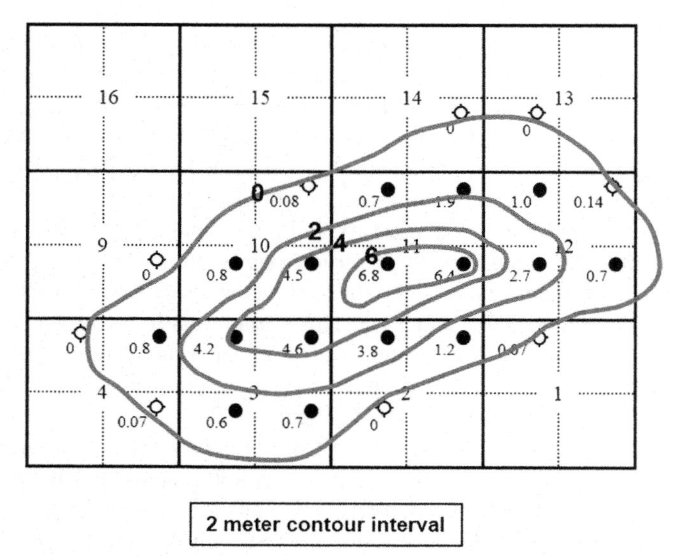

Figure 10.2.5 Porosity-pay map from example thickness and porosity data. Grid lines are section lines and are one mile apart.

Permeability-pay: Permeability or permeability-pay (kh) can be mapped directly if sufficient permeability data are available from cores and build-up tests. This is usually not the case. Instead, an initial permeability-pay map is obtained from a porosity-pay map using a k-ϕ correlation. To apply this method, there must be a reasonably good fit

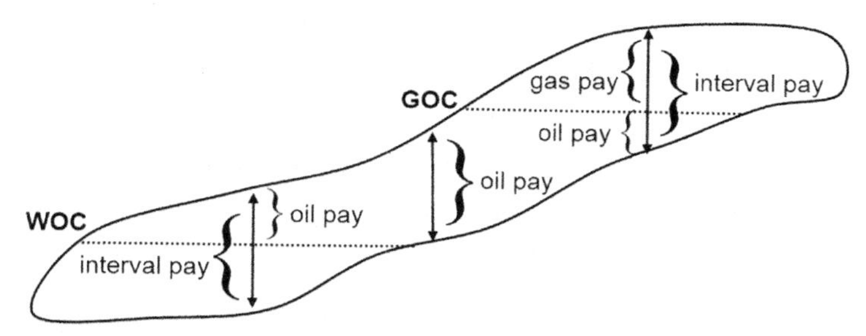

Figure 10.2.6 Schematic reservoir profile showing oil, gas, and interval pay thicknesses.

of log(k) to core ϕ and there must be a good correlation of log ϕ to core ϕ. Because there is usually large uncertainty in the k-ϕ correlation, the indirect kh mapping is only useful as a guideline or interpretive tool. Permeability and kh maps are useful for identifying:

- High-permeability regions where high production rates and good recovery are expected;
- Low-permeability regions where low production rates and poor recovery are expected;
- Some potential large-scale permeability barriers.

A check of the initial well productivity versus mapped kh should also be performed to validate the kh mapping.

Geological Contour Maps for Simulation

The maps used for volumetric calculations are not necessarily the most convenient maps for input into reservoir simulators. For example, in a reservoir with variable porosity, a porosity-pay map is used for volumetrics. However, a net pay and a porosity map are required to initialize a simulator. The following contour maps are required for a simulator when the given property is nonuniform:

- Net-to-gross ratio;
- Porosity;
- Water saturation.

Usually a porosity map is required. Water saturation maps are only rarely used.

Gross, net, porosity-, and permeability-pay can be mapped for the total porous formation (interval) or for the hydrocarbon zone only (oil or gas pay), as shown in Figure 10.2.6. Mappings of the hydrocarbon zone are used for volumetrics and the determination of average reservoir properties. Mappings of the net interval are required for simulation because a simulator requires a continuous flow unit that includes the oil, gas, and water zones whenever they are present. Mapping requirements for various applications are summarized in Table 10.2.2.

10.2.2 Geological Cross Sections

A cross section is a vertical cross-sectional view of the reservoir prepared from well data. Wells are arranged in a line corresponding to their location, and common features such as formation tops, fluid contacts, shale breaks, and regions of contrasting

Table 10.2.2 Maps required for various applications

Application	Required map(s)	Other data
Reservoir interpretation	Structure Gross hydrocarbon pay NGR (if variable) Porosity (if variable)	Average NGR Average porosity Average water saturation
Volumetrics uniform reservoir	Gross oil pay Gross gas pay	Average NGR Average porosity Average water saturation
Volumetrics nonuniform NGR	Net oil pay Net gas pay	Average porosity Average water saturation
Volumetrics nonuniform porosity	Oil porosity-pay Gas porosity-pay	Average water saturation
Volumetrics nonuniform water saturation	Hydrocarbon porosity-pay	–
Reservoir simulation uniform reservoir	Top structure Gross oil pay Gross gas pay	Average NGR Average porosity Average water saturation
Reservoir simulation nonuniform reservoir	Top structure NGR Net interval pay Porosity	Average water saturation

porosity are connected between the wells. There are two types of cross-section: structural and stratigraphic.

Structural Cross Section: The wells are aligned according to their elevation or true vertical depth. A structural cross-section shows the reservoir structural profile. They are used to identify WOC, GOC, and pool boundaries. If the reservoir is permeable across the interval, the GOC and the WOC are usually level. If there are large changes in reservoir quality or ground water movement, there can be variation in the WOCs. Faulting, lack of continuity, and development of secondary gas caps can cause lack of agreement of GOCs (Figure 10.2.7). An example of a structural cross-section with-GOC and water-oil (shown as oil-water or OWC) contacts is given in Figure 10.2.8. Note that, both the GOC and the WOC vary, although it is a structural cross section.

Stratigraphic Cross Section: The wells are aligned according to a regional geological marker, such as a coal seam or formation top. Stratigraphic cross sections attempt to show the reservoir profile as it was originally deposited. They are used to correlate features such as shale breaks or regions of contrasting porosity between wells. Stratigraphic cross sections are particularly useful in assessing pool continuity, that is, identifying connected and isolated regions in a given formation. An example of a stratigraphic cross section is given in Figure 10.2.9.

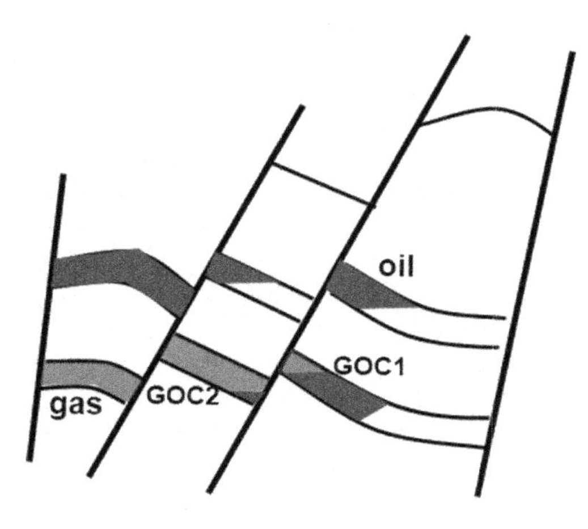

Figure 10.2.7 Schematic cross section of reservoirs created through faulting. Note how the faulting allows gas to reside in a structurally lower position than oil.

10.2.3 Mapping More Complex Reservoirs

The hypothetical mapping example discussed in Section 10.2.1 was a straightforward case with continuous reservoir and sufficient well control to completely define the reservoir. Most real reservoirs are more complex and have more limited data. Space does not permit a discussion of all the geological considerations required for mapping complex reservoirs. However, three issues of relevance to reservoir engineering are addressed:

- Identifying major pool boundaries;
- Mapping reservoir continuity within a single horizon;
- Mapping when net and gross pay are different.

Identifying Major Pool Boundaries

One of the challenges in reservoir mapping is to identify the areal extent of a reservoir, that is, the pool boundaries. Pools can be bounded by faults, porosity pinch-outs, facies changes, or fluid contacts.

Faults can occasionally be identified through geological cross sections and changes in fluid contacts between wells. The locations of faults can also be determined from seismic data, provided the faults are sufficient in size (throw). Usually, a geology and geophysics team prepares initial maps for reservoirs bounded by faults, and the engineer has little involvement unless long-term production testing is used to help identify barriers to flow. The mapped location of the faults can sometimes be tested once production data are available. For example, different pressure profiles and water breakthroughs are expected in different reservoir compartments isolated by faults. A hypothetical cross section of reservoirs created by faulting is shown in Figure 10.2.7.

Porosity pinch-outs and facies changes can be identified from seismic maps or from cross sections in which a porous zone or rock type disappears between wells (see Figure 10.2.10). Regional geological mapping may identify erosional discontinuities that bound some reservoirs. For example, in Western Canada, layers of carbonates

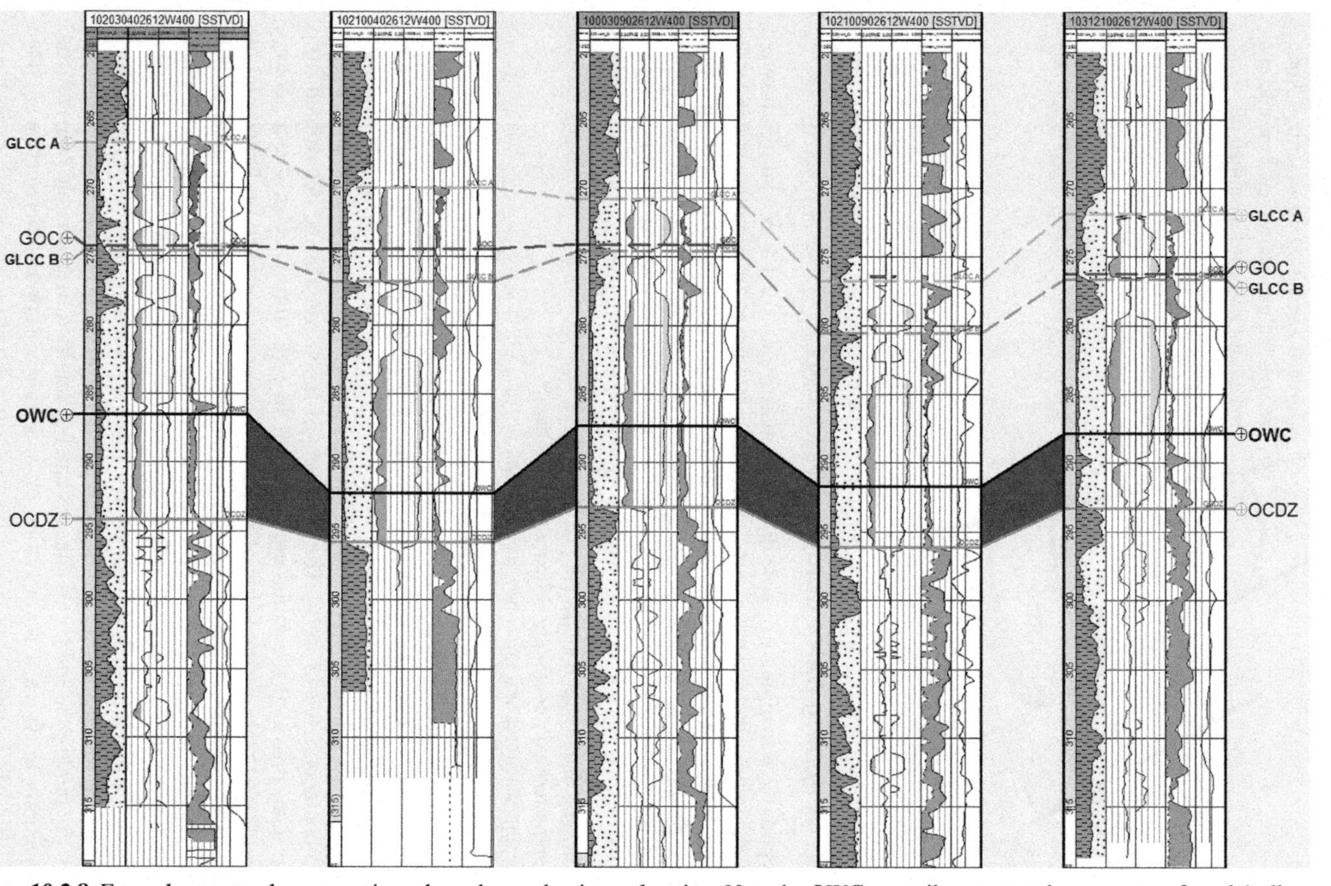

Figure 10.2.8 Example structural cross section where the marker is an elevation. Note the OWC, gas-oil-contact and occurrence of sand (yellow shading (Light gray in print versions)) and shale (grey shading). The dark red shading (Dark gray in print version) indicates gas effect on the density-neutron logs.

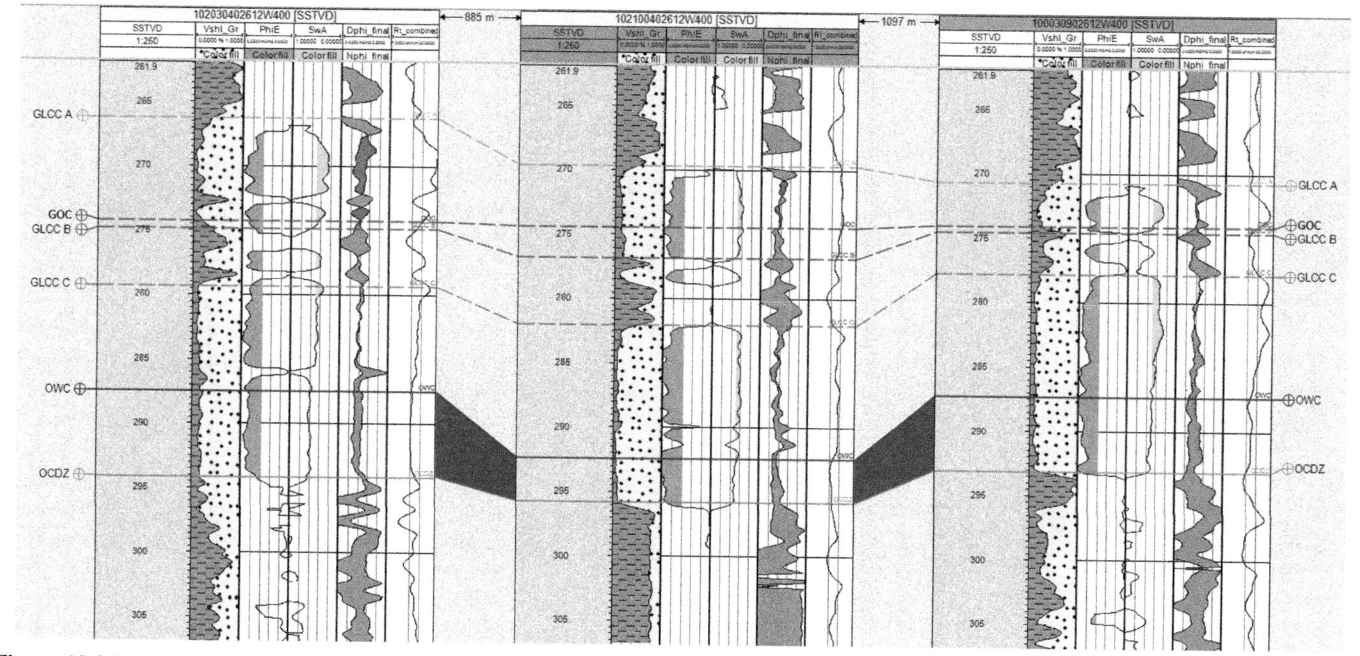

Figure 10.2.9 Example stratigraphic cross section. The marker is the top of the Glauconitic A zone (GLCCA). Note WOC is the same as OWC.

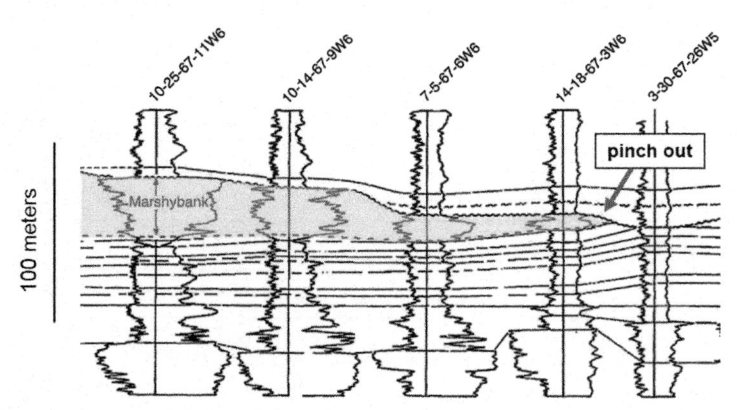

Figure 10.2.10 Porosity pinch-out of a regional sand in Western Canada. Adapted from Leckie et al. (2012).

were partly eroded and overlain with sandstones. The transition from carbonate to sandstone or from porous to nonporous rock marks the erosional discontinuity, as shown in Figure 10.2.11.

Structure maps and cross sections can help identify fluid contact boundaries. The elevation of the contact is determined from structural contacts. The boundaries of the fluid contacts are determined from structure maps. In dome-shaped reservoirs, there is a single WOC line and a single GOC line, as shown in Figure 10.2.12. However, in tilted reservoirs, fluid contacts intersect both the top and bottom of a reservoir structure, as shown in Figure 10.2.13. The line where the WOC intersects the top of the reservoir structure is the outside WOC and marks the outer extent of the oil zone. The line where the WOC crosses the bottom of the structure is the inside WOC and marks the inner extent of the water zone. The zero pay boundary of an oil zone is the outer WOC. Water is under the oil between the outside and inside WOCs. There is no underlying water within the inside WOC.

Similar logic applies to the GOC. The inside GOC (top structure) marks the outer extent of the gas zone, that is, the zero pay boundary for gas. The outside GOC (bottom structure) marks the furthest extent of the oil zone, that is, the zero pay boundary for

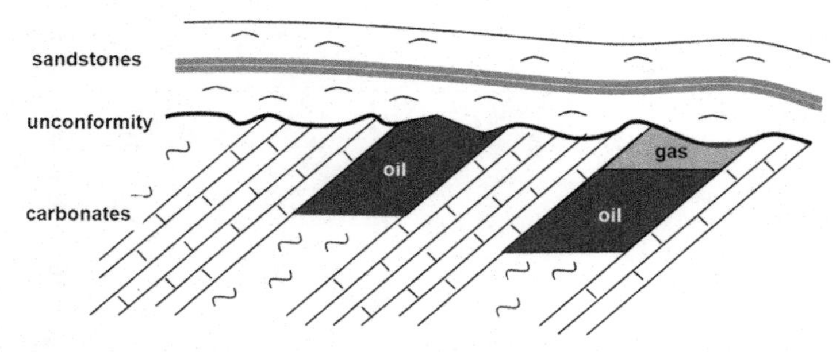

Figure 10.2.11 Traps formed from erosional discontinuity.

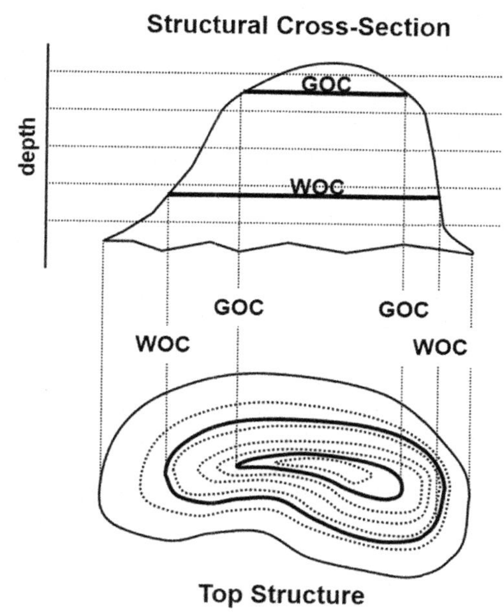

Figure 10.2.12 Projection of WOC and GOC on a top structure map of a map of a dome-shaped reservoir.

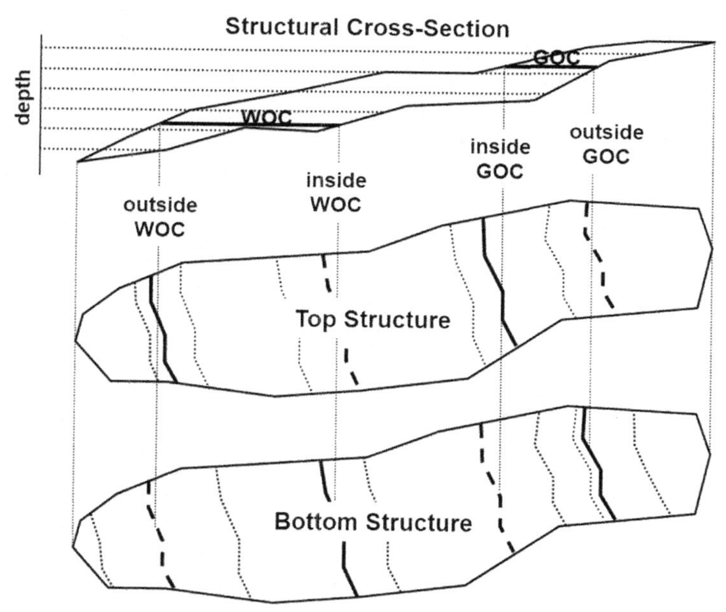

Figure 10.2.13 Projection of WOC and GOC on the top and bottom structure of a tilted reservoir.

the oil zone. Gas lies on the oil zone between the inside and outside GOCs. No gas is found down dip of the inside GOC. No oil is found up dip of the outside GOC.

Mapping Reservoir Continuity and Flow Units

Another challenge in mapping and reservoir analysis is to locate isolated pools, zones of high or low permeability, and barriers to flow within a single geological horizon. Some examples of flow barriers are shale layers, faults, or low-permeability reservoir rock. Faults can sometimes be identified from seismic data. Shale layers and low-porosity zones can sometimes be identified on logs and correlated between wells using cross sections. However, in many cases, flow barriers can only be inferred from production and pressure data. If a reservoir has regions of different pressure, it is an indication of poor communication within the reservoir. Poor communication may result from low average permeability or the presence of flow barriers. The movement of a displacement fluid can also indicate whether flow barriers are present. For example, if injection water moves to some nearby wells but not others, a flow barrier may be present, as shown schematically on Figure 10.2.14.

When flow barriers exist, it can be difficult to determine whether a horizon consists of a single reservoir with poor internal communication or several isolated pools. Sometimes isolated pools will have different fluid compositions. If fluid contacts are present, they may occur at different elevations in each pool. Usually, a combination of geological analysis and reservoir performance analysis is required to assess whether pools are completely isolated or in partial communication. When there is more than one isolated pool in a single geological horizon, each pool should be mapped and analyzed individually.

It is common practice to divide a reservoir with flow barriers into flow units. A flow unit is a portion of the reservoir that has good internal communication, but is partially isolated from other parts of the reservoir. In some cases, each flow unit may be mapped separately. The division of a reservoir into flow units is one of the most difficult challenges in characterizing the reservoir because the barriers that create the flow units are not easily detected. However, the number, size, and shape of the flow units has a large

Figure 10.2.14 Schematic illustrating the effect of a flow barrier on waterflood performance.

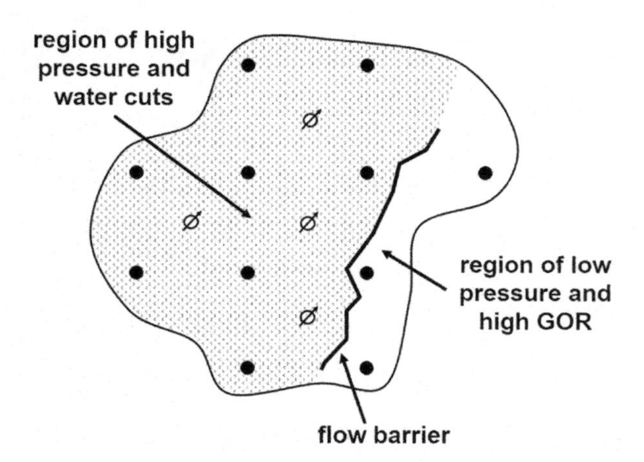

effect on reservoir performance. For this reason, the division of a reservoir into flow units is often iterated many times until a satisfactory match with pool performance is obtained.

A method to determine flow units is to evaluate cross sections of the log and core data to identify shales within the reservoir interval. Effective barriers are formed if the shales have very low permeability ($k_{air} < 0.001$ mD), are thicker than approximately two feet, and are laterally continuous (extending to at least half the interwell distance). In this case, the porous units above and below the shale can likely be divided in separate flow units, as shown in Figure 10.2.15. If the shale break does not meet the above criteria, the two porous units could be treated as a single flow unit.

Another method to identify flow units is to drill wells in stages, so that early wells can be put on production and provide some project revenue. Then pressure measurements can be made in the later wells, such as using repeat formation testers (Chapter 6), to determine intervals at which pressure has dropped (indicating some communication) and other intervals at which pressure is still at original levels (indicating limited or no communication).

The choice of flow units also depends on the direction of the bulk flow. A relatively large number of flow units may be required to model flow parallel to bedding, that is, parallel to the direction of the permeability contrast, such as a shale break. Fewer flow units are required when the flow is perpendicular to the bedding.

Mapping When Net and Gross Pay are Different

When net and gross pay are different, it is straightforward to map net pay and gross pay separately. However, a net pay contour map provides no information on the distribution of the net pay within the gross pay zone. A map of NGR shows the aerial distribution of the net pay, but does not show the vertical distribution. The vertical distribution of net pay can have a significant impact on the movement of fluid in a

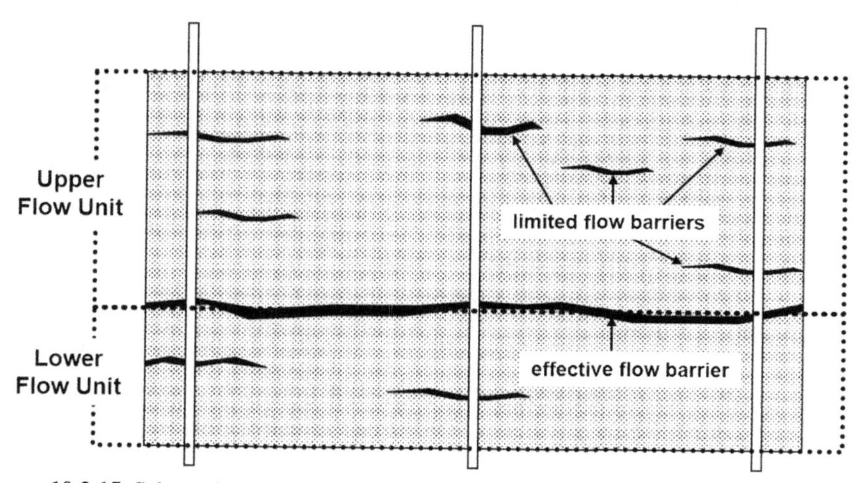

Figure 10.2.15 Schematic of the division of a reservoir into two flow units based on the type of shale breaks.

reservoir. For example, a laterally extensive shale break above the WOC can inhibit coning. Therefore, some judgment is required when mapping a reservoir with an NGR less than unity.

The appropriate choice of mapping depends on the size and distribution of the zero net pay zones (permeability barriers). The following mappings are recommended for the indicated permeability distributions:

- Randomly distributed and limited extent permeability barriers, use
  - Gross pay
  - Average NGR
- Randomly distributed but laterally extensive barriers, use
  - Gross pay
  - Average NGR
  - Reduced vertical permeability
  - Individually adjusted vertical permeability around well bores as required to model coning
- Nonrandom permeability barriers
  - Gross pay
  - Permeability map if only areal effects are important
  - Permeability for each reservoir layer if areal and vertical effects are important
  - Transmissibility barriers if a fault or shale layer is identified

Some Consistency Checks for Reservoir Mapping

When preparing or evaluating a set of reservoir maps, there are some consistency checks to consider. The first set of checks applies if there are fluid contacts in the reservoir. When mapping an oil zone, the pool boundary must coincide with the location of the (outside) WOC and the (outside) GOC, if present. Similarly the boundary of the gas cap should coincide with the location of the (inside) GOC. The gross pay thickness of an oil zone above the WOC should equal the difference between the top of the structure and the elevation of the WOC. If there is gas above and water below, the gross pay should equal the difference between the elevations of the GOC and WOC. The thickness of a gas cap should equal the difference between the top structure and the elevation of the GOC.

The second set of checks applies when the NGR is less than unity. The net pay must always be less than or equal to the gross pay. If an NGR map is prepared, the net pay thickness over the gross pay thickness should match the mapped NGR at all locations in the reservoir.

The final check applies if multiple zones are mapped. The total gross (or net) pay must equal the sum of the individual gross (or net) pays.

10.3 Volumetrics

Volumetrics is the determination of the reservoir bulk, pore, or hydrocarbon-pore volume from a contour map of net pay, porosity-pay, or hydrocarbon porosity-pay, respectively. As was discussed in Section 10.2, bulk volume is used when the porosity and water saturation are uniform, pore volume is used when the porosity varies, and hydrocarbon pore volume is used when water saturation varies.

Volume is determined from a contour map based on the area of each contour. For example, consider a contour map of a pyramid that has a base of 50 by 50 m and a height of 40 m, Figure 10.3.1. Each of the contoured layers is a trapezoid. The volume of a trapezoid is given by:

$$V_{trap} = 0.5\left(A_{base} + A_{top}\right)\Delta h \tag{10.3.1}$$

in which A_{base} and A_{top} are the areas of the base and top of the trapezoid, respectively, and Δh is the contour thickness. If the volumes of the four layers of the pyramid in Figure 10.3.1 are summed, the following expression is obtained:

$$\begin{aligned} V_{pyr} &= 0.5(A_1 + A_2)\Delta h + 0.5(A_2 + A_3)\Delta h + 0.5(A_3 + A_4)\Delta h \\ &\quad + 0.5(A_4)\Delta h \\ &= \left(0.5A_1 + \sum_{2}^{3} A_i + 0.5A_4\right)\Delta h \end{aligned} \tag{10.3.2}$$

Note that the top layer has zero top area.

The volumetric calculations for the entire pyramid are summarized in Table 10.3.1. The volume from the volumetric calculation is 33,590 m^3, slightly higher than the exact solution of 33,333 m^3 ($A_{base} \cdot h/3$). In this case, the error occurs in the top layer,

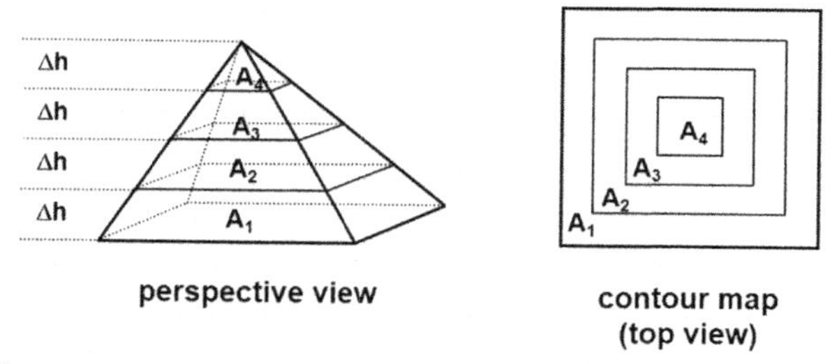

Figure 10.3.1 Contour mapping of a pyramid for volumetrics.

Table 10.3.1 Volumetric calculations for 50×50×40 m pyramid. The contour interval is 10 m

Layer	Base area (m^2)	Coefficient	Volume (m^3)
1	2500	0.5	12,500
2	1406	1	14,060
3	625	1	6250
4	156	0.5	780
Total	–	–	33,590

which is a pyramid, not a trapezoid. A more accurate volumetric formula used the volume of a pyramid for the top layer:

$$V_{pyr} = \left(0.5A_1 + \sum_{2}^{3} A_i + 0.333A_4\right)\Delta h \tag{10.3.3}$$

This formula provides an exact solution for the pyramid example.

The same procedure is followed for any contour map, with the assumption that the top layer is a pyramid and all other layers are approximately trapezoidal. The mapped volume is then given by:

$$V = \left(0.5A_1 + \sum_{2}^{n-1} A_i + 0.333A_n\right)\Delta X \tag{10.3.4}$$

in which X is h, ϕh, or $\phi h(1 - Sw)$ for net pay, porosity-pay, and hydrocarbon porosity-pay maps, respectively.

Reservoir Volumetrics

If a geological model or a reservoir simulation is to be constructed, the appropriate set of maps is digitized or scanned and input into the software. The program then calculates the reservoir bulk, pore, and hydrocarbon pore volumes, as well as the original oil and gas in place. It is not always convenient to use such software. In this case, the reservoir volumes are determined using the trapezoidal method described above and the following procedure:

1. Measure the areas of each contour on the appropriate map. The area within a contour can be determined with a planimeter. This instrument provides an area measurement for the space traced by a movable arm. If a planimeter is not available, the contour area can be divided into small squares of known area. The number of squares is counted. The total area is the product of the number of squares and the area of each square.
2. Convert the measured area to a mapped area. The measured area is in some arbitrary set of units. The measurement must be scaled to the units of the given map. When using a planimeter, the simplest approach is to measure a known area, such as a section (1 mi^2 or 259 ha), on the map. The contour area is then given by:

$$A_c = A_{c,m}\frac{A_s}{A_{s,m}} \tag{10.3.5}$$

   in which A_c and A_s are the actual areas of the contoured space and the known area (e.g., a section), respectively, and $A_{c,m}$ and $A_{s,m}$ are the measured areas of the same respective areas. A similar approach can be used with the counting squares method or the length of an edge of the square can be scaled to the length scale of the map.
3. Apply the trapezoidal method using the scaled areas and the contour interval of the map.

Once the reservoir volume is known, the hydrocarbon-in-place is determined. If uniform porosity and water saturation are assumed the original *OOIP* is given by:

$$OOIP = \frac{BV\phi(1 - S_{wi})}{B_{oi}} \tag{10.3.6}$$

in which *BV* is the bulk volume determined from a net pay map, ϕ is the average porosity, S_{wi} is the average initial water saturation, and B_{oi} is the initial oil formation volume factor. The oil formation volume factor converts a reservoir volume to a surface volume; hence, *OOIP* is in surface units.

If the porosity is variable, then, *OOIP* is given by:

$$OOIP = \frac{PV(1 - S_{wi})}{B_{oi}} \tag{10.3.7}$$

in which PV is the pore volume determined from a ϕh map.

If the water saturation is variable, then, *OOIP* is given by:

$$OOIP = \frac{HCPV}{B_{oi}} \tag{10.3.8}$$

in which HCPV is the hydrocarbon pore volume determined from a $\phi h(1 - S_{wi})$ map.

It is conventional when reporting reserves to use the following formulation:

$$OOIP = \frac{hA\phi(1 - S_{wi})}{B_{oi}} \tag{10.3.9}$$

in which h, ϕ, and S_{wi} are the average pay thickness, porosity, and initial water saturation respectively, and A is the area of the zero net pay contour. The average net pay is given by:

$$h = \frac{BV}{A} \tag{10.3.10}$$

If porosity is nonuniform, the average porosity is determined as follows:

$$\phi = \frac{PV}{BV} \tag{10.3.11}$$

and if water saturation is nonuniform, the average water saturation is given by:

$$S_{wi} = 1 - \frac{HCPV}{PV} \tag{10.3.12}$$

A similar approach is used to determine the *OGIP* for a gas cap. The conventional formulation is:

$$OGIP = \frac{hA\varphi(1 - S_{wi})}{B_{gi}} \tag{10.3.13}$$

in which B_{gi} is the initial gas formation volume factor. It is also useful to determine the original solution gas in place (*OSGIP*), given by:

$$OSGIP = R_{si}OOIP \tag{10.3.14}$$

in which R_{si} is the initial solution gas-oil ratio.

Example

Determine the reservoir pore volume and *OOIP* for the example problem using the net pay and porosity-pay maps, Figures 10.2.4 and 10.2.5, respectively. Assume an average water saturation of 12% and a B_{oi} of 1.25 m^3/scm.

First, let us use the net pay map to determine the bulk volume of the reservoir. The volumetric calculations are summarized in Table 10.3.2. The bulk volume of the reservoir is 241×10^6 m^3. The thickness weighted average porosity was determined from the data in Table 10.2.1 as follows:

$$\phi = \frac{\sum h_i\phi_i}{\sum h_i} = \frac{41.84 \text{ m}}{381.0 \text{ m}} = 0.110$$

The *OOIP* is then given by:

$$OOIP = \frac{BV\varphi(1 - S_{wi})}{B_{oi}} = \frac{241 \times 10^6 \text{m}^3(0.110)(1 - 0.12)}{1.25} = 18.6 \times 10^6 \text{m}^3$$

Table 10.3.2 Volumetric calculations for example problem based on net pay map of Figure 10.2.4. The contour interval is 10 m

Contour	Base area (ha)	Coefficient	Volume (10^6 m^3)
0	1360	0.5	68.0
10	696	1	69.6
20	486	1	48.6
30	324	1	32.4
40	194	1	19.4
50	81	0.333	2.7
Total	–	–	240.7

Now let us determine the pore volume and *OOIP* using the ϕh map. The volumetric calculations are summarized in Table 10.3.3. The pore volume of the reservoir is 29.1×10^6 m^3. The *OOIP* is then given by:

$$OOIP = \frac{PV(1 - S_{wi})}{B_{oi}} = \frac{29.1 \times 10^6 \text{m}^3 (1 - 0.12)}{1.25} = 20.5 \times 10^6 \text{m}^3$$

The *OOIP* based on the ϕh map is 10% greater than the *OOIP* based on the net pay map. The difference may result from the accuracy of the contouring, the number of contour intervals, or the accuracy of the average porosity calculation. The net pay map has more contour intervals and, therefore, will give a more accurate volumetric calculation. However, the thickness weighted average porosity is less accurate than the volume weighted porosity implicit in the ϕh map. In general, with accurate contour maps, the *OOIP* based on a ϕh map is more accurate than one based on a net pay map and an average porosity. However, it is easier to make contouring errors with a ϕh map because the ϕh product is not a physical parameter that can be checked by inspection in the same way a porosity or pay thickness can be checked.

Let us assume that the bulk volume and pore volume are correct but the thickness weighted average porosity is inaccurate. The zero-pay contour area of the reservoir is 1360 ha. The average pay thickness is:

$$h = \frac{BV}{A} = \frac{241 \times 10^6 \text{ m}^3}{1360 \times 10^4 \text{ m}^2} = 17.7 \text{ m}$$

and the average porosity is:

$$\varphi = \frac{PV}{BV} = \frac{29.1 \times 10^6 \text{ m}^3}{241 \times 10^6 \text{ m}^3} = 0.121$$

Table 10.3.3 Volumetric calculations for example problem based on porosity-pay map of Figure 10.2.5. The contour interval is 2 m

Contour	Base area (ha)	Coefficient	Volume (10^6 m^3)
0	1360	0.5	13.6
2	518	1	10.3
4	243	1	4.9
6	49	0.333	0.3
Total	–	–	29.1

The reservoir volumetrics for the example problem are summarized below:

A	1360 ha
h	17.7 m
ϕ	12.1 %
S_{wi}	12 %
B_{oi}	1.25 m^3/scm
$OOIP$	20.5×10^6 scm

Uncertainty in Volumetric Reserves

The main source of error in volumetric reserves is the error in mapping the reservoir. Usually, there are insufficient data to accurately determine the pool boundaries. (See Section 10.2.1 and Figure 10.2.3 for an example of this issue.) There is also uncertainty in the cutoffs chosen to determine net pay. The choice of average porosity and water saturation may also introduce error. The oil formation volume factor is usually known with reasonable accuracy.

The net pay and therefore *OOIP* of a reservoir is a function of the cutoffs (such as porosity, shale content, and water saturation). This sensitivity of the *OOIP* to the choice of cutoffs is different for each reservoir. For example, in very clean sands with a distinct boundary between the sand and shale, the net pay is often insensitive to the choice of cutoff. In shaly sands and carbonates, the net pay is often very sensitive to the choice of cutoff. Therefore, it is useful to assess the sensitivity of the *OOIP* to the cutoffs, particularly when there is difficulty reconciling the volumetric *OOIP* with a material balance *OOIP*.

To determine the sensitivity of the volumetric *OOIP* to the cutoffs, the net pay is determined from a series of different cutoffs such as 3%, 6%, and 9% porosity or 40%, 50%, and 60% water saturation. In many cases, the type of cutoff most likely to affect the net pay can be determined from inspection of the well logs, and only sensitivities on that particular cutoff need be examined. Then the reservoir is remapped, and the *OOIP* determined for each choice of cutoff. To illustrate the sensitivity, the calculated *OOIP* or percent change in *OOIP* can be plotted against the cutoff values of percent change in cutoff values, as shown in Figure 10.3.2. Figure 10.3.2(a) is a spider diagram, and Figure 10.3.2(b) is a tornado chart of the same data. In Figure 10.3.2, the *OOIP* is most sensitive to the porosity cutoff and least sensitive to the water saturation cutoff.

In general, if the pool boundaries can be clearly established, volumetric reserves are accurate to within approximately ±20%. In well-developed fields with a large number of wells and limited "edge effects," the volumetric reserves can be determined more precisely. If pool boundaries cannot be accurately determined, then there may be order of magnitude errors in the volumetric reserves.

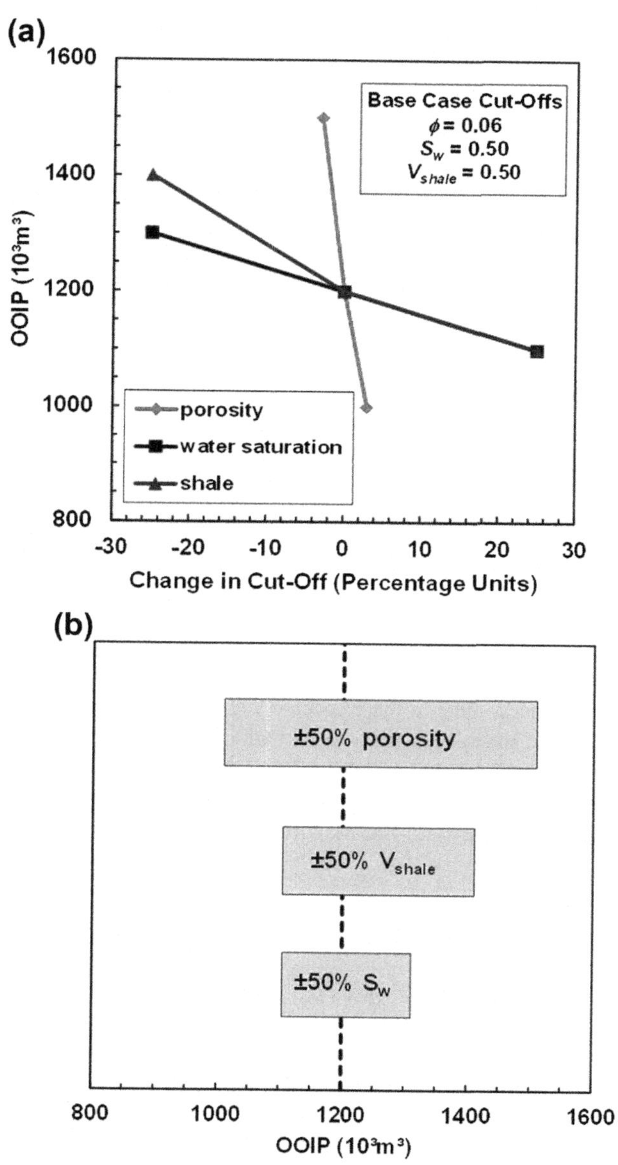

Figure 10.3.2 Example of sensitivity of OOIP to cutoffs: (a) spider diagram; (b) tornado chart. Note that change in cutoffs presented in percentage units (i.e., 6 ± 3% units) in the spider diagram and as a percentage difference (i.e., 6% porosity ±50%) in the tornado chart.

Example 1: Effect of Uncertainty of Reservoir Properties
Let us reevaluate the volumetric reserves of the above example, assuming that the average porosity is accurate to ±2% and the water saturation is accurate to ±4%. Also assume that the area and net pay are accurate. The lower limit on the *OOIP* is then:

$$
\begin{aligned}
OOIP &= \frac{hA\varphi(1 - S_{wi})}{B_{oi}} \\
&= \frac{(17.7\ \text{m})(13.6\cdot 10^6\text{m}^2)(0.0121 - 0.02)(1 - 0.12 - 0.04)}{1.25\ \text{m}^3/\text{scm}} \\
&= 16.3 \times 10^6\ \text{m}^3
\end{aligned}
$$

and the upper limit is:

$$
\begin{aligned}
OOIP &= \frac{hA\varphi(1 - S_{wi})}{B_{oi}} \\
&= \frac{(17.7\ \text{m})(13.6\cdot 10^6\text{m}^2)(0.0121 + 0.02)(1 - 0.12 + 0.04)}{1.25\ \text{m}^3/\text{scm}} \\
&= 22.8 \times 10^6\ \text{m}^3
\end{aligned}
$$

Hence, a plausible range of uncertainty in the reservoir properties leads to approximately ±10–20% uncertainty in the expected *OOIP*.

At this point we can start to address the first and possibly the second question asked at the start of the chapter (1) What is the *OOIP*?; (2) What is remaining recoverable oil in place (*RROIP*)? However, given the possible uncertainty in the volumetric numbers, they must also be reconciled with the dynamic pressure and production data.

10.4 Analysis of Well, Production, and Pressure History

An examination of the production, pressure, and development history of a reservoir can often aid in characterizing the reservoir, including the drive mechanism, the permeability distribution, and the presence and likely location of flow barriers. There are characteristic production profiles that correspond to different drive mechanisms. The permeability distribution and presence of flow barriers can sometimes be inferred from fluid breakthroughs, the distribution of water cuts and GORs between individual wells, and the distribution of pressure in a reservoir. Here, we partially address the third question from the start of the chapter: What is the permeability distribution or reservoir architecture?

10.4.1 Characteristic Production Profiles

While each reservoir has a unique production history, there are characteristic profiles that depend largely on the reservoir drive mechanisms. The production profile also

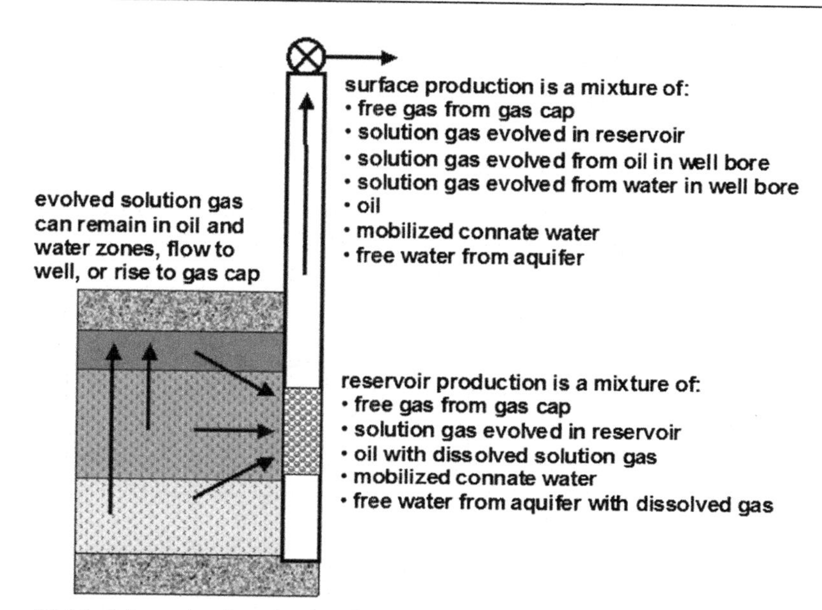

Figure 10.4.1 Schematic of production fluids in reservoir and wellbore and their sources.

depends on the reservoir architecture and the extent of the segregation of flow of evolved solution gas.

When interpreting production profiles, it is helpful to recall that the produced gas can be a mixture of gases from different sources. Consider the reservoir and wellbore schematic shown in Figure 10.4.1. As a reservoir is produced and its pressure drops below the bubble point, solution gas evolves in the reservoir. The evolved solution gas may flow with the oil to the wellbore, or it may rise to join an existing gas cap or form a new secondary gas cap. The fluid that flows into the wellbore can then be a mixture of oil, water, solution gas evolved in the reservoir, and gas cap gas. As the fluid rises up the wellbore and pressure decreases further, more solution gas evolves from the oil in the wellbore. Hence, the gas at the surface can be a mixture of solution gas that evolved in the reservoir, solution gas that evolved in the wellbore, and gas cap gas. Note that there is also dissolved solution gas in reservoir water, but the amount is usually negligible. As will be discussed later, examination of the producing GOR can sometimes help to identify the drive mechanism and whether gas flow is segregated or nonsegregated.

The rate at which oil production declines is also an indication of the drive mechanism. To analyze a production profile, it is sufficient to recognize that there are three types of decline curves used in production analysis, exponential, hyperbolic, and harmonic, as shown in Figure 10.4.2. The equations for each type of decline are given by:

Exponential:

$$\frac{dq_o}{dt} = -cq_o \tag{10.4.1}$$

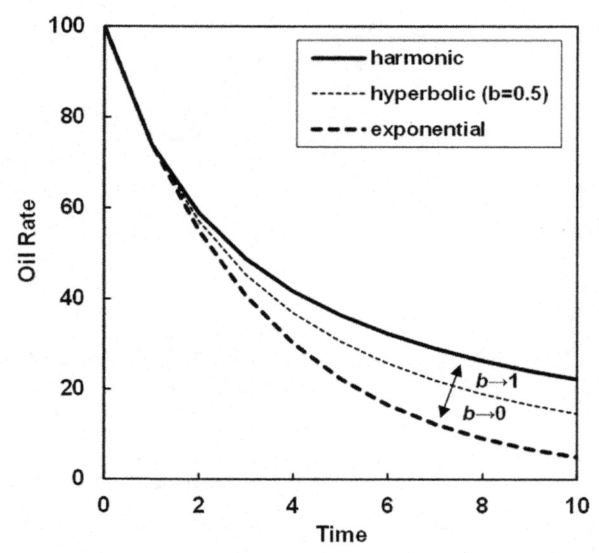

Figure 10.4.2 Comparison of harmonic, hyperbolic, and exponential decline profiles. The decline coefficients (c) were adjusted so that the declines in the first interval were identical.

Harmonic:

$$\frac{dq_o}{dt} = -cq_o^2 \tag{10.4.2}$$

Hyperbolic:

$$\frac{dq_o}{dt} = -cq_o^{b+1}, \quad where \ 0 < b < 1 \tag{10.4.3}$$

in which q_o is the oil rate, t is time, and c is a constant. The exponent constant, b, is sometimes referred to as the Arps exponent (Arps, 1945). Exponential decline has the steepest decline rate, and harmonic has the shallowest. Hyperbolic decline approaches exponential decline as b goes to zero and approaches harmonic decline as b goes to unity.

Now let us consider the production histories that can be expected from different reservoir drive mechanisms. We will consider plots of pressure, oil rate, GOR, and water cut versus time or cumulative oil production.

10.4.1.1 Undersaturated Oil Drive

A typical production profile under undersaturated oil drive is shown in Figure 10.4.3(a) and (b) versus cumulative production and time, respectively. When oil is above the bubble point and there is no aquifer support, the reservoir energy is supplied

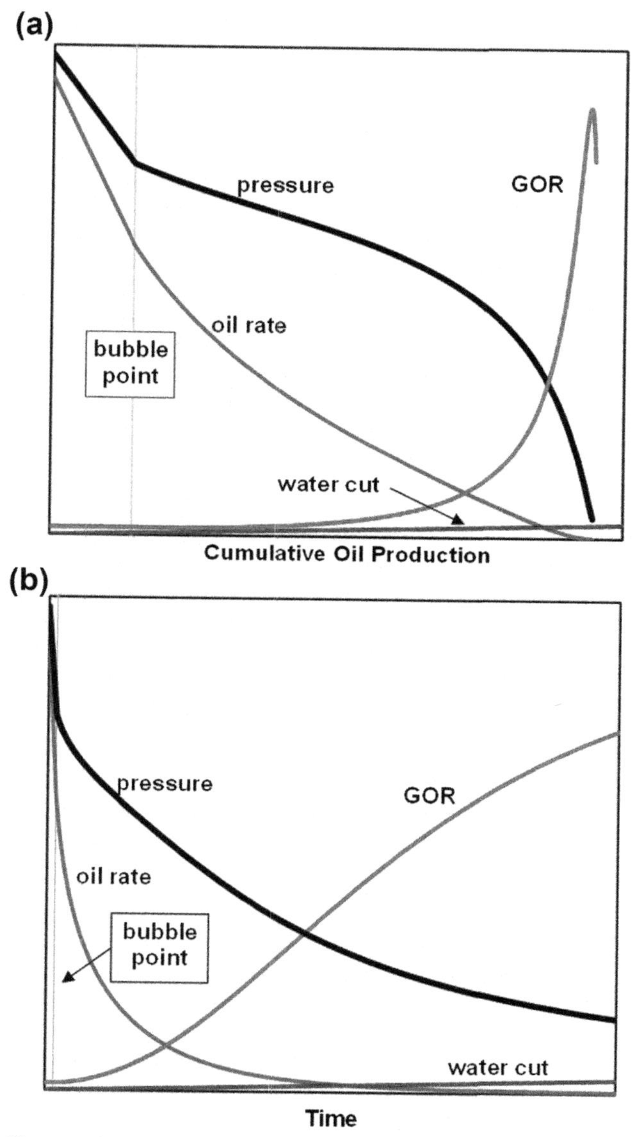

Figure 10.4.3 Characteristic production profile versus cumulative oil production (a) and time (b) of an undersaturated oil reservoir—undersaturated oil drive, followed by solution gas drive.

by formation compaction, connate water expansion, and oil expansion. The overall compressibility is low, and therefore, a small expansion of the fluid results in a large pressure decrease. Therefore, reservoir pressure declines rapidly. Because the oil rate is proportional to the reservoir pressure, the oil rate also declines rapidly. No solution gas evolves in the reservoir because the reservoir is above the bubble point. The only gas produced is solution gas from the produced oil; hence, the producing GOR is constant and equal to the initial solution gas-oil ratio. A small amount of connate water may be produced as the pressure drops, but the water cut is expected to be near zero.

Individual well performance depends on local permeability and pay thickness. The initial flow rates may differ from well to well, but the decline is usually exponential with a similar decline rate for each well. The late-stage decline rates are similar because the oil rate is directly proportional to reservoir pressure and, in many reservoirs at pseudo steady-state, the pressure declines uniformly in different regions. Note that permeability-pay maps can often be derived from the oil rates at a given time because there is relatively little uncertainty in the single-phase inflow equations.

The ultimate recovery factor for undersaturated systems is usually less than 5% unless significant rock compaction occurs. However, in many cases, the bubble point pressure is high enough that significant additional recovery is attained under solution gas drive.

10.4.1.2 Solution Gas Drive

A typical production profile under solution gas drive (SGD) with nonsegregated gas flow is shown in Figure 10.4.3 As the pressure falls below the bubble point, solution gas evolves in the reservoir. Gas is highly compressible, and therefore the reservoir pressure decreases more gradually than under undersaturated oil drive. The reservoir performance strongly depends on gas segregation. Production characteristics of SGD reservoirs are summarized in Figure 10.4.4. A field example of a solution drive reservoir is shown in Chapter 11.

If the solution gas is segregated, it will tend to remain in the reservoir and form a secondary gas cap. The producing GOR equals the solution gas-oil ratio and decreases at first as the pressure drops below the bubble point. The GOR eventually

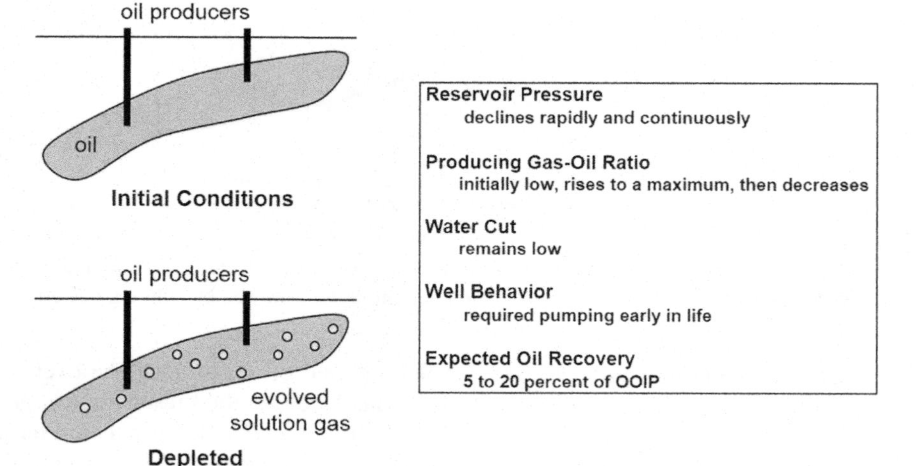

Figure 10.4.4 Production characteristics of SGD reservoirs.

increases dramatically when gas is coned from the secondary gas cap. The reservoir pressure and oil rate both decrease slowly until gas is coned. Water cuts remain low. The reservoir performance with segregated gas flow is similar to that of a gas cap reservoir.

If the solution gas is nonsegregated, a secondary gas cap does not form and pressure support is less. The producing GOR increases, reaches a plateau, and ultimately decreases when the evolved gas supply is exhausted. The pressure and oil rate both decline more rapidly than with segregated gas flow. Water cuts remain low.

With nonsegregated gas flow, the individual well performance will usually follow similar hyperbolic decline profiles with Arps exponent $b$in the range of 0.2–0.4. In tight formations, the initial decline rates may differ, but after gas evolves in the reservoir, reservoir pressure tends to equalize and similar profiles are observed. With segregated gas flow, individual well performance will not only depend on local permeability, but on the distance between the perforations and the GOC. The decline is usually hyperbolic.

Ultimate oil recovery factors for SGD systems typically vary between 5% and 20%. Ultimate solution gas recovery factors are usually between 85% and 95%. An estimate of the ultimate oil recovery can be obtained from a Cartesian or semi-log plot of cumulative oil production versus cumulative gas production (or gas recovery factor). Oil recovery usually changes little at high gas recovery factors, as shown on Figure 10.4.5. Hence, the oil recovery estimated from a linear extrapolation to a 90% gas recovery factor is often a reasonable estimate of the ultimate oil recovery. Note the gas recovery factor depends on the producing wells effectively draining the reservoir pressure down to a low pressure, which may not be the case in tighter reservoirs.

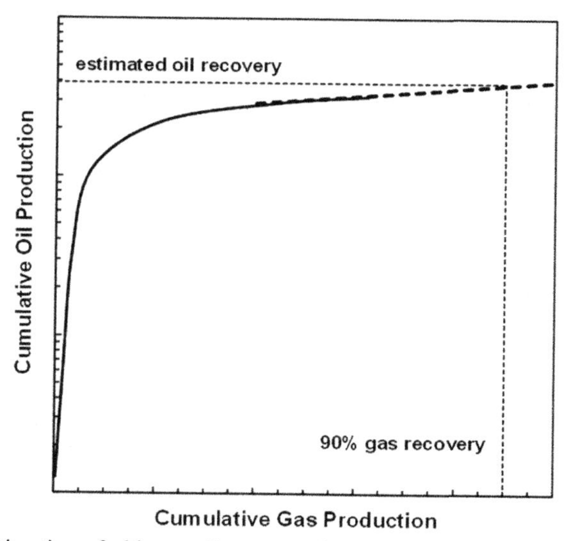

Figure 10.4.5 Estimation of ultimate oil recovery from a semi-log plot of cumulative oil production versus cumulative gas production.

10.4.1.3 Gas Cap Drive

A typical production profile under gas cap drive with segregated gas flow is shown in Figure 10.4.6. The gas cap is a volume of highly compressible fluid and can provide considerable pressure support depending on its size. Reservoir performance again depends strongly on gas segregation. In general, more gradual pressure and oil rate decline than for solution gas drive can be expected. Once gas breakthrough occurs, the GOR is expected to continue rising. Water cuts are low. An example of a field with gas cap drive reservoir is shown in Chapter 11.

Individual well decline is usually hyperbolic with an Arps exponent of approximately 0.5 (Lefkovits and Matthews, 1958). However, individual well performance is strongly dependent on the structure of the reservoir and the distance between the well and the GOC. The permeability distribution is also an important factor because: (1) drawdown pressure is related to local permeability and coning is influenced by the drawdown pressure; (2) the extent of gas migration depends on the permeability distribution. When there is a laterally extensive gas cap blanketing the oil zone, pressure interference is common and can also affect the individual well decline rates. Production characteristics of gas cap reservoirs are summarized in Figure 10.4.7.

Recovery factors for gas cap reservoirs vary over a large range depending on pay thickness, the relative thickness of the gas and oil columns, permeability, and the ability to control the GOR. Generally, for reservoirs with small gas caps with low permeability, ultimate recovery factors are in the range of 5–20% because solution gas drive dominates the production profile. For reservoirs with large gas caps (m>0.3), thick oil columns (h>10 m), and high permeability, the ultimate recovery factor is usually

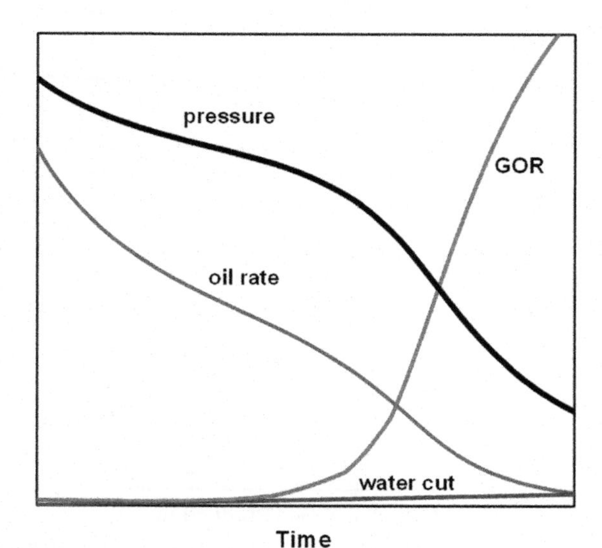

Figure 10.4.6 Characteristic production profile of a gas cap drive reservoir with segregated gas flow.

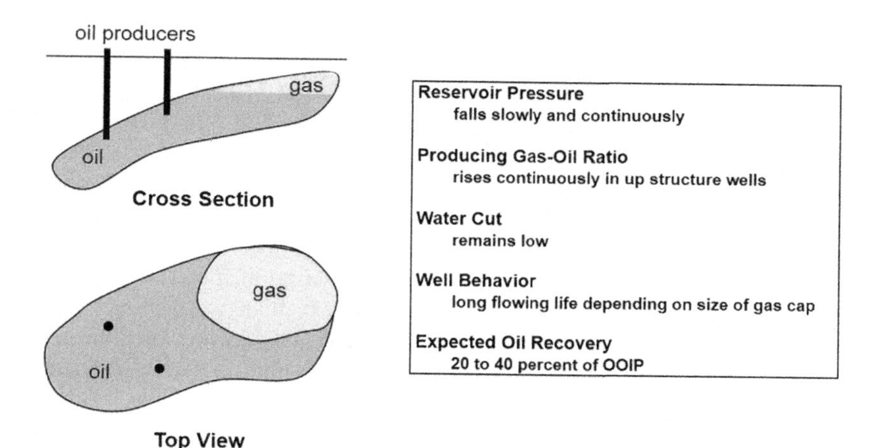

Figure 10.4.7 Production characteristics of gas cap drive reservoirs.

greater than 30%. For high permeability and thick reservoirs in which gravity drainage dominates, the ultimate recovery factor can exceed 50%.

10.4.1.4 Water Drive

A typical production profile under strong water drive is shown in Figure 10.4.8. Reservoir performance depends on the strength of the aquifer relative to the reservoir fluid withdrawal rate. If the aquifer is strong enough to maintain the reservoir pressure above the bubble point, then a nearly constant oil rate can be maintained until water

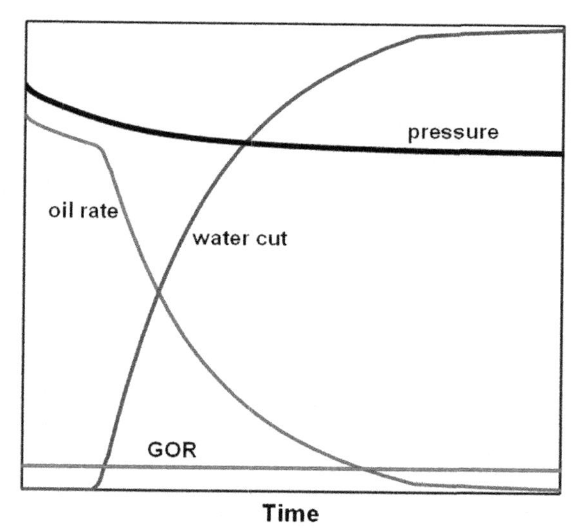

Figure 10.4.8 Characteristic production profile of a strong water drive reservoir.

breakthrough. After water breakthrough, the water cut rises continuously and the oil rate usually declines hyperbolically or sometimes harmonically. With a strong aquifer, the decline rate depends mainly on the mobility ratio (oil viscosity and the oil and water relative permeabilities) and the permeability distribution. The producing GOR equals the solution gas-oil ratio.

If the aquifer support is weaker, the reservoir pressure dips below the bubble point, and the drive mechanism becomes a mixture of water drive and solution gas drive. Above the bubble point, the pressure and oil rate decline more slowly than for an undersaturated oil reservoir. Below the bubble point, pressure and oil rate decline more slowly and GOR increases more slowly than for a purely solution gas drive reservoir, at least until water breakthrough. After water breakthrough, a more rapid decline in oil rate is expected due to three-phase relative permeability effects and coning.

Individual well performance is strongly dependent on the structure of the reservoir and the distance between the well and the water-oil contact. As mentioned above, the decline rate is typically hyperbolic to harmonic (Arps exponent between 0.2 and 1.0) depending mainly on the aquifer strength, voidage replacement, the mobility ratio, and the permeability distribution. Ultimate recovery factors for reservoirs under water drive are usually high, ranging from 20% to 80%. Production characteristics of water drive reservoirs are summarized in Figure 10.4.9. An example of a water drive reservoir is shown in Chapter 11.

There are many diagnostic plots for water drive reservoirs, including semi-log WOR versus cumulative oil production and semi-log oil cut versus cumulative oil production. Extrapolation of these plots can provide a good estimate of the ultimate oil recovery. The log WOR and log oil cut trends are often linear in moderate to high oil viscosity reservoirs. In light oil water drive reservoirs, the water cuts often rise so rapidly that clear trends are not observed.

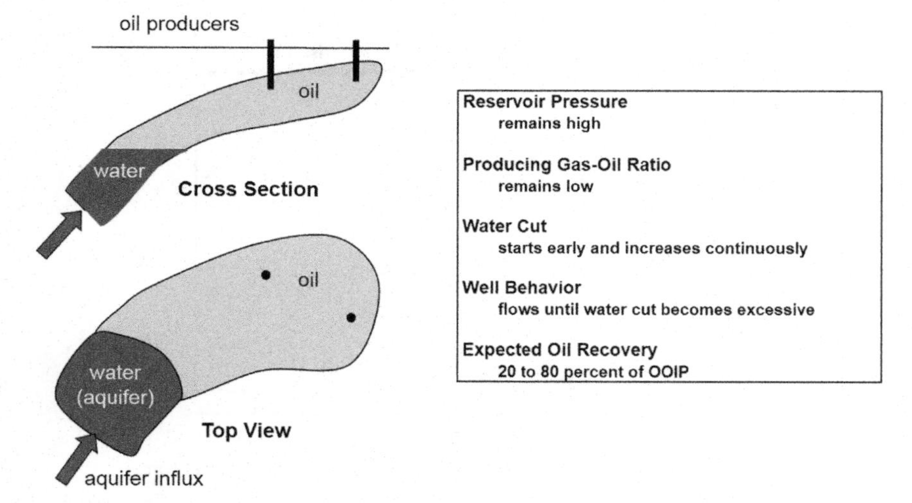

Figure 10.4.9 Production characteristics of water drive reservoirs.

10.4.1.5 Combination Drive

A reservoir under both gas cap and water drive will usually have good pressure maintenance and slow decline in the oil rates until gas and/or water breakthrough occurs. After gas breakthrough, the GOR rises continuously. After water breakthrough, the water cut rises continuously. Breakthrough of one fluid often leads to pressure decline and quick breakthrough of the second fluid. In a well-managed reservoir, breakthrough of either fluid is delayed as long as possible and occurs nearly simultaneous for both fluids. Rapid decline is expected after breakthrough. Individual well performance depends on the permeability distribution, the structure of the reservoir, and the distance between the well and the WOC and GOC.

10.4.2 Analyzing Production and Pressure Data

Production and pressure data can be used to determine the drive mechanism, determine the bubble point of a solution gas drive reservoir, assess the permeability distribution, and estimate the *OOIP*. Each type of analysis is discussed below.

10.4.2.1 Determination of Drive Mechanism

In the previous section, characteristic production profiles were described. In practice, it can be difficult to distinguish one profile from another, particularly early in the production history of the reservoir. If the *OOIP* is known from volumetrics or a material balance, then production profiles can be plotted versus the oil recovery factor. In many cases, the distinction between drive mechanisms becomes clearer.

Figure 10.4.10 shows normalized reservoir pressure (pressure/initial pressure) versus oil recovery factor. A strong water drive stands out with the small pressure decrease with oil recovery. Undersaturated oil and SGD also stand out with rapid

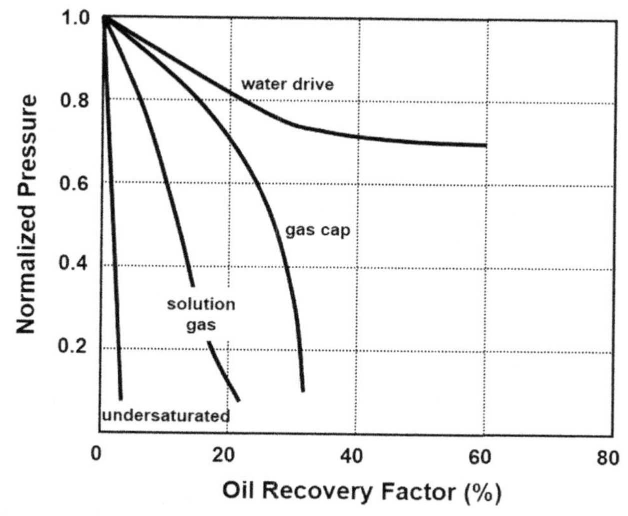

Figure 10.4.10 Comparison of pressure trends for different drive mechanisms. Adapted from Allen and Roberts (1968).

pressure decline. Of course, the distinctions are not always so clear cut. Weak water drive may have a similar pressure profile to a gas cap drive reservoir or a SGD reservoir with gas segregation. A gas cap drive reservoir with rapid coning may have a similar pressure profile to a SGD reservoir. Combination drive may be indistinguishable from gas cap or water drive depending on the strength of the aquifer.

Sufficient pressure data to analyze trends are not always available. However, a plot of normalized fluid rate (fluid rate/initial fluid rate) versus recovery factor is expected to show the same trends as the pressure plot. Oil decline rates are also an indicator of drive mechanism. Undersaturated oil drive usually results in exponential decline. SGD exhibits hyperbolic decline with an Arps exponent between 0.2 and 0.4. Gas cap drive also exhibits hyperbolic decline with an Arps exponent of approximately 0.5. A strong water drive usually results in hyperbolic decline with an Arps exponent greater than 0.5 or even harmonic decline. Unfortunately, it is often impossible to distinguish decline profiles until far into the productive life of a reservoir, see Figure 10.4.2.

Figure 10.4.11 shows GOR versus oil recovery factor. Again, it is relatively easy to diagnose a strong water drive since the GOR remains low. SGD usually exhibits high GORs at relatively low oil recovery. The decrease in GOR at the end of the life of an SGD is a clear diagnostic, but is often not observed in practice because the reservoir is abandoned when the GOR begins to decrease due to uneconomic production rates. Gas segregation, weak water drive, premature water, or gas breakthrough can all obscure the differences between GOR profiles from different drive mechanisms.

Figure 10.4.12 shows water cut versus cumulative oil recovery factor. This plot is the most obvious diagnostic for water drive reservoirs, because water cuts remain low for both solution gas and gas cap drive reservoirs, but increases continuously for water drives. A combination drive reservoir is expected to show increasing water cuts as

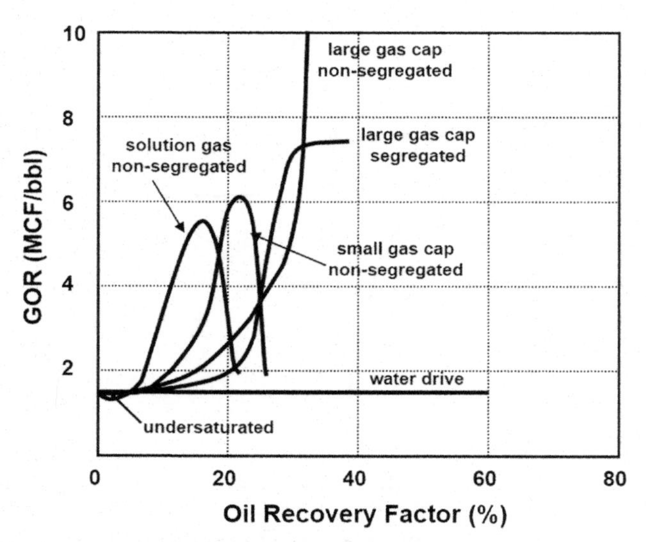

Figure 10.4.11 Comparison of GOR trends for different drive mechanisms. Adapted from Allen and Roberts (1968).

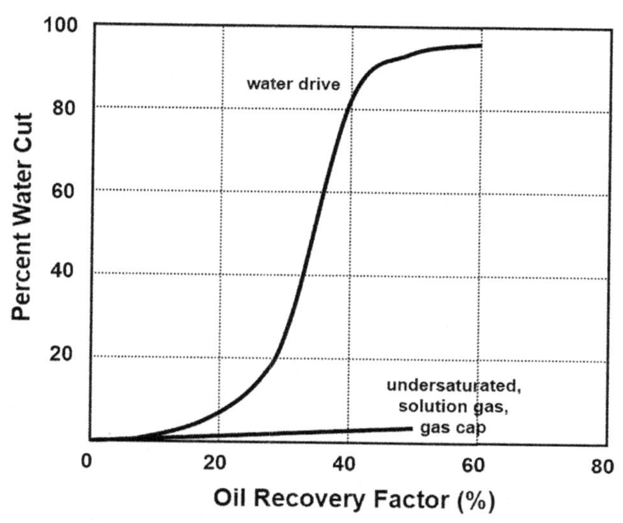

Figure 10.4.12 Comparison of water cut trends for different drive mechanisms.

well. However, a combination drive reservoir is likely to have higher GORs than a water drive reservoir at least in the late stages of depletion.

Often, a combination of production plots is required to make a diagnosis. Diagnostic indicators are listed in Table 10.4.1. These indicators are guidelines only because production profiles also depend on the reservoir architecture.

10.4.2.2 Estimation of Bubble Point Pressure from Field Data

The bubble point pressure of a gas cap reservoir is identical to the initial pressure. Undersaturated reservoirs are initially above the bubble point. As production depletes the reservoir, the pressure will eventually drop below the bubble point unless there is a strong water drive. There can be distinctive changes in production profiles as the pressure drops below the bubble point. Above the bubble point, the GOR is constant at the solution gas-oil ratio. Below the bubble point, the GOR increases with production. The oil rate and pressure decline less rapidly below the bubble point than above.

To illustrate how to determine a bubble point from pressure and production data, consider the pressure, GOR, and oil rate data for the Bilbo A Cardium A Pool, presented in Figures 10.4.13–10.4.15. Some reservoir properties are summarized below:

Oil gravity	45 API	P_i	15,700 kPa
h	1.23 m	*OOIP*	1217 $10^3 m^3$
A	1538 ha	R_s	75 scm/scm
ϕ	12%	SGIP	91×10^6 m^3
S_w	33%	m	0
B_{oi}	1.25 m^3/scm	*OGIP*	0 $10^6 m^3$

Table 10.4.1 Diagnostic indicators for drive mechanism

Drive mechanism	Indicators
Undersaturated oil	• Ultimate recovery <5% • Very rapid pressure decline • Exponential oil decline • Low GOR • Low water cut
Solution Gas drive	• Ultimate recovery 5−20% • Rapid pressure decline • Hyperbolic oil decline (0.2<b<0.4) • Rapid increase in GOR • Decrease in GOR at high recovery factor • Low water cut
Gas cap drive	• Ultimate recovery 20−40% • Moderate pressure decline • Hyperbolic oil decline (b ≈ 0.5) • Continuous increase in GOR • GOR sensitive to perforation location • Low water cut
Water drive	• Ultimate recovery 20−80% • Low pressure decline • Hyperbolic oil decline (b>0.5) • Sometimes near harmonic oil decline • Low GOR • Continuously increasing water cut • Water cut sensitive to perf location • Pressure gradient from edge aquifer
Combination drive	• Ultimate recovery 20−80% • Moderate to low pressure decline • Hyperbolic oil decline (b>0.5) • Continuously increasing GOR • Continuously increasing water cut • GOR and water cut sensitive to perf location • Pressure gradient from edge aquifer

This reservoir was initially undersaturated. Figure 10.4.13 shows that pressure declined rapidly during the first year of production and then more gradually in later years. There is some scatter in the production data, but the bubble point was likely reached in 1988 and possibly as late as 1991. The bubble point pressure is approximately 10,500 kPa.

Figure 10.4.14 shows that the GOR was stable at the solution gas-oil ratio until 1991. Based on GOR alone, one would conclude that the bubble point was reached

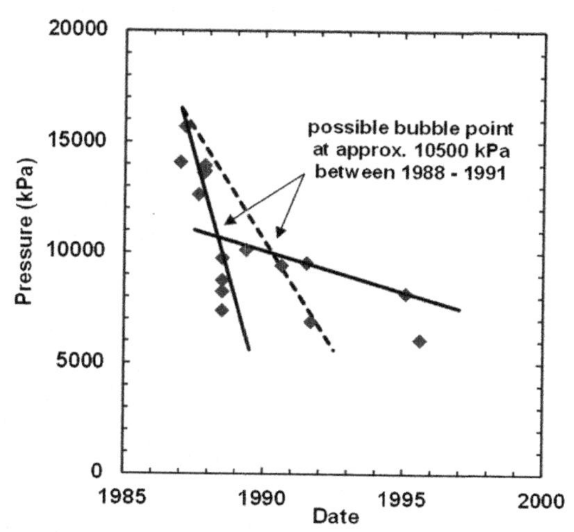

Figure 10.4.13 Pressure history of the Bilbo A Cardium A Pool. Dashed and solid lines represent two possible interpretations of the pressure-time behavior prior to reaching the bubble point pressure.

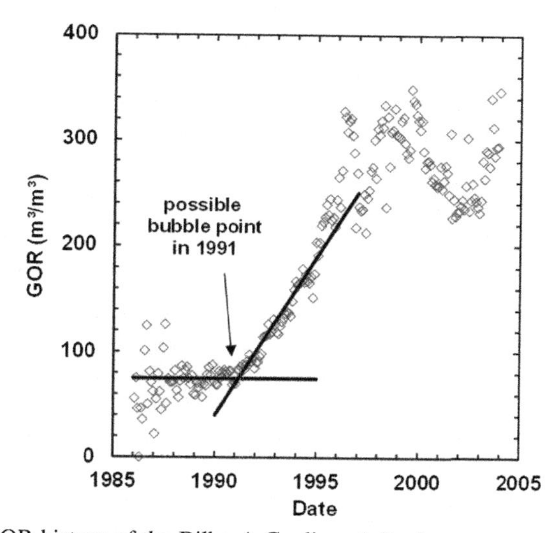

Figure 10.4.14 GOR history of the Bilbo A Cardium A Pool.

in 1991. However, GOR may not be the most reliable indicator for the bubble point for two reasons:

1. Gas may evolve in the reservoir without reaching the wellbore due to a critical gas saturation or to gas segregation;
2. Gas may evolve locally near the wellbore because pressure is lowest near the wellbore.

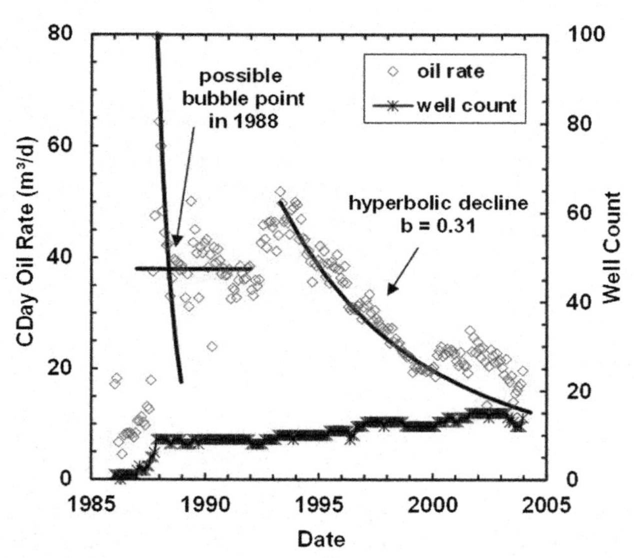

Figure 10.4.15 Oil production history of the Bilbo A Cardium A Pool. CD oil rate is calendar day oil rate.

Therefore, GOR data are best used in combination with other data. In this case, it is likely that the bubble point was reached in 1988, but evolved gas was not produced until 1991. GOR data should only be used to confirm other data such as the base solution gas-oil ratio (R_S). Figure 10.4.14 shows a reasonable estimate of Rs from the field data is 75–80 m^3/m^3.

Figure 10.4.15 shows that the pool oil rate declined rapidly from the initial rate of 80 m^3/d in 1987. In 1988, oil production stabilized at a rate of approximately 38 m^3/d. The oil production data are consistent with the pressure data, strongly suggesting that the bubble point was reached in 1988. Note also that the early production declines exponentially, consistent with undersaturated drive. The late production declines hyperbolically with an Arps exponent of 0.35, consistent with solution gas drive.

10.4.2.3 Assessing the Permeability Distribution

Perhaps the best indicators of a reservoir's permeability distribution are the fluid rates, fluid movement, and pressure distribution in the reservoir. These factors can be analyzed using bubble maps, breakthrough maps, isobaric maps, aggregate well plots, and pressure versus time plots. An aggregate well plot is simply a set of individual well production histories from various wells plotted together to see if the decline in oil rates and rise in water cuts or gas-oil ratios occur at the same time. A bubble map is a well location map on which a given production variable is plotted for each well as a circular area centered on the well. The area of the circle is proportional to the magnitude of the variable.

Production variables include: rate, cumulative production, GOR, water cut, and WOR. A breakthrough map is a well location map on which the date that water or gas breakthrough occurred is plotted for each well. An isobaric map is a contour map of the reservoir pressure at a given time. Fluid rate, fluid movement, pressure distribution, and the appropriate mapping to analyze each factor are discussed below. The analysis of permeability distribution is then illustrated with an example.

Fluid Rate and Cumulative Fluid Production: Fluid rates depend largely on the permeability-thickness product *kh* and the reservoir pressure, unless otherwise constrained, e.g., partial well penetration. If the reservoir pressure is nearly uniform and GORs are stable, the producing fluid rate at each well is proportional to the *kh* of the well. Hence, regions with high producing rates are likely regions of relatively high *kh* and vice versa. Similarly, regions with high cumulative fluid production are likely regions of relatively high permeability-pay. Some caution is required when considering cumulative production, because regions with older wells may have higher cumulative production simply because they have been producing for a longer time and possibly when reservoir pressure was higher and water or gas breakthrough had not occurred. Bubble maps of oil rate, fluid rate, cumulative oil production, or cumulative fluid production can be used to examine the distribution of productivity in the reservoir.

Fluid Movement: Fluid flows through the path of least resistance or, in other words, through the path of highest permeability. If water or gas is moving through an oil zone, the changes in GOR and water cut indicate where the fluid is moving. Higher permeability pathways can then be inferred from the distribution of GORs and water cuts. Fluid movement can be examined using water or gas breakthrough time maps as well as bubble maps of GOR, water cut, WOR, cumulative water production, and cumulative gas production. Note that the initial distribution of GORs and water cuts can also help to locate the original fluid contacts.

Reservoir Pressure Distribution: Reservoir pressure becomes nonuniform when fluid withdrawal rates are different in different parts of a reservoir. In a high permeability reservoir with no flow barriers, the reservoir pressure equalizes quickly, and nearly uniform reservoir pressure is observed. In low permeability reservoirs, pressure may take considerable time to equalize and regions of different pressure may develop. The lowest pressures are observed in regions where the greatest proportion of the *OOIP* and OGIP has been withdrawn. In other words, pressure will be lowest where relatively large oil and gas volumes have been produced from a region with a relatively low pore volume. In reservoirs with flow barriers, the reservoir pressure may not equalize during the productive life of the reservoir. The reservoir pressure distribution can be assessed using isobaric maps or comparing the pressure versus time trends of individual wells.

Example: The Medicine River Jurassic A Pool is a carbonate reservoir under a primarily edge water drive. This example will show how pressure data can be used to confirm the drive mechanism and to address question 3: What is the permeability distribution or reservoir architecture?

Oil gravity	28 API	P_i	16,036 kPa
h	4.69 m	$OOIP$	$5750\times10^3\ m^3$
A	1429 ha	R_s	147 scm/scm
ϕ	14%	$SGIP$	$844\times10^6\ m^3$
S_w	25%	m	0.017
B_{oi}	$1.25\ m^3$/scm	$OGIP$	$157\times10^6\ m^3$

There is a small gas cap that had little impact on the reservoir performance. The aquifer is weak and has been supplemented with water injection, as shown on the injection bubble map, Figure 10.4.16. The response to the water flood has been uneven (Figure 10.4.17). Let us examine the pressure and water breakthrough patterns for the

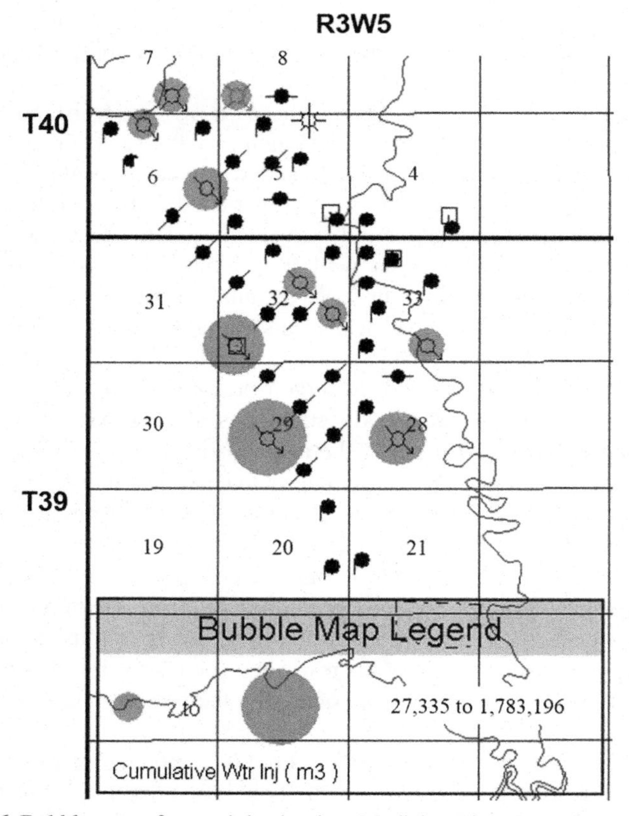

Figure 10.4.16 Bubble map of water injection into Medicine River Jurassic A Pool. Grid lines are section lines and are one mile apart.

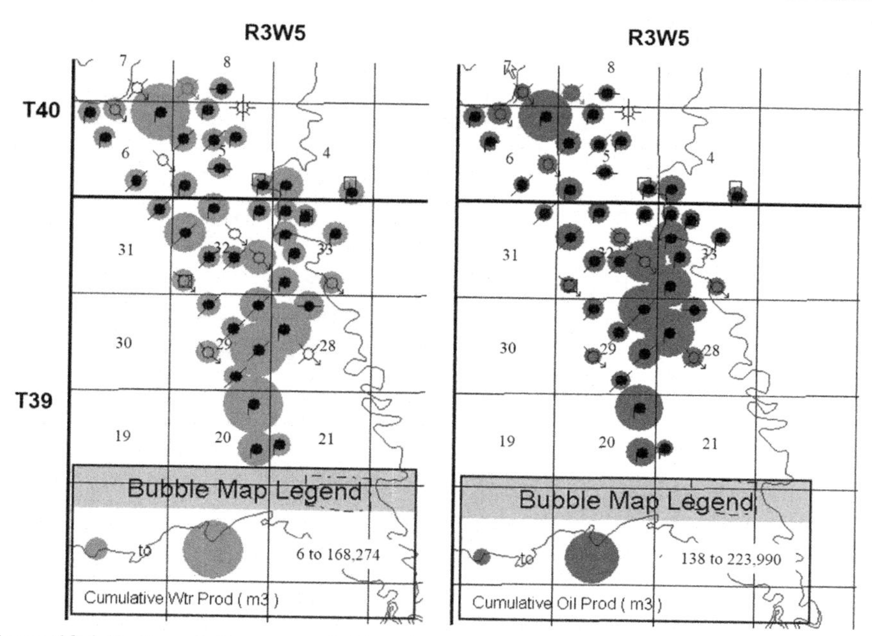

Figure 10.4.17 Cumulative water (left) and oil (right) production bubble maps for Medicine River Jurassic A Pool as of January 2004. Grid lines are section lines and are one mile apart.

reservoir and attempt to identify the permeability distribution and explain the nonuniform waterflood response.

Most of the available pressure data were obtained prior to 1996. Therefore, only the injection response prior to 1996 is evaluated. The injectors active during this period are summarized in Table 10.4.2.

Table 10.4.2 Medicine River Jurassic A Pool water injectors and date injection started

Injector	Start date
06-29-39-3W5	1966
04-32-39-3W5	1966
06-28-39-3W5	1972
02-07-40-3W5	1972
08-06-40-3W5	1975
02-33-29-3W5	1988
04-08-40-3W5	1993

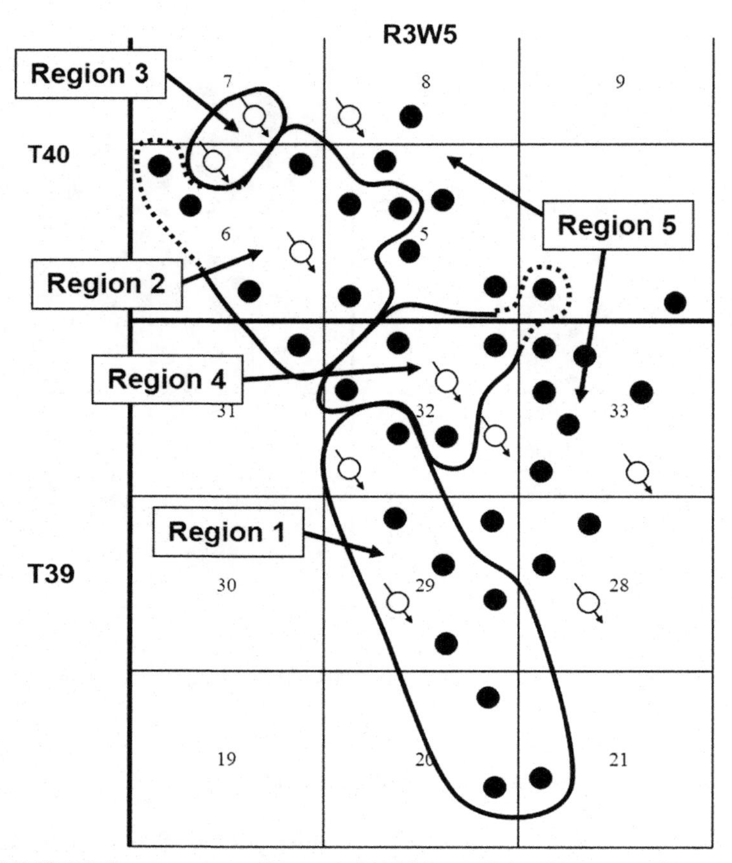

Figure 10.4.18 Distinct pressure regions as of 1996 in Medicine River Jurassic A Pool. The dotted lines are areas where pressure data were not available and wells were assigned to the appropriate region based on water breakthrough data.

An examination of the individual well pressure histories showed that there were five distinct pressure profiles in the reservoir. The reservoir regions corresponding to each pressure profile are identified in Figure 10.4.18, and the profiles are shown in Figure 10.4.19. Region one is the part of the reservoir responding to injection at 06-29 and 4-32-39-3W5. Region two is responding to injection at 08-06-40-3W5. Region three appears to be an isolated part of the reservoir responding to injection at 02-07-40-3W5. Region four appears to be receiving injection support in the 1980s. It is not clear whether the support is from the 04-32 injector or from the 08-06 injector. Region five does not receive strong pressure support.

A series of isobaric maps, Figure 10.4.20, shows how the pressure regions have developed over time. Note that in years in which pressure data were sparse, pressures from several years was including in the same map. This approach is reasonable as long as pressure does not change significantly over the course of a few years. The shape of

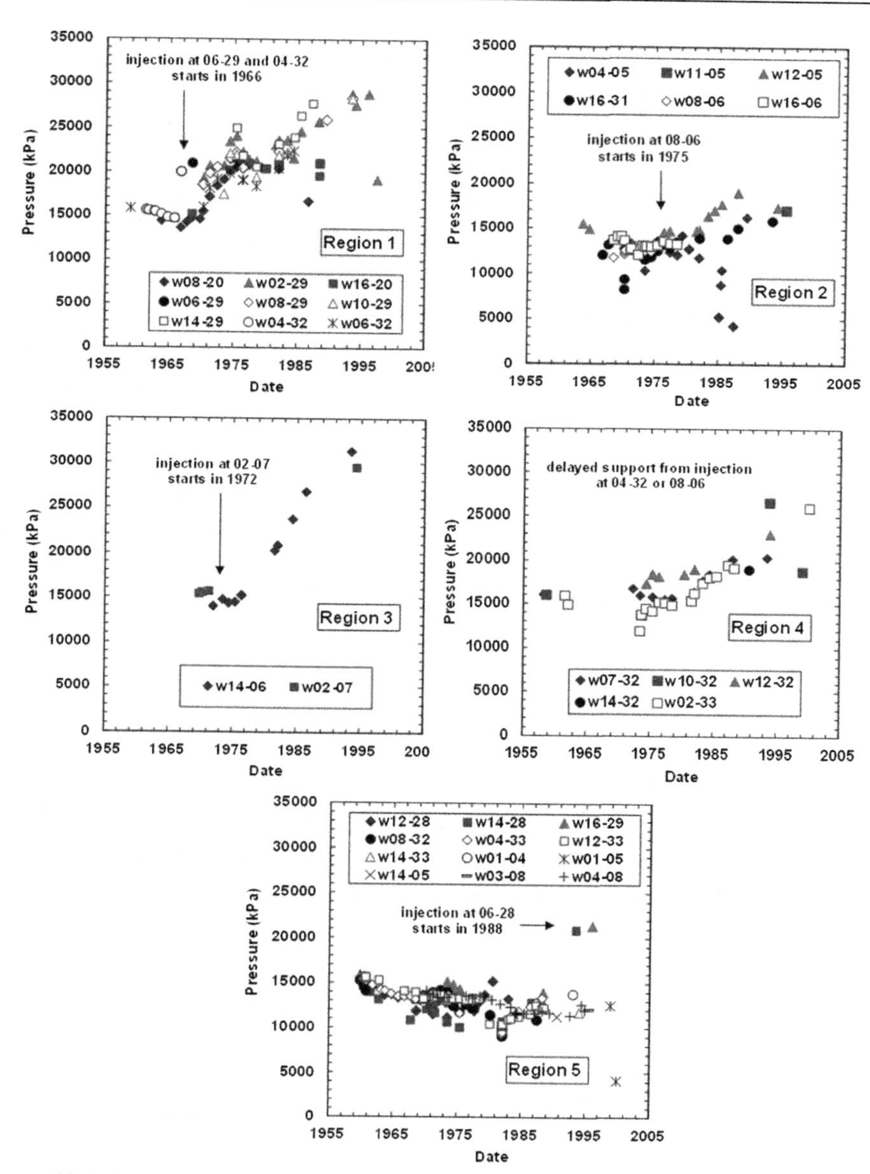

Figure 10.4.19 Individual well pressure histories of Medicine River Jurassic A Pool grouped into five common profiles.

the isobars suggest that region four is receiving delayed pressure support from the 04-32-39-3W5 injector.

A water breakthrough map, shown in Figure 10.4.21, confirms the regional trends observed from the pressure data. The reservoir appears to be a low

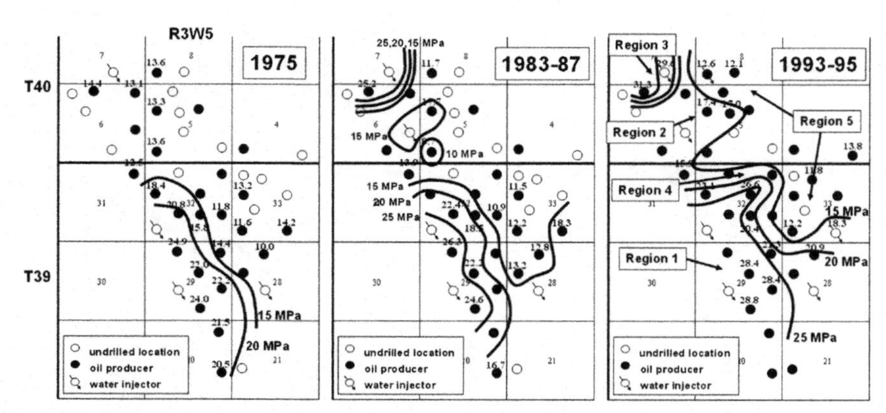

Figure 10.4.20 Isobaric maps illustrating the development of distinct pressure regions in the Medicine River Jurassic A Pool. Note that well status is shown as it was at the time the pressure data were collected.

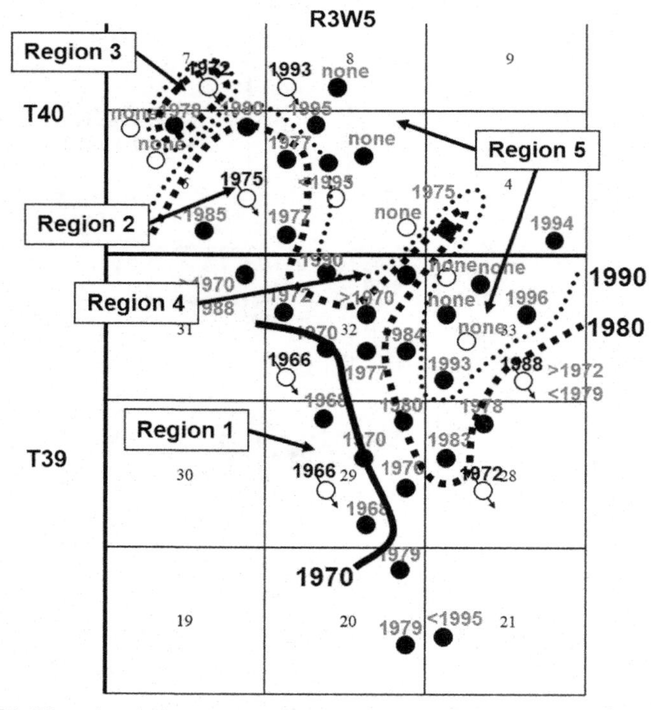

Figure 10.4.21 Water breakthroughs as of 1996 for Medicine River Jurassic A Pool. The lines indicate the movement of the flood front over time. The dates are the date injection commenced for injectors or the date of water breakthrough for producers.

permeability reservoir with limited communication throughout the reservoir. The initial injection in 1996 provided immediate pressure support to region one, with water breakthroughs occurring within 5–10 years. Similarly, injection at 08-06 provided local pressure support in region two, with water breakthroughs also occurring within 5–10 years of the start in injection in 1975. Region three appears to be completely isolated, with rapid but localized pressure response and water breakthrough from injection at 02-07. Region four appears to be a narrow segment of reservoir in direct communication with region one. The pressure response and water breakthroughs are more rapid in this region than in region five. Region five is in partial communication with region one and has benefited from partial pressure maintenance and delayed water breakthroughs. As a result, the highest cumulative oil production has been from the portion of region five adjacent to region one, as was shown in Figure 10.4.17.

The pressure data and water breakthrough data indicate that the overall permeability is likely low. There is clearly a permeability barrier between region three and the rest of the pool. There is likely a partial permeability barrier or a permeability contrast between regions one and four and the rest of the reservoir. Note that the interpretation should be further refined through the examination of structure maps, isopach maps, and permeability data.

10.4.2.4 Estimating Original Oil in Place From Production Data

Cumulative oil and gas production are easily obtained from production data. The expected ultimate cumulative oil production (recoverable oil in place or *ROIP*) and ultimate cumulative gas production (recoverable gas in place or RGIP) can be estimated using the forecasting methods described in Chapter 5. Then the *OOIP* can be inferred from the *ROIP* or RGIP. Because a recovery factor must be assumed, this method is not very accurate. Nonetheless, the method provides some constraints on the possible *OOIP*. To present these constraints, let us assume that the *ROIP* and RGIP have been estimated from an analysis and extrapolation of the production history.

Constraint 1: Order of Magnitude Estimate Based on Mobile Oil in Place

The *ROIP* can be used to find a lower limit on the possible *OOIP*. First recall that only a fraction of the oil in a reservoir is mobile. For water displacement, the mobile oil in place (*MOIP*) is given by:

$$MOIP = \frac{PV(1 - S_{wi} - S_{orw})}{B_{oi}} \tag{10.4.4}$$

The original oil in place is given by:

$$OOIP = \frac{PV(1 - S_{wi})}{B_{oi}} \tag{10.4.5}$$

Therefore, *OOIP* is related to *MOIP* as follows:

$$OOIP = \frac{(1 - S_{wi})}{(1 - S_{wi} - S_{orw})} MOIP \tag{10.4.6}$$

If we assume that the *ROIP* is the same as the *MOIP*, then:

$$OOIP_{(1)} = \frac{(1 - S_{wi})}{(1 - S_{wi} - S_{orw})} ROIP \tag{10.4.7}$$

In reality, the *ROIP* is less than the *MOIP* because a perfect volumetric sweep is never attained. Therefore, the *OOIP* estimated from this method is a lower limit. A better estimate can be obtained if the volumetric sweep efficiency, E_{vol}, is included as follows:

$$OOIP_{(1)} = \frac{(1 - S_{wi})}{(1 - S_{wi} - S_{orw})} \frac{ROIP}{E_{vol}} \tag{10.4.8}$$

Volumetric sweep efficiencies for successful conventional oil waterfloods are typically in the range of 50–70%. Sweep efficiencies for other drive mechanisms vary, but are typically less than for a water flood. Given the uncertainty in the volumetric sweep efficiency, this method can only provide an order of magnitude estimate of the *OOIP*.

Example: What is the lower limit of the *OOIP* for a reservoir with an *ROIP* of 500,000 m^3, an initial water saturation of 25% and a residual oil saturation of 30%? The lower limit for the *OOIP* is:

$$OOIP_{(1)} = \frac{(1 - 0.25)}{(1 - 0.25 - 0.30)} 500{,}000\ \text{m}^3 = 833{,}000\ \text{m}^3 \tag{10.4.9}$$

If a volumetric sweep efficiency of 50% is assumed, the estimated *OOIP* is then:

$$OOIP_{(1)} = \frac{(1 - 0.25)}{(1 - 0.25 - 0.30)} \frac{500{,}000\ \text{m}^3}{0.50} = 1{,}660{,}000\ \text{m}^3 \tag{10.4.10}$$

Constraint 2: Order of Magnitude Estimate Based on Expected Recovery Factor
The *OOIP* can be found from the *ROIP* and the recovery factor as follows:

$$OOIP_{(2)} = \frac{ROIP}{RF} \tag{10.4.11}$$

Of course, the recovery factor for a given pool is not known and is usually calculated from the *ROIP* and some other estimate of the *OOIP*. However, typical ranges of recovery factors are known for given drive mechanisms and geological characteristics. Recovery factors may also be known for analogous reservoirs. Hence, the *OOIP* can be estimated from an educated guess of the recovery factor. This method provides an order of magnitude estimate of the *OOIP*.

Example: What is the likely *OOIP* for an SGD reservoir with an *ROIP* of 500,000 m^3? The recovery factor for an SGD reservoir is expected to be between 5% and 20%. Therefore, the estimated *OOIP* is between 2.5×10^6 m^3 and 10×10^6 m^3. If the reservoir architecture is well understood and/or data from analogous reservoirs are available, the recovery factor and *OOIP* estimate can be refined further.

Constraint 3: Recovery of Solution Gas

For solution gas drive reservoirs, the solution gas in place can be determined as follows:

$$OSGIP = \frac{RGIP}{RF_{sg}}$$

The *OOIP* is then given by:

$$OOIP_{(3)} = \frac{OSGIP}{R_{si}} = \frac{RGIP}{R_{si}RF_{sg}}$$

The advantage of this method is that solution gas recovery factors usually fall within a narrow range between 85% and 95%. Therefore, a more accurate estimate of *OOIP* is possible. The disadvantage of this method is that is only applies to solution gas or weak water drive reservoirs.

Example: What is the likely *OOIP* for an SGD reservoir with an RGIP of 520×10^6 m^3 and an R_{si} of 90 m^3/m^3? The recovery factor for an SGD is expected to be between 85% and 95%. Therefore, the estimated *OOIP* is between 6.1×10^6 m^3 and 6.8×10^6 m^3.

10.5 Material Balance

The primary objectives of a material balance analysis are to:

- Estimate the original hydrocarbons in place;
- Determine the amount (if any) of water influx;
- Predict future reservoir pressures and production (when coupled with inflow equations).

The material balance technique can also be used as a diagnostic tool to determine the drive mechanism and assess the strength of an aquifer.

Material balance is a statement of the conservation of mass, but expressed in terms of volume. No spatial variation or transient effects are considered. The advantages of the material balance approach are that:

- It provides a mathematically simple set of equations;
- It provides insight into the mechanisms and physics of hydrocarbon production and reservoir performance;
- It is an excellent tool for sensitivity studies during early stages of depletion.

Material Balance Equation:

In Chapter 2, the following general material balance was derived for black oil reservoirs:

$$N_p\left(B_o + \left(R_p - R_s\right)B_g\right) + W_pB_w - G_iB_{g,inj} - W_iB_{w,inj}$$

$$= N\left[\begin{array}{l}(B_o - B_{oi}) + (R_{si} - R_s)B_g + \dfrac{mB_{oi}}{B_{gi}}\left(B_g - B_{gi}\right) \\ \\ +\dfrac{(1+m)B_{oi}}{1-S_{wi}}(c_f + S_{wi}c_w)\Delta P\end{array}\right] + W_eB_w \tag{10.5.1}$$

in which:

N_p = cumulative oil production (scm or stb)
R_p = cumulative producing gas-oil ratio (scm/scm or SCF/stb)
R_p = cumulative gas production/cumulative oil production
G_i = cumulative gas injection (scm or SCF)
W_p = cumulative water production (scm or stb)
W_i = cumulative gas injection (scm or stb)
B_o = oil formation volume factor (m^3/scm or bbl/stb)
B_{oi} = initial oil formation volume factor (m^3/scm or bbl/stb)
R_s = solution gas-oil ratio (scm/scm or SCF/stb)
R_{si} = initial solution gas-oil ratio (scm/scm or SCF/stb)
B_g = gas formation volume factor (m^3/scm or bbl/SCF)
B_{gi} = initial gas formation volume factor (m^3/scm or bbl/SCF)
$B_{g,inj}$ = injection gas formation volume factor (m^3/scm or bbl/SCF)
B_w = water formation volume factor (m^3/scm or bbl/stb)
$B_{w,inj}$ = water formation volume factor (m^3/scm or bbl/stb)
N = original oil in place, *OOIP* (scm or stb)
m = ratio of gas cap pore volume to oil pore volume (m^3/m^3 or bbl/bbl)
W_e = aquifer influx (scm or stb)
c_f = formation compressibility (kPa^{-1} or psi^{-1})
c_w = water compressibility (kPa^{-1} or psi^{-1})
S_{wi} = initial water saturation

The left hand side of the equation represents the reservoir withdrawals (production and injection). The right hand side represents the reservoir response, including rock compaction and fluid expansion. Note that when gas injection occurs, the B_g of the produced gas may not be the same as the B_g of the solution gas at the same pressure.

For convenience, let us define the following terms:

$$F = N_p\left(B_o + \left(R_p - R_s\right)B_g\right) + W_pB_w - G_iB_{g,inj} - W_iB_{w,inj} \tag{10.5.2}$$

in which F is the reservoir withdrawals in m^3 or stb;

$$E_o = (B_o - B_{oi}) + (R_{si} - R_s)B_g \tag{10.5.3}$$

in which E_o is the oil shrinkage and solution gas expansion in m^3 or stb;

$$E_g = B_{oi}\left(\frac{B_g}{B_{gi}} - 1\right) \tag{10.5.4}$$

in which E_g is the gas cap expansion in m^3 or stb;

$$E_f = \frac{(1+m)B_{oi}}{1-S_{wi}}(c_f + S_{wi}c_w)\Delta P \tag{10.5.5}$$

in which E_f is the compaction of the formation and the expansion of formation water in m^3 or stb. The general material balance can then be expressed as:

$$F = N(E_o + mE_g + E_f) + W_eB_w \tag{10.5.6}$$

To apply a general material balance, it is necessary to determine values for F, E_o, E_g, E_f, and W_e as appropriate for the particular drive mechanism. The starting point is always pressure data, as shown in Table 10.5.1. The cumulative production values are then determined at the date of each pressure test. The pressure, volume, and temperature (PVT) properties are determined at the test pressure, and then the withdrawal and expansion terms are calculated.

Applicability and Limitations:

The applicability of the material balance technique requires that the following conditions be satisfied:

1. Thermodynamic equilibrium exists, particularly between oil and its solution gas. If there is a supersaturation of a liquid phase with gas as reservoir pressure declines, the reservoir pressure is lower than it would be at equilibrium (Craft and Hawkins, 1991). In East Texas reservoirs, Wieland and Kennedy (1957) showed that supersaturation effects were about 20 psi in magnitude.
2. PVT data should be obtained using a gas liberation process that closely duplicates the conditions in the reservoir. If not, the predicted fluid properties will not be correct and the material balance will be in error.

Table 10.5.1 Summary of data and equations required for a reservoir material balance

											Equation			
Date	P	N_p	R_p	W_p	G_i	W_i	B_o	R_s	B_g	B_w	F	E_o	E_g	E_f
Pressure test data		Production data at date of pressure test					Fluid properties at test pressure				10.5.2	10.5.3	10.5.4	10.5.5

3. Pressure data must accurately represent the average reservoir pressure values. There are three potential sources of errors in such data: instrument errors, difficulties in obtaining true static build-up pressures, and problems of correctly weighting or averaging the individual well pressures. The effect of pressure error on the material balance depends on its magnitude relative to the reservoir pressure decline; the smaller the pressure decline, as might be the case with a strong aquifer or if a large gas cap is associated with the oil zone, the more that pressure errors will affect the material balance estimate. "Errors in bottomhole pressure (both measurement and interpretation) are among the major contributors to errors in material balance calculations" (Cronquist, 2001). Extrapolation of pressure transient results to the static reservoir pressure is a large error source.
4. Usually, there must be a good initial estimate of the ratio of the free gas volume to the initial reservoir oil volume (m) for gas cap drive reservoirs. The potential error in *OOIP* increases with the size of this ratio.
5. All the production data must be accurate. While the cumulative oil production is usually known accurately, the reported gas and, particularly, water production are usually less accurate.
6. The reservoir drive mechanisms must be determined correctly. Erroneous reserve estimates are obtained if drives involved in the production process are ignored or grossly underestimated in the material balance calculations.

In general, the material balance is best suited for moderate to high permeability reservoirs with good built up pressure data. It is important to corroborate material balance results from other sources whenever possible.

Several limitations to the application of the material balance method are listed below:

1. A material balance is a zero dimensional mathematical model in which fluid properties and pressures are averaged over the entire reservoir. Variations in initial fluid properties (for instance, a gradient in bubble point either laterally or with depth) cannot be accounted for. We can determine average oil, gas, and water saturations, but no conclusion can be drawn on how those saturations are distributed. For example, we may calculate a 10% average gas saturation, but we do not know whether the gas has risen to the top of the reservoir or whether it is dispersed throughout the oil column.
2. A material balance is difficult to apply when there are large pressure gradients in the reservoir, for example, in low permeability or low continuity reservoirs and especially in water floods. Suppose the pressure of a reservoir that has dropped to 500 psia on primary production and it is desired to repressure it to 2000 psia. After injecting water, the actual reservoir pressure is 3000 psia near injectors and 700 psia near producers. In this case, the calculated "average" reservoir pressure strongly depends on the selection of wells and the shut-in time of pressure build-up tests. Similarly, it may be difficult to apply a material balance to areally extensive low transmissibility reservoirs (low kh/μ) at different stages of pressure depletion due to timing of drilling and field development. Also, in heterogeneous situations in which layers of high permeability are interbedded with low permeability layers or in fractured reservoirs with low matrix permeability, it may not be possible to determine an "average" reservoir pressure.
3. For combination drives, it is theoretically possible to solve the material balance for the water influx, *OOIP*, and gas cap size. However, in practice, due to the data scatter, it is difficult to determine all three unknowns with any degree of accuracy.

4. It may not be possible to apply a material balance to a strong water drive or gas cap drive in which the decrease in reservoir pressure is small. If the pressure decline has a magnitude comparable to the error in pressure measurements, there is too high an uncertainty in the calculations. A rule of thumb is that there must be at least a 5–10% pressure decline from the original reservoir pressure to apply a material balance.
5. It is challenging to accurately model aquifer influx. Initially, a finite aquifer may appear to be infinitely large, but later the pressure disturbance reaches the outer boundary of the aquifer. When the boundary is encountered, the rate of water influx decreases. It is possible to match early data with an infinite aquifer model and overestimate future influx when the boundary is reached. It is therefore prudent reservoir engineering practice to achieve a match with the smallest possible aquifer that still matches historical data.
6. Combining the material balance with unsteady state flow equations for the case of a water drive system is a weakness of the method. The material balance may not be a good tool for calculating water influx in such a situation because water influx is determined by subtracting the reservoir fluid expansions from the reservoir fluid withdrawals. In other words, when there are slight errors in reservoir pressure, small errors in production measurement, or errors in extrapolation to fully built-up pressures, there may be large errors in the calculated water influx. Obviously, multiple calculations at different time periods reduce this error somewhat.
7. "The material balance does not and cannot reflect the mechanism of production except to the extent that the values inserted reflect it. If an erroneous value is inserted, one or more other values will change to maintain the equality, regardless of what is happening in the reservoir." When using the above equations, these limitations should be remembered (Campbell and Campbell, 1978).
8. Injection of fluids may result in re-pressurization of a reservoir. The material balance implicitly assumes that the gas goes back into solution. Two extreme cases may exist in the reservoir: (1) free gas goes back into the oil according to the solution gas-oil ratio versus pressure relationship as determined by PVT analysis; or (2) almost no free gas re-dissolves, because all the gas in the main reservoir may have migrated to the top of structure and therefore may not be available to go back into solution. The expansion calculation is different in each case, and therefore a material balance error can occur if the gas does not redissolve.
9. Muskat's (1945) statement still holds true: "The material balance method is by no means a universal tool for estimating reserves. In some cases it is excellent. In others it may be grossly misleading. It is always instructive to try it, if only to find out that it does not work and why. It should be a part of the 'stock in trade' of all reservoir engineers. It will boomerang if applied blindly as a mystic hocus-pocus to evade the admission of ignorance. The algebraic symbolism may impress the 'old timer' and help convince a Corporation Commission, but it will not fool the reservoir. Reservoirs pay little heed to either wishful thinking or libelous misinterpretation. Reservoirs always do what they 'ought' to do. They continually unfold a past with an inevitability that defies all 'man-made' laws. To predict this past while it is still the future is the business of the reservoir engineer. But whether the engineer is clever or stupid, honest or dishonest, right or wrong, the reservoir is always 'right'."

Initial steps:

Prior to starting the material balance, the assumptions of the method should be verified. The following steps are recommended:

1. Check for the adequacy and quality of the production and PVT data (for example, sometimes gas production is missing or there is poor pressure build-up data or interpretation).

2. Verify the uniformity of pressure decline throughout the reservoir by plotting static reservoir pressure data for all wells versus time on the same graph; observe the degree of scatter and whether or not well pressures pass through the bubble point simultaneously; that is, at least within a few weeks of each other. The above plot will frequently provide valuable knowledge regarding the degree of communication in the reservoir. The plot may also show groups of wells that are in separate reservoirs.
3. It is good practice to determine and/or confirm the nature of the drive mechanisms involved in the case under consideration. Production performance data such as oil, gas, and water rates as well as GOR and watercut trends are usually good indicators of prevailing reservoir mechanisms.

In the following sections, the general material balance is simplified for each type of drive mechanism. The graphical techniques developed by Havlena and Odeh (1963) to obtain "best-fit" solutions for each drive mechanism are presented. The limits and potential errors of the material balance technique for each drive mechanism are also discussed, along with some examples.

10.5.1 Expansion Drive (Undersaturated Oil)

In an expansion drive reservoir, there is no gas cap or aquifer, and therefore m and W_e are zero. Also, the solution gas-oil ratio must always equal the initial solution gas-oil ratio. We will only consider the case in which there is no gas injection and therefore G_i is zero, and thus the cumulative producing gas-oil ratio must equal the initial solution gas-oil ratio. Hence, the general material balance equation (Eqn (10.5.1)) simplifies to:

$$N_pB_o + W_pB_w - W_iB_{w,inj} = N\left[(B_o - B_{oi}) + \frac{B_{oi}}{1 - S_{wi}}(c_f + S_{wi}c_w)\Delta P\right] \tag{10.5.7}$$

Now recall that the oil compressibility is defined as:

$$c_o = \frac{B_o - B_{oi}}{B_{oi}\Delta P}. \tag{10.5.8}$$

Equation (10.5.8) is substituted into Eqn (10.5.7) to obtain:

$$N_pB_o + W_pB_w - W_iB_{w,inj} = NB_{oi}\left[\frac{(1 - S_{wi})c_o + c_f + S_{wi}c_w}{1 - S_{wi}}\right]\Delta P \tag{10.5.9}$$

Equation (10.5.9) can be written as:

$$F' = NB_{oi}c_e\Delta P \tag{10.5.10}$$

in which F' is the net reservoir withdrawal assuming no gas injection and is given by:

$$F' = N_pB_o + W_pB_w - W_iB_{w,inj} \tag{10.5.11}$$

and c_e is the effective compressibility, given by:

$$c_e = \frac{(1 - S_{wi})c_o + c_f + S_{wi}c_w}{1 - S_{wi}}. \tag{10.5.12}$$

The data required to perform a material balance on an undersaturated reservoir are summarized in Table 10.5.2. Note that Eqn (10.5.9) is only valid above the bubble point, and therefore only pressures above the bubble point should be used for this type of material balance. An SGD material balance is used below the bubble point.

As a first approximation, a plot of the net reservoir withdrawals versus the change in reservoir pressure is expected to be linear with a slope of $NB_{oi}c_e$ and a zero intercept. If c_e and B_{oi} are known, the *OOIP*, N, can be determined from the slope of a best fit line through the origin, as shown in Figure 10.5.1(a). If there is a substantial change in reservoir pressure and the compressibility is not constant, then N can be determined from a plot of F' versus $c_e\Delta P$, as shown in Figure 10.5.1(b). In this case, the slope is NB_{oi}, and the intercept is again zero.

The advantages of this approach are:

1. All the data points on the pressure production plot are used, which increases the accuracy of calculation; at the same time, any erroneous data can be detected.
2. The data points form a straight line when plotted in terms of the coordinates defined by the various combinations of the new variables. A requirement of the linearity imposes an additional constraint on the relationship, which should increase the accuracy of data analysis.

The disadvantages of the Havlena–Odeh approach are that:

1. It can too highly weight early data points (low pressure drop points), and these early points are usually less reliable.
2. It may not be as easy to relate each term to a physical process.

Table 10.5.2 Summary of data and calculations required for a material balance on an undersaturated reservoir

Date	P	ΔP	N_p	W_p	W_i	c_o	c_f	c_w	c_e	F'
Pressure test data		From pressure data	Production data at date of pressure test			Fluid properties at test pressure			Equation (10.5.12)	Equation (10.5.11)

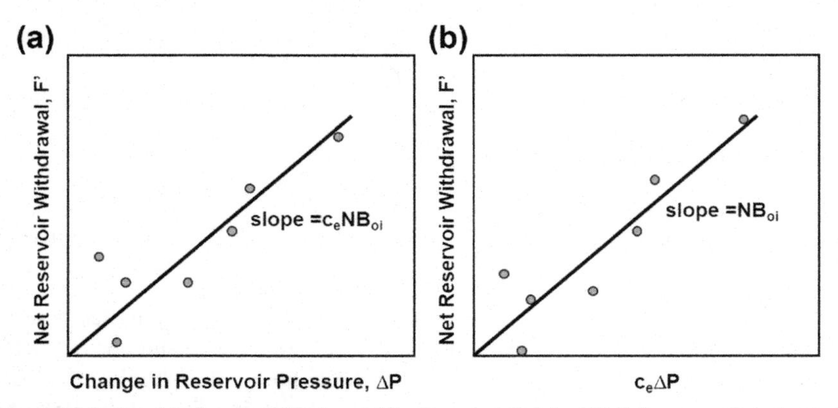

Figure 10.5.1 Graphical material balance (Havlena and Odeh, 1964) for an undersaturated reservoir: (a) constant compressibility; (b) compressibility varies with pressure.

In theory, the material balance for expansion drive can give the least ambiguous measure of the oil in place because only one drive mechanism is operating. In practice, it can be difficult to obtain an accurate *OOIP* for the following reasons:

1. The change in reservoir pressure above the bubble point may be small relative to the error in the pressure measurements.
2. The effective compressibility (especially rock compressibility) may only be known to an order of magnitude accuracy.
3. The bubble point may be unknown or incorrect, and pressure data below the bubble point may inadvertently be used in the material balance.
4. The pressure data may not be representative of the average reservoir pressure.
5. Another drive mechanism such as water drive is operating.

In general, when pressure data are scattered, the greatest weight should be given to the data points with the largest ΔP. An example expansion drive material balance is provided in Section 10.5.3.

10.5.2 Solution Gas Drive

In an SGD reservoir, there is no initial gas cap or aquifer, and therefore m and W_e are zero. Also, because the gas compressibility is much higher than the water or formation compressibilities, $E_o \gg E_f$ so that E_f, can be neglected. Hence, the general material balance equation simplifies to:

$$N_p\left(B_o + \left(R_p - R_s\right)B_g\right) + W_pB_w - G_iB_{g,inj} - W_iB_{w,inj} = N\left[\left(B_o - B_{oi}\right) + \left(R_{si} - R_s\right)B_g\right] \quad (10.5.13)$$

or

$$F = NE_o \quad (10.5.14)$$

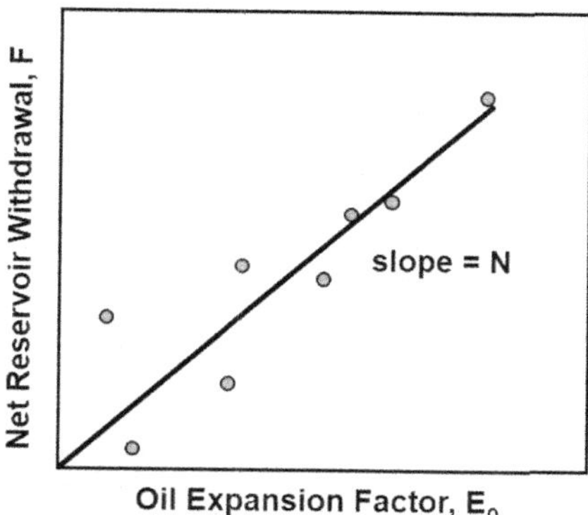

Figure 10.5.2 Graphical material balance for an SGD reservoir.

A plot of F versus E_o is expected to be linear with an intercept of zero and a slope of N, as shown in Figure 10.5.2.

Once sufficient reservoir pressure decrease has occurred, the SGD material balance can give an accurate determination of the *OOIP*. Possible sources of error are:

1. The change in reservoir pressure is small relative to the error in the pressure measurements.
2. The withdrawal and gas expansion is so small that the assumption that $E_o \gg E_f$ is not valid
3. The bubble point may be unknown or incorrect, and pressure data above the bubble point may be used in the material balance.
4. The measured well pressure data are not representative of the average reservoir pressure.
5. Another drive mechanism such as gas cap or water drive is operating.

In general, when pressure data are scattered, the greatest weight should be given to the data points with the largest net withdrawal.

If there is a large amount of scatter in the pressure data, it can be helpful to plot the value of N calculated from each pressure point versus pressure, cumulative oil production, or net withdrawal. Such a plot can demonstrate if the scatter results from low withdrawal and a high sensitivity to errors in the pressure data. If so, the values of N will tend to converge at higher net withdrawals (or lower pressure or higher cumulative oil production), as shown in Figure 10.5.3. Fig. 10.5.3 demonstrates why there is value in collecting pressure data throughout a reservoir's producing life. Examples of SGD material balances are given in Sections 10.5.3 and 10.7.

10.5.3 Transition through the Bubble Point of an Initially Undersaturated Reservoir

Unless pressure is maintained, an undersatured reservoir will eventually reach and pass its bubble point. The drive mechanism changes from expansion drive to SGD. Both drive mechanisms can be graphically represented based on the following equation:

$$F = N(E_o + E_f) \tag{10.5.15}$$

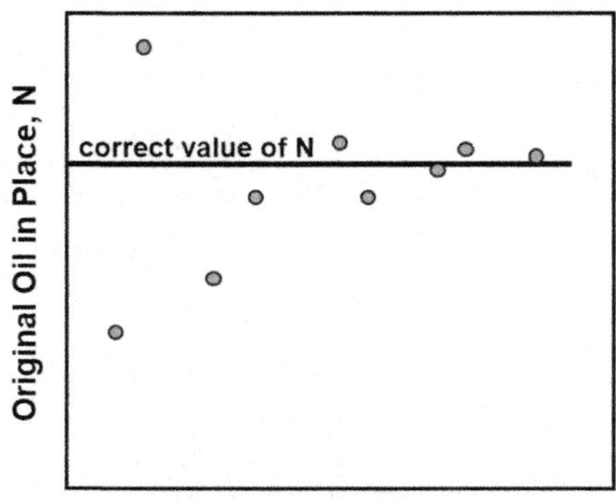

Figure 10.5.3 Convergence of calculated OOIP as net withdrawal increases.

A plot of F versus $E_o + E_f$ is expected to be linear with an intercept of zero and a slope of N, as shown in Figure 10.5.4(a).

However, in practice, there is often a change in slope at the bubble point. The expansion drive is dominated by the effective compressibility. The SGD is dominated by the compressibility of the evolved solution gas (more compressible by up to a factor of 100). Small errors in each compressibility or in the measured cumulative gas-oil ratio can result in a different slope in the regions in which each compressibility dominates, as shown in Figure 10.5.4(b). The change in slope is sometimes confused with water influx.

A plot F versus $E_o + E_f$ is best used as a method to determine the bubble point. In general, it is recommended to divide the production into production above and below the bubble point. The above-bubble point data are analyzed with the expansion drive material balance, and the below-bubble point data are analyzed with the SGD material

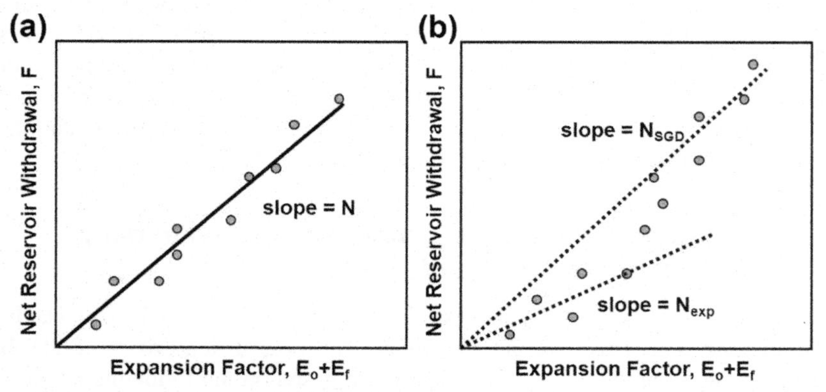

Figure 10.5.4 Graphical material balance for an expansion drive reservoir passing through the bubble point: (a) perfect match of slopes in expansion drive and SGD regions; (b) imperfect match.

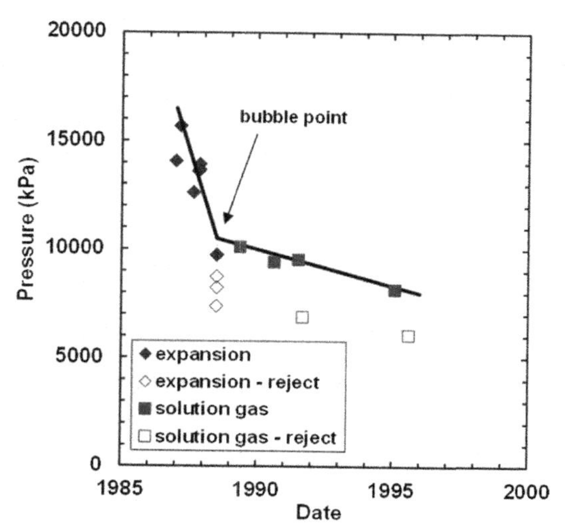

Figure 10.5.5 Bilbo A Cardium A Pool pressure history under expansion drive and then SGD.

balance. Note that for the SGD material balance, the cumulative production at the bubble point must be zero.

Example: Material Balance On an Initially Undersaturated Reservoir
The Bilbo A Cardium A Pool was presented in Section 10.4.2.2. It is an undersaturated reservoir, as shown in the pressure versus time plot in Figure 10.5.5. The pressure data have been divided into data in which expansion drive applies (diamonds) and data in which SGD applies (squares). Judgment is needed with interpretation of reservoir pressure. In highly compartmentalized reservoirs (labyrinth or jigsaw types), small compartments with large withdrawal volumes will have very low pressures compared with the field average. If this is the case, then material balance is not meaningful. Usually, this will be observed in the measured build-up pressures by little or no pressure increase after a 72-h buildup test for wells located within the small compartments.

Separate material balances must be performed for the time the reservoir is under expansion drive and SGD. Let us first consider expansion drive. The reservoir properties required for a material balance are given in Table 10.5.3. The material balance calculations are summarized in Table 10.5.4, and the net reservoir withdrawal is plotted versus reservoir pressure drop in Figure 10.5.6. The slope of the plot was determined from the final point with the greatest pressure change and is 4.57 m^3/kPa. The slope is equal to c_eNB_{oi}, assuming constant compressibility. The effective compressibility was determined from Eqn (10.5.12) as follows:

$$c_e = \frac{(1 - S_{wi})c_o + c_f + S_{wi}c_w}{1 - S_{wi}}$$

$$= \frac{((1 - 0.33)1.73 + 0.764 + (0.33)1.26)10^{-6}}{1 - 0.33} = 3.49 \times 10^{-6}\text{kPa}^{-1}$$

Table 10.5.3 Reservoir properties required for material balance

Property	Value
Initial water saturation, S_{wi}	33%
Oil compressibility, c_o	1.73 10^{-6} kPa^{-1}
Water compressibility, c_w	1.26 10^{-6} kPa^{-1}
Formation compressibility, c_f	7.64 10^{-7} kPa^{-1}
Initial oil volume factor, B_{oi}	1.271 m^3/scm

Table 10.5.4 Summary of expansion drive material balance calculations

Date	Pressure (kPa)	ΔP (kPa)	N_p (10^3m^3)	W_p (10^3m^3)	B_o (m^3/scm)	F' (10^3m^3)
initial	15,700	–	–	–	–	–
1987.0	14,079	1621	3.16	0.00	1.263	3.99
1987.1	15,690	10	3.81	0.01	1.259	4.81
1987.6	12,623	3077	6.10	0.04	1.266	7.76
1987.8	13,607	2093	8.70	0.04	1.264	11.03
1987.9	13,689	2011	8.70	0.04	1.264	11.03
1987.9	13,935	1765	8.70	0.04	1.263	11.02
1988.5	9742	5958	21.34	0.04	1.272	27.20

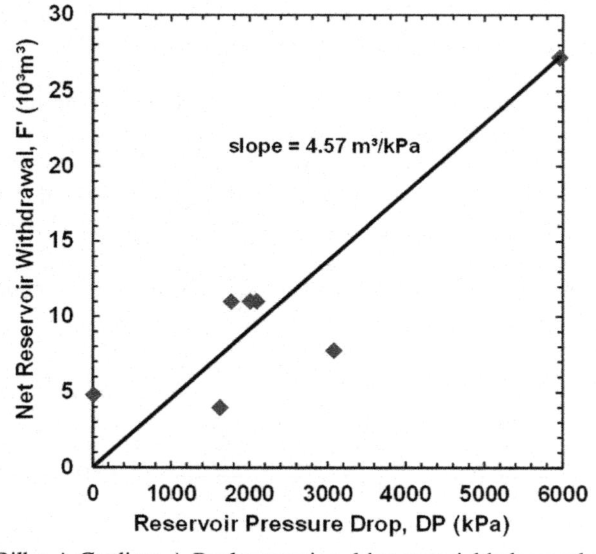

Figure 10.5.6 Bilbo A Cardium A Pool expansion drive material balance plot.

The *OOIP* is then given by:

$$N = \frac{slope}{c_e B_{oi}} = \frac{4.57\ \text{m}^3/\text{kPa}}{3.49\cdot 10^{-6}\text{kPa}^{-1}1.271\ \text{m}^3/\text{scm}} = 1030 \times 10^3\text{scm}$$

Now let us consider SGD. The material balance calculations are summarized in Tables 10.5.5 and 10.5.6. Note that SGD begins at the bubble point, and the net withdrawal for the SGD material balance must be zero at the bubble point. In this case, the bubble point was reached in mid-1988. Therefore, the cumulative production as of mid-1998 was subtracted from all the subsequent cumulative production values.

Table 10.5.5 Production data zeroed at bubble point for SGD material balance

Date	N_p (10^3m^3)	G_p (10^3m^3)	W_p (10^3m^3)	$N_p{}^a$ (10^3m^3)	$G_p{}^a$ (10^3m^3)	$R_p{}^a$ (10^3m^3)	$W_p{}^a$ (10^3m^3)
1988.5	21.34	1.53	0.04	0.00	0.00	75.18	0.00
1989.4	32.63	2.34	0.05	11.28	0.81	71.72	0.01
1990.6	52.30	3.83	0.06	30.95	2.30	74.26	0.02
1991.5	63.19	4.70	0.07	41.85	3.17	75.64	0.03
1991.7	65.50	4.90	0.11	44.16	3.37	76.38	0.07
1995.1	118.68	12.21	0.11	97.34	10.68	109.70	0.07
1995.6	125.71	13.78	0.12	104.37	12.25	117.34	0.08

G_p is cumulative gas production.
[a]indicates zeroed values.

Table 10.5.6 Summary of remaining SGD material balance equations

Date	Pressure (kPa)	B_o (m^3/scm)	R_s (scm/scm)	B_g (m^3/scm)	E_o (m^3/scm)	F (10^3m^3)
1988.5	9742	1.264	70.9	0.0141	0.0545	0.0
1989.4	10,112	1.267	73.0	0.0137	0.0270	14.1
1990.6	9417	1.261	69.0	0.0145	0.0803	41.4
1991.5	9533	1.262	69.7	0.0144	0.0709	56.4
1991.7[a]	6881	1.240	54.6	0.0183	0.3474	72.5
1995.1	8168	1.251	61.9	0.0162	0.1950	197.0
1995.6[a]	6064	1.233	49.9	0.0200	0.4691	269.8

[a]insufficient duration of build-up test.

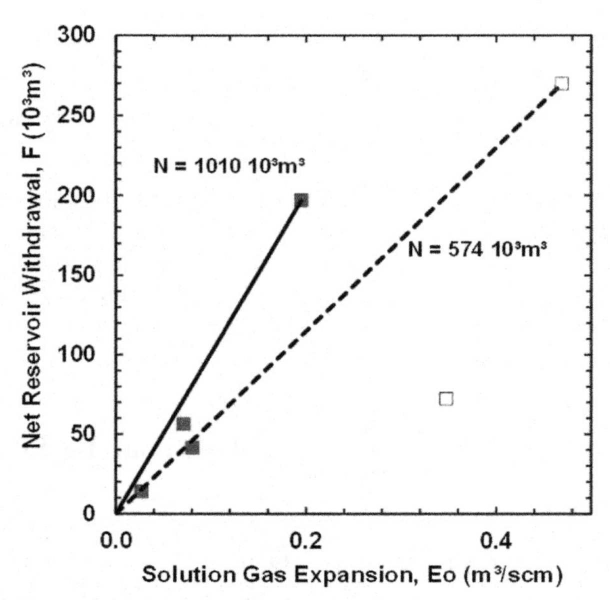

Figure 10.5.7 Bilbo A Cardium A Pool SGD material balance plot.

The net reservoir withdrawal, F, is plotted versus the solution gas expansion factor, E_o, in Figure 10.5.7. Based on the data point with sufficient build-up and the greatest net withdrawal, the slope is $1010\times10^3\text{m}^3$. The two data points with insufficient build-up are also plotted to demonstrate the significant effect of including poor pressure data. In this case, the slope is approximately halved when these data points are included in the analysis.

The slope of $1010\times10^3\text{m}^3$ is the *OOIP* based on the SGD material balance. This value is in remarkably good agreement with the *OOIP* of $1030\times10^3\text{m}^3$ from the expansion drive material balance. Such an exact agreement is rare and is probably coincidental given the uncertainties in pressures, compressibilities, and fluid properties. Note that the volumetric *OOIP* for this reservoir is reported at $1217\times10^3\text{m}^3$, which is also in reasonable agreement with the material balance results.

10.5.4 Gas Cap Drive

In a gas cap drive reservoir, there is no aquifer, and therefore W_e is zero. Also, because the gas compressibility is much higher than the water or formation compressibilities, E_f, can be neglected. Hence, the general material balance equation simplifies to:

$$N_p\left(B_o + \left(R_p - R_s\right)B_g\right) + W_pB_w - G_iB_{g,inj} - W_iB_{w,inj}$$
$$= N\left[\left(B_o - B_{oi}\right) + \left(R_{si} - R_s\right)B_g + \frac{mB_{oi}}{B_{gi}}\left(B_g - B_{gi}\right)\right] \tag{10.5.16}$$

or

$$F = N(E_o + mE_g) \tag{10.5.17}$$

A plot of F versus $E_o + mE_g$ is expected to be linear with an intercept of zero and a slope of N, as shown in Figure 10.5.8(a). It is necessary to assume a value of m to apply this method. If the assumed m is too small, the trend on the plot will curve upward. If the assumed value is too large, the trend will curve downward. Hence, the correct value of m is found when the data fall on a straight line.

An alternate expression for Eqn (10.5.17) is given by:

$$\frac{F}{E_o} = N + mN\frac{E_g}{E_o} \tag{10.5.18}$$

In this case, a plot of F/E_o versus E_g/E_o is expected to be linear with an intercept of N and a slope of mN, as shown in Figure 10.5.8(b). No trial and error is required for this approach.

In theory, either of the two aforementioned approaches can be used to determine both m and N. In practice, the compressible energy of the gas cap often dominates the pressure response, and the contribution of the oil zone is often negligible. Therefore, it is often not possible to determine m or N. However, the size of the gas cap can be determined from the slope of the plot of F/E_o versus E_g/E_o. The slope is mN and the surface volume of the gas cap is given by:

$$OGIP = \frac{(mN)B_{oi}}{B_{gi}} \tag{10.5.19}$$

Note that it is not necessary to specify N or m to determine the $OGIP$.

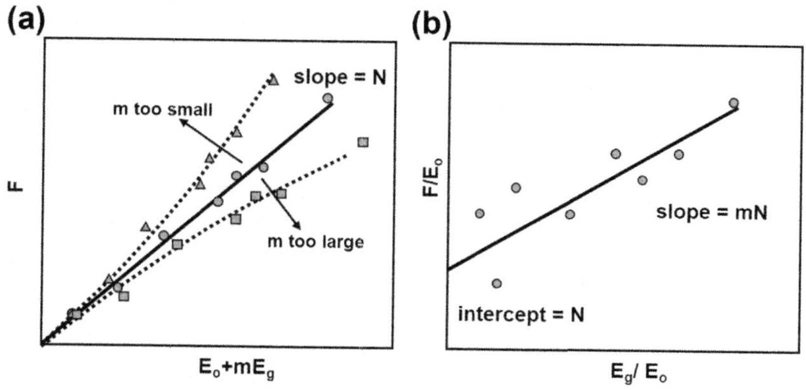

Figure 10.5.8 Graphical material balance for a gas cap drive.

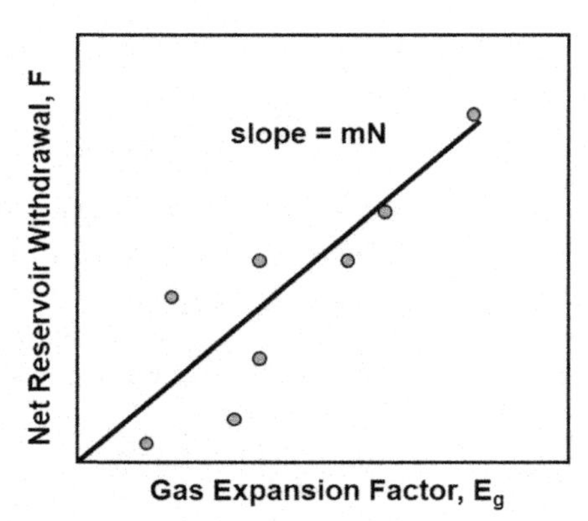

Figure 10.5.9 Graphical material balance to determine size of gas cap when oil zone influence is negligible.

The size of the gas cap can also be estimated if it is assumed that $NE_o \ll mNE_g$. Eqn (10.5.17) then becomes:

$$F \approx mNE_g \tag{10.5.20}$$

A plot of F versus E_g is expected to be linear with a zero intercept and a slope of mN, as shown in Figure 10.5.9. As before, the OGIP is determined from Eqn (10.5.19). It is strongly recommended to check the assumption that $NE_o \ll mNE_g$ using the volumetric *OOIP* and OGIP.

Once sufficient reservoir pressure decrease has occurred, the gas cap drive material balance can usually give an accurate determination of the OGIP and sometimes of the *OOIP* as well. Possible sources of error are:

1. The change in reservoir pressure is small relative to the error in the pressure measurements.
2. The contribution of the oil and solution gas expansion is too small relative to the gas cap expansion to accurately calculate N.
3. The pressure data are not built up or representative of the average reservoir pressure.
4. Another drive mechanism, such as water drive, is operating.

In general, when pressure data are scattered, the greatest weight should be given to the data points with the largest net withdrawal.

If there is a large amount of scatter in the pressure data, it can be helpful to plot the value of mN calculated from each pressure point versus pressure, cumulative oil production, or net withdrawal. Such a plot can demonstrate whether the scatter results from low withdrawal and a high sensitivity to errors in the pressure data. If so, the values of mN will tend to converge at higher net withdrawals (or lower pressure or higher cumulative oil production), as shown in Figure 10.5.10.

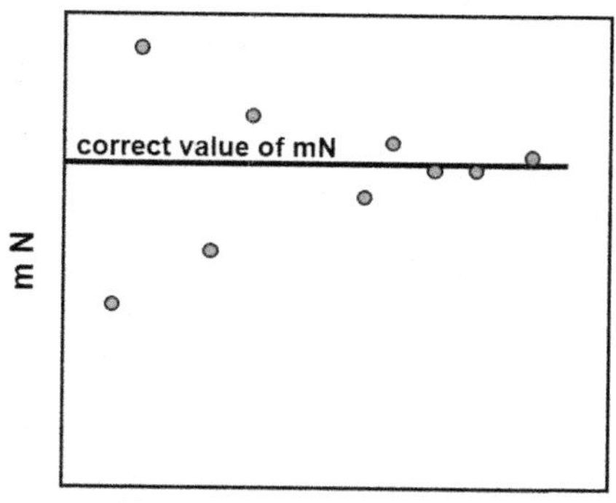

Figure 10.5.10 Convergence of calculated mN as net withdrawal increases (valid only when oil zone influence is negligible).

10.5.5 Water Drive

In a water drive reservoir, there is no gas cap, and therefore m is zero. Also, because the water influx plus solution gas expansion is much higher than the formation compaction, E_f, can be neglected. Hence, the general material balance equation simplifies to:

$$N_p\left(B_o + \left(R_p - R_s\right)B_g\right) + W_pB_w - G_iB_{g,inj} - W_iB_{w,inj} = N\left[(B_o - B_{oi}) + (R_{si} - R_s)B_g\right] + W_eB_w \tag{10.5.21}$$

or

$$F = NE_o + W_eB_w \tag{10.5.22}$$

The application of the water drive material balance depends on the strength of the aquifer.

If a reservoir experiences significant pressure drop but remains above the bubble point, the material balance including formation compaction becomes:

$$N_p\left(B_o + \left(R_p - R_s\right)B_g\right) + W_pB_w - G_iB_{g,inj} - W_iB_{w,inj} = c_eNB_{oi}\Delta P + W_eB_w \tag{10.5.23}$$

or

$$F = c_eNB_{oi}\Delta P + W_eB_w \tag{10.5.24}$$

Usually, an accurate value of N cannot be obtained from this material balance because $W_eB_w \gg c_eNB_{oi}$.

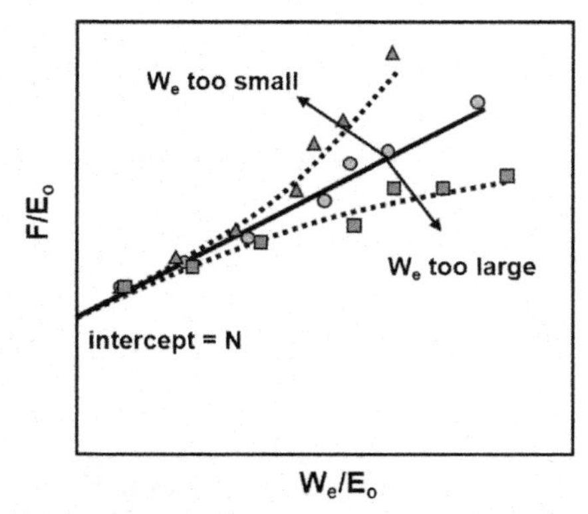

Figure 10.5.11 Graphical material balance for a weak or moderate water drive.

Strong Water Drive

In a strong water drive, there is relatively little pressure decline and $NE_o \ll W_eB_w$. Eqn (10.5.22) reduces to:

$$F \approx W_eB_w \tag{10.5.25}$$

It is not possible to determine the *OOIP* of a reservoir under strong water drive using a material balance. Only the water influx can be determined.

Even under a strong water drive, there must be some pressure decline to establish flow. If the pressure decline is greater than the error in the pressure measurements, a relationship between the water influx and the change in reservoir pressure can be determined. This relationship can be used to fit an appropriate aquifer model, as discussed in Section 10.5.8.

Weak or Moderate Water Drive

With a weak or moderate water drive, there is a significant decline in reservoir pressure, and both solution gas expansion and water influx occur. Eqn (10.5.22) is rearranged as follows:

$$\frac{F}{E_o} = N + \frac{W_eB_w}{E_o} \tag{10.5.26}$$

A plot of F/E_o versus W_eB_w/E_o is expected to be linear with an intercept of N and a unit slope of 45°, as shown in Figure 10.5.11. It is necessary to predict W_e as a function of reservoir pressure to apply this method. If the predicted W_e is too small, the trend on the plot will curve upward. If the predicted values are too large, the trend will curve downward. Hence, the correct value of W_e is found when the data fall on a straight line.

The prediction of aquifer influx is discussed in Section 10.5.9. The case of a small, permeable aquifer is included here to illustrate the approach. For a small, permeable

Table 10.5.7 Summary of data and calculations required for a material balance on a weak water drive reservoir

Date	P	ΔP	N_p	R_p	W_p	W_i	B_o	R_s	B_g	c_f	c_w	W_e	F
Pressure test data		From pressure data	Production data at date of pressure test				Fluid properties at test pressure					Equation (10.5.27)	Equation (10.5.2)

aquifer, the pressure response is assumed to be instantaneous throughout the aquifer. The influx is then given by:

$$W_e = (c_w + c_f)W_i\Delta P \tag{10.5.27}$$

in which W_i is the volume of water in the aquifer, ΔP is the decrease in reservoir pressure, and the total compressibility of the water-filled pore space is simply the sum of the water and formation compressibilities, c_w and c_f.

The data required to perform a material balance for a small aquifer drive are summarized in Table 10.5.7. The volume of the aquifer water, W_i, is adjusted until the slope of a plot of F/E_o versus W_eB_w/E_o is linear with unit slope. If the simple aquifer model is incorrect, a trend in the data also deviates from linearity as shown in Figure 10.5.7.

In practice, it is usually only possible to perform a material balance to determine water influx from a strong aquifer or to analyze a reservoir supported by a small, permeable aquifer. For a strong aquifer, only a poor estimate of the *OOIP* may be possible because the water influx dominates the reservoir energy. For a small aquifer, the water drive material balance can usually give a good determination of the *OOIP*. If the *OOIP* is known from the volumetrics, then the water influx can be determined from the material balance as follows:

below bubble point:

$$W_eB_w = F - NE_o \tag{10.5.28}$$

above bubble point:

$$W_eB_w = F - c_eNB_{oi}\Delta P. \tag{10.5.29}$$

Possible sources of error are:

1. The change in reservoir pressure is small relative to the error in the pressure measurements.
2. The contribution of the oil and solution gas expansion is too small relative to the water influx to accurately calculate N.
3. The pressure data are not built up or representative of the average reservoir pressure.
4. Another drive mechanism, such as gas cap drive, is operating.

In general, when pressure data are scattered, the greatest weight should be given to the data points with the largest net withdrawal.

10.5.6 Combination Drive

In a combination drive reservoir, there is a gas cap and an aquifer. Only the formation compaction, E_f, can be neglected. The general material balance equation simplifies to:

$$N_p\left(B_o + \left(R_p - R_s\right)B_g\right) + W_pB_w - G_iB_{g,inj} - W_iB_{w,inj} = N\left[\left(B_o - B_{oi}\right) + \left(R_{si} - R_s\right)B_g + \frac{mB_{oi}}{B_{gi}}\left(B_g - B_{gi}\right)\right] + W_eB_w \quad (10.5.30)$$

or

$$F = N\left(E_o + mE_g\right) + W_eB_w \quad (10.5.31)$$

It is rarely possible to apply a material balance to a combination drive reservoir and determine the *OOIP*, the gas cap size, and the aquifer influx. There are three sources of reservoir energy and only one material balance equation, so the problem is usually indeterminate. If the gas cap is large and the water influx is small, then the influx can be neglected and the reservoir analyzed as a gas cap reservoir. If the influx is large and the gas cap is small, the reservoir can be analyzed as a water drive reservoir. Finally, if the *OOIP* and OGIP are known from volumetrics, then the material balance can be used to determine the water influx as follows:

$$W_eB_w = F - N\left(E_o + mE_g\right) \quad (10.5.32)$$

The calculated influx and measured decrease in reservoir pressure can be used to tune an aquifer model as described in Section 10.5.8.

10.5.7 Material Balance Diagnostic Plots to Determine Drive Mechanism

The role of the diagnostic plot is to determine the drive mechanism and to assess the strength of the aquifer, if present. Diagnostic plots are usually a plot of calculated *OOIP* versus time or net withdrawal.

Testing for Gas Cap or Water Influx:
If the reservoir is expected to be a solution gas or expansion drive reservoir, then plot the *OOIP* (equivalent to N) calculated at each measured pressure from a material balance excluding water influx and gas cap expansion. The equation for the *OOIP* is given by:

$$N = \frac{F}{E_o + E_f} \quad (10.5.33)$$

The calculated value of N is plotted versus time (or net withdrawal) as shown in Figure 10.5.12. If there is a gas cap or water influx, the calculated values of N will deviate upward over time as the neglected effects of gas cap expansion and water

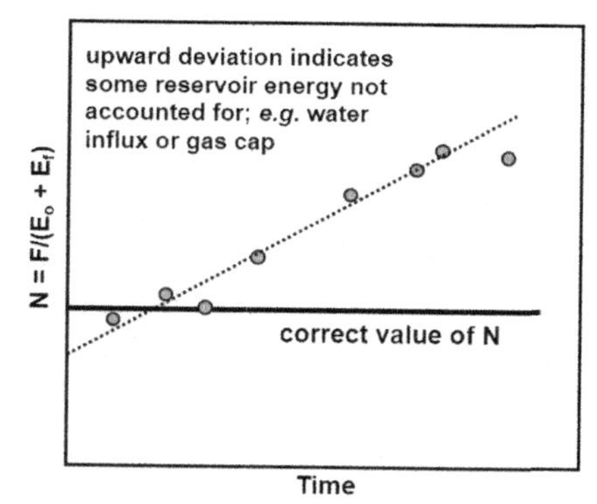

Figure 10.5.12 Effect of neglected gas cap expansion or water influx on calculated OOIP versus time. The horizontal line shows the calculated value of OOIP(N) before external energy starts to affect the calculation of N. Often in water drives there is a delay in the external support mechanism, especially if the aquifer is long and of lower permeability compared to the oil reservoir. In gas cap drives similar delay in external support from gas cap effects can occur, but usually in long thin low permeability gas caps, because of the low gas viscosity.

influx increase. Note that early pressure points will often have some large percentage errors due to lower reservoir pressure drops. This may distort early pressure trends and lead to an incorrect assessment of the drive mechanism.

Testing for Water Influx:

If the size of a reservoir's gas cap is known, the *OOIP* is calculated from a material balance with no water influx:

$$N = \frac{F}{E_o + mE_g + E_f}$$

The calculated value of N is then plotted versus time, in the same manner as was shown in Figure 10.5.12. The assumption of zero water influx is reasonable for the initial times in which little influx is expected. If there is influx, the calculated values of N will increase over time and are expected to be larger than the volumetric *OOIP* (Campbell and Campbell, 1978). The expected effect of different aquifer strengths is shown in Figure 10.5.13. Note that if the estimated gas cap size is too small, a similar effect may be observed.

Drive Indices

The complete material balance can be written as:

$$N(E_o + E_f) + mNE_g + W_eB_w = F \tag{10.5.34}$$

If we divide through by F, the following expression is obtained:

$$\frac{N(E_o + E_f)}{F} + \frac{mNE_g}{F} + \frac{W_eB_w}{F} = 1 \tag{10.5.35}$$

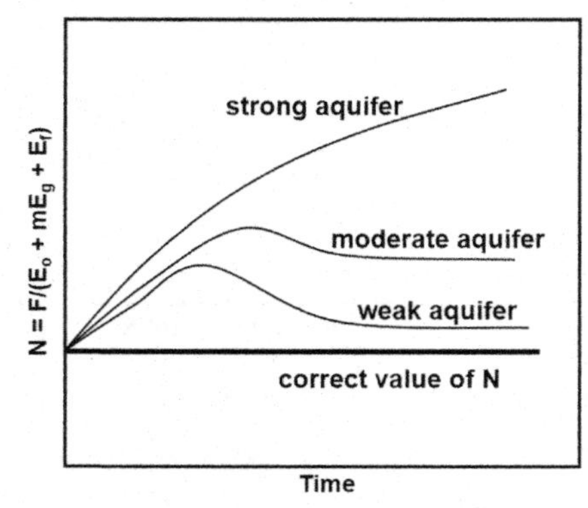

Figure 10.5.13 Effect of neglected water influx on calculated OOIP versus time.

In effect, the contribution of each major drive mechanism has been normalized so that the contributions sum to unity. Each normalized contribution is defined as a drive index, and the usual abbreviations are:

- DDI (depletion drive index) for the combined solution gas and expansion drive contributions;
- SDI (segregation drive index) for the gas cap drive contribution;
- WDI (water drive index) for the water drive contribution.

The drive index form of the material balance is then given by:

$$DDI + SDI + WDI = 1 \tag{10.5.36}$$

The drive indices indicate the relative magnitude of each drive mechanism. They are used as a guide to determine when a material balance can be used to obtain an accurate *OOIP*, OGIP, or water influx volume. For example, if DDI>0.8, then it is usually possible to determine an accurate N (*OOIP*) after sufficient withdrawals. If SDI>0.8, then it may be possible to determine an accurate mN (OGIP for the gas cap), but it is unlikely an accurate N can be obtained. If WDI>0.8, then it may be possible to determine the water influx, but probably not an accurate N. If all the indices are less than 0.8, it may be impossible to solve the material balance without fixing either m, N, or W_e before applying the material balance.

10.5.8 Aquifer Models

Whenever a water zone is in hydraulic communication with a hydrocarbon-filled reservoir, the water will expand in response to pressure decline in the reservoir. If the water zone is large or fed from another source (an aquifer), the water expansion will be a significant contribution to the pressure support of the reservoir and must be accounted for in a material balance.

Aquifers are usually considered to be radial or linear. The water volume in a linear aquifer is given by:

$$W_i = hwL\phi \qquad (10.5.37)$$

in which W_i is the water volume, w is the width, and L is the length of the aquifer, as shown on Figure 10.5.14.The water volume in a radial aquifer is given by:

$$W_i = h\pi r_w^2\theta\phi \qquad (10.5.38)$$

in which h is the thickness, r_w is the radius of the aquifer, θ is the encroachment angle in radians, and ϕ is the porosity, as shown on Figure 10.5.14. Note the oil zone volume is assumed to be small compared with the aquifer volume ($r_w \gg r_o$).

Aquifer models can be used to predict water influx based on reservoir pressure changes over time. Aquifer modeling is one of the most challenging material balance applications. Often various aquifer models will be able to match the water influx (or pressure versus time) of the reservoir. Therefore, incorporating regional geological and production data (water cuts, water rates) is usually very important. For example, there may be a large, strong aquifer, but shales or other geological features limit the influx.

Aquifer models can be classified into steady-state and unsteady state models. Steady-state models assume a constant pressure at the water-oil contact. Unsteady state models allow the pressure to vary with time. Figure 10.5.15 compares the pressure profiles in steady-state and unsteady state aquifers over time. In both cases, the reservoir pressure is assumed to be uniform. The aquifer pressure is constant in the steady-state model, but decreases with time in the unsteady state model. Steady-state aquifers must be large compared to the size of the hydrocarbon reservoir, for example, regional aquifers in extensive high permeability sands. Unsteady state aquifers are either small in comparison with the hydrocarbon reservoir or have relatively low permeability.

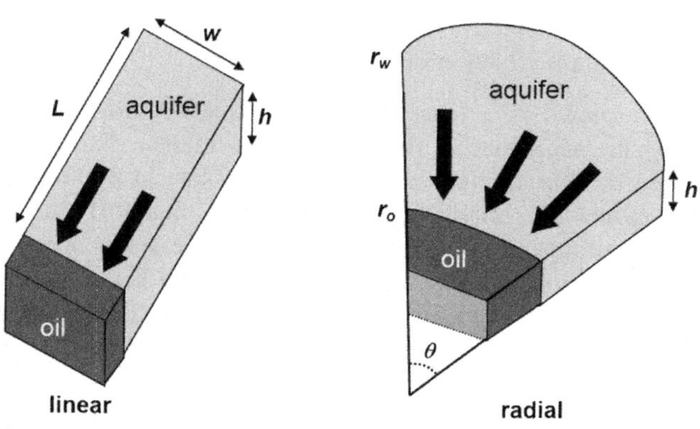

Figure 10.5.14 Aquifer geometries.

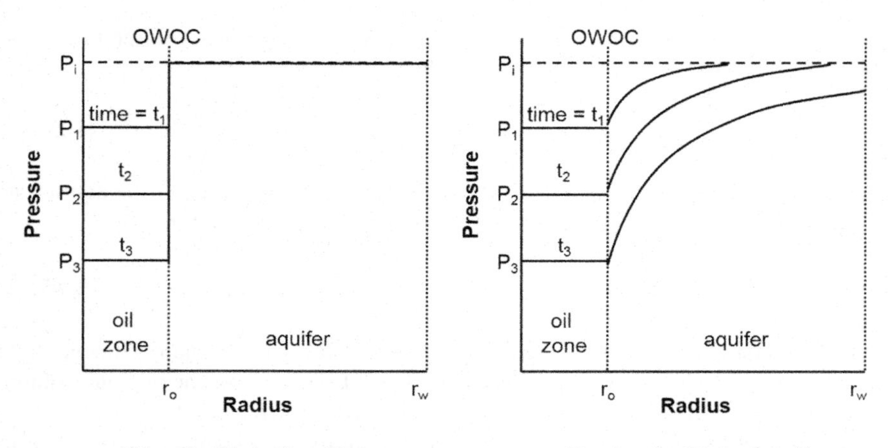

Figure 10.5.15 Pressure profiles in steady and unsteady state radial aquifer models. The pressure at the original WOC is constant in the steady-state model, but decreases as the reservoir pressure declines in the unsteady state model.

In general, equations for water influx can be written as a product of an aquifer constant, U, and a driving force, $S(P,t)$:

$$W_e = US(P,t) \tag{10.5.39}$$

The driving force is a pressure gradient that is a function of time, and the aquifer constant is the "expansivity" of the aquifer and depends on the size, compressibility, and permeability of the aquifer. A number of aquifer models are derived from Eqn (10.5.39) depending on the choice of boundary conditions, including the following:

- Pot aquifer model (unsteady state);
- Schilthuis (1936) model (infinite acting steady-state);
- Fetkovitch (1971) model (unsteady state);
- van Everdingen and Hurst (1949) model (unsteady state).

The pot aquifer model is a reasonable approximation for small aquifers that are approximately the same size as the reservoir of interest. The Schilthuis model applies to large permeable aquifers in which the pressure at the WOC is constant. The Fetkovitch model is suitable for aquifers less than 10 times the size of the reservoir. The infinite acting model is a special case in which the aquifer boundary pressure is constant. The van Everdingen and Hurst model is suitable for aquifers of any size. The procedure for the Hurst van Everdingen model is involved and is beyond the scope of this book. The interested reader is referred to Dake (1978).

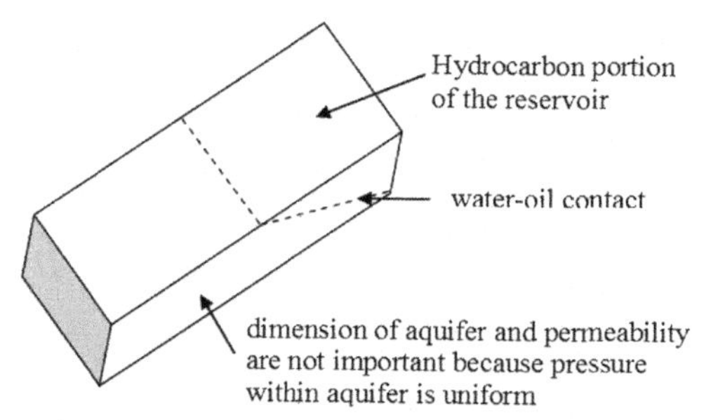

Figure 10.5.16 Schematic diagram of pot (small) aquifer.

Pot Aquifer

A pot aquifer, Figure 10.5.16, is small enough and has high enough permeability that the pressure is assumed to equalize instantaneously throughout the aquifer. The change in average aquifer pressure is then identical to the change in average reservoir pressure at any time. The influx from the aquifer is simply the expansion of the water volume for the given decrease in reservoir pressure:

$$W_e = c_t W_i (P_i - P_a) = c_t W_i (P_{ri} - P_r) \tag{10.5.40}$$

in which W_e is the water influx, c_t is the total compressibility of the aquifer, W_i is the water volume in the aquifer, P_i is the initial aquifer pressure, P_a is the current aquifer pressure, P_{ri} is the initial reservoir pressure, and P_r is the current reservoir pressure. Here, the aquifer constant, $U = c_t W_i$ and the driving force, $S(P,t) = P_{ri} - P_r$.

If the initial reservoir pressure and initial aquifer pressure are both defined at the WOC, then the pressures at any time will also be identical. Note that the total compressibility of an aquifer is simply the sum of the rock compressibility, c_f, and the water compressibility, c_w, because the water saturation is unity.

The pot aquifer model is only valid for aquifers of a similar size to the reservoir and with moderate to high permeability (approximately 100 mD or more). Otherwise, the pressure will not equalize rapidly, and the aquifer pressure will not equal the reservoir pressure. A pot aquifer solution does not involve flow calculations; therefore the shape or permeability of the reservoir is not needed.

Schilthuis Model—Infinite Acting Steady-State Aquifer

The Schilthuis model applies to systems in which aquifer volume is assumed to be much larger than the reservoir volume. It is assumed that the aquifer's pressure is constant and equal to the initial pressure value. The rate of water influx is determined at the water-oil boundary and is given by:

$$\frac{dW_e}{dt} = U(P_i - P_r) \tag{10.5.41}$$

in which P_i is the constant aquifer pressure, and P_r is the time-dependent reservoir pressure. The reservoir pressure changes in time with production, but not necessarily in a way that can be represented with a straightforward equation. Therefore, the water influx is determined numerically:

$$W_e = U\sum_{j-1}^{n}\left[P_i - \frac{P^{[j]} + P^{[j-1]}}{2}\right]\left[t^{[j]} - t^{[j-1]}\right] \tag{10.5.42}$$

in which j is a time step, and n is the total number of steps. The aquifer constant is usually derived from history matching the pressure versus withdrawal plots.

Fetkovitch Model—Finite Pseudo Steady-State Aquifer

When the aquifer pressure does not equalize instantaneously, the water influx is given by:

$$W_e = c_t W_i(P_i - P_a) \neq c_t W_i(P_{ri} - P_r) \tag{10.5.43}$$

Here, P_a is the average pressure in the aquifer, and P_r is the average pressure in the reservoir. The change in aquifer pressure is no longer equal to the change in reservoir pressure. Eqn (10.5.43) can be rearranged to solve for the average pressure as follows:

$$P_a = P_i - \frac{W_e}{c_t W_i} \tag{10.5.44}$$

Let us assume that c_t is constant and note that W_i is constant by definition. The change in average aquifer pressure with time is then given by:

$$\frac{dP_a}{dt} = -\frac{1}{c_t W_i}\frac{dW_e}{dt} \tag{10.5.45}$$

Fetkovitch assumed that the rate of aquifer influx, dW_e/dt, could be expressed in terms of a productivity index, J, as follows:

$$\frac{dW_e}{dt} = J(P_a - P_r) \tag{10.5.46}$$

Equation (10.5.44) is substituted into Eqn (10.5.45) to obtain:

$$\frac{dP_a}{dt} = -\frac{J}{c_t W_i}(P_a - P_r) \tag{10.5.47}$$

or, after some rearrangement:

$$\frac{dP_a}{P_a - P_r} = -\frac{J}{c_t W_i}dt \tag{10.5.48}$$

If J, c_t, and P_r are constant, Eqn (10.5.48) can be integrated from the initial aquifer pressure to the current aquifer pressure with the following solution:

$$\ln\left\{\frac{P_a - P_r}{P_i - P_r}\right\} = -\frac{J}{c_t W_i}t \tag{10.5.49}$$

or

$$P_a - P_r = (P_i - P_r)\exp\left\{-\frac{J}{c_t W_i}t\right\} \tag{10.5.50}$$

Now Eqn (10.5.50) is substituted into Eqn (10.5.46) to obtain an expression for the rate of water influx as a function of time:

$$\frac{dW_e}{dt} = J(P_i - P_r)\exp\left\{-\frac{J}{c_t W_i}t\right\} \tag{10.5.51}$$

Equation (10.5.51) is then integrated to obtain:

$$W_e = c_t W_i (P_i - P_r)\left(1 - \exp\left\{-\frac{J}{c_t W_i}t\right\}\right) \tag{10.5.52}$$

The term $P_i - P_r$ is an instantaneous pressure drop applied at the boundary between the reservoir and the aquifer. The term $c_t W_i$ determines the maximum possible aquifer influx for a given pressure drop, and the term $J/c_t W_i$ determines the rate at which the aquifer response approaches the maximum. As time goes to infinity, the aquifer influx becomes:

$$W_e = c_t W_i (P_i - P_r) \tag{10.5.53}$$

Hence, the ultimate response of a Fetkovitch aquifer to an instantaneous pressure drop followed by constant reservoir pressure, Eqn (10.5.53), is identical to the instantaneous response of a pot aquifer, Eqn (10.5.40).

In reality, reservoir pressure usually changes continuously, and Eqn (10.5.52) is not valid. Instead, Fetkovich devised a step-wise scheme in which water influx is determined for small time increments. The reservoir pressure is assumed to be constant over each time increment. The average reservoir pressure is calculated at the end of each time step, and the new pressure is used for the next time step. The numerical scheme is as follows:

$$\Delta W_e^{[j]} = c_t W_i\left(P_a^{[j-1]} - P_r^{[j]}\right)\left(1 - \exp\left\{-\frac{J}{c_t W_i}\Delta t^{[j]}\right\}\right) \tag{10.5.54}$$

$$W_e^{[j]} = \sum_{k=1}^{j} \Delta W_e^{[k]} \tag{10.5.55}$$

$$P_a^{[j-1]} = P_i - \frac{W_e^{[j-1]}}{c_t W_i} \tag{10.5.56}$$

$$P_r^{[j]} = \frac{P_r^{[j-1]} + P_r^{[j]}}{2} \tag{10.5.57}$$

If the pressure history is known, the aquifer influx is solved explicitly at each pressure point. The aquifer size and productivity index can be used as history match parameters to match water influx determined independently from a material balance or contact measurements. If the pressure history is not known, the Fetkovitch model can be combined with a material balance to solve implicitly for the pressure and aquifer influx.

The Practice of Reservoir Characterization

Chapter Outline

The purpose of reservoir characterization is to capture the features that determine the fluid flow in the reservoir and affect production profiles and hydrocarbon recovery. As discussed in Chapter 1, the reservoir characterization and classification are used to determine suitable development strategies and production forecasting methods. In other words, reservoir characterization attempts to answer the following questions:

- What are the limiting factors in oil recovery?
- What can be done to improve recovery and or oil rates?
- Where do we make changes? (Which wells can I convert or drill? Which wells do I convert to injectors? What should be the injection or production rates?)
- Are the changes proposed economically viable?

Reservoir characterization involves combining all of the reservoir data discussed in Chapters 3–9 and the methods of Chapter 10 to form a one-, two-, or three-dimensional representation of the reservoir. This model can be a mental model, analytical, geological, or flow simulation model. The components include the:

- Fluid type (heavy oil, conventional oil, volatile oil, retrograde condensate, gas);
- Reservoir architecture (size and structure, porosity, permeability, and saturation distributions); sometimes this is called the rock model, whereas the structural model characterizes only the structural aspects;
- Drive mechanism (expansion, solution gas, gas cap, water, combination);
- Flow mechanism characterization.

Practical Reservoir Engineering and Characterization. http://dx.doi.org/10.1016/B978-0-12-801811-8.00011-0

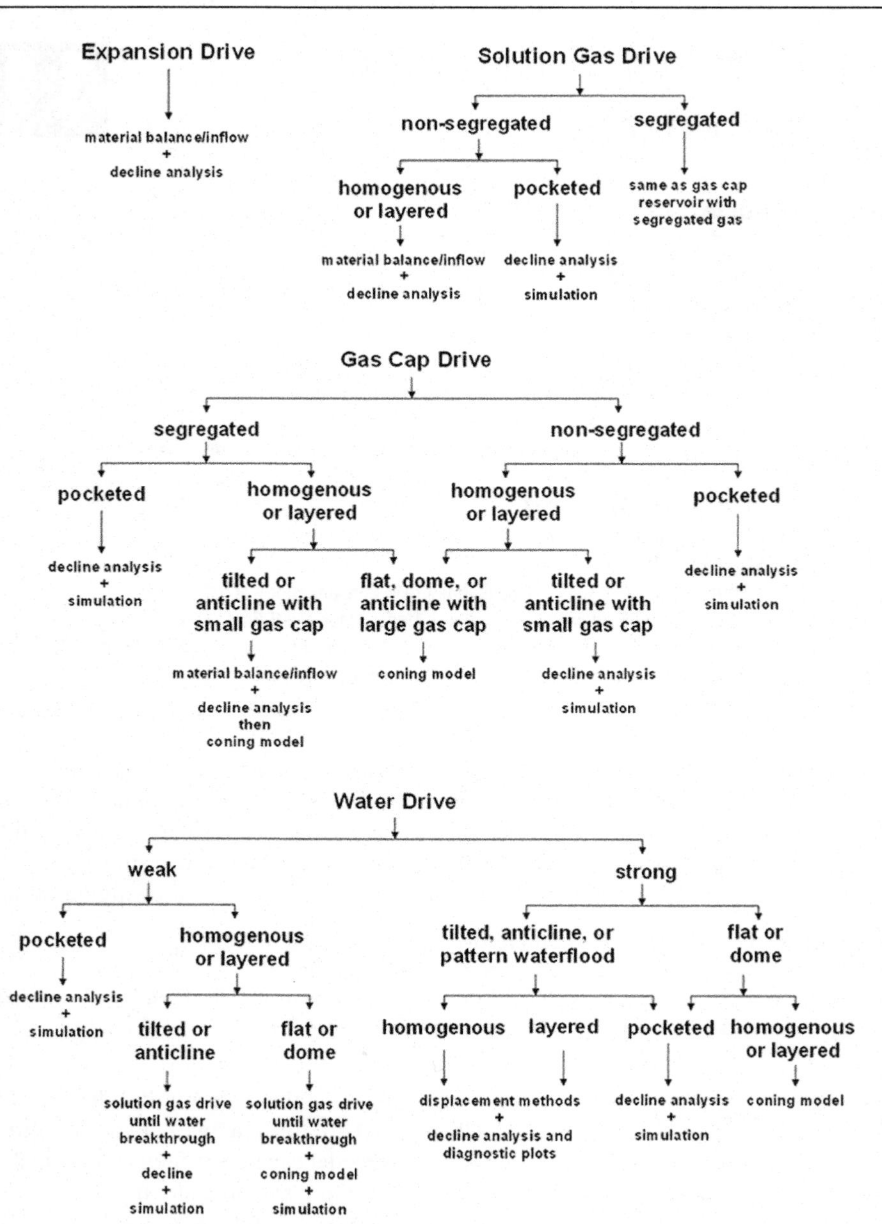

Figure 11.0.1 Simplified guidelines for appropriate choice of analytical production history matching and forecasting techniques. Simulation is listed when analytical approaches are not likely to apply.

These classifications are then used to select the appropriate modeling and forecasting methods, Figure 11.0.1. Note the choices of forecast model and material balance method are determined by drive mechanism, flow mechanism, and reservoir architecture. For example, material balance-based methods, inflow calculations, and decline analysis are appropriate methods for homogeneous solution gas drive reservoirs with nonsegregated gas flow. Chapter 11 focuses on how to characterize and classify the reservoir prior to forecasting. The forecasting methods themselves are beyond the scope of this book. A general overview of characterization is provided in Section 11.1, a characterization workflow is introduced in Section 11.2, and classification into fluid type, architecture, drive mechanism, and flow mechanism is discussed in Section 11.3. This approach to characterization is illustrated with several case studies provided in Section 11.4.

11.1 Overview of Characterization

Characterizing a reservoir from what are usually partial and inconsistent data is probably the greatest challenge in reservoir engineering, as well as geology and geophysics. While data and techniques used often vary from one field to the next, one or more of the following methods are usually used to characterize a reservoir (Chapter 10):

- Volumetrics;
- Material balance;
- Analogy;
- Decline analysis;
- Analytical;
- Data mining;
- Simulation.

However, any one method is likely to provide a large range of possible solutions for the oil in place, expected production rates, and recoverable reserves. It is often necessary to apply several methods to obtain a more accurate solution, as illustrated in Figure 11.1.1. The main problem is that we need dynamic data to determine the drive mechanism, reservoir architecture, and flow mechanisms in the reservoirs. Hence, reservoir characterizations are periodically updated as new data come in over time and increasingly more accurate solutions are realized, Figure 11.1.2. Also, the results from one method are often used to refine the inputs into another method. In a sense, the reservoir model spirals into a narrower range of solutions as each method is refined based on new data and analysis.

The reservoir characterization team must then judge which interpretation is most likely, but also must be flexible in its thinking. In most, if not all cases, several iterations are required to obtain a satisfactory characterization. Often, there is insufficient data to perform the full step-by-step analysis, and the reservoir engineer and reservoir characterization team must adjust their approach according to the data available. Often, the quality of reservoir characterization depends on the amount, type, and quality of the measurements. Good engineering and geoscience work will not fully compensate for lack of data.

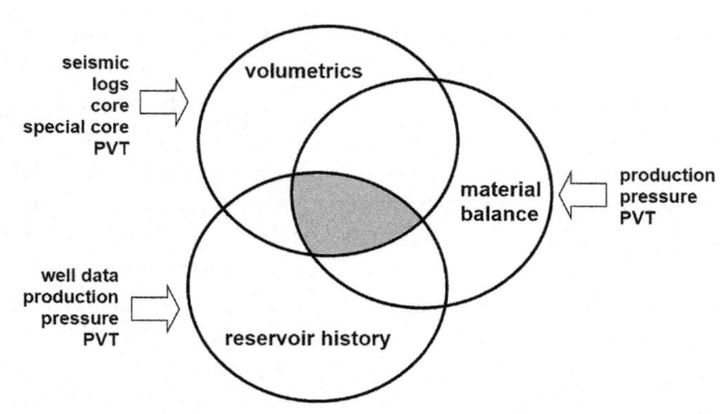

Figure 11.1.1 Overlapping range of solutions (e.g., original oil in place (OOIP), original gas in place (OGIP)) from several techniques can provide significantly more accurate solutions than any one technique.

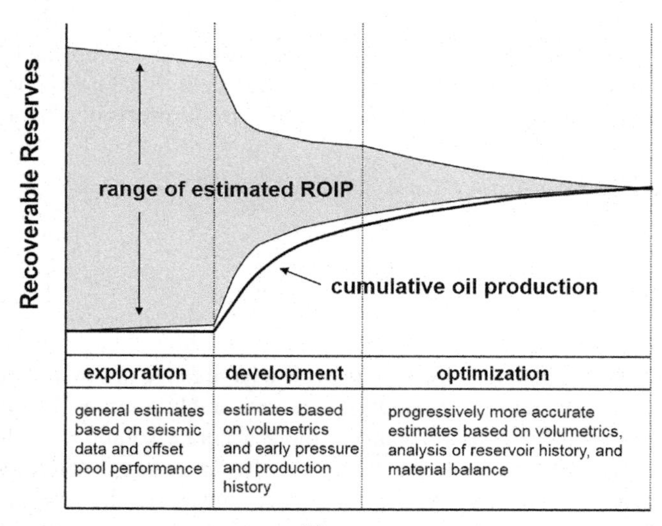

Figure 11.1.2 Improved accuracy of reservoir charaterization (recoverable oil in place (ROIP), for this case) over time as more data become available.
Adapted from Arps (1956).

We strongly recommend using material balance (Chapter 10) and analytical techniques to range-find or identify key variables, as discussed in Chapter 1. As pointed out by Dake (1994), a material balance allows an independent view of OOIP and drive mechanism. Similar analytical equations, such as Buckley Leverett (1942) or Dietz (1953) analysis, will allow us to identify waterflood or injectant swept regions, and the comparison with volumteric analysis will allow us to understand reservoir continuity. These methods are not perfect, but they often will allow the team to better conceptualize flow in the reservoir.

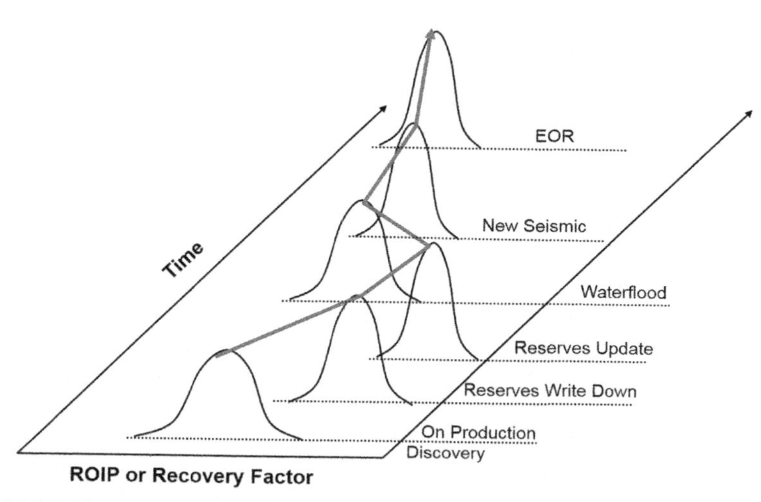

Figure 11.1.3 Changes to apparent reserves probability distribution due to operational changes and updated reservoir characterization view over time.

The above reservoir characterization analysis can be complicated by infill drilling programs or fluid injection projects that markedly change the interwell flow regimes and sometimes drive mechanisms in a reservoir. Particularly for large fields, there may be changes in recovery factor due to manmade factors, as shown in Figure 11.1.3. Therefore, it is sometimes necessary to break the history into segments to extract information from the different periods corresponding to different production strategies. For example, during primary production, we can evaluate the total OOIP during the pressure depletion stage. If water is injected into the reservoir, a new set of signals is introduced into the reservoir. The water and oil movement provides useful data on areal heterogeneity, but little information about total OOIP. Instead, injection processes usually give information on the swept oil volume in place, connected OOIP, and an updated estimate of the ROIP. Enhanced oil recovery (EOR) will further increase the recovery factor (and ROIP).

11.2 Characterization Workflow

Figure 11.2.1 shows the workflow for long-term reservoir characterization. Generally, for most new reservoirs (green fields), an initial static model is constructed from the geological model, seismic data, and petrophysical information. Later, a dynamic model is built. At this green field stage, the characterization is "model driven" rather than data driven because of the limited sampling of the reservoir. As more and more dynamic data become available, we history match or tune the dynamic model's behavior to match well and field performance. It is critical to gather dynamic data because most initial static and dynamic models will be an oversimplification of reality. For older fields (brown fields), the challenge is to integrate the dynamic data as early as possible to avoid needless iterations. The geoscience team should be guided by both

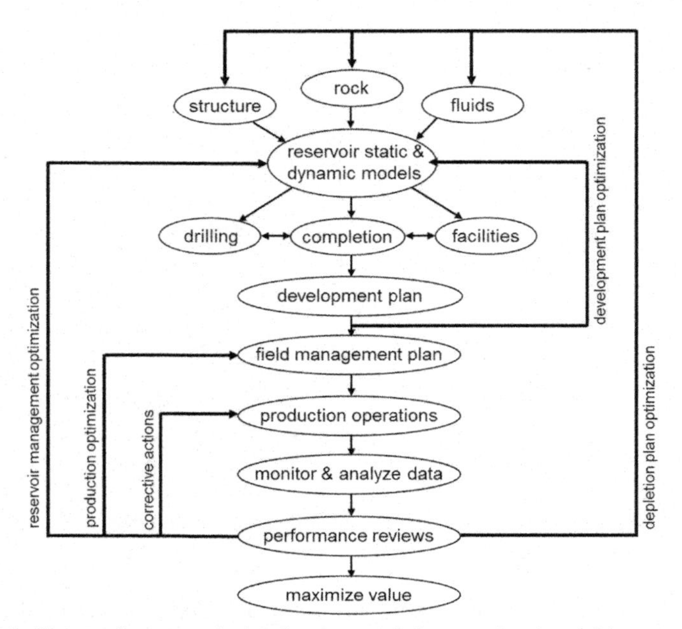

Figure 11.2.1 Stages of a reservoir development and the associated activities.

the static and the dynamic data, especially in assessing the degree of continuity or degree of pressure communication both vertically and areally.

A data-oriented approach to characterization at any point in the reservoir history was introduced at the beginning of part 3 of this book, Figure 11.2.2. Reconciling core and log, porosity and water saturation is usually one of the first steps. Then reconciling drive mechanisms with fluid types and production as well as pressure history is next. Reconciling volumetric values of oil- or gas-in-place with material balance is a third step. Finally, reconciling production data with flow mechanisms and completion data or well trajectory is done.

The following workflow is recommended for brownfields:

1. Determine fluid type from fluid analysis and PVT studies, as discussed in Chapter 5.
2. Determine the type of reservoir drive mechanism by examining field or area production and pressure profiles if available, as shown in Section 10.4.1 and Table 10.4.1.
3. Develop a structural model from geological discussions and modeling/mapping concentrating on top of pay structure, faulting, and the location of gas-oil contact (GOC) and water–oil contact (WOC), as discussed in Chapter 10.
4. Develop a permeability model from geological analysis and modeling/mapping concentrating on net pay, permeability maps, or porosity maps.
5. Determine the fluid flow patterns by examining aggregated individual well performances:
   - **a.** Communication is indicated if wells in close proximity have similar behavior in terms of GOR or watercuts or oil decline behavior, as shown in Figure 11.2.3;
   - **b.** If wells or groups of wells behave differently in terms of GOR or watercuts or oil decline behavior, check if the behavior is related to top of structure or distance from gas cap or aquifer.

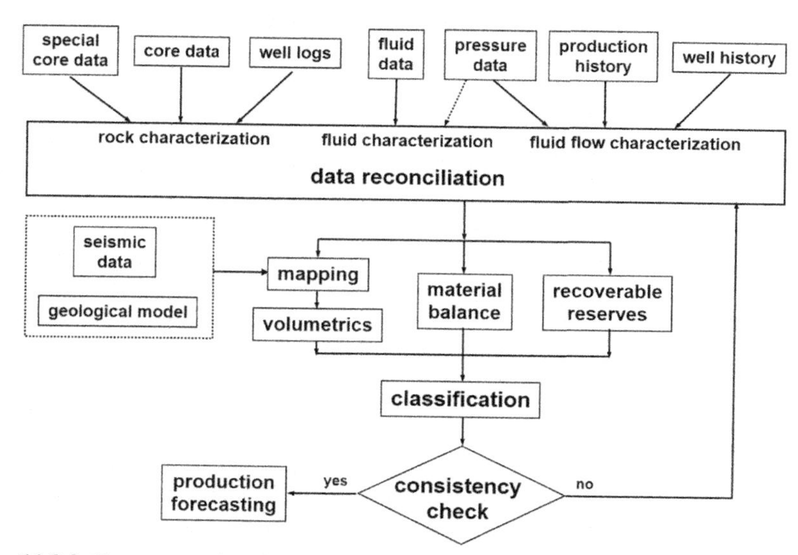

Figure 11.2.2 Components involved in the reservoir characterization process.

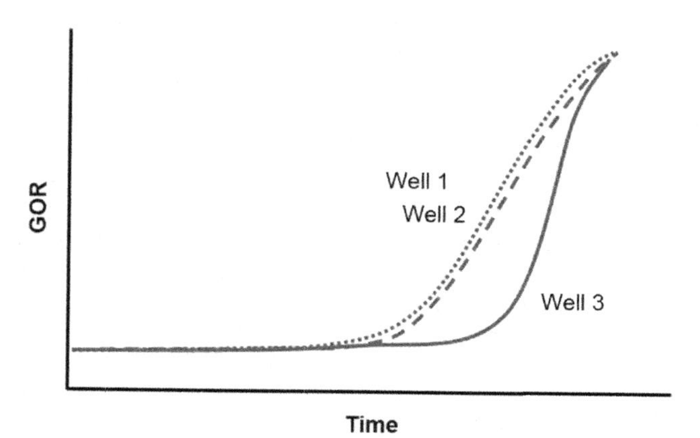

Figure 11.2.3 Well aggregate plot. If there is good communication in the reservoir, the GORs of structurally similar wells drilled at the same time will track each other (wells 1 and 3). A new drill (well 2) may not initially follow the same trend, but once free gas flow is established near the well, it will reach the same trends if it is in communication with the other wells.

6. Determine how operational effects (changes in flowing pressure) and rate changes have affected the production profile.
7. Select the development strategy (recovery technique) to be used, for example, continue on primary solution gas drive, infill drill, water flood.
8. Select an analogous reservoir, if possible, and use as a guide for field performance.
9. Update the reservoir characterization with new production data, new infill well openhole log data, and seismic data. The engineer may need to request particular measurements that have

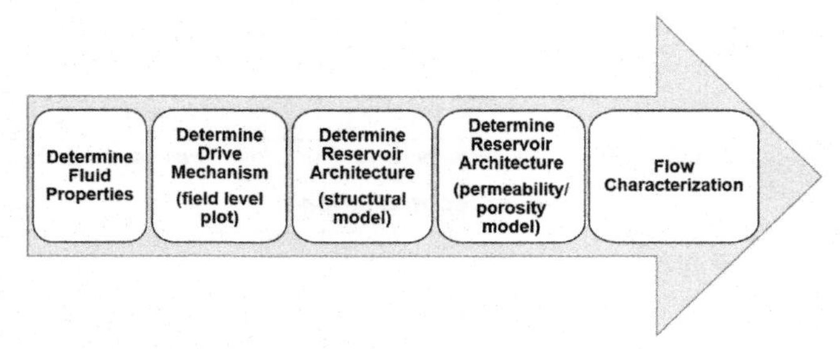

Figure 11.2.4 Reservoir classification workflow.

not been acquired yet. This may involve providing an economic justification to show the benefits of the data requested exceeds the cost of their acquisition.

10. Conduct the development and operation of the reservoir so to ensure the maximum possible efficiency in oil recovery.

Often, some of the steps above will be done concurrently. As the characterization progresses, the workflow shifts to classification and modeling according to fluid properties, drive mechanism, architecture, and flow characterization, Figure 11.2.4, to the determination of reservoir flows. Reservoir flow mechanism is critical to the determination of remaining reserves, optimization strategy, and EOR potetential. Each of these components is discussed in the next section.

Note that we have focussed on individual well analysis in production for reservoir characterization. Analysing individual well production and restoring a problem well is also a critical task of reservoir engineers. A well must be considered from a wellbore and reservoir point of view to intelligently diagnose its behaviour.

This chapter presents classification as black and white logic. It is rarely that simple. As the geoscience team becomes more skilled and more knowledgeable about a reservoir, the black and white logic presented here can be replaced with shades of grey. However, initially for the sake of learning, we will present a yes or no classification system.

11.3 Reservoir Classification

Reservoir classification includes four components: fluid type, reservoir architecture, drive mechanism, and flow mechanism. Each component is discussed below.

11.3.1 Fluid Type

There are four types of oil and three types of gas, listed below in order of increasing density:

- Dry gas;
- Wet gas;
- Retrograde condensate gas;
- Volatile oil;

- Black oil;
- Heavy oil;
- Bitumen.

A dry gas remains completely in the gas phase even at surface conditions, while a wet gas forms a two-phase mixture at surface or wellbore conditions. Dry and wet gas will not drop out liquid in the reservoir as pressure decreases, but a retrograde condensate gas does. Therefore, the distinction between retrograde condensate gas and the other gas types is important for reservoir engineering. Volatile oils require compositional models to accurately predict phase behavior. Both conventional and heavy oils are considered to be "black oils." Heavy oils often require specialized production practices because their viscosity is high. Bitumens are nearly solid at ambient conditions and shallow deposits are usually mined as an ore. The bitumen is then extracted from the ore in surface processes that are not related to reservoir engineering. Deeper bitumen are usually extracted using thermal processes such as steam injection.

The fluid type is determined from fluid properties, including oil and gas analyses and fluid studies. The producing GOR or the liquid content of a gas is also a good indicator of fluid type. Typical properties of black (nonvolatile) oils, volatile oils, retrograde condensates, wet gases, and dry gases are listed in Tables 11.3.1–11.3.5. The Unitar definition for heavy oils is provided in Table 11.3.6. These tables are guidelines only, and a fluid study may be required to identify a volatile oil or to distinguish between a retrograde condensate gas and a wet gas.

Table 11.3.1 Black (nonvolatile) oil typical composition and classification criteria

Parameter	Schlumberger (1982)	Smith et al. (1992)	Amyx et al. (1960)
Methane, mol%	44	–	57
Hexane plus, mol%	43	–	35
Oil gravity, °API	30–40	20–32	<45
GOR, SCF/STB	100–2500	50–500	<1000

Table 11.3.2 Volatile oil typical composition and classification criteria

Parameter	Schlumberger (1982)	Smith et al. (1992)	Amyx et al. (1960)
Methane, mol%	60–65	60	–
Hexane plus, mol%	15–20	21	–
Oil gravity, °API	40–50	30–50	45–60
GOR, SCF/STB	≈3000	500–6000	1000–8000

Table 11.3.3 Retrograde condensate gas typical composition and classification criteria

Parameter	Schlumberger (1982)	Smith et al. (1992)	Amyx et al. (1960)
Methane, mol%	75	75	–
Hexane plus, mol%	8	8.5	–
Liquid gravity, °API	50–70	50–70	50–60
Liquid content, STB/MMSCF	10–200	67–500	15–125

Table 11.3.4 Wet gas typical composition and classification criteria

Parameter	Schlumberger (1982)	Smith et al. (1992)	Amyx et al. (1960)
Methane, mol%	–	–	87
Hexane plus, mol%	–	–	3.7
Liquid gravity, °API	50–70	>60	>60
Liquid content, STB/MMSCF	<10	10–67	10–17

Table 11.3.5 Dry gas typical composition and classification criteria

Parameter	Schlumberger (1982)	Smith et al. (1992)	Amyx et al. (1960)
Methane, mol%	90	–	91
Hexane plus, mol%	1	–	0.4
Liquid content, STB/MMSCF	0	0	0

Table 11.3.6 Unitar definition of crude oils (Gray, 1994)

Oil type	Viscosity (mPas)	Density (kg/m^3)	API gravity
Conventional oil	<100	<934	>20
Heavy oil	100–100,000	934–1000	20–10
Bitumen	>100,000	>1000	<10

11.3.2 Reservoir Architecture

Geology and reservoir architecture are usually a key controlling factor in reservoir performance, processes involving injection. Having said this, it is important to realize that there are considerable limitations to the initial knowledge of a reservoir because core and logs represent only about one 10 billionth of the reservoir. The challenge of reservoir complexity has been appreciated for many years; for example, Rowan observes that,

> *It is no insult to the geologist to say that the subsurface picture is frequently uncertain, for no one knows better than he the difficulties involved in the creating this picture and the constant need for revision as new information is obtained from wells that are drilled. The geological picture, however, only depicts the broad features and broad features in themselves are insufficient to explain detailed behavior, from within the large scale pattern there are much smaller scale variations which have a pronounced effect on observed behavior.*

To simplify the challenge of interpreting reservoir architecture, we approach the problem from two aspects: structure and permeability/porosity distribution (rock model).

11.3.2.1 Reservoir Architecture; Structural Model

First in general terms, there are five main types of structure in the structural model, as described in Chapter One.

- Flat;
- Tilted;
- Anticline;
- Dome;
- Heavily faulted.

The structure model is identified using structural contour maps and structural cross-sections. These maps are based on seismic data and well logs, as discussed in Sections 1.2.3.1, 4.2.1, and 10.2.3. Anticlines are the dominant structure worldwide, but reservoirs may have components of all the above structures. In large reservoirs, there may be mixed structural types.

Reservoir structure has significant impact on flow characterization. For example, fluid flow in strongly tilted or anticline reservoirs, with dips $>10°$, will usually be controlled by both viscous and gravity forces. Therefore, the location of a well within the structure in relation to the fluid contacts may strongly affect its performance. If gravity dominates, such as a gravity-dominated gas flood, then higher recovery factors are expected. In flat or dome-shaped reservoirs, pay thickness and vertical permeability will dictate the degree of segregation of fluids in the reservoir. In general, the average thickness and dip angle in a reservoir structure are usually critical in reservoirs in which gravity plays an important role, such as gas cap drive or bottom/edge water drive. Finally, the structural model is usually critical in reservoirs that are heavily

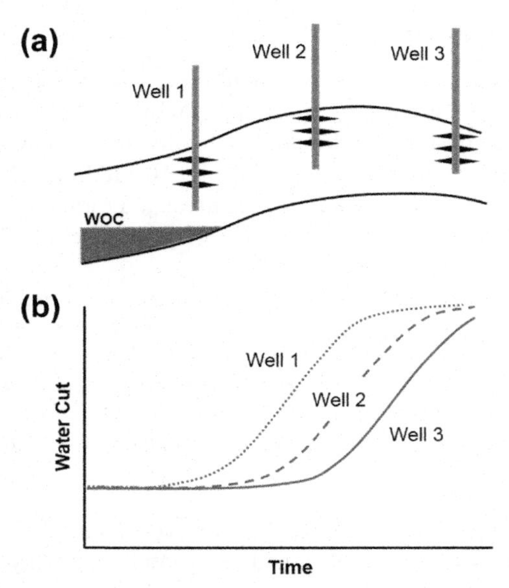

Figure 11.3.1 (a) Cross-sectional view of three wells showing their proximity to a water-oil contact (WOC); (b) water cut production profile of the three wells.

faulted, and those faults may cause compartments when the faults have small permeability or may act as conduits if the faults act as flow pathways.

When developing a structural model, it is important to examine the production data in terms of water cut/water rate (and GOR/gas rate) behavior compared to distance (standoff distance) from the WOC (and GOC), as shown in Figure 11.3.1. Breakthrough times of water or free gas production provide a confirmation of the structural model and communication within the reservoir.

11.3.2.2 Reservoir Architecture; Porosity and Permeability Model

The second component of reservoir architecture is the permeability/porosity distribution. There are five main types of permeability/porosity distribution (rock model), Figure 11.3.2:

- Homogeneous;
- Layered;
- Pocketed (jigsaw and labyrinth);
- Fractured;
- Tight (low permeability $k < 0.1$ mD).

The permeability distribution is assessed based on core data, well logs, production and pressure history, and well test analysis. Guidelines for determining each type of permeability distribution and some common diagnostic problems are given below.

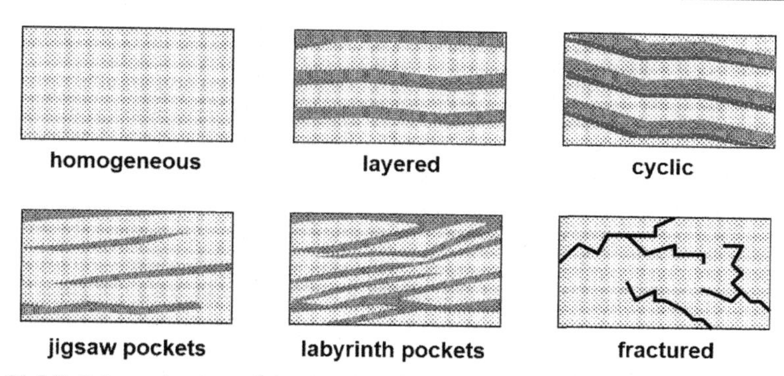

Figure 11.3.2 Schematic view of the six major types of permeability and porosity distribution.

Most reservoirs may not perfectly fit into one of the categories, but the categories are useful, especially initially, because they allow us to communicate more effectively with the geoscience team.

Homogeneous Permeability and Porosity

A homogeneous reservoir has the following characteristics:

- Less than an order of magnitude variation in permeability;
- Porosity and permeability variations are random, e.g., no cycles or trends;
- No extensive shale breaks or other permeability barrier with in the layer;
- Nearly uniform productivity indices;
- Uniform pressure and production profiles;
- No flow barriers detected through pressure transient analysis or static pressures.

In general, material balance and most analytical methods work well in homogeneous high permeability ($k > 100$ mD) reservoirs. The production profiles for individual wells are similar to other wells in the same area or structural position, at least in flat reservoirs (Figure 11.2.3). Material balance and simulation work well in these types of reservoirs because wellbore pressure equalizes to reservoir pressure after only short shut-in periods, and large pressure gradients do not usually occur, particularly in high permeability single-layered reservoirs. The exception is heavy oil, in which, even for high permeability reservoirs ($k > 1000$ mD), measured well pressures are not likely to equilibrate with a short shut-in period ($t < 1$ week).

In distributed drive mechanism systems, such as solution gas or compressible drive systems, individual well oil rate declines will be similar, Figure 11.3.3. For tilted or anticlinal system with higher dips and gas cap or water drive mechanisms, individual well performance will usually be consistent for wells in a similar structural position. Gas cap drive or water drive can yield fairly high primary recovery factors in such reservoirs if they are thick or have substantial dip angle. Usually fluid injection projects and EOR are successful in these types of reservoirs.

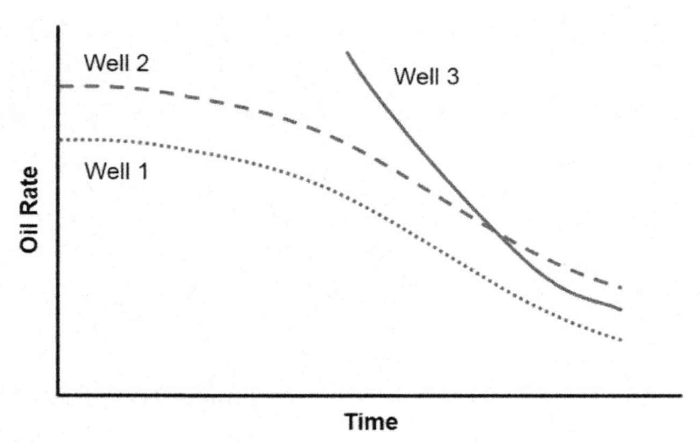

Figure 11.3.3 Aggregate oil rate analysis example for a homogenous reservoir. Because oil rate is a function of reservoir pressure, at later times the oil decline rates of all three wells are similar.

Heterogeneous Permeability and Porosity

Most reservoirs have heterogeneous permeability and porosity distributions. We must assess whether the permeability is random, lenticular, or strongly layered, Figure 11.3.4. The continuity of the permeability and porosity features dictates the classification. The continuity is determined through openhole log and core data, production and pressure data, and seismic or remote sensing data. All three types of permeability distribution may look similar on a single well log or core, Figure 11.3.5. However, in a strongly layered (layered cake) reservoir, alternating layers of higher and lower permeability and porosity are continuous between wells. Cyclic heterogeneity is a special form of layering, where the characteristics of alternating layers are similar.

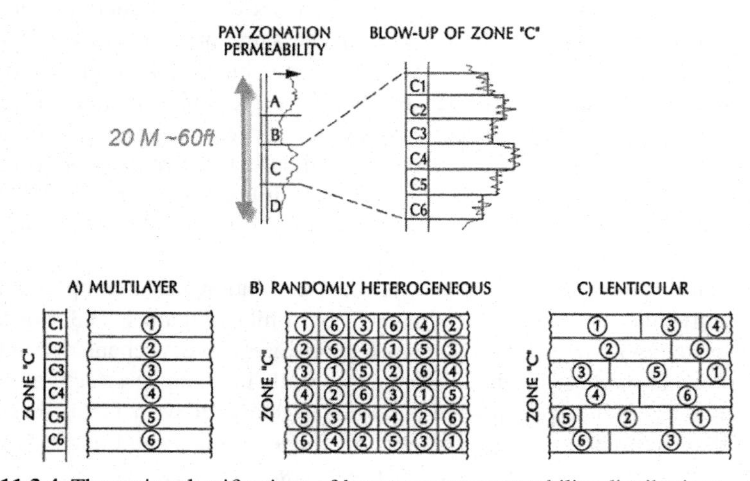

Figure 11.3.4 The major classifications of heterogenous permeability distributions (Baker, 1994).

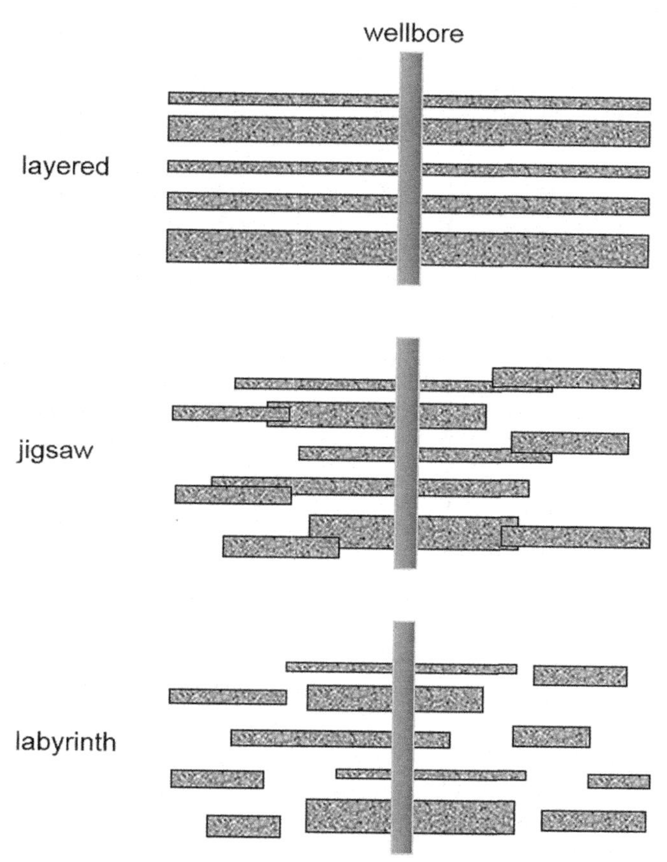

Figure 11.3.5 Schematic view of layered-, jigsaw-, and labyrinth-type permeability and porosity distributions at a single well. Note that although the well data is the same for all three models, the continuity of beds is dramatically different.
Adapted from Baker (1994).

For jigsaw and labyrinth reservoirs, the log characteristics are different from well to well, Figures 11.3.6–11.3.8. (For more discussion of the geological characteristics of these reservoir types, see Weber and van Geuns, 1990.) Production data are essential to distinguish between these more complex permeability distributions, as is discussed for each type below.

Layered and Cyclic Reservoirs

A layered reservoir has the following characteristics:

- Zones of contrasting permeability or porosity correlate from well to well in stratigraphic cross sections;
- Initial pressures of infill wells (using methods discussed in Chapter 6) will show large differences between wells;
- Early breakthrough of water or gas in fluid injection projects;

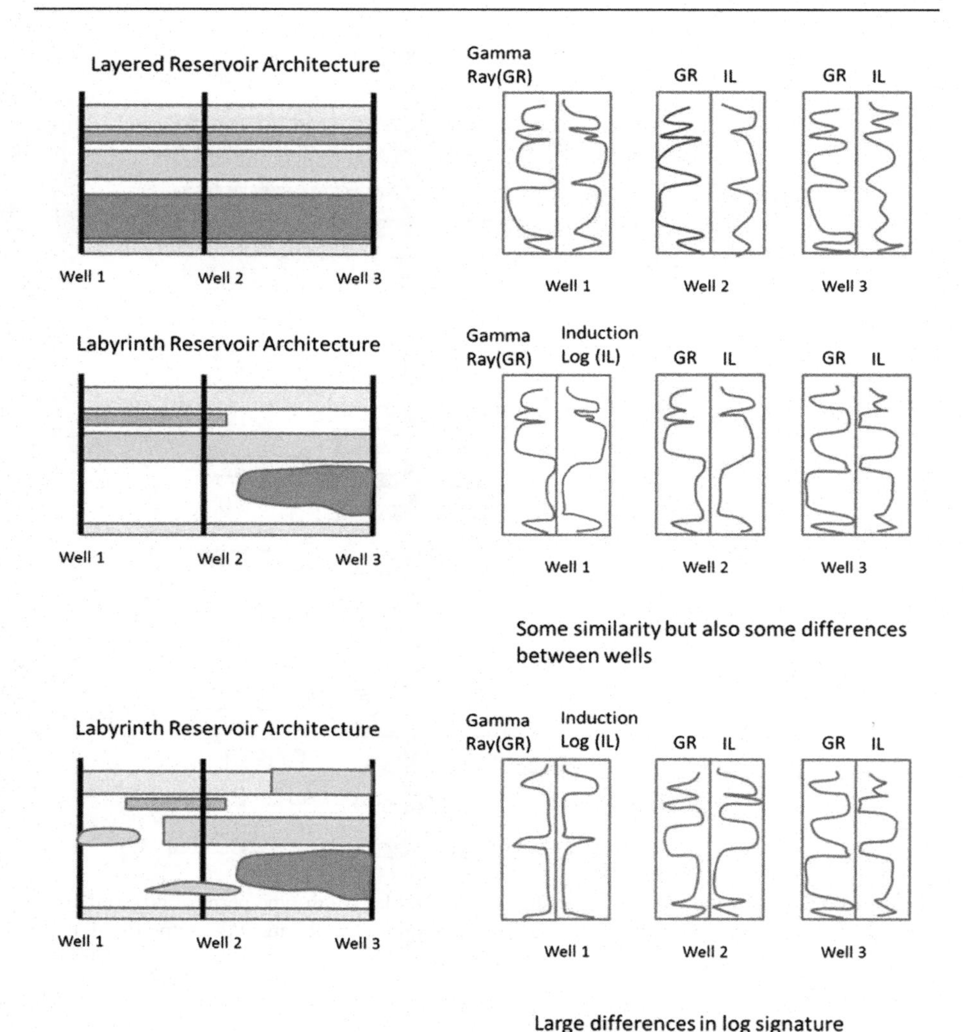

Figure 11.3.6 Schematic and openhole log cross-sectional views of layered-, jigsaw-, and labyrinth-type reservoirs *(adapted from James, 1999)*. The layer cake reservoir has consistent log signatures between wells, but the labyrinth reservoir has poor correlation from well to well.

- Early breakthrough of water or gas in water drive or gas cap drive reservoirs;
- Relatively high GORs as a high permeability layer acts as a conduit for gas for gas-injection or solution gas drives or similarly high water production for waterdrive or waterfloods;
- Stepwise changes in water cut for waterdrive or waterfloods for individual wells.

Depending on how layers are connected and the permeability of individual layers, the production profiles for individual wells may be similar for wells in the same area or structural position. Figure 11.3.7 shows for a single layer reservoir or multilayered

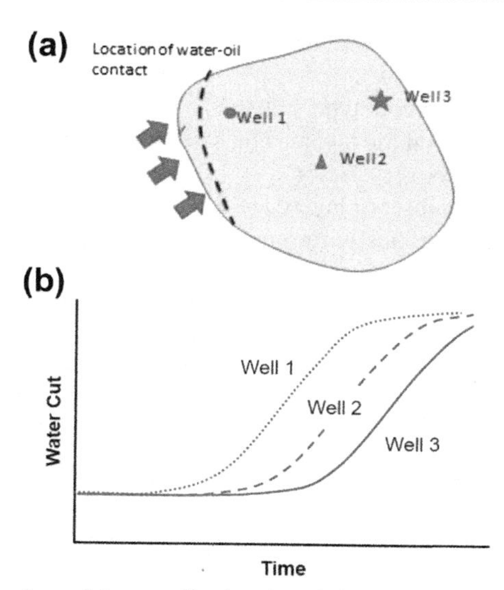

Figure 11.3.7 (a) Top view of three wells showing their proximity to a water-oil contact (WOC); (b) water cut production profile of the three wells. Note, the sequence of water cut increases is related to distance from WOC and can give important clues to both the permeability distribution and drive mechanism.

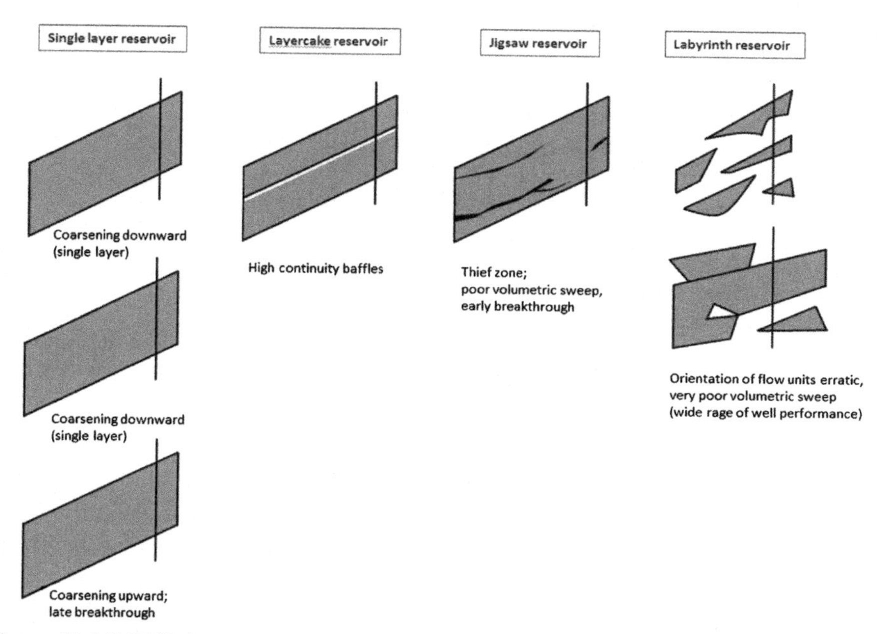

Figure 11.3.8 Differing water sweep efficiency according to the type of reservoir.

reservoir with low permeability contrast, the influence of gravity on water cut production profiles.

Performance of this reservoir type is usually strongly dependent on the average permeability contrast between the layers. The lower the permeability contrast between layers, the higher is the recovery factor.

Cyclic reservoirs are a subset of layered reservoirs in which the properties of sets of two or more layers repeat vertically throughout the reservoir. Cyclic heterogeneity is common in reservoirs because many depositional processes have one or more repetitive aspects. For example, floods may occur each spring, and sea levels rise and fall according to the Earth's orbit around the Sun. The advantage to recognizing cyclicity is that layers in similar parts of cycles tend to have similar characteristics, making reservoir analysis and characterization simpler. That is, if one cycle is adequately measured and evaluated, similar characteristics may be assumed for other cycles. Geological guidance is helpful for identifying the source(s) of cyclicity and the validity of the assumption about neighboring cycles.

A material balance can be difficult to apply to a layered reservoir because the depletion rate maybe different in each layer. Material balance and simulation are more challenging in these types of reservoirs because wellbore pressure in a shut-in well with commingled zones may not equalize to reservoir pressure within short shut-in periods. Large pressure gradients may occur in a multilayered reservoir. It can also be difficult to predict inflow performance if there is cross-flow between reservoir layers near the wellbore. Therefore, vertical profile logs in vertical wells and vertical allocation factors are critical for developing good reservoir characterizations for layered reservoirs. If layer flow allocation factors can be determined, fluid injection projects and EOR can be successful in these types of reservoirs. The key challenge is to get recovery from lower permeability layers.

Pocketed (Jigsaw and Labyrinth) Reservoirs

A pocketed reservoir has the following characteristics:

- Often a carbonate or an interbedded low permeability sandstone;
- There are zones of contrasting permeability or porosity, but they do not correlate from well to well throughout the reservoir based on stratigraphic cross sections;
- Large regional variations in pressure;
- Regional variations in production profiles;
- If injection processes are implemented, there are nonuniform water or gas breakthrough times; continuity and permeability of beds will control which wells show injectant breakthrough;
- Boundaries are detected in pressure transient analysis.

In a jigsaw permeability distribution, the pockets are formed from different sand or lithology units that overlap and are interconnected through large areas within the reservoir structure. In a labyrinth permeability distribution, the different sand bodies or units have only local, three-dimensional connections.

It is difficult to apply a material balance to a jigsaw reservoir and almost impossible to apply a material balance to a labyrinth reservoir because the permeability pockets may deplete at different rates and even have different dominant drive mechanisms. This causes material balance estimates to underpredict reserves. Analytical methods often do not apply to pocketed reservoirs, and reservoir simulation is usually required to obtain a forecast other than a decline analysis. As a result of the differential drainage areas, individual well production profiles will not have much coherence.

Infill drilling is usually the only way to increase recovery from these fields. The key challenge with these reservoirs is to increase the well connectivity with various pockets of the reservoir. Often these reservoirs will be approached in a statistical manner in terms of development because of the large degree of uncertainty.

Heterogeneity also influences aquifer influx, Figure 11.3.8. In layered systems, the influx may be strong depending on the aquifer size and permeability. In labyrinth- and jigsaw-type reservoirs, aquifer water influx is generally less of an issue due to the lack of continuity. Figure 11.3.8 shows how aquifer influx propagates through different permeability distributions For a homogeneous single-layer reservoir, the water movement is influenced by displacement efficiency and gravity. Internal reservoir architecture porosity and permeability distribution also influence influx and production behavior. In a coarsening upward sequence, the permeability increases as the depth decreases, and similarly in a fining upward sequence, the permeability decreases as the depth decreases. So in either case, the water can breakthrough earlier compared to a more homogenous rock sequence. However, gravity and a fining upward sequence can act together to cause earlier water breakthrough.

For a layer cake reservoir, often one layer will have preferential water breakthrough in one of the layers. For jigsaw and labyrinth reservoirs, often discontinuity of layers is an issue and water drive or water flood performance may vary greatly from one well to the next.

Fractured

A fractured reservoir has the following characteristics:

- Sometimes associated with faulted reservoirs;
- Fractures may show up in cores (although many times vertical core will miss seeing fractures if fracture spacing is too wide);
- Highly variable production rates with the best wells producing much more than expected based on matrix porosity;
- Very rapid water or gas breakthroughs if fractures are connected;
- Fractures detected in production logs or interference tests.

Fractured reservoirs have a huge range of behaviors depending on the fracture spacing, fracture aperture, and matrix permeability and drive mechanism. (Nelson, 2001) Simulation and material balance may or may not be useful as reservoir characterization tools depending on the fracture type and distribution of fractures. Usually, the main limitation with characterizing naturally fractured reservoirs is a lack of sufficient information on fracture distributions and fracture connectivity. Fractured reservoirs require specialized analytical and simulation methods that are beyond the scope of this book.

Tight

A tight oil reservoir has the following characteristics (Clarkson et al., 2012):

- Air permeability is less than approximately 1 mD based on core data;
- Permeability values are often stress-sensitive;
- Reservoirs may be self-sourcing; that is, the hydrocarbons may come from organic material within the rock matrix so that well logs overestimate available porosity;
- Effective permeability is less than approximately 0.1 mD based on pressure transient analysis;
- Pore and pore throat geometry and size are particularly important in controlling storage and flow because:
  - Flow mechanisms besides Darcy flow, e.g., diffusion, become important;
  - Pores and throat sizes approach the sizes of the fluid molecules.

Usually in these types of reservoirs, pay continuity and the value of permeability are controlling factors for recovery factor.

A material balance can be difficult to apply because there are often large pressure gradients in the reservoir. The flow is often in the transient region. Material balance and simulation are more challenging in these types of reservoirs because wellbore pressure in a shut-in well (particularly with commingled zones) may not equalize to the reservoir pressure within short shut-in periods. Large pressure gradients may occur in a tight reservoir. The analytical methods used for this type of permeability are rate and pressure transient analyses.

11.3.3 Drive Mechanism

There are five main drive mechanisms:

- Expansion drive;
- Solution gas drive;
- Gas cap drive;
- Water drive;
- Combination drive.

The drive mechanism is determined from a combination of fluid contact data, production profiles, fluid properties, and material balance (Chapter 10).

There are seven indicators to diagnose the type of drive mechanism:

- Pressure profile;
- Oil rate decline or decline exponent;
- GOR or gas production profile;
- Water cut or water rate production profile;
- Ultimate recovery levels;
- Material balance diagnostic plots;
- Examination of the reservoir architecture through seismic or geological/petrophysical models.

Well interference and operational aspects (back pressure and skin) can also have a large impact on production profiles. It is therefore important to identify the drive mechanism of the reservoir early, before a significant decline in pressure has taken place. Drive mechanisms will also be a critical factor in determining whether a well can have a successful workover or not. The characteristics, influential parameters, and appropriate analytical tools for each drive mechanism are discussed below and summarized in Table 11.3.7.

Table 11.3.7 Critical parameters for different drive mechanisms (controlling factors) and recommended techniques

Drive mechanism	Critical parameters	Analytical tool used
Expansion drive	OOIP, total compressibility, total permeability, pay thickness	Material balance and single-phase inflow performance equation
Solution gas drive	OOIP, critical gas saturation, gas-oil relative permeability ratio (particularly near S_{gc}), horizontal permeability, vertical permeability, pay thickness	Tracy Tarner material balance and multiphase phase inflow performance equation (Vogel type curves)
Gas cap drive	OOIP, OGIP (gas cap size), mobile oil volume, standoff distance (distance from gas cap to top of perforations), vertical distribution of OOIP, horizontal permeability, vertical permeability, thickness	Gas cap material balance and multiphase phase coning equations (Muskat, 1935; Schols, 1972)
Water drive	OOIP, size and permeability thickness of aquifer, mobile oil volume, mobility ratio, vertical distribution of OOIP, permeability distribution, standoff distance (distance from water leg to top of perforations), boundary conditions of aquifer; boundary conditions; lateral edge drive versus bottom water drive	Water drive oil material balance and multiphase phase coning equations (these references deal with near wellbore flow: Kuo, 1983; Lefkovits and Mathews, 1958; Muskat, 1935; Schols, 1972; Singhal, 1993; Sobocinski, 1965; Yang and Wattenbarger, 1991; these references deal with aquifer influence on reservoir: Hurst, 1960; Pletcher, 2000; Slider, 1983; Tehrani, 1992; Walsh, 1999)
Combination drive	Critical parameters depend on relative energy that each mechanism contributes	

Expansion Drive

Expansion drive is the most straightforward case because only single-phase flow occurs. Production history matches and forecasts can be made from coupled material balance and inflow equations or using decline analysis. Reservoir architecture is not usually an issue, except if the reservoir has permeability pockets. In this case, it may be necessary to model each pocket individually. Gas segregation is not a factor because there is no gas flow. The low total compressibility of the system is usually the limiting factor; therefore values of rock and fluid compressibility are important.

Characteristics of Expansion Drive

- the initial pressure is greater than bubble point pressure from a fluid study or an analysis of production data
- no GOC is observed and if WOC is present there is no evidence of water encroachment (small water drive index)
- the GOR is constant and equal to the initial solution gas-oil ratio
- the water cut is always less than 5%
- pressure decline is rapid
- the oil decline rate is exponential ($b = 0$, Section 10.4)
- the ultimate oil recovery is less than 5%
- the OOIP from an expansion drive material balance is in good agreement with the volumetric OOIP

Solution Gas Drive

Gas segregation strongly impacts solution gas drive performance. If the gas flow is segregated, the solution gas drive can be treated in the same manner as gas cap drive with segregated flow. If gas flow is non-segregated and the reservoir is homogeneous, a material balance coupled with inflow equations can be employed. This approach may also work for layered reservoirs as long as the highest permeability layer does not act as a conduit for evolving solution gas. Decline analysis can usually be applied for any reservoir architecture under solution gas drive with non-segregated gas flow. Reservoir simulation is usually required if the reservoir has permeability pockets.

Characteristics of Solution Gas Drive

- the pressure has dropped below the bubble point pressure
- no initial GOC is observed and if a WOC is present, there is no evidence of water encroachment (small water drive index)
- high GOR's develop in wells near the top of the reservoir structure
- the pool GOR rises rapidly and may reach a maximum at low reservoir pressures
- the water cut is always less than 5%
- the oil decline rate is hyperbolic ($0.2 < b < 0.4$)
- the ultimate oil recovery is between 5% and 20%
- the OOIP from a solution gas drive material balance is in good agreement with the volumetric OOIP

Gas Cap Drive

Gas cap drive can only be forecast with decline analysis or simulation with two exceptions.

1. If the gas flow is highly segregated, the GOR can be forecast to remain near the solution-gas-oil ratio until gas coning occurs. If there is a small overlap between the oil and the gas zone (such as tilted or some anticlinal reservoirs), gas coning can be delayed until late in the life of the reservoir. In this case, a material balance can be coupled with inflow equations to predict oil rates until gas breakthrough. This approach is usually only applicable to homogeneous reservoirs. After breakthrough decline analysis or coning models are required.
2. When there is a large overlap between the oil and gas zones (flat, dome, and some anticlinal reservoirs), coning usually dominates. If the reservoir is homogenous or layered, production can be forecast with coning models. In some cases, it may be necessary to model the contact movement as well.

Characterisitcs of Gas Cap Drive

- the initial pressure is equal to the bubble point pressure
- a GOC is observed on structural cross-sections or the reservoir structure extends above the uppermost well log data
- if a WOC is present, there is no evidence of water encroachment (small water drive index)
- high GOR's are observed in wells near the top of the reservoir structure
- the pool GOR rises continuously
- the water cut is always less than 5%
- the oil decline rate is hyperbolic ($b \approx 0.5$)
- the ultimate oil recovery is between 20% and 40%
- the OGIP from a gas cap drive material balance is in good agreement with the volumetric OGIP

Water Drive

The approach for water drive reservoirs depends on the strength of the water drive. For weak water drives, solution gas drive will play a significant role. The same techniques used for solution gas drive reservoirs are often applicable until water breakthrough. After water breakthrough coning models are appropriate for flat or dome reservoirs and decline analysis for tilted or anticlinal reservoirs. Simulation is often required for weak water drives.

For moderate or strong water drives, reservoir pressure will be constant. Usually there are three components of the problem: (1) the direction of flow bottom water drive and edge water drive; (2) the amount of pressure support given by the aquifer; and (3) the near-wellbore coning.

There are two main categories of strong water drives: bottom water drive and edge water drive. Bottom water drive applies to flat or dome reservoirs. Oil is displaced vertically and production is forecast using coning models. Edge water drive applies to tilted or anticlinal reservoirs or flat reservoirs undergoing a pattern waterflood. Oil is displaced horizontally. Production is forecast using displacement methods, decline analysis, and diagnostic plots. The choice of displacement method depends on the permeability distribution.

Characteristics of Water Drive

- the initial pressure is sometimes greater than the bubble point pressure
- the reservoir pressure declines slowly
- a WOC is observed on structural cross-sections or the reservoir structure extends below the lowermost well log data
- no GOC is observed
- high water cuts are observed in wells at the bottom of the reservoir structure
- the pool water cut rises continuously
- the pool GOR remains low
- the oil decline rate is hyperbolic ($b > 0.5$)
- the ultimate oil recovery is between 20% and 80%
- the material balance is dominated by water influx

Combination Drive

Combination drive reservoirs can usually only be forecast with decline analysis or simulation with the following exceptions:

1. Homogenous dome or anticlinal reservoirs with a large overlap between the water, oil, and gas zones can be forecast using coning models. It may be necessary to model the contact movement as well.
2. A material balance can be coupled with inflow equations for reservoirs where coning is delayed, the GOR remains low, and the water cut is zero. There must be little overlap between the water, oil, and gas zones (tilted or some anticlinal reservoirs). The reservoir must also be homogeneous and the gas flow must be segregated. This approach applies until water or gas breakthrough occurs.

Characteristics of Combination Drive

- the initial pressure is equal to the bubble point pressure
- the reservoir pressure declines slowly
- a WOC is observed on structural cross sections or the reservoir structure extends below the lowermost well log data
- a GOC is observed on structural cross sections or the reservoir structure extends above the uppermost well log data
- high water cuts are observed in wells at the bottom of the reservoir structure
- the pool water cut rises continuously
- high GORs are observed in wells near the top of the reservoir structure
- the pool GOR rises continuously
- the oil decline rate is hyperbolic ($b > 0.5$)
- the ultimate oil recovery is between 20% and 80%

11.3.4 Flow Characterization

Four main factors to consider in obtaining a consistent flow interpretation in the reservoir and near wellbore flow are: (1) the primary direction of fluid flow (horizontal or vertical) relative to the major reservoir heterogeieities; (2) the direction of pressure support (outer boundary conditions); (3) the uniformity of the flow; and (4) the extent of gas or water segregation. These factors were discussed in Chapter 1 and are reviewed briefly here.

1. Horizontal (flow along layers or bedding) flow usually involves radial or linear inflow and horizontal displacement of oil by gas or water. Vertical (flow across layers or bedding) fluid flow leads to coning of gas or water. The production profiles and required analytical methods are different for the two flow directions.
2. Gas caps, water legs, and injectors often give reservoir pressure support to producing wells. Estimation of the location and degree of communication is important to understand flow.
3. Uniform flow occurs in homogeneous reservoirs. Many analytical methods apply to uniform flow reservoirs. Nonuniform flow occurs in layered, pocketed, or fractured reservoirs. Nonuniform flow often results in poorer reservoir performance and sometimes requires simulation to model.
4. The extent of gas segregation has a significant impact on solution gas drive, gas cap, and combination drive reservoirs. If gas flow is segregated, the gas rises to form a secondary gas cap or to join an existing gas cap. Pressure support and oil recovery tend to be high. If gas flow is nonsegregated, the gas is produced with the oil. Pressure depletes more rapidly, and oil recovery is lower. Different analytical methods may be required for different types of gas flow.

11.3.5 Consistency Checks and Choice of Analytical Techniques

Once a preliminary classification is complete, it is highly recommended to check that the volumetric OOIP and OGIP are consistent with the material balance results and the ultimate recovery factors forecast from the production history. It is also recommended to compare the expected decline rates and recovery with the performance of analogous pools.

If the results are not consistent, then it is necessary to re-evaluate the data and the interpretation. The general approach is to assess the relative accuracy and completeness of all the data. The interpretation is adjusted to give greater weight to the data with the greatest confidence. This is an iterative process in which judgment is required and universal guidelines do not exist. Some examples are provided below to illustrate the concept.

1. A reservoir is well delineated with wells, but has limited pressure data. The volumetric OOIP and material balance OOIP do not agree. Here, a greater emphasis should be given to the volumetric OOIP.
2. The volumetric OOIP and OGIP of a reservoir are insufficient to account for the observed pressure support. Here, an increased OOIP, a gas cap, or an aquifer should be considered.
3. The predicted oil recovery factor is significantly lower than analogous reservoirs. In this case, the estimated OOIP may be too large, or the reservoir architecture may be limiting recovery. For example, there may be permeability contrasts that lead to high local GORs and pressure depletion, or the reservoir structure may lead to early coning.

Finally, model error must be considered. As mentioned in Chapter 1, errors in models are a function of three components (Poeter and Hill, 1997):

1. Errors in data caused by inaccuracy and imprecision by the measuring device or human error;
2. Error as a result of the model being a simplification of the physical system;
3. Error due to neglecting physical phenomena.

Usually in both the geological (static) model and in the flow simulation (dynamic) model, the second type of error occurs. The other problem can be that we are missing in our models some phenomena that partially control recovery and/or flow rates. In some cases, waxes, fines migration, chemical reactions, or fracture opening/closure may dramatically affect production profiles. Sometimes routine flow simulation neglects these effects and attributes poor performance incorrectly to reservoir effects.

11.4 Case Studies

Below are field examples that illustrate drive mechanisms and production diagnostics. Like most field examples, the data are noisy, but the authors intentionally did not filter the data. Field and well behaviors are controlled by multiple variables, so some of the scatter on plots is due to changes in operational procedure, and some is due to incorrect measurements. In the hundreds of fields that we have worked on, this type of scatter is typical.

Note that it is important to examine general trends of measure data such as oil, gas and water rates, and cumulatives, as well as GOR, water cuts, or resesvoir pressure. Individual measurement of GOR can vary significantly due to transient effects in the wellbore or near wellbore region. These effects add noise and scatter to the data; therefore, it is useful to look at cumulatives or averages or periods and look for general trends first. Start with the big picture, then move to the details after you have a good reconfirmation of drive mechanism, reservoir architecture, and reservoir flow mechanisms.

The first nine chapters of this book are mainly concerned with solving the direct problem; that is, if the components of the system are known, how does the system behave? When analyzing field production data, we are trying to solve the inverse problem; that is, if a system behaves in a certain way, what are the components of the system? Unfortunately, the inverse problem in a reservoir rarely has a unique solution, and therefore interpretation is subjective.

However we can infer much from examining the data systematically. Confirming OOIP, ROIP, drive mechanisms, continuity, and flow mechanism are products of analyzing production data. The examples below show various drive mechanism and associated production characteristics. They also show different reservoir architectures (permeability distributions) and their impact on production characteristics.

11.4.1 Solution Gas Drive Field Example: Pine Creek Cardium Pool

Figure 11.4.1 shows the well locations for the Pine Creek Cardium Pool, and Table 11.4.1 summarizes the reservoir properties. This reservoir is a thin sandstone (average thickness ~ 12 ft) with light oil (44.3 API) and very low initial water saturation. Pine Creek is one of several high-permeability sandstone reservoirs that are

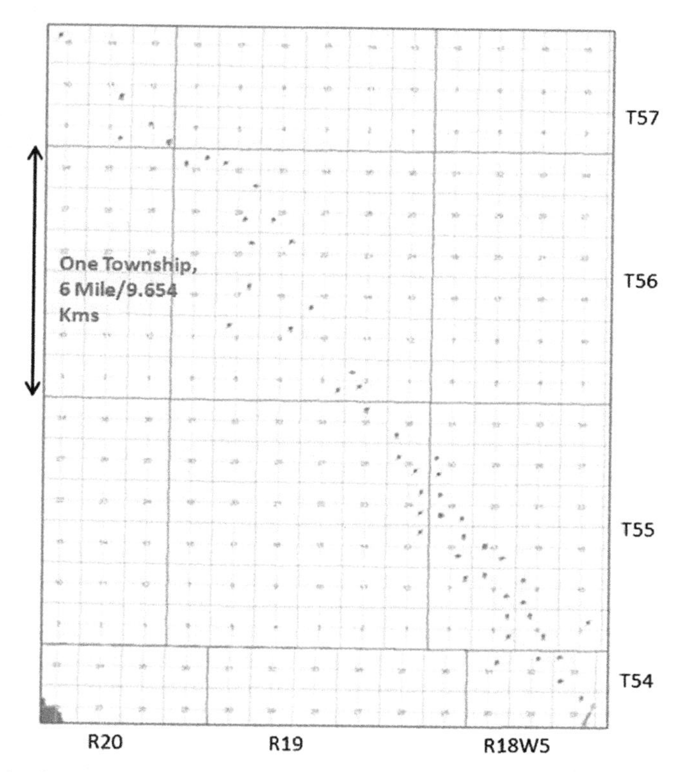

Figure 11.4.1 Pine Creek Cardium in Alberta, Canada. Grid lines are section lines and are one mile apart.

aligned parallel to an ancient shoreface in the Cardium Formation. The high oil gravity and formation volume factor (1.43 m^3/m^3) suggest a near-volatile oil. There was no evidence of a gas cap or water zone. The reservoir is orientated in an NW–SE direction. As of 2012, there were 25 active wells in the pool.

Figure 11.4.2 shows the field production characteristics of this pool. There was an infilling drilling program between 1995 and 2000 with obvious effects on the gas production, Figure 11.4.3. The water cuts remain low throughout the production history of the pool. It is not unusual to produce some water production at the late stages of a field, but this water production is not associated with a large aquifer.

The GOR is erratic, but trending upward, as shown in Figure 11.4.4. Pine Creek is one of several high-permeability sandstone reservoirs that are aligned parallel to an ancient shore face in the Cardium Formation. The GOR spiked from 1997 to 2000 as a result of increased gas production rather than reduced oil production, Figure 11.4.5. The spike in gas production was associated with the perforation of high structure wells. In 2000, the GOR decreased when the oil production rate increased from wells perforated lower in the structure. The GORs are indicative of a rise in free gas saturation and possibly formation and growth of secondary gas cap.

Table 11.4.1 Pine Creek Cardium (Alberta, Canada) properties (ERCB, 2013)

Property	Value
Area	4201 ha
Thickness (h)	3.71 m (12.1 ft)
Porosity (ϕ)	11%
Initial water saturation (S_{wi})	15%
API gravity	44.3°
Oil formation volume factor (B_{oi})	1.43
Solution Gas-Oil ratio (R_{si})	167 m^3/m^3
OOIP	10.2 10^6 m^3 (64.2 MMSTB)
SGIP	1.70 10^9 m^3 (60.1 BCF)
Datum elevation	893.8 m (2932 ft)
Initial pressure (P_i)	22,093 kPa (3204 psi)
Temperature	68 °C
Active well count	25 producers

The reservoir pressure was already depleted by 1997, Figure 11.4.6. As discussed in Section 11.3.3, as the GOR rises in these solution gas fields the pressure decline increases.

At early time, the oil rates are mainly a function of well lift capability and government regulation of maximum rates rather than a function of reservoir parameters. Figures 11.4.7–11.4.9 show a classic solution gas drive oil rate decline in which the oil rate has a hyperbolic decline characteristic or "double exponential" feature with steep initial decline, but a flattening of oil rate as pressure decreases and the compressibility of the system increases. Note the decline analysis must be performed before and after but not during the infill program. The decline in oil rate is smooth with later wells adding incremental oil. Decline analysis supports an Arp's exponent in the range of 0.2–0.3, but the decline signature pick is highly subjective, as shown in Figures 11.4.8 and 11.4.9. To reduce the subjectivity of the decline analysis, it is recommended that the calibration interval have constant well count and have if possible a smooth decline. Obviously picking a longer calibration interval will characterize the decline better. The infill drilling program in the late 1990s seems to have added substantial reserves (increase of 14%). Recovery factor has improved from 3.7% to 4.2% OOIP by the infill drilling program; however, low reservoir pressure becomes the limiting factor. By 2002, reservoir pressure is only about ~25% of original reservoir pressure at least for many of the wells (Figure 11.4.6).

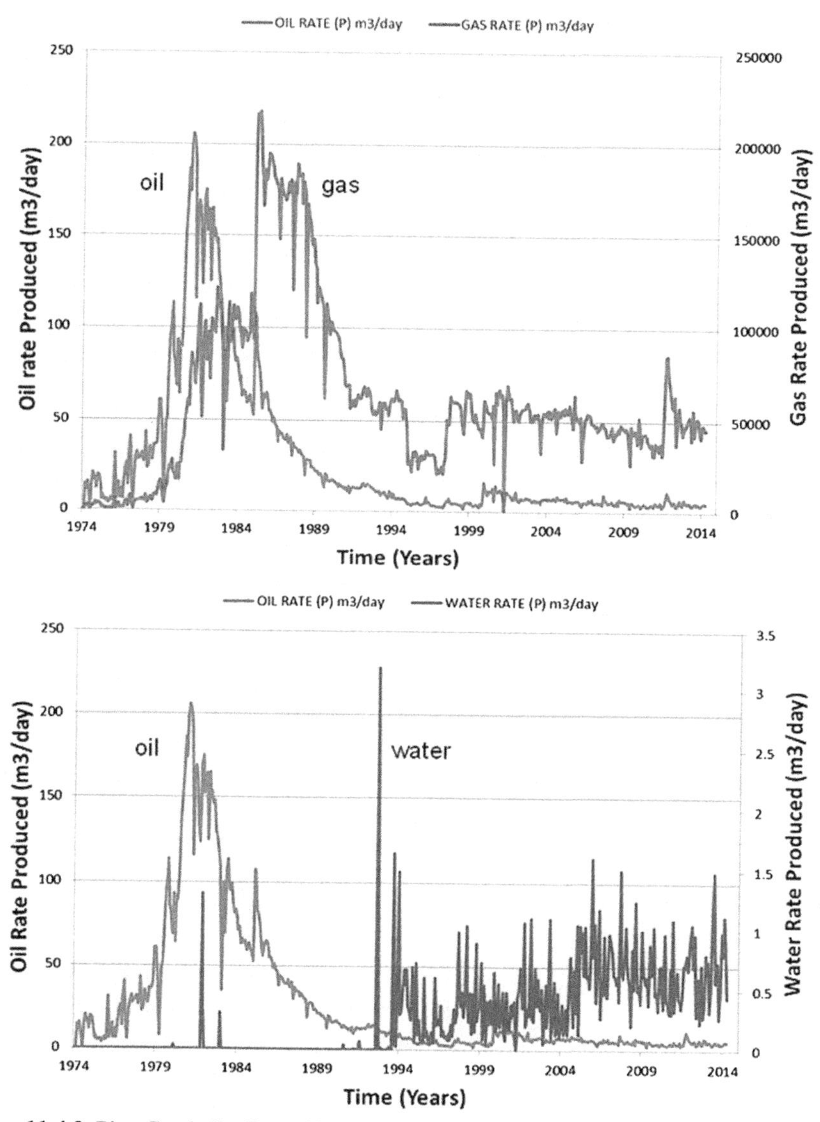

Figure 11.4.2 Pine Creek Cardium oil and gas rate profiles (top), and oil and water rate profiles (bottom).

In analyzing the GOR and the oil rate decline plots, as in many production data plots, there may be considerable scatter when analyzing the data. It is important to look for trends and disregard putting too much emphasis on single data point. Oil, gas, and water production rates are a function of many variables, including infill drilling, changing choke sizes, and workovers, as well as reservoir parameters. Usually

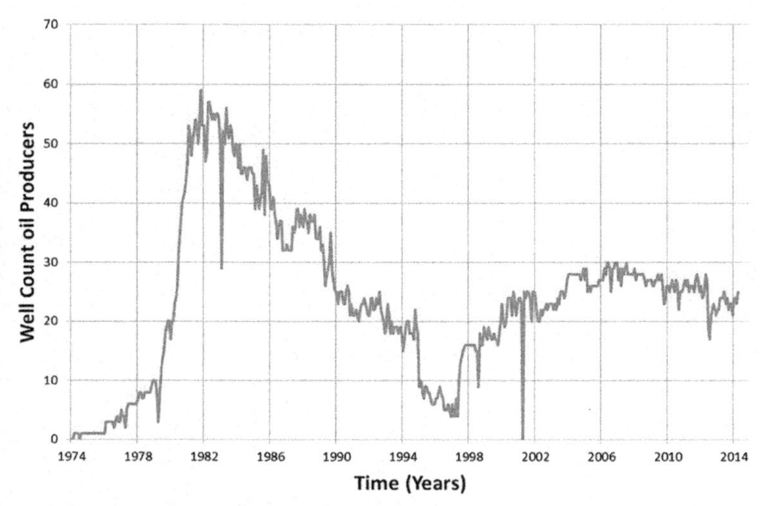

Figure 11.4.3 Pine Creek Cardium well count profile.

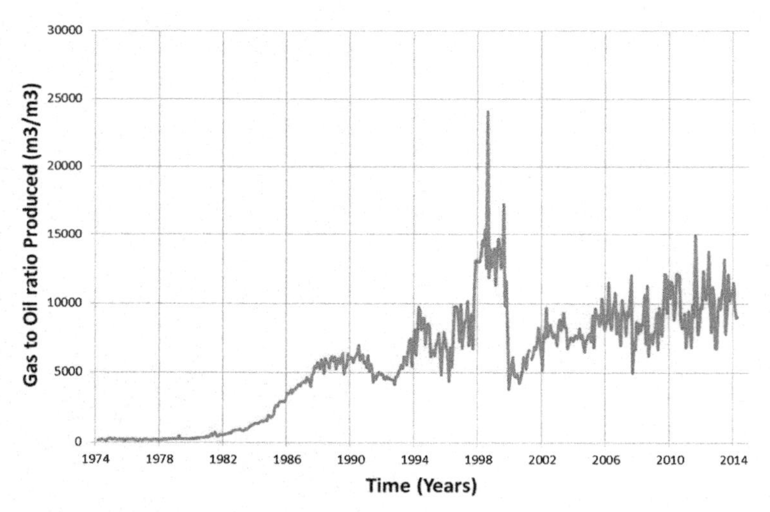

Figure 11.4.4 Pine Creek Cardium GOR profile.

reservoir phenomena such as decreasing pressure or increasing gas saturation will be revealed by long-term trends, but we also expect some scatter in the data.

Figures 11.4.10–11.4.13 show individual well performance. As stated above, at early time the oil rates are mainly a function of well lift capability and government regulation of maximum rates rather than a function of reservoir parameters. However, at late stages, most wells strongly track each other, especially during 1980–1988, indicating good reservoir continuity between the wells and sufficient permeability to allow fluids to move easily. The decline rates of many of the wells

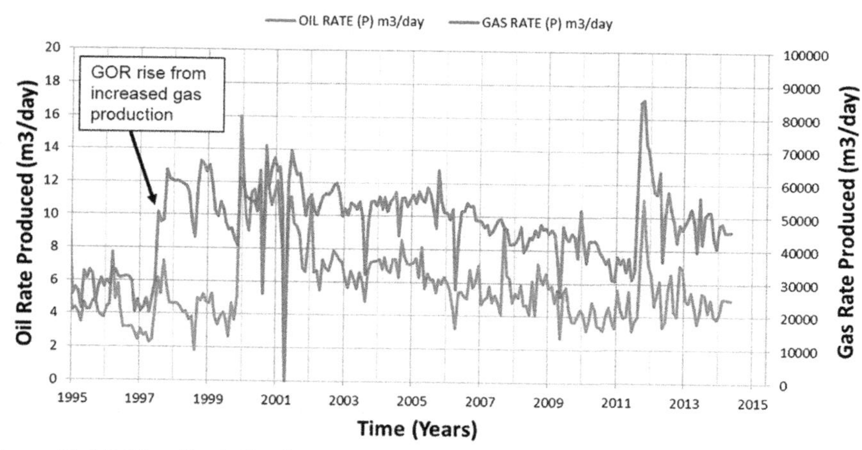

Figure 11.4.5 Pine Creek Cardium oil rate and gas rate profiles since 1995.

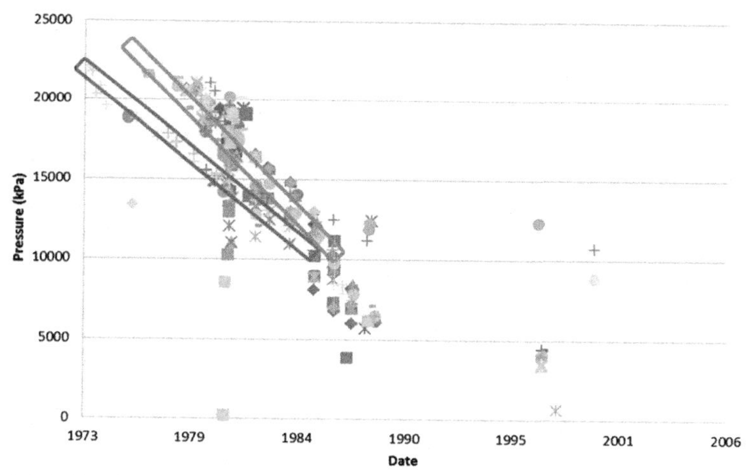

Figure 11.4.6 Pine Creek Cardium oil pressure profile (individual wells have different symbols; reservoir pressure versus date). The wells in the lower rectangle are the ones located to the NW of the field, while wells in the upper rectangles are located SE.

are similar in the late stages because reservoir pressure and formation of gas saturation control the decline rate. In small but reasonably high permeability fields such as Pine Creek, the pressure difference, (Figure 11.4.6), between wells is small and the gas saturation is similar; therefore we see similar oil rate decline rates for the wells that are well connected. On the other hand, a few wells show some differences in oil rate signatures and GORs over time. For example, the oil rate decline and GOR behavior of Well 01-02-57-20W5 both have different shapes compared to

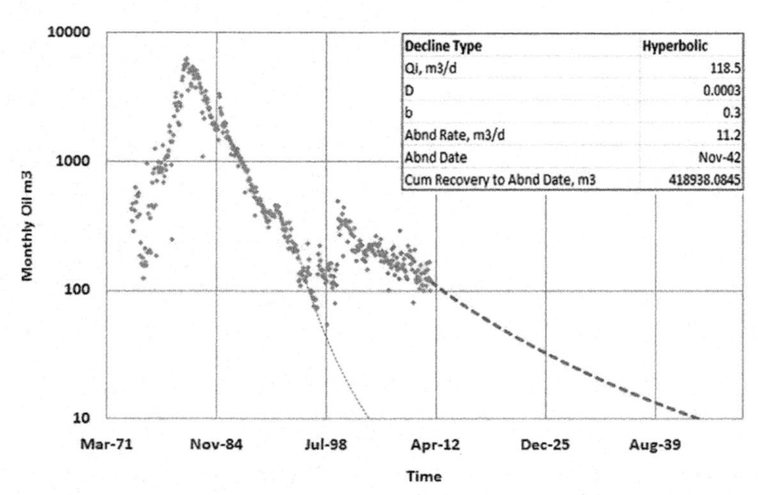

Figure 11.4.7 Pine Creek Cardium Pool oil rate decline on semi-log plot (hyperbolic decline with $b = 0.3$).

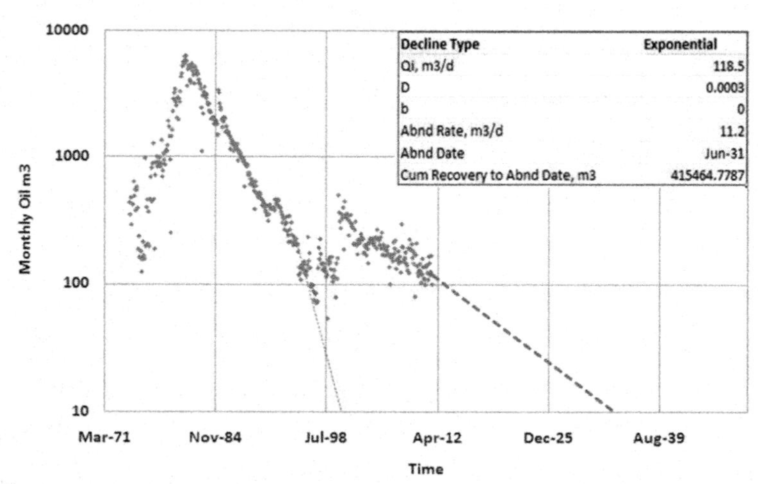

Figure 11.4.8 Pine Creek Cardium Pool oil rate decline on semi-log plot (exponential decline, $b = 0$).

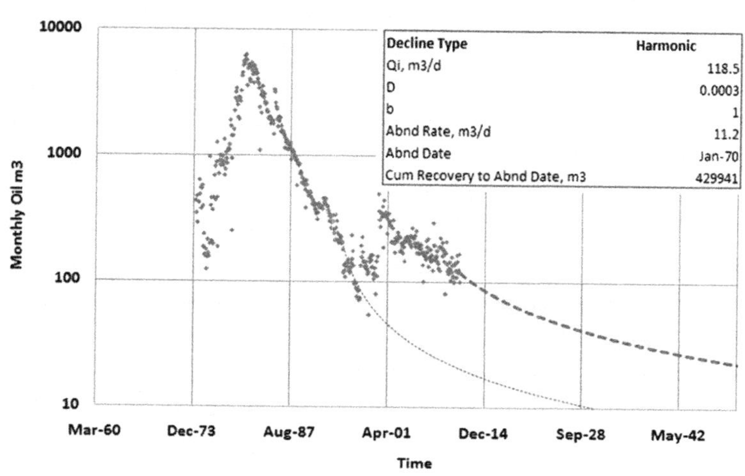

Figure 11.4.9 Pine Creek Cardium Pool oil rate decline on semi-log plot (harmonic decline, $b = 1$).

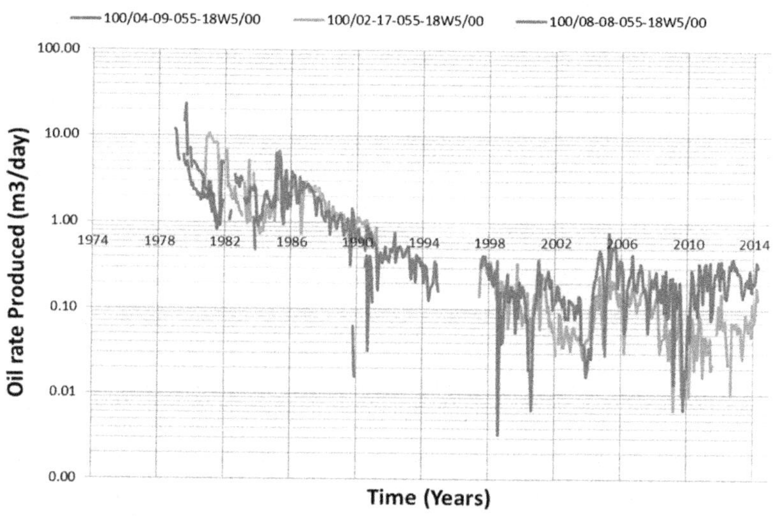

Figure 11.4.10 Pine Creek Cardium aggregate oil decline analysis (group 1).

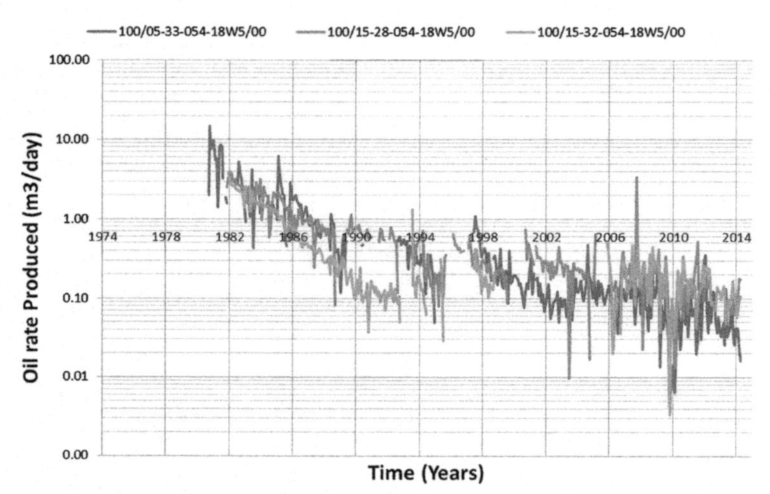

Figure 11.4.11 Pine Creek Cardium aggregate oil decline analysis (group 2).

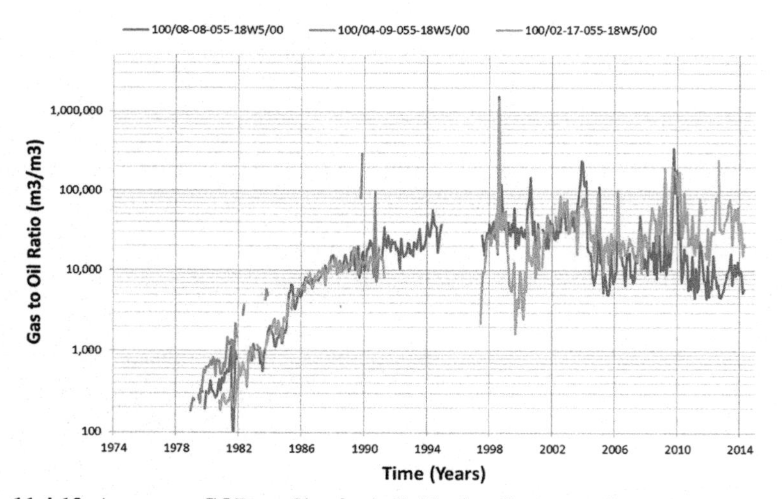

Figure 11.4.12 Aggregate GOR profiles for individual wells (group 1).

the other wells. There may be poor continuity and permeability between this well and the others.

Figure 11.4.6 is a good illustration of what we mean by similar trends, but it also shows noisy data as well as some differences in trends. In general, all the wells appear to have similar decreases in reservoir pressure. However, there is a small but consistent regional difference. The wells in the blue rectangle, located to the NW of the field, have lower pressures than the wells in the red rectangles located in the SE. Hence, the pressure trends show pool-wide communication with a regional deviation likely arising from the later development of the SE part of the pool.

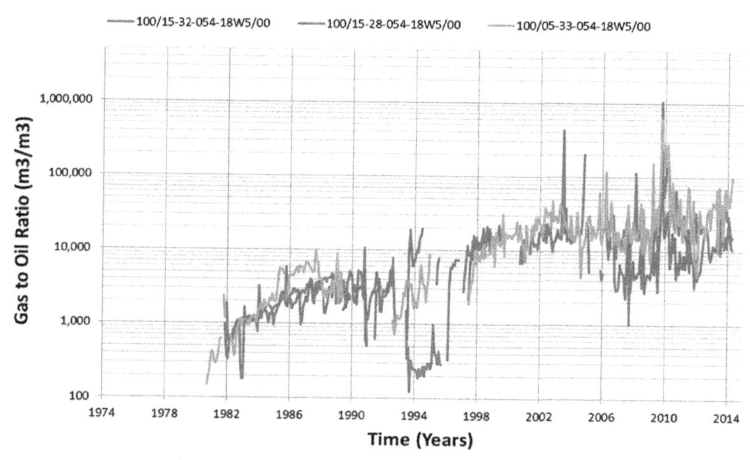

Figure 11.4.13 Aggregate GOR profiles for individual wells (group 2).

Figure 11.4.14(a) and (b) show oil bubble plots that show the field focused into areas. The reservoir itself is one of several following an NW to SE trend. Understanding the reason for this trend (that it follows a shoreline) makes for a useful exploration strategy of drilling along the trend to locate other reservoirs deposited under similar conditions.

Cumulative oil production plots show initially the NW portion of field has the highest oil production. But then a second lobe of high cumulative oil production, one in the NW and a second in the SE occur. The time lapse plots show two factors; one is the variability of rock quality as shown in Figure 11.4.14(a) and (b), and the effects of structure are shown in Figure 11.4.14(a) and (b). In general, the wells on the eastern edge of the pool have higher structure. Also, Figure 11.4.14(c) and (d) show cumulative gas-to-oil ratio bubble plots. The highest gas production occurs at the south end of the field, suggesting that a secondary gas cap formed here. Also note that the high GOR wells initially are also the highest structure wells.

We can use the trends in GOR as an indictor of reservoir connectivity as well as the degree of segregation and the development of secondary gas caps. The early-time bubble plots (1974 and 1979) indicate limited gas production, suggesting that no gas cap existed initially. Later-time plots (1989–2004), however, show significant gas production, particularly in the southeast. The bubble plots show that GORs are highest in the south end of the pool, and so we would expect a situation like Figure 11.4.15. On the other hand, the individual GOR aggregate plots, Figures 11.4.12 and 11.4.13, indicate that there is good communication in the reservoir.

The plot of cumulative gas versus cumulative oil plot, as shown Figure 11.4.16, is a useful plot for forecasting solution gas systems because the amount of produced gas is an indicator of the energy in the system. When the solution gas is depleted, there will be no further oil recovery. Figure 11.4.16 shows an asymptotic profile, with a maximum oil recovery under solution gas drive of approximately 400,000 m^3. This

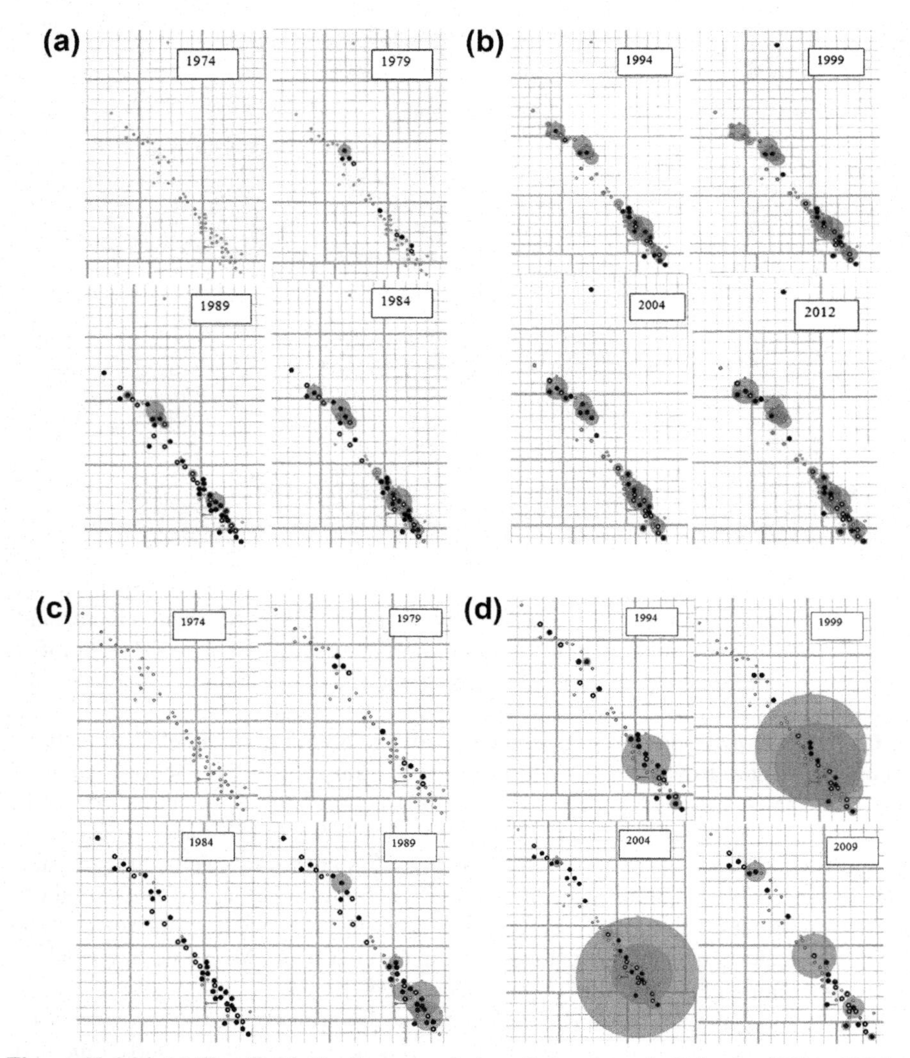

Figure 11.4.14 (a) Pine Creek Cardium cumulative oil time lapse bubble plot (1974–1984). (b) Pine Creek Cardium cumulative oil time lapse bubble plot (1994–2012). (c) Pine Creek Cardium cumulative gas-oil ratio time lapse bubble plot (1974–1994). (d) Pine Creek Cardium cumulative gas-oil ratio time lapse bubble plot (1994–2012).

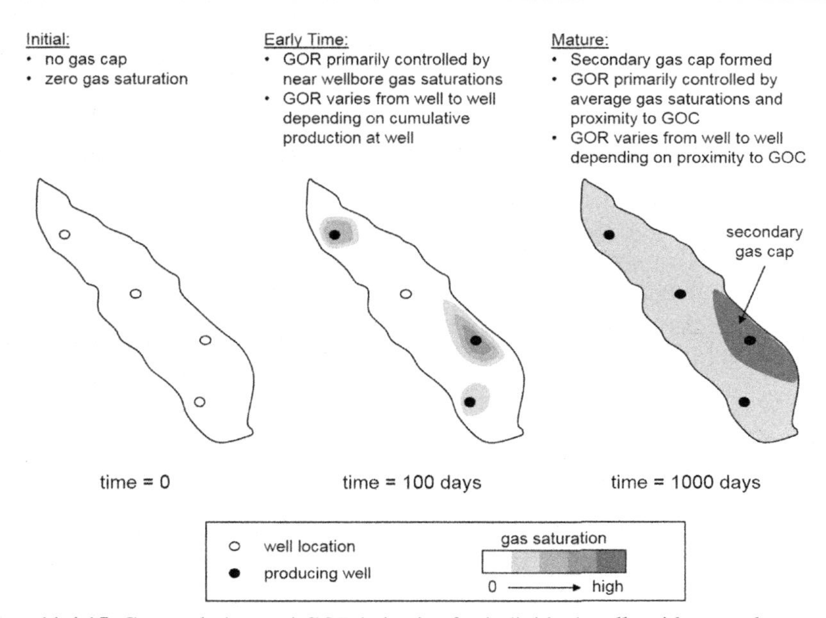

Figure 11.4.15 Gas evolution and GOR behavior for individual wells with secondary gas cap formation.

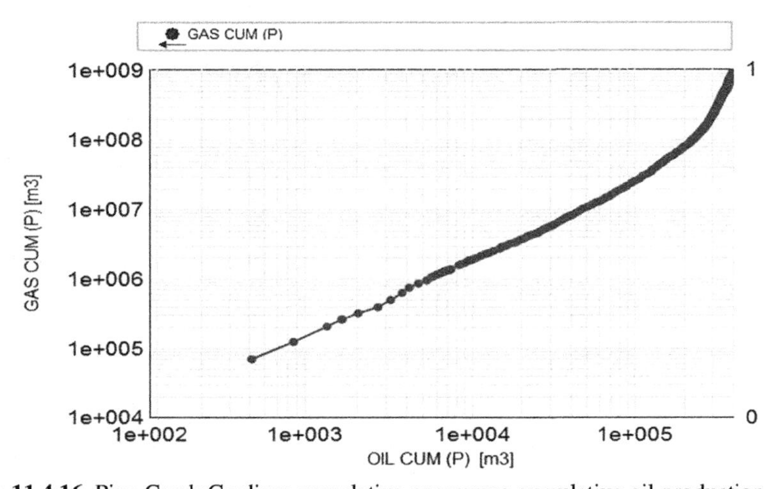

Figure 11.4.16 Pine Creek Cardium cumulative gas versus cumulative oil production.

value is similar to values obtained using conventional decline curve analysis, as shown in Figures 11.4.7 and 11.4.9.

11.4.2 Gas Cap Drive Field Example: Jumpbush Upper Mannville B Pool

Figure 11.4.17 shows the well locations for the Jumpbush Upper Mannville B Pool, and Table 11.4.2 summarizes the reservoir properties. The Jumpbush Upper Mannville B Pool is a small sandstone reservoir containing a light oil with a medium-sized gas cap. As of 2012, there was only one active well in the pool.

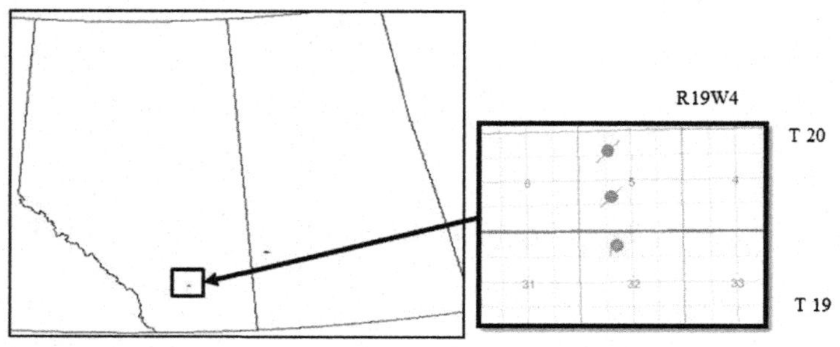

Figure 11.4.17 Jumpbush Upper Mannville B (Alberta, Canada) location map. Grid lines are section lines and are one mile apart.

Table 11.4.2 Jumpbush Upper Mannville B (Alberta, Canada) properties (ERCB, 2013)

Property	Value
Area	142 ha
Thickness (h)	2.5 m (8.2 ft)
Porosity (ϕ)	19%
Initial water saturation (S_{wi})	25%
API gravity	36°
Oil formation volume factor (B_{oi})	1.205
Solution gas-oil ratio (R_{si})	73 m^3/m^3
OOIP	420 10^3 m^3 (2.6 MMSTB)
SGIP	31.0 10^6 m^3 (1.1 BCF)
Initial pressure (P_i)	10,960 kPa (1590 psi)

Figures 11.4.18–11.4.21 show the oil production, liquid production, well count, and pressure history of this pool. As is typical for gas cap reservoirs, the water cut is near zero, and the GOR increased continuously with production. The GOR spiked dramatically at the end of the pool life as gas breakthrough occurred and the gas cap was blown down. The pressure declined steadily with production, with the decline rate accelerating late in the pool life when the GOR increased significantly.

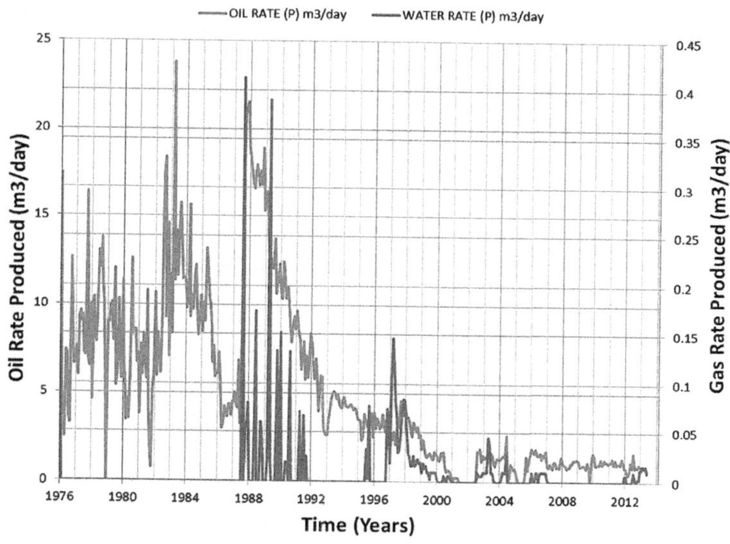

Figure 11.4.18 Jumpbush Upper Mannville B Pool oil and water rate profile.

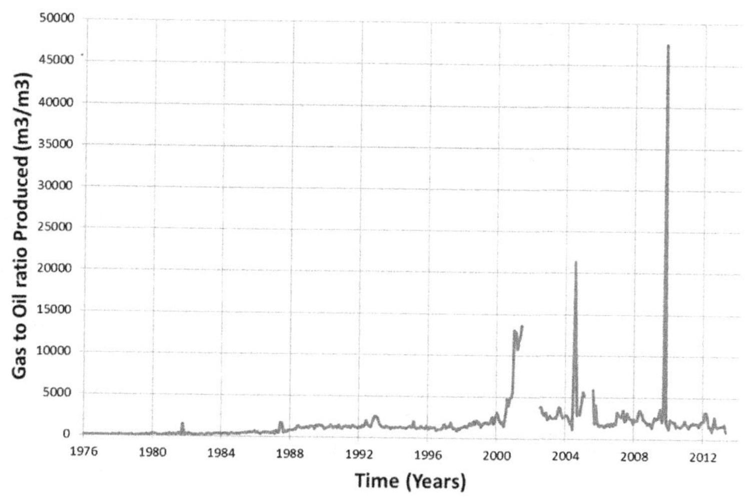

Figure 11.4.19 Jumpbush Upper Mannville B Pool GOR profile.

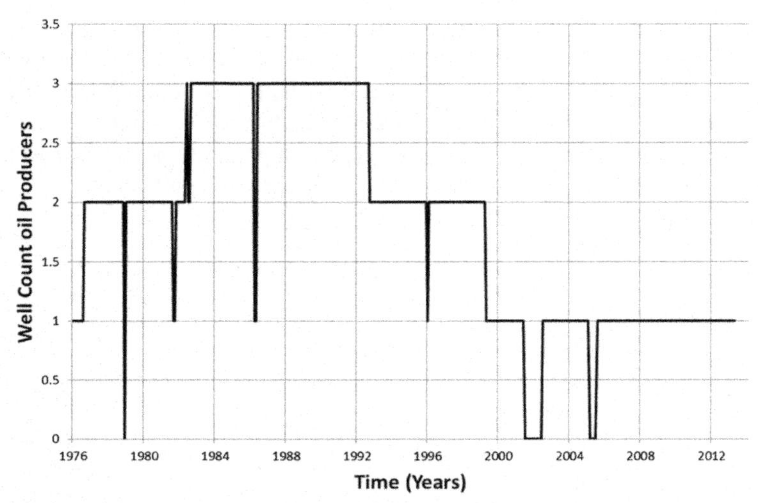

Figure 11.4.20 Jumpbush Upper Mannville B Pool field well count.

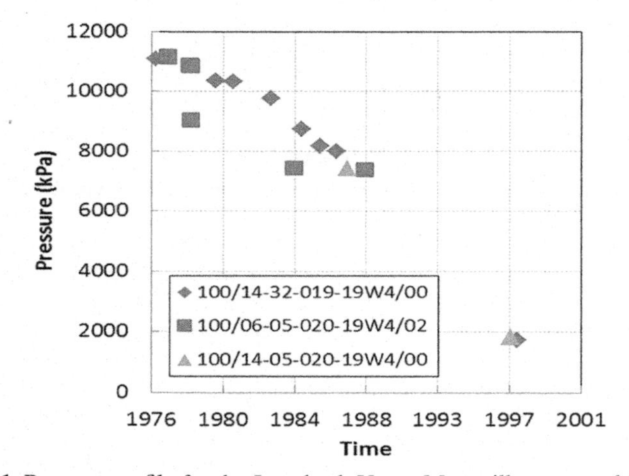

Figure 11.4.21 Pressure profile for the Jumpbush Upper Mannville gas cap drive reservoir.

Initially, oil rates seem to be constrained by government maximum rate regulations or facility controlled production. Note that there is constant oil production from 1977 to 1984, with a rise in oil production in 1985–1986 due to the addition of one well, as shown in Figure 11.4.22. Normally, we would expect if the wells were controlled by reservoir phenomena in a reservoir at bubble point and a gas cap with pressure depletion we should see noticeable oil decline rate with a thinning oil column. Figure 11.4.23 shows that there was reservoir pressure decline for the period of 1977–1986. The lack

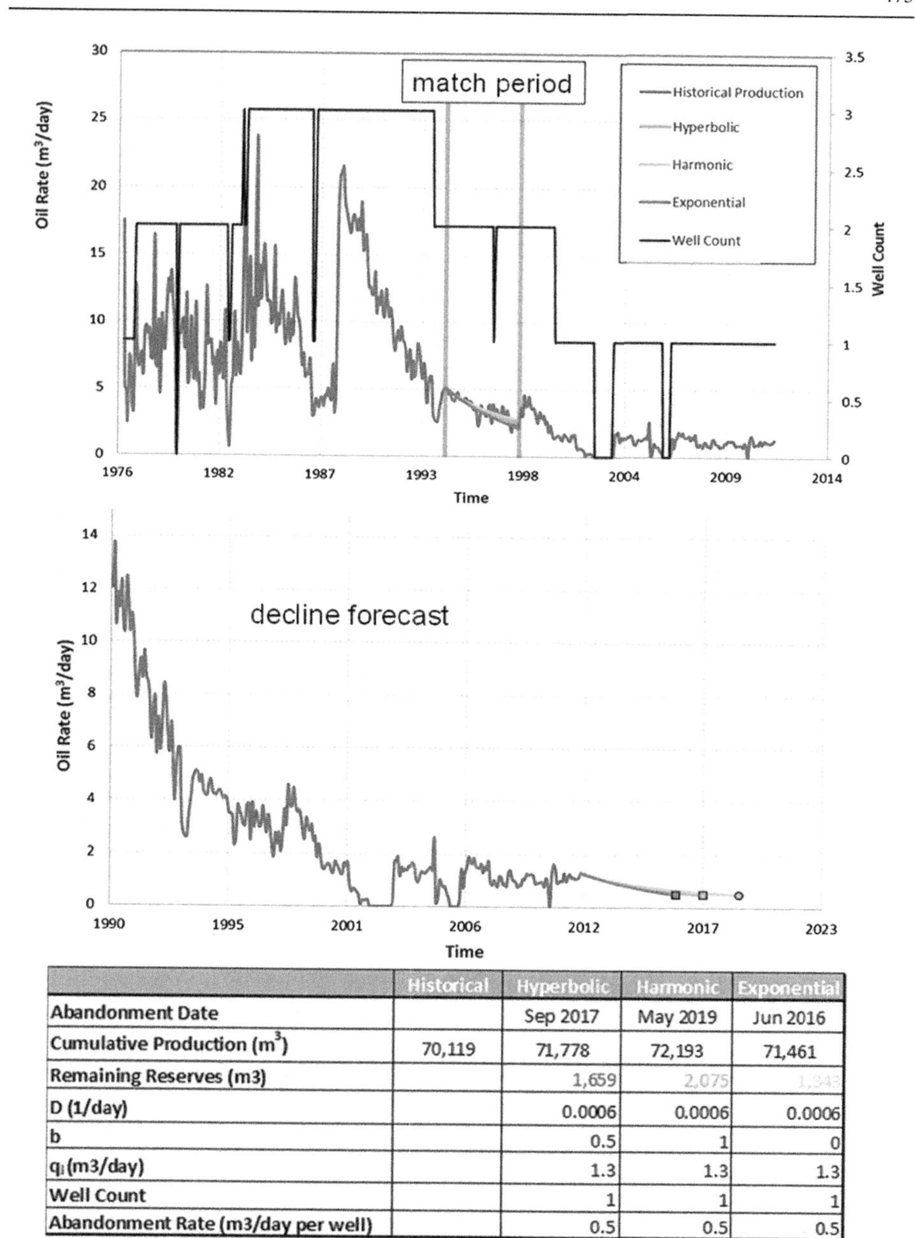

	Historical	Hyperbolic	Harmonic	Exponential
Abandonment Date		Sep 2017	May 2019	Jun 2016
Cumulative Production (m^3)	70,119	71,778	72,193	71,461
Remaining Reserves (m3)		1,659	2,075	1,343
D (1/day)		0.0006	0.0006	0.0006
b		0.5	1	0
q_i (m3/day)		1.3	1.3	1.3
Well Count		1	1	1
Abandonment Rate (m3/day per well)		0.5	0.5	0.5

Figure 11.4.22 Jumpbush Upper Mannville B oil rate decline: upper plot shows match period; lower plot shows decline forecast; table shows decline parameters.

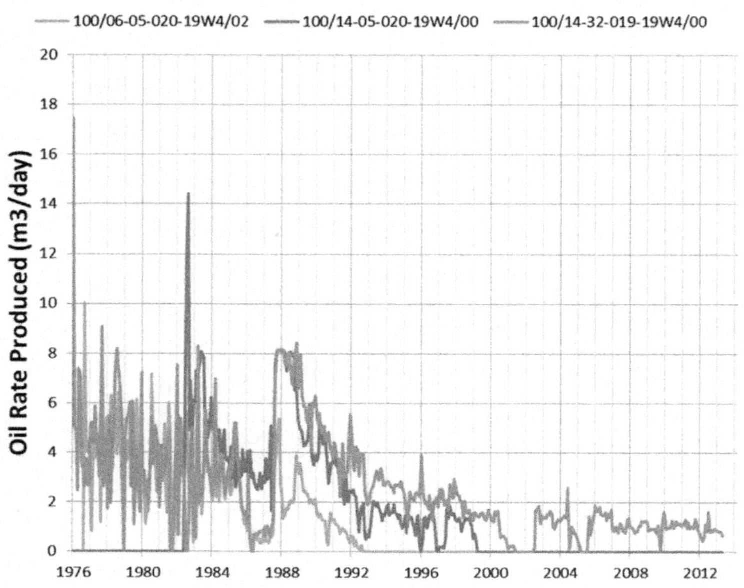

Figure 11.4.23 Jumpbush Mannville B Pool oil rate profiles for individual wells.

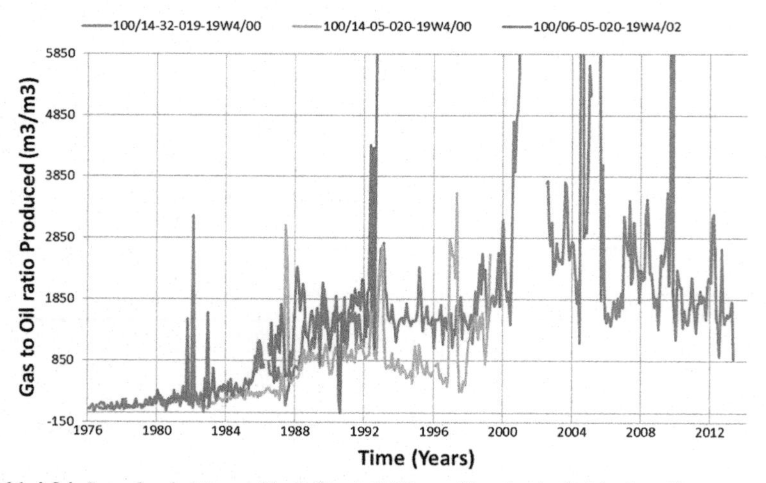

Figure 11.4.24 Jumpbush Mannville B Pool GOR profiles for individual wells.

of decline during this period is probably indicative of low production compared to reserve size, as a result of either government maximum rate regulations or facility controlled production. Many fields initially are controlled by the government until a certain point of time to understand the drive mechanism and reservoir architecture and their effect on recovery. After 1988, the oil wells appear to operate at maximum rates, and as the GOR increased and pressure declined rapidly, the oil rate declined steeply at a hyperbolical rate with an Arp's exponent of approximately 0.5, as shown in Figure 11.4.24.

Aggregate individual well analysis shows that, after the rate restrictions were removed, the oil decline and GOR signature of the wells were similar, Figures 11.4.23 and 11.4.24, respectively. Note that in 1989–1993 the oil decline rate on all wells was similar. The rise in GOR in all three wells in the same period, shown in Figure 11.4.24, confirms that there is good continuity and permeability within this reservoir.

In a gas cap reservoir with a relatively thin oil column like this field, oil decline rate is a function of two components: the development of the gas cone and the pressure of the reservoir. Because the reservoir is thin, the gas cap pressure and oil column pressure will be approximately equal, with a slight difference to fluid head. We see little difference in pressure between wells, as shown in Figure 11.4.22. We see a steep decline on this pool from 1988 to 1993, due to rapidly declining oil pressure and coning behavior both initally decreasing oil production. Decline rates at this point can vary from well to well due to differences in standoff or local permeability variation. However, in this pool, the decline rates are similar for all wells. At late stages, we expect that the oil decline rate of individual wells to be roughly similar in these types of fields because the decrease in reservoir pressure is the controlling factor at that time.

Figure 11.4.23 shows that the decline correlation period we used is between 1994 and 1998. This correlation period was selected because of its constant well count. From the pressure plot, Figure 11.4.22, we can see this is a period of rapid pressure decline. Thus in this period the decline rate should be a strong function of gas cap and oil leg reservoir pressure. GOR plots indicate that communication with the gas cap happened before this period. Examination of individual well performance support the grouping of the wells, as shown in Figures 11.4.23 and 11.4.24. Figure 11.4.23 shows that in the correlation period, oil decline rate is similar between the two remaining producing wells.

The ultimate oil recovery factor of approximately 17% is low for a gas cap reservoir, probably because the oil zone is only 2.5 m thick and because the gas cap overlies much of the oil zone. It is not usually possible to effectively sweep thin oil layers before significant gas breakthrough occurs.

11.4.3 Combination Drive (Predominately Gas Cap Drive) Field Example: Leduc D3 Reef

Figure 11.4.25 shows the well locations for the Leduc D3 Pool (Alberta, Canada), and Table 11.4.3 summarizes the reservoir properties. Leduc is a carbonate reef and has an average porosity of 10%. It contains fractured vuggy dolomite with excellent permeability, with permeabilities in the darcy range. The pay thickness (for the initial oil column) is ~11.6 m. It contains light oil with 40 API gravity oil, but also a large gas cap. Like many highly permeable reservoirs worldwide, it has a natural waterdrive plus reinjection of produced water. This pool is an example of a combination drive, with gas cap expansion being the major drive mechanism. There was a gas injection into the gas cap to recover liquids from the gas cap. This combination drive field is an excellent example of how drive mechanism, reservoir architecture, well workovers, well control, and blowdown strongly affect the production profile.

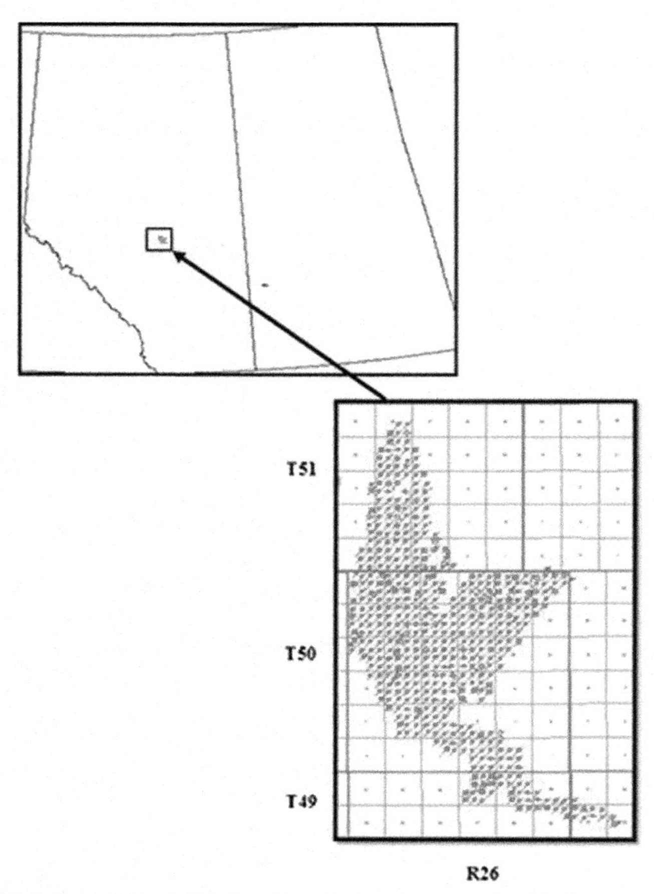

Figure 11.4.25 Leduc D3 Pool (Alberta, Canada) location map. Grid lines are section lines and are one mile apart.

The oil rates, GOR and watercut, gas rates, and well count profiles are shown in Figures 11.4.26–11.4.29. Water breakthrough occurred in 1970 and reached watercuts of 80% by 1980. Some government rate restrictions were removed in the 1973, and the oil rates declined continuously as the watercuts increased. There was also an aggressive workover program starting in 1970s. Therefore, before 1970, the oil rates and decline are not functions of reservoir depletion or changes in saturations.

After 1973, the wells were no longer on government maximum oil rate limitations for the pool, although there were still GOR restrictions on wells. There was a dramatic rise in oil rate in 1973. There is then consistent decline in oil rate from 1973 to 1990, as shown in Figure 11.4.26. The oil rates and oil decline at this time are functions of reservoir depletion and of changes in saturations and are due to the thinning of the oil column.

The pool was managed to minimize GORs until 1990, as shown Figure 11.4.27. Minimizing GORs from individual wells allows minimum gas production from the

Table 11.4.3 **Leduc D-3A Pool properties (ERCB, 2013)**

Property	Value
Area	8854 ha
Thickness (h)	9.91 m (32.5 ft)
Porosity (ϕ)	10.8%
Initial water saturation (S_{wi})	14%
API gravity	40°
Oil formation volume factor (B_{oi})	1.33
Solution gas-oil ratio (R_{si})	98 m^3/m^3
OOIP	61.2 10^6 m^3 (385 MMSTB)
SGIP	6.0 10^9 m^3 (212 BCF)
Datum elevation	915.5 m (3003 ft)
Initial pressure (P_i)	13,500 kPa (1960 psi)
Active well count	27 producers, 2 injectors

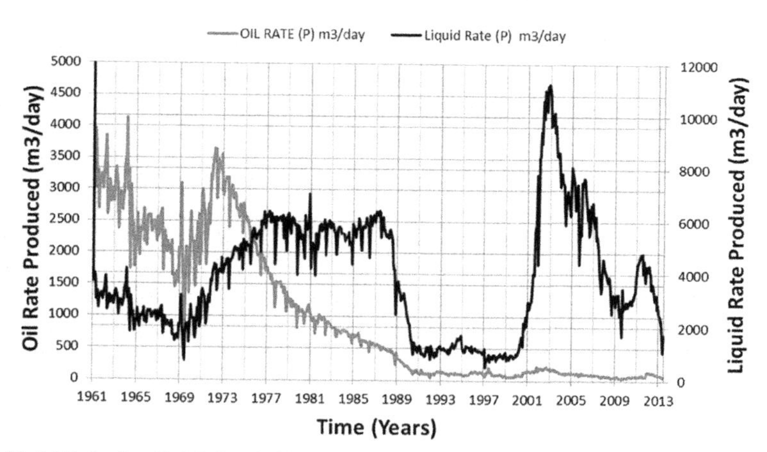

Figure 11.4.26 Leduc D-3A Pool oil and liquid rate profiles.

gas cap and therefore maximizes reservoir pressure and oil recovery. Figure 11.4.27 shows a spike in GOR associated with gas cap blowdown. The large spike in gas production from 1990 to 1996 was the blowdown of the gas cap (blowdown in Canada is regulated by the government). The pressure profile shows a small and slow decline until blowdown occurs in 1990, Figure 11.4.30. Note how the reservoir pressure is

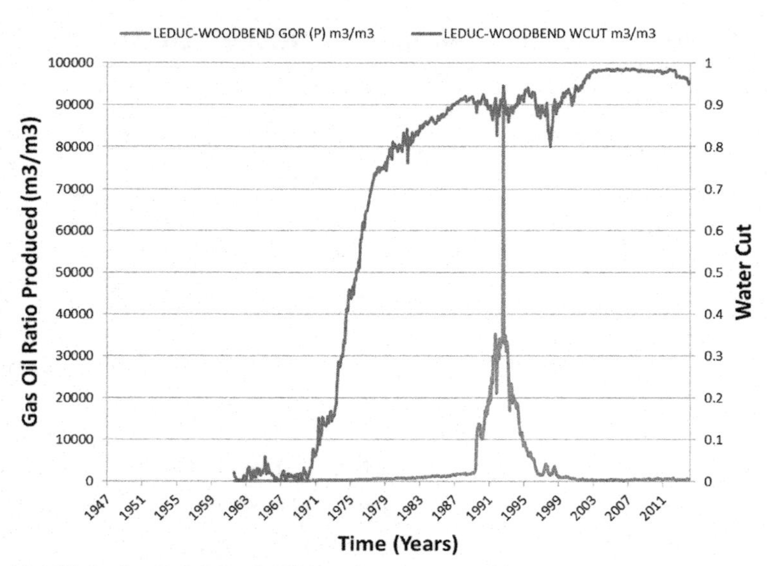

Figure 11.4.27 Leduc D-3A Pool GOR and watercut profiles.

uniform across the reservoir until blowdown, consistent with a high permeability reservoir and an extensive gas cap. However, note there is scatter in the measured pressure of certain wells. We believe that the data scatter is a reflection of the dynamic well flowing pressure and is not indicative of static reservoir pressure.

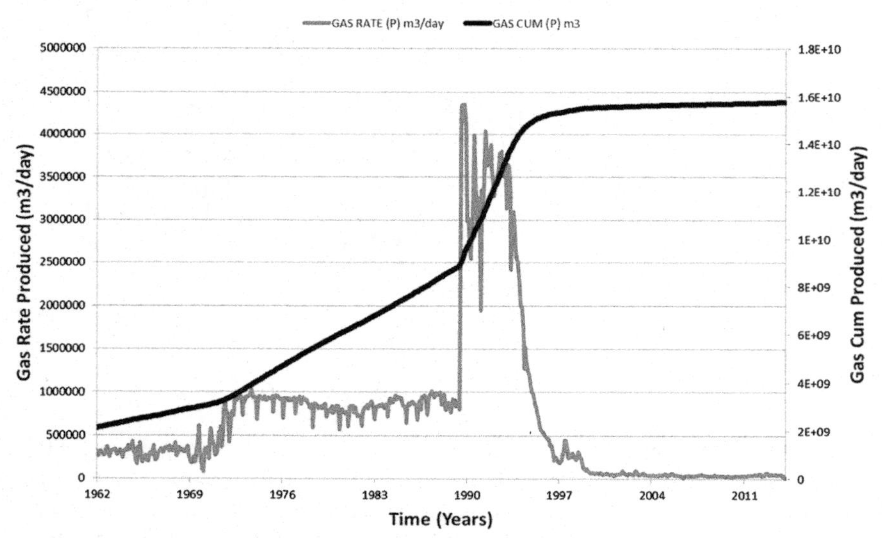

Figure 11.4.28 Leduc D-3A Pool gas rate and cumulative gas profiles.

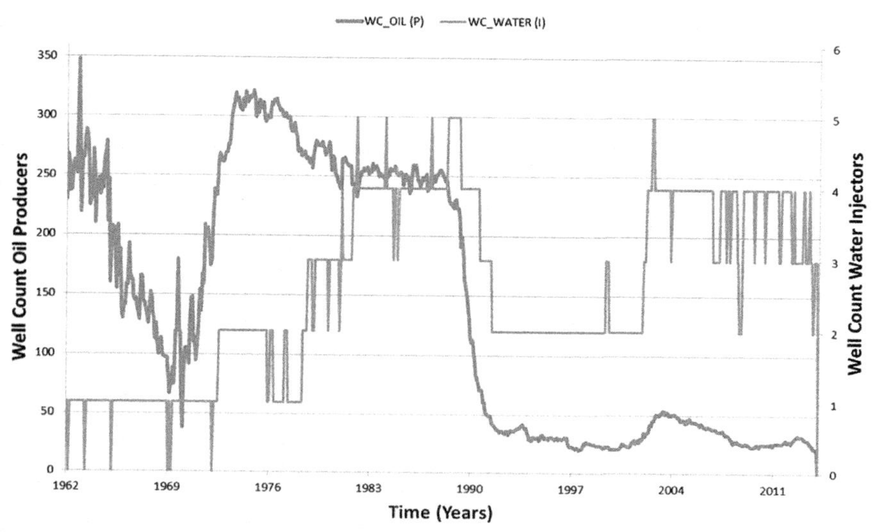

Figure 11.4.29 Leduc D-3A Pool injection (right axis) and production (left axis) well counts.

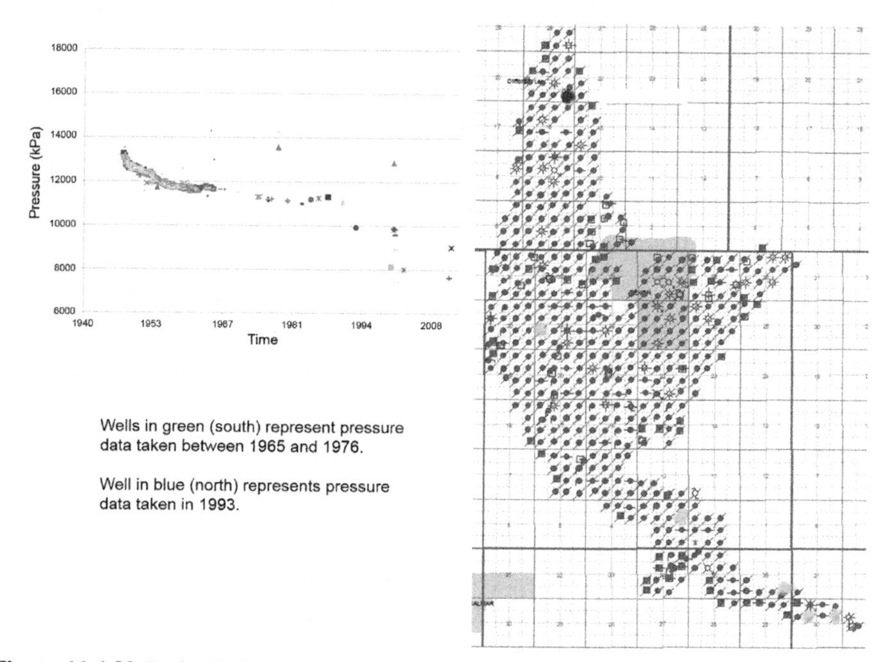

Figure 11.4.30 Leduc D-3A Pool oil pressure profile (individual wells have different symbols).

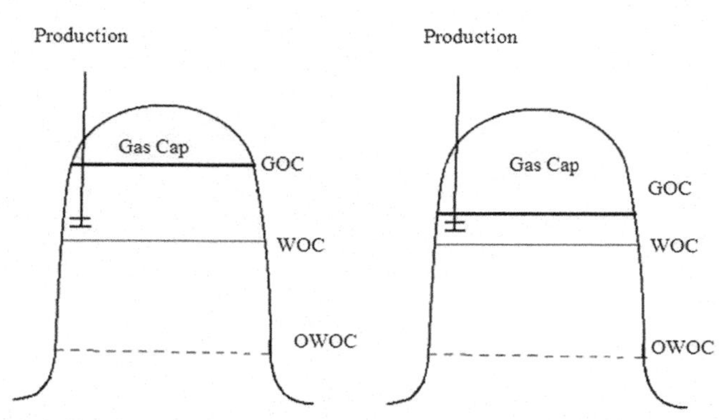

Figure 11.4.31 Schematic showing how GOR was managing by controlling the location of contacts relative to the perforations. OWOC is the original (preproduction) water-oil contact. Note in most pools the perforations are relocated as the contacts move. In the Leduc, the operator adjusted the withdrawal and injection rates and shut-in wells (for GOR and watercut control) to keep the oil zone centered on the completion intervals.

WOC and GOC were measured, and they show good consistency between wells, thus confirming good permeability. This consistency is typical of fractured and vuggy carbonate reservoirs. The oil column "sandwich" was managed well. The oil column started at 11.6 m and decreased to an estimated thickness of only a 1.5 m.

In 2002, oil rates increased as a result of a combination of a workovers, infill drilling, and an increase in fluid handling capacity. Oil rates began to decline again by 2004. As of 2012, there were 50 active producers in the pool. The productivity of the pool is then a function of the thinness of the oil column or "sandwich" thickness, as shown on Figure 11.4.31. The overall recovery factor as of November 2012 is high, at 64.1% (cumulative oil to date of $40\,10^3\text{m}^3$ from an OOIP of $61.2\,10^6\,\text{m}^3$). The excellent recovery factor is a function of the high reservoir permeability, the high displacement efficiency of gravity drainage, and the moderately thick oil column.

Figure 11.4.38 shows that water injection in this pool is dispersed but, because the water in injected into the water zone, and the permeability is high, the static reservoir pressure and the contact level remain uniform. The field oil production profile is smooth and has an exponential to hyperbolic ($b = 0.5$) decline signature after 1973 and before 1990, as shown in Figure 11.4.32. The main decline in oil well production occurs in the late 1970s. Similarly, the GOR is low for all wells until the 1980s, when the oil column thins. Oil rate signatures were affected by blowdown gas. Note that the change in flow mechanism and rapid pressure depletion caused by blowdown dramatically changes the decline exponent and the nature of oil decline. During 1990 to 1991, a large number of wells were shut in, as shown in Figure 11.4.29. Total fluid rates and oil rates dropped off in this period. After 1990, using decline analysis to forecast production is suspect.

We can see from Figure 11.4.33 that the oil production varies considerably from well to well, mainly depending on the well's location in the structure (distance of

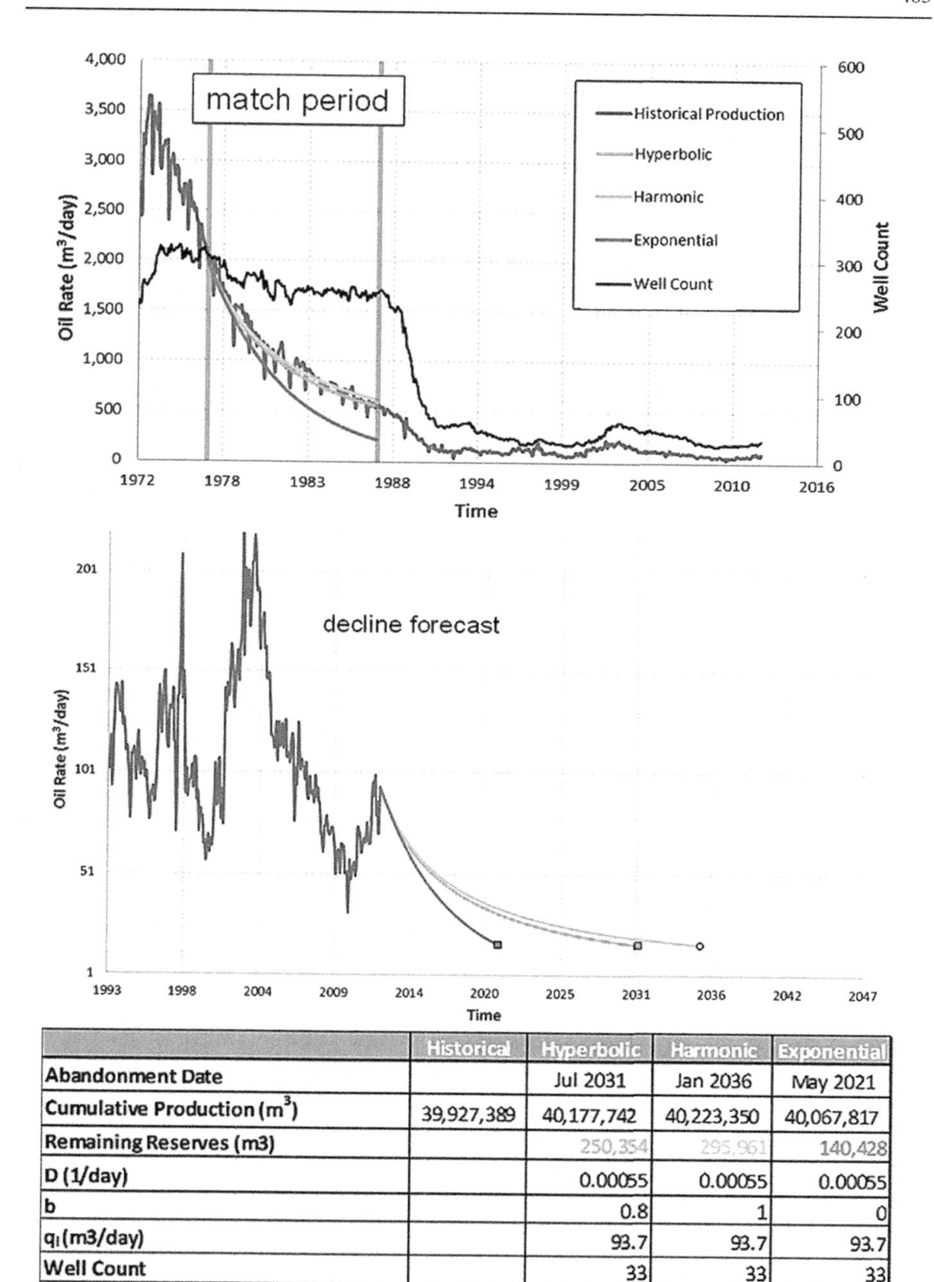

	Historical	Hyperbolic	Harmonic	Exponential
Abandonment Date		Jul 2031	Jan 2036	May 2021
Cumulative Production (m^3)	39,927,389	40,177,742	40,223,350	40,067,817
Remaining Reserves (m3)		250,354	295,961	140,428
D (1/day)		0.00055	0.00055	0.00055
b		0.8	1	0
q_i (m3/day)		93.7	93.7	93.7
Well Count		33	33	33
Abandonment Rate (m3/day per well)		0.5	0.5	0.5

Figure 11.4.32 Leduc D-3A Pool oil rate decline with match period before gas cap blowdown.

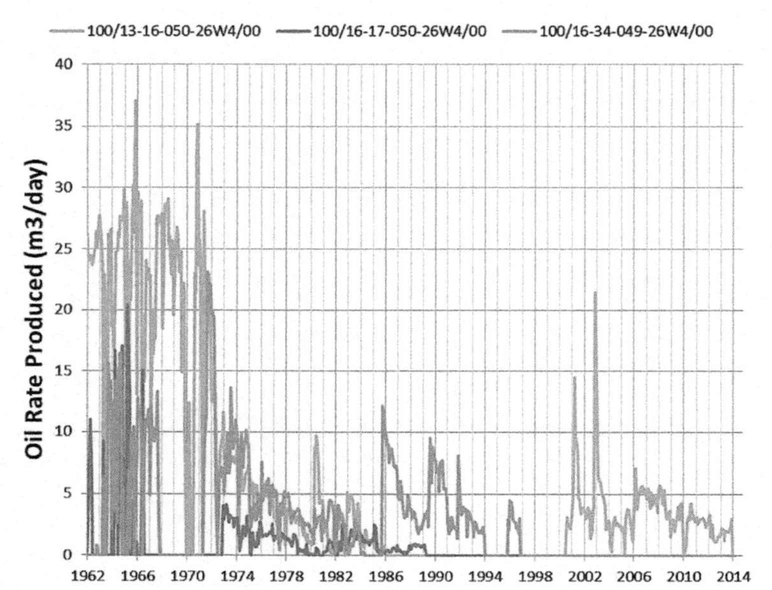

Figure 11.4.33 Leduc D-3A Pool aggregate oil rate profile for individual wells, Cartesian coordinates.

perforations from contacts). At the late stages, the location of the active oil producers was sporadic. For pools like the Leduc D3A in the late stages of field life after pressure blowdown, the wells now have a thinner oil sandwich (column) and a rapidly changing interface, so oil well productivity is sporadic and GORs are erratic.

11.4.4 Water Drive Bottom Water Drive Field Example: Provost Dina N Pool

Figure 11.4.35 shows the well locations for the Provost Dina N Pool, and Table 11.4.4 summarizes the reservoir properties. The Provost Dina N Pool is a sandstone reservoir with 6.5 m of pay containing a medium-heavy oil located above an aquifer. The pool is under a moderate to strong bottom water drive, supplemented by produced water injection.

Figures 11.4.36–11.4.40 show the oil rates, water injection rates, GOR and water cut, gas production rates, well count, and pressure profiles. The pool was put on production in 1984, and new wells were added until 1996. As of 2012, there were 150 active wells in the pool (Figure 11.4.40). Water breakthrough occurred rapidly, and GORs remained low and relatively stable (Figure 11.4.38). The pressure declined with production, but was still at approximately 75% of the original pressure at an

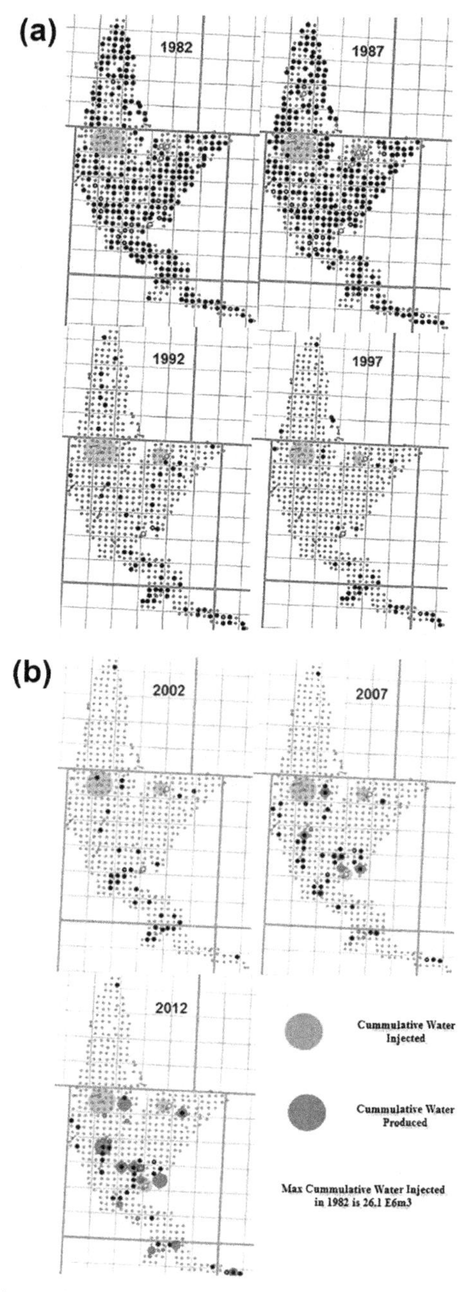

Figure 11.4.34 (a) Leduc D-3A bubble plots of yearly injection volumes from 1982 to 1997 (scale: water injected to water produced: 10:1). (b) Leduc D-3A bubble plots of yearly injection volumes from 2002 to 2012 (scale: water injected to water produced: 10:1).

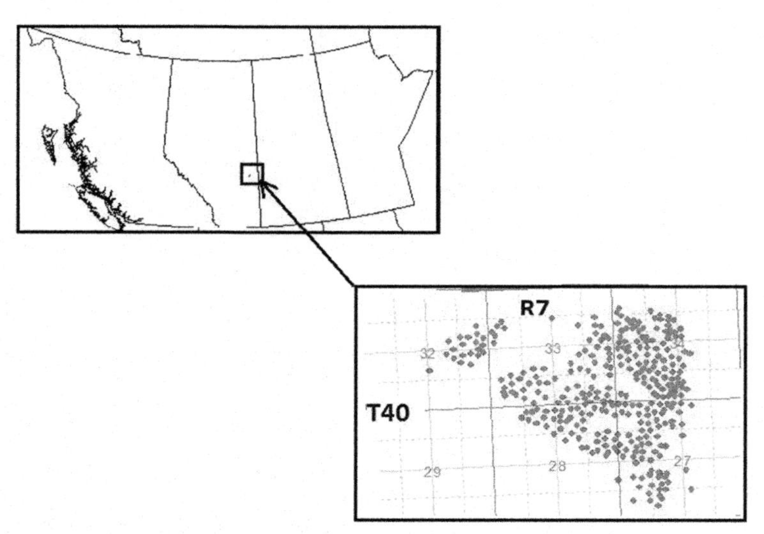

Figure 11.4.35 Provost Dina N Pool location map.

Table 11.4.4 **Provost Dina N Pool properties (ERCB 2013)**

Property	Value
Area	517 ha
Thickness (h)	6.45 m (21.1 ft)
Porosity (ϕ)	29%
Initial water saturation (S_{wi})	14%
API gravity	20°
Oil formation volume factor (B_{oi})	1.03
Solution gas-oil ratio (R_{si})	10 m^3/m^3
OOIP	8.07 10^6 m^3 (50.8 MMSTB)
SGIP	80.7 10^6 m^3 (2.8 BCF)
Initial pressure (P_i)	5969 kPa (866 psi)

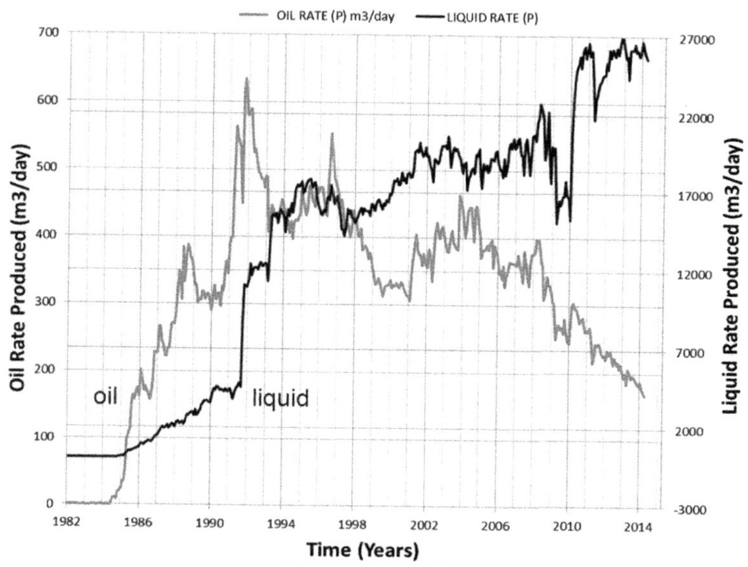

Figure 11.4.36 Provost Dina N Pool oil and liquid rate profiles.

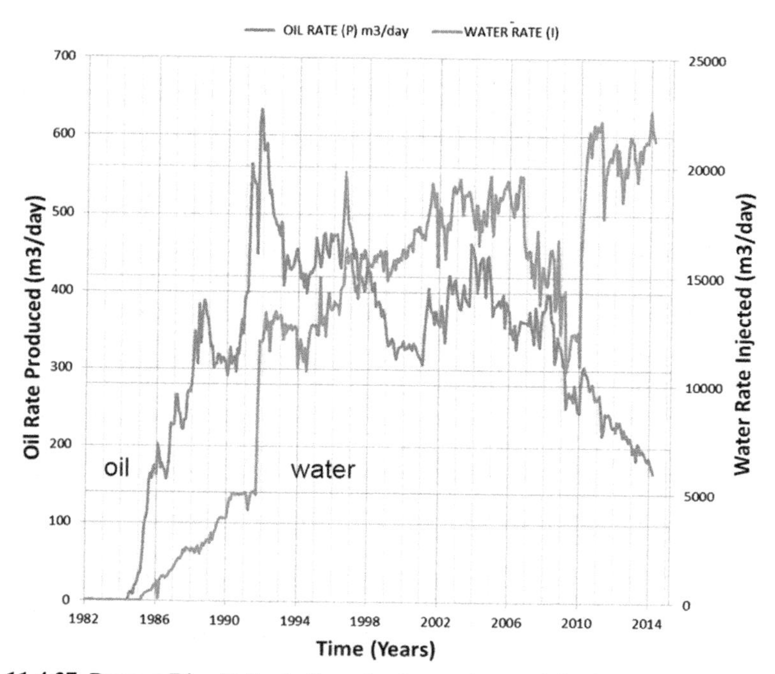

Figure 11.4.37 Provost Dina N Pool oil production and water injection rate profiles.

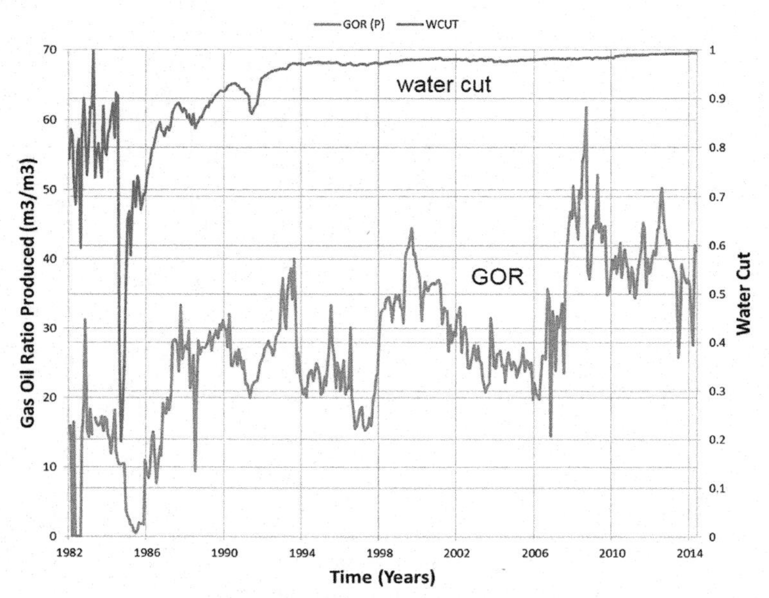

Figure 11.4.38 Provost Dina N Pool GOR and water cut profiles.

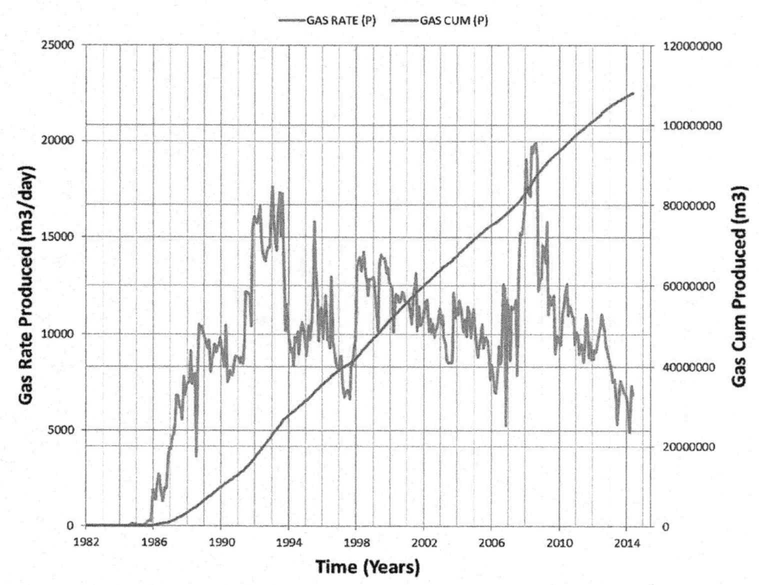

Figure 11.4.39 Provost Dina N Pool gas rate and cumulative gas production profiles.

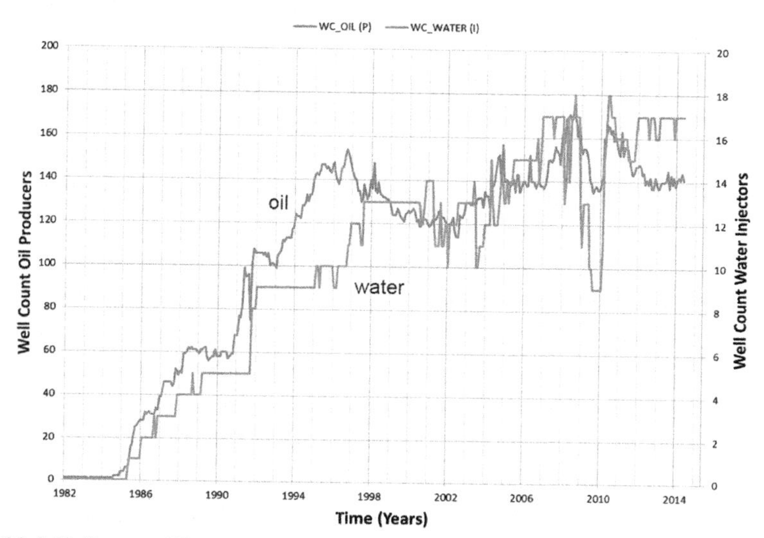

Figure 11.4.40 Provost Dina N Pool well counts.

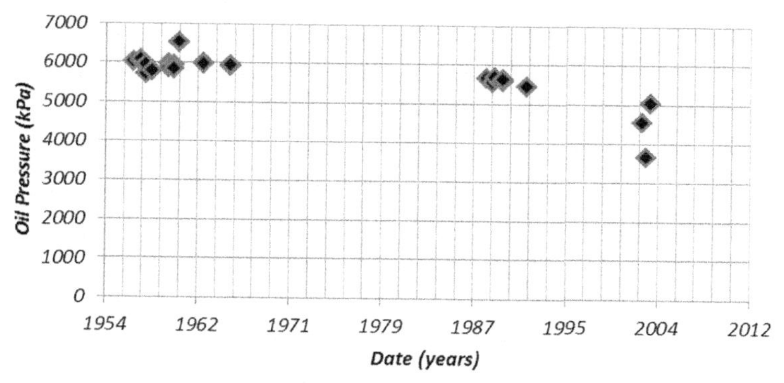

Figure 11.4.41 Provost Dina N Pool oil pressure profile.

oil recovery factor of 32% (Figure 11.4.41). These profiles are characteristic of a moderate to strong water drive. Recovery factor for these types of reservoirs (medium-heavy oil pool under a moderate to strong bottom water drive) are heavily dependent on economic water cut limitations both from a facility and from a well point of view.

The oil rate declined gradually to approximately 50% of the maximum rate (Figure 11.4.42). In this case, the type of pool (field) early decline is not a useful diagnostic because breakthrough occurs at different times in different wells. In cases in

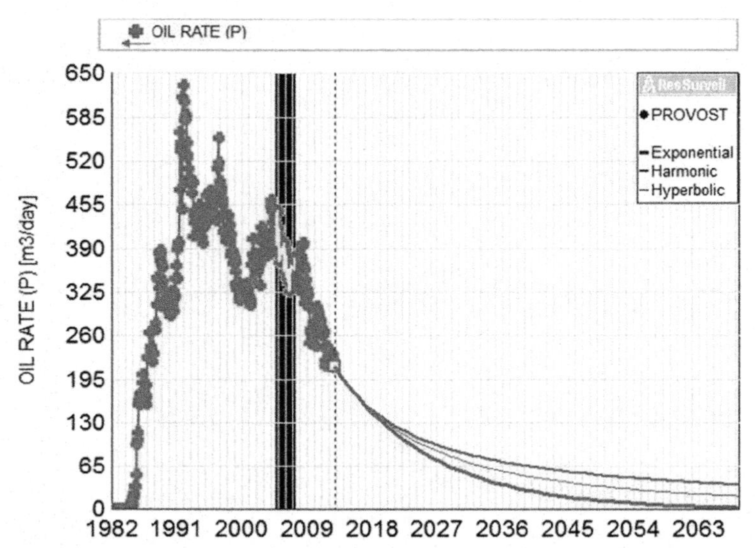

Figure 11.4.42 Provost Dina N Pool oil rate decline. Yellow (white in print versions) highlighted points show dates ~2007–2008 were used in the decline analysis.

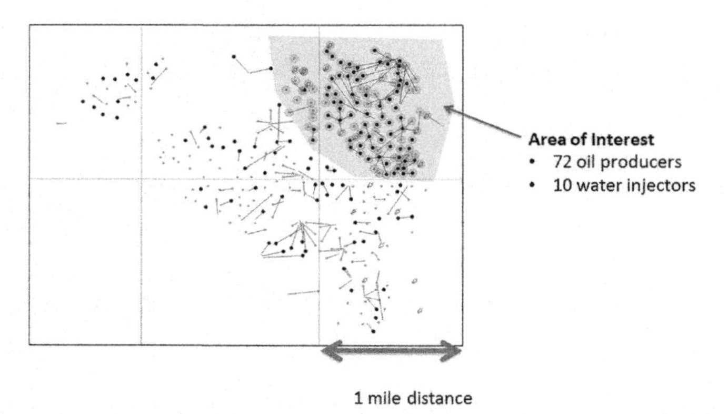

Figure 11.4.43 Provost Dina N Pool map showing location of injectors and producers in area of interest (part of pool shown in Figure 11.4.35).

which the water breakthrough is more consistent, a pool decline analysis can be useful. An Arp's constant greater than 0.5 is a partial indicator of a strong water drive. Sometimes, coning models, or possibly the summation of individual well declines, is required to forecast future production.

Figure 11.4.43 shows that most of the water injection is at the edges of the pool and most of the water production comes from the north and south areas. The current injector

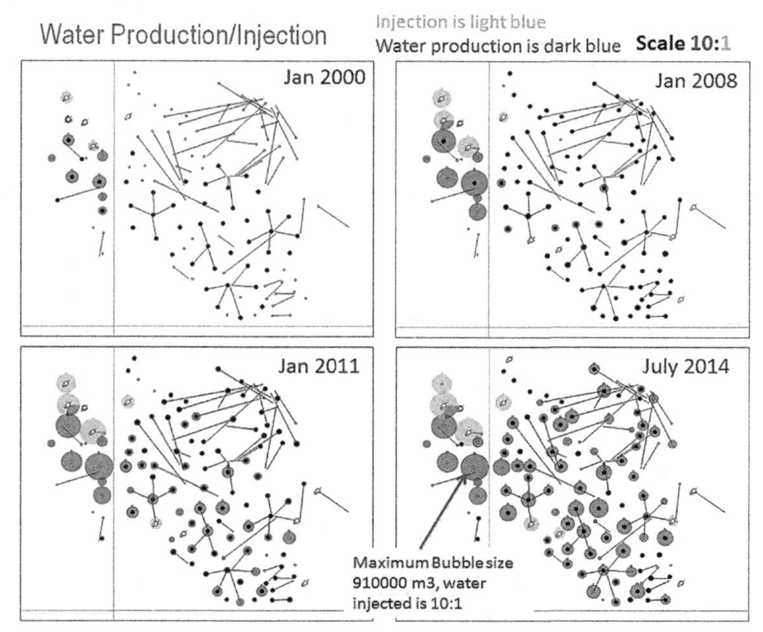

Figure 11.4.44 Time lapse bubble maps of cumulative water produced and injected (active producing wells are solid black dots; scale: water injected to water produced is 10:1).

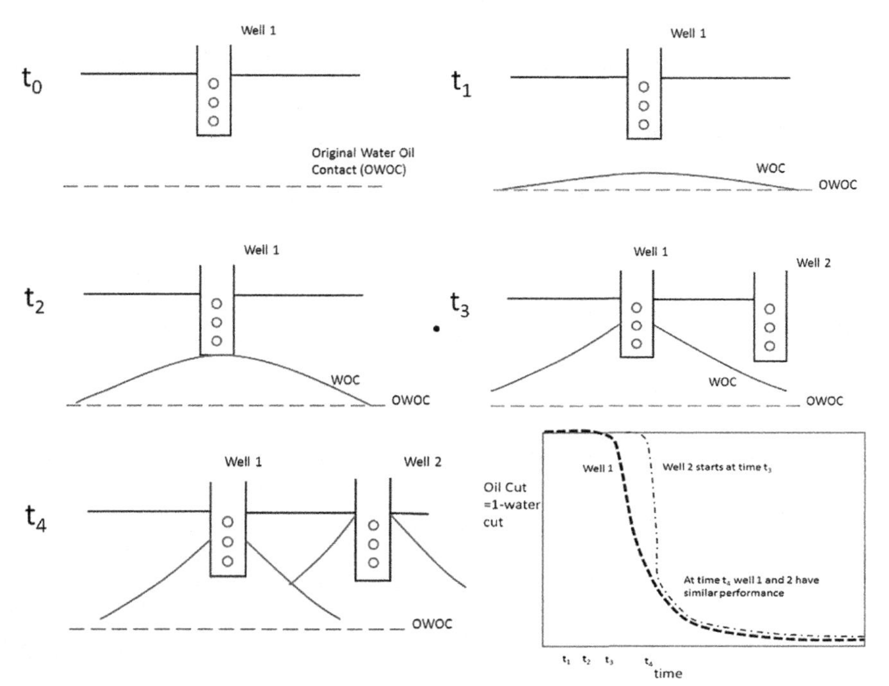

Figure 11.4.45 Typical water coning development in individual wells and associated oil cut profile.

wells are often from producer-to-injector conversions. Typically, watered out producers are selected for conversion, and these wells are found near the edge of the pool where limited top structure forces the perforations to be placed near the WOC.

Water breakthrough at each well depends on its standoff from the underlying aquifer, Figure 11.4.44. Hence, water breakthrough or oil cut decreases are often different from one well to the next, if the wells in close proximity to each other, as shown in Figures 11.4.45–11.4.47. The performance after breakthrough is similar

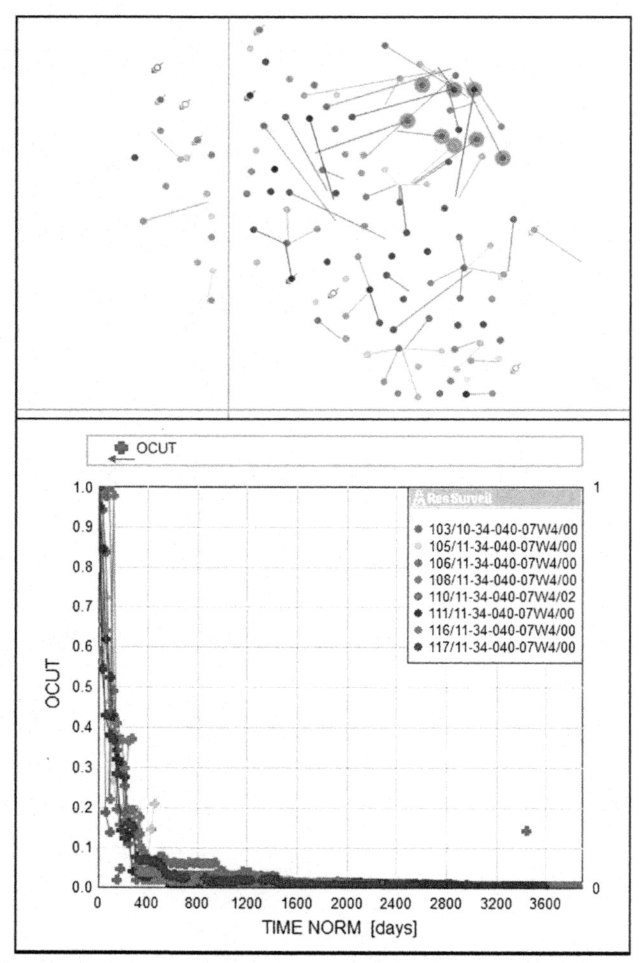

Figure 11.4.46 Aggregate plot of oil cut profile (1 - watercut) for individual wells (highlighted in purple on map) in Provost Dina N Pool. The wells in this group follow similar trends, but they are not identical. Also note the contrast of oil cut behavior with the wells on Figure 11.4.47. The differences in performance are due to differences in distances to the water-oil contact or water injectors.

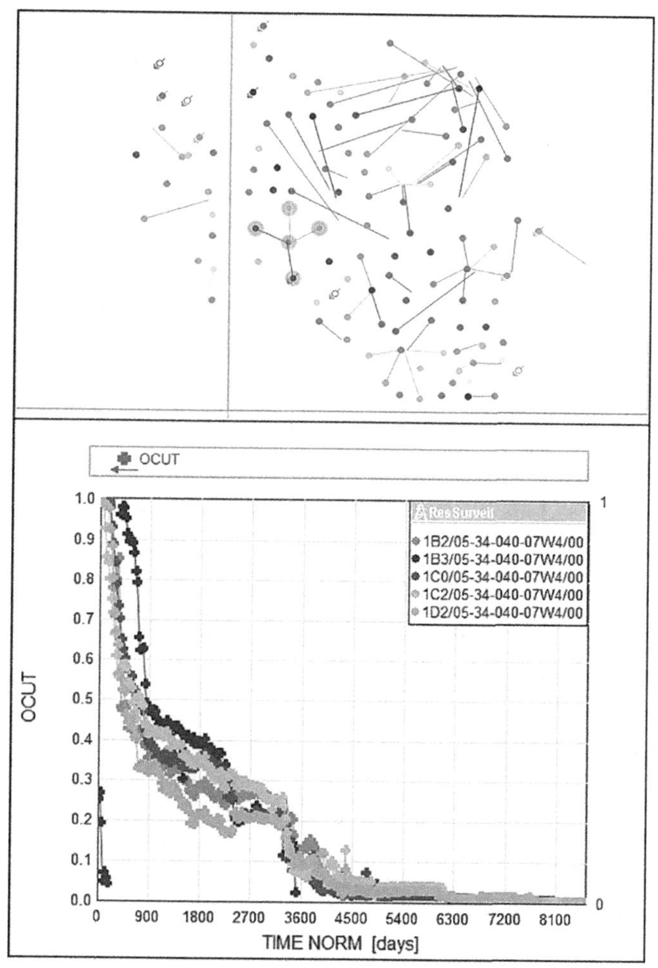

Figure 11.4.47 Aggregate plot of oil cut profile (1 - watercut) for individual wells highlighted in blue on map) in Provost Dina N Pool. The wells in this group follow similar trends, but they are not identical. Note the contrast of oil cut behavior with the wells on Figure 11.4.46. The difference in well performance between the two groups is like due to proximity to original WOC or water injectors.

for a group of wells. Often, new infill wells will initially have low water cuts, but as the water cones develop near the wells, their water cut behavior will be similar to the original "legacy wells" because the rise of the average water-oil contact for the wells in the area will affect rise of watercut for individual wells. This pattern is characteristic of water coning.

References

General References

Geology:

Chapman, R.E., 1983. Petroleum Geology. Elsevier, Amsterdam.

Selley, R.C., 1998. Elements of Petroleum Geology, second ed. Academic Press, San Diego.

Slatt, R.M., 2006. Stratigraphic Characterization for Petroleum Geologists, Geophysicists, and Engineers. Elsevier, Amsterdam.

Reservoir Engineering:

Ahmed, T., 2000. Reservoir Engineering Handbook. Gulf Publishing Co., Houston.

Amyx, J.M., Bass, D.M., Whiting, R.L., 1960. Petroleum Reservoir Engineering. McGraw-Hill, New York.

Craft, B.C., Hawkins, M., 1991. Applied Petroleum Reservoir Engineering, second ed. Prentice Hall, Englewood Cliffs. Revised by R.E. Terry.

Cronquist, C., 2001. Estimation and Classification of Reserves of Crude Oils, Natural Gas and Condensate. SPE, Richardson, TX.

Dake, L.P., 1978. Fundamentals of Reservoir Engineering. Elsevier, New York.

Slider, H.C., 1983. Worldwide Practical Petroleum Reservoir Engineering Methods. PennWell Books, Tulsa.

Towler, B.F., 2002. Fundamental Principles of Reservoir Engineering. Society of Petroleum Engineers, Richardson.

Production and Processing:

Allen, T.O., Roberts, A.P., 1982. Production Operations 1 and 2 Well Completions, Workovers and Stimulation. Oil & Gas Consultants International Inc., Tulsa.

Arnold, K., Stewart, M., 1986. Design of Oil Handling Systems and Facilities, vol. 1. Gulf Publishing Co., Houston.

Chilingarian, G.V., Robertson, J.G., Kumar, S., 1987. Surface Operations in Petroleum Production, 1. Elsevier, New York.

Golan, M., Whitson, C.H., 1991. Well Performance, second ed. Prentice Hall, Eaglewood Cliffs, NJ.

Fluid Properties:

Butler, R.M., 1987. Thermal Recovery of Oil and Bitumen. GravDrain Inc., Calgary.

Gas Processors Suppliers Association, 1980. GPSA Engineering Data Book. Gas Processors Suppliers Association, Tulsa.

McCain, W., 1990. The Properties of Petroleum Fluids. Pennwell Publishing, Tulsa.

Pedersen, K.S., Fredenslund, A., Thomassen, P., 1989. Properties of Oils and Natural Gases. Gulf Publishing Co., Houston.

Riazi, M.R., 2005. Characterization and Properties of Petroleum Fractions. ASTM International, West Conshohocken, USA.

Whitson, C.H., Brulé, M.R., 2000. Phase Behavior. In: Monograph, vol. 20. Henry L. Doherty Series, Society of Petroleum Engineers Inc., USA.

Pressure and Flow Testing:

Earlogher Jr., R.C., 1977. Advances in Well Test Analysis. In: SPE Monograph, vol. 7. Henry Doherty Series, New York.

Lee, J., 1982. Well Testing. Society of Petroleum Engineers, New York.

Core Analysis:

Bass Jr., D.M., 1987. Properties of reservoir rocks (Chapter 26). In: Bradley, H.B. (Ed.), Petroleum Engineering Handbook. SPE, Richardson, TX.

Rose, W., 1987. Relative permeability (Chapter 28). In: Bradley, H.B. (Ed.), Petroleum Engineering Handbook. SPE, Richardson, TX.

Core Laboratories Inc, 1973. The Fundamentals of Core Analysis. Course Notes, Dallas, TX.

Log Interpretation:

Baker Hughes, 1995. Introduction to Wireline Log Analysis. Baker Hughes, Houston, Texas.

Dresser Atlas, 1995. Log Review 1. Dresser Industries, Inc., Houston, Texas.

Schlumberger Educational Services, 1987. Log Interpretation Principles/Applications, second ed. Houston.

Welex (Halliburton), 1987. An Introduction to Log Analysis. Houston.

Chapter References

Ahr, W.M., 2008. Geology of Carbonate Reservoirs. Wiley, New Jersey.

Alberta Oil and Gas Conservation Regulations, 2004. Standards of Accuracy for Oil, Gas, and Water Production. Schedule 9.

Allen, T.O., Roberts, A.P., December 1968. Guidelines to problem well diagnosis. Pet. Engineer 43–46.

Amyx, J.M., Bass, D.M., Whiting, R.L., 1960. Petroleum Reservoir Engineering. McGraw-Hill, New York.

Anderson, W.G., October 1986. Wettability literature survey Part 1: rock/oil/brine interactions and the effects of core handling on wettability. J. Pet. Technol. 1125.

Hepler, L.G., His, C. (Eds.), 1989. AOSTRA Technical Handbook on Oil Sands, Bitumens, and Heavy Oils. AOSTRA Technical Publication Series #6, Edmonton.

Araya, A., Ozkan, E., 2002. An account of decline-type-curve analysis of vertical, fractured, and horizontal well production data. In: SPE Annual Technical Conference and Exhibition. Society of Petroleum Engineers, San Antonio, Texas.

Arps, J.J., 1945. Analysis of decline curves. Trans. AIME 160, 228–247.

Arps, J.J., 1956. Estimation of primary oil reserves. Trans. AIME 207, 182–191.

Baker, R.O., July 1994. Effect of Reservoir Heterogenities and Flow Mechanisms on Numerical Simulation Requirements (MSc thesis). University of Calgary.

Baker Hughes, 1995. Introduction to Wireline Log Analysis. Baker Hughes, Houston, Texas.

Bass Jr., D.M., 1987. Properties of reservoir rocks (Chapter 26). In: Bradley, H.B. (Ed.), Petroleum Engineering Handbook. SPE, Richardson, TX.

Begg, S., King, P.R., 1985. Modelling the Effects of Shales on Reservoir Performance: Calculation of Effective Vertical Permeability. SPE Paper 13529.

Beggs, H.D., Robinson, J.R., 1975. Estimating viscosity of crude oil systems. J. Pet. Technol. 27 (9), 1140–1141.

Belfield, W.C., October 1988. Characterization of a Naturally Fractures Carbonate Reservoir; Lisburne Field, Prudhoe Bay, Alaska. In: SPE paper 18174.

Beliveau, D., December 1995. Heterogeneity, geostatistics, horizontal wells, and blackjack poker. J. Pet. Technol. 1068.

Bennett, F., Geoghegan, J.G., 1992. Monitoring the performance of Pembina Nisku miscible floods, relative permeability calculations from pore size distribution data. J. Can. Pet. Technol. 31 (3), 16–21.

Bennion, D.B., Bietz, R.F., Thomas, F.B., November 1994. Reductions in the productivity of oil and low permeability gas reservoirs due to aqueous phase trapping. J. Can. Pet. Technol. 33 (9), 45–54.

Berge, F., Ruyter, E., Gronas, T., August 31–September 2, 2006. Enhanced oil recovery by utilization of multilateral wells on the Troll West Field. SPE 103912. In: Intl. Oil Conf., Cancun.

Bergman, D.F., Sutton, R.P., November 11–14, 2007. A consistent and accurate dead oil viscosity method. SPE Paper 110194. In: SPE Annual Technical Conference.

Bobek, J.E., Mattax, C.C., Denekas, M.O., July 1958. Reservoir rock wettability, - its significance and evaluation. J. Pet. Technol. 155–160.

Brooks, R.H., Corey, A.T., 1964. Hydraulic Properties of Porous Media. Hydrological Papers. Colorado State University.

Buckles, R.S., 1965. Correlating and averaging connate water saturation data. J. Can. Pet. Technol. 4, 42–52.

Buckley, S.E., Leverett, M.C., 1942. Mechanism of fluid displacements in sands. Trans. AIME 146, 107–116.

Burdine, N.T., 1953. Relative permeability calculations from pore size distribution data. J. Pet. Technol. 5 (3), 71–78.

Butler, R.M., 1997. Thermal Recovery of Oil and Bitumen. GravDrain Inc., Calgary.

Camacho-Velazquez, R.G., Raghavan, R., 1989. Boundary dominated flow in solution gas drive reservoirs. SPE Reservoir Eng. 4 (4), 503–512.

Campbell, R.A., Campbell Sr., J.M., 1978. Mineral Property Economics. In: Petroleum Property Evaluation, vol. 3. Campbell Petroleum Series, Norman, OK.

Canadian Well Logging Society (CWLS), 1987. Formation Water Resistivities of Canada.

Capen, E., 1976. The difficulty of assessing uncertainty. J. Pet. Technol. 28 (8), 843–850.

Chabris, C., Simons, D., 2011. The Invisible Gorilla: How Our Intuitions Deceive Us. Random House, New York.

Chaperon, I., October 5–8, 1986. Theoretical study of coning toward horizontal and vertical wells in anisotropic formations: subcritical and critical rates. SPE 15377. In: SPE Annual Conf., New Orleans.

Clarkson, C.R., Jensen, J.L., Chipperfield, S., 2012. Unconventional gas reservoir evaluation: what do we have to consider? J. Nat. Gas Sci. Eng 8, 9–33.

Clavier, C., Coates, G., Dumanoir, J., 1977. Theoretical and experimental bases for the dual water model for interpretation of shaly sands. SPE J. 24, 153–168.

Cobb, W.M., Marek, F.J., September 27–30, 1998. Net pay determination for primary and waterflood depletion mechanisms. SPE Paper 48952. In: SPE Annual Technical Conference, New Orleans.

Corbett, P.W.M., 1993. Reservoir Characterization of a Laminated Sediment, the Rannoch Formation, Middle Jurassic, North Sea (Ph.D. dissertation). Heriot-Watt University, Edinburgh, Scotland, p.145.

Core Laboratories Inc, 1973. The Fundamentals of Core Analysis. Course Notes, Dallas, TX.

Corey, A.T., Rathjens, C.H., Henderson, J.H., Wyllie, M.R.J., 1956. Three phase relative permeability. Trans. AIME 207, 349–351.

Cowan, G., Bradney, J., 1997. Regional diagenetic controls on reservoir properties in the Millom accumulation: implications for field development. In: Meadows, N.S., Trueblood, S.R., Hardman, M., Cowan, G. (Eds.), Petroleum Geology of the Irish Sea and Adjacent Areas, vol. 124, pp. 373–386.

Craft, B.C., Hawkins, M., 1991. Applied Petroleum Reservoir Engineering. Revised by R.E. Terry, second ed. Prentice Hall, Englewood Cliffs.

Craig Jr., F.F., 1971. The Reservoir Engineering Aspects of Waterflooding. In: SPE Monograph, vol. 9. Henry Doherty Series, New York.

Cronquist, C., 2001. Estimation and Classification of Reserves of Crude Oils, Natural Gas and Condensate. SPE, Richardson, TX.

Dake, L.P., 1978. Fundamentals of Reservoir Engineering. Elsevier, New York.

Dake, L.P., 1994. The Practice of Reservoir Engineering. Elsevier Science Ltd., Amsterdam.

Dietz, D.N., 1953. A theoretical approach to the problem of encroaching and by-passing edge water. Proc. Akad. Van Wtenschappen vol. 56B, 83. Amsterdam.

Dietz, D.N., August 1965. Determination of average reservoir pressure from build-up surveys. J. Pet. Technol. 955–959.

Doll, S.P., 1949. Log theoretical analysis and principles of interpretation. Trans. AIME 179, 146.

Dresser Atlas, 1995. Log Review 1. Dresser Industries, Inc., Houston, Texas.

Earlogher Jr., R.C., 1977. Advances in Well Test Analysis. In: SPE Monograph, vol. 7. Henry Doherty Series, New York.

ERCB (Energy Resources Conservation Board), 2013. Alberta Pool Properties as Reported in Geoscout^{TM}.

Ershaghi, I., Omoregie, O., February 1978. A method for extrapolation of cut vs. recovery curves. J. Pet. Technol. 203–204.

Ewens, S., 2012. Personal Communication, Calgary.

Lo, K.K., Warner, H.R., Johnson, J.B., A study of the post-breakthrough characteristics of waterfloods, In: SPE Paper 20064, SPE California Regional Meeting, Ventura, April 4-6, 1990.

van Everdingen, A.F., Hurst, W., 1949. The application of the Laplace transformation to flow problems in reservoirs. Trans. AIME 186, 305.

van Everdingen, A.F., 1953. The skin effect and its influence on the productive capacity of wells. Trans. AIME 198, 71.

Felsenthal, M., October 1979. A statistical study of some waterflood parameters. J. Pet. Technol. 31, 1303–1304.

Fetkovich, M.J., 1971. A simplified approach to water influx calculations—finite aquifer systems. J. Pet. Technol. 23 (7), 814–828.

Gawith, D., Gutteridge, P., 1999. Decision-driven reservoir modelling: the next big thing. In: SPE Reservoir Simulation Symposium. Society of Petroleum Engineers, Houston.

Gaynor, G.C., Sneider, R.M., 1993. Effective pay determination. In: Morton-Thompson, D., Woods, A.M. (Eds.), Development Geology Reference Manual. Amer. Assoc. Petroleum Geologists.

Georgi, D., February 2013. Personal Communication.

Gilchrest, R.E., Adams, J.E., July 1993. How to best utilize PVT reports. Pet. Eng. Int. 22, 38–41.

Golan, M., Whitson, C.H., 1991. Well Performance. Prentice Hall, New York.

Gas Processors Suppliers Association, 1980. GPSA Engineering Data Book. Gas Processors Suppliers Association, Tulsa.

Gray, M.R., 1994. Upgrading Petroleum Residues and Heavy Oils. CRC Press, Boca Raton.

Hale, B.W., 1986. Analysis of tight gas well production histories in the Rocky Mountains. SPE Production Eng. 1 (4), 310–322.

Haldorsen, H.H., 1989. On the modelling of vertical permeability barriers in single-well simulation models. SPE Formation Eval. 2, 349–358.

Haldorsen, H.H., 1986. Simulator parameter assignment and the problem of scale in reservoir engineering. In: Lake, L.W., Carroll, H.B. (Eds.), Reservoir Characterization. Academic Press, pp. 293–340.

Hall, H.N., 1953. Compressibility of reservoir rocks. Trans. AIME 198, 309.

Havlena, D., Odeh, A.S., August 1963. The material balance as an equation of a straight line. J. Pet. Technol. 15 (8), 896–900.

Havlena, D., Odeh, A.S., July 1964. The material balance as an equation of a straight line, Part II – field cases. J. Pet. Technol. 815–822.

Heffer, K., Zhang, X., Koutsabeloulis, N., Main, I., Li, L., June 11–14, 2007. Identification of activated (therefore potentially conductive) faults and fractures through statistical correlations in production and injection rates and coupled flow-geomechanical modeling. SPE Paper 107164. In: SPE European/EAG Annual Conf., London.

Honarpour, M., Koederitz, L.F., Harvey, A.H., December 1982. Empirical equations for estimating two-phase relative permeability in consolidated rock. J. Pet. Technol. 2905–2908.

Honarpour, M., 1986. Relative Permeability of Petroleum Reservoirs. CRC Press Inc., Boca Raton, Fl.

Horner, D.R., 1951. Pressure build up in wells. In: Proc. 3rd World Petr. Congress. The Hague.

Hurst, W., 1960. Interference between Oil Fields, vol. 219. Pet. Trans. AIME. Paper #1335-G.

James, S.J., 1999. Brent field reservoir modeling: laying the foundations of a brown field redevelopment. SPE Reservoir Eval. Eng. 2 (1), 104–111.

Jensen, J.L., Lake, L.W., Corbett, P.W.M., Goggin, D.J., 2000. Statistics for Petroleum Engineers and Geoscientists, second ed. Elsevier, Amsterdam.

Jensen, J.L., Menke, J.Y., 2006. Some statistical issues in selecting porosity cutoffs for estimating net pay. Petrophysics 47, 315–320.

Jerauld, G.R., February 1997. Prudhoe bay gas/oil relative permeability. SPE Reservoir Eng. 12 (4), 255–263.

Jones, F.O., Owens, W.W., September 1980. Laboratory study of low permeability gas sands. J. Pet. Technol. 1631–1640.

Joshi, S.D., June 1988. Augmentation of well productivity with slant and horizontal wells. J. Pet. Technol. 729–739.

Joshi, S.D., 1991. Horizontal Well Technology. Pennwell, Tulsa.

Kay, J., 2011. Obliquity: Why Our Goals Are Best Achieved Indirectly. Profile Books Ltd., London.

Keelan, D.K., April–June 1972. A critical review of core analysis techniques. J. Can. Pet. Technol. 11 (2), 42–55.

Keushnig, H., July 1976. Imbibition Characteristics of Reservoir Rock, Petrophysics. Technical Note No. 16. Energy Resources Conservation Board, Calgary.

Khazaeni, Y., Mohaghegh, S.D., 2011. Intelligent production modeling using full field pattern recognition. SPE Reservoir Eval. Eng. 14 (6), 735–749.

Klinkenberg, L.J., 1941. The Permeability of Porous Media to Liquids and Gases. API Drilling Production Pract., 200–213.

Knutsen, C.F., September 1954. Definition of water table. Am. Assoc. Pet. Geol. Bull. 38, 2020–2027.

Kuo, M.C.T., Desbrisay, C.L., 1983. A simplified method for water coning predictions. SPE Paper 12067. In: 58th Annual Fall Meeting, San Francisco.

Lachance, D.P., Anderson, M.A., October 5–8, 1983. Comparison of uniaxial strain and hydrostatic stress test pore-volume compressibilities in the Nugget sandstone. SPE Paper 11971. In: SPE Annual Technical Conf. and Meeting, San Francisco.

Land, C.S., June 1968. Calculation of imbibition relative permeability for two and three phase flow from rock properties. SPE J. 149–156.

Lange, E.A., April 1998. Correlation and prediction of residual oil saturation for gas-injection-enhanced-oil-recovery processes. SPE Reservoir Eval. Eng. 127.

Larionov, V.V., 1969. Radiometry of Boreholes (In Russian). NEDRA, Moscow.

Larter, S., Adams, J., Gates, I.D., Bennett, B., Huang, H., 2008. The origin, prediction and impact of oil viscosity heterogeneity on the production characteristics of tar sand and heavy oil reservoirs. J. Can. Pet. Technol. 47 (10), 52–61.

Lasater, J.A., 1958. Bubble point pressure correlation. Trans. AIME 213, 379–381.

Lee, A.L., Gonzalez, M.H., Eakin, B.E., August 1966. Viscosity of natural gases. J. Pet. Technol. 997–1000. Trans. AIME, 1966, 234.

Leckie, D.A., Bhattacharya, J.P., Bloch, J., Gilboy, C.F., 2012. Chapter 20 Cretaceous Colorado/Alberta group of the Western Canada sedimentary basin. In: Geological Atlas of the Western Canada Sedimentary Basin. Government of Alberta. www.ags.gov.ab.ca/publications/wcsb_atlas/a_ch20/ch_20.html.

Lee, J., 1982. Well Testing. Society of Petroleum Engineers, New York.

Lefkovits, H.C., Matthews, C.S., 1958. Application of decline curves to gravity-drainage reservoirs in the stripper stage. Trans. AIME 213, 275–280.

Lo, K.K., Warner, H.R., Johnson, J.B., 1990. A study of the post-breakthrough characteristics of waterfloods. SPE paper 20064.

Lucia, F.J., 1999. Carbonate Reservoir Characterization. Springer-Verlag, Berlin.

Masoner, L., 1998. Decline analysis' relationship to relative permeability in secondary and tertiary recovery. In: SPE Rocky Mountain Regional/Low-Permeability Reservoirs Symposium. Society of Petroleum Engineers, Denver, Colorado.

Mattar, L., McNeil, R., February 1998. The 'flowing' material balance. J. Can. Pet. Technol. 37 (2), 52–55.

McCain, W.D., 1988. The Properties of Petroleum Fluids, second ed. Petroleum Publishing Co., Tulsa.

McCain, W.D., 1990b. The Properties of Petroleum Fluids. Pennwell Publishing, Tulsa.

Memon, A.I., Gao, J., Taylor, S.D., Davies, T.L., Jia, N., October 19–21, 2010. A systematic workflow process for heavy oil characterization: experimental techniques and challenges. SPE Paper 137006. In: SPE Canadian Unconventional Resources and International Petroleum Conference, Calgary.

Mijnssen, F.C.J., Weber, K., Rozendal, S., Kool, H., 1992. The effect of horizontal shales on vertical permeability in clastic reservoir rocks. In: Spencer, A.M. (Ed.), Generation, Accumulation and Production of Europe's Hydrocarbons; II, vol. 2. Special Publication of the European Association of Petroleum Geoscientists, pp. 279–289.

Miller, K., Nelson, L.A., Almond, R.M., 2006. Should you trust your heavy oil viscosity measurement? J. Can. Pet. Technol. 45, 42–48.

Motahhari, H., Yarranton, H.W., Satyro, M.A., 2013. Extension of the expanded fluid viscosity model to characterized oils. Energy Fuels 27, 1881–1898.

Muskat, M., Wyckoff, R.D., 1935. An approximate theory of water coning in oil production. Trans. AIME 114 (1), 161.

Muskat, M., 1945. The production histories of oil producing gas-drive reservoirs. J. Appl. Phys. 16, 147.

Nelson, R.A., 2001. Geologic Analysis of Naturally Fractured Reservoirs, second ed. Gulf Professional Publishing, Elsevier.

Newman, G.H., Febuary 1973. Pore volume compressibility of consolidated, friable, and unconsolidated reservoir rocks under hydrostatic loading. J. Pet. Technol. 129–134.

Osif, T.L., September 16–19, 1984. The effects of salt, gas, temperature, and pressure on the compressibility of water. SPE Paper 13174. In: SPE Annual Technical Conference, Houston.

Pedersen, K.S., Fredenslund, A., Thomassen, P., 1989. Properties of Oils and Natural Gases. Gulf Publishing Co., Houston.

Peneloux, A.E., Rauzy, E., Freze, R., 1982. A consistent correction for Redlich-Kwong-Soave volumes. Fluid Phase Equil. 8, 7–23.

Pirson, S.J., 1958. Oil Reservoir Engineering, Second ed. McGraw Hill, New York.

Pletcher, J.L., October 1–4, 2000. Improvements to reservoir material balance methods. SPE Paper 62882. In: SPE Annual Technical Conference and Exhibition Dallas.

Poeter, E.P., Hill, M.C., 1997. Inverse models: a necessary next step in ground-water modeling. Ground Water 35 (2), 250–260.

Purvis, R.A., July–August 1987. Further analysis of production-performance graphs. J. Can. Pet. Technol. 26, 74–79.

Van Poollen, H., April 1966. How decline curves help to evaluate treatments. Oil Gas J. 177–180.

Raymer, L.L., Hunt, E.R., Gardner, J.S., 1980. An improved sonic transit time-to-porosity transform. In: Trans. Annual Logging Symp. SPWLA, pp. 1–12.

Raza, S.H., Treiber, L.E., Archer, D.L., April 1968. Wettability of reservoir rocks and its evaluation. Producers Monthly 32, 2–7.

Rowan, G., Clegg, M.W., 1963. The cybernetic approach to reservoir engineering. In: SPE 38th ATCE, Oct 6-9 New Orleans, LA.

da Silva, L.C.F., Portella, R.C.M., Ebecken, N.F.F., Emerick, A.A. Predictive data-mining technologies for oil-production prediction in petroleum reservoir (SPE 107371). In: Latin American & Caribbean Petroleum Engineering Conference, 15–18 April 2007, Buenos Aires, Argentina.

Da Sie, W.J., Guo, D.S., May 1990. Assessment of a vertical hydrocarbon miscible flood in the Westpem Nisku D reef. SPE Reservoir Eng. 147–154.

Saleri, N.G., 1998. Re-engineering simulation: managing complexity and complexification in reservoir projects. SPE Reservoir Eval. Eng. 1 (1), 5–11.

Saleri, N.G., Toronyi, R.M., October 2–5, 1988. Engineering control in reservoir simulation: Part I. In: SPE Annual Technical Conference, Houston.

Schilthuis, R.J., 1936. Active oil and reservoir energy. Trans. AIME 118, 33–52.

Schlumberger Educational Services, 1987. Log Interpretation Principles/Applications, second ed. Houston.

Schlumberger Publication, 1982. Reservoir and Production Fundamentals.

Schols, R.S., January 1972. An Empirical Formula for the Critical Oil Production Rate. Erdoel Erdgas 88 (1), 6–11.

Selley, R.C., 1998. Elements of Petroleum Geology, second ed. Academic Press, San Diego.

Shen, C., Batycky, J., April 1999. Some observations of mobility enhancement of heavy oils flowing through sand pack under solution gas drive. J. Can. Pet. Technol. 38, 46–50.

Singhal, K.A., October 18–20, 1993. Water and gas coning/cresting: a technology overview. In: Petroleum Recovery Institute Paper, 5th Petroleum Conference of South Saskatchewan, Regina.

Slatt, R.M., 2006. Stratigraphic Reservoir Characterization for Petroleum Geologists, Geophysicists, and Engineers. Elsevier, Amsterdam.

Slider, H.C., 1983. Worldwide Practical Petroleum Reservoir Engineering Methods. PennWell Books, PennWell Publishing Company, Tulsa, Oklahoma, pp. 486–501.

Smith, C.R., Tracy, G.W., Farrar, R.L., 1992. Applied Reservoir Engineering. OGCI Publications.

Sobocinski, D.P., Cornelius, A.J., May 1965. A correlation predicting water coning time. J. Pet. Technol. 594–600.

Springer, S.J., Singhal, A.K., Sadal, S., November 4–7, 2002. To drill or not to drill a horizontal well. In: SPE Intl. Thermal Operations Heavy Oil Symp. Horizontal Well Technol. Conf., Calgary. SPE Paper 79086.

Standing, M.B., 1947. A Pressure-volume-temperature Correlation for Mixtures of California Oils and Gases, Drilling and Production Practices. API, 275–287.

Startzman, R.A., Wu, C.H., December 1984. Discussion of empirical prediction technique for immiscible processes. J. Pet. Technol. 2192–2194.

Steiber, S.J., October 4–7, 1970. Pulse neutron capture log evaluation in the Louisiana Gulf Coast. In: SPE Annual Meeting, Houston. SPE Paper 2961.

Stoian, E., Telford, A.S., February 1966. Determination of natural gas recovery factors. J. Can. Pet. Technol. 115–129.

Stone, H.L., October–December 1973. Estimation of three-phase relative permeability and residual oil data. J. Can. Pet. Technol. 12 (4), 53–61.

Tang, G.-Q., Firoozabadi, A., 2003. Gas and Liquid Phase Relative Permeabilities for Cold Production from Heavy Oil Reservoirs. SPE Paper 83667. SPE, Richardson, TX.

Tehrani, D.H., July 1992. An analysis of volumetric balance equation for calculation of oil-in place and water influx. SPE Paper 5990. J. Pet. Technol. 1664–1670.

The Petroleum Society of the Canadian Institute of Mining, Metallurgy and Petroleum, 1994. Determination of Oil and Gas Reserves.

Treiber, L.E., Owens, W.W., 1972. Laboratory evaluation of the wettability of fifty oil-producing reservoirs. Soc. Pet. Eng. J. 12 (6), 531–540.

Twu, C.H., 1985. An internally consistent correlation for predicting liquid viscosities of petroleum fractions. Ind. Eng. Chem. Process Des. Dev. 24 (4), 1289–1293.

Vasquez, M., Beggs, H.D., 1980. Correlations for fluid physical property prediction. J. Pet. Technol. 968–970.

Vogel, H., 1921. The law of the relation between the viscosity of liquids and the temperature. Physik. Z. 22, 645–646.

Vogel, J.V., January 1968. Inflow performance relationships for solution gas drive wells. J. Pet. Technol. 83–92.

Wahl, J.S., Tittman, J., Johnstone, C.W., Alger, R.P., 1964. The dual spacing formation density log. J. Pet. Technol. 16, 1411–1417.

Walsh, M.P., October 3–6, 1999. Effect of pressure uncertainty on material-balance plots. SPE Paper 56691. In: SPE Annual Technical Conf., Houston.

Wattenbarger, R.A., Ramey Jr., H.J., August 1968. Gas well testing with turbulence, damage and wellbore storage. SPE Paper 1835 J. Pet. Technol. 877.

Weber, K.J., van Geuns, L.C., 1990. Framework for constructing clastic reservoir simulation models. J. Pet. Technol. 42, 1248–1253.

Welge, H.J., Bruce, W.A., 1947. The Restored State Method for Determination of Oil in Place and Connate Water, Drilling and Production Practices. American Petroleum Institute, Dallas, 166–174.

Whitson, C.H., Brulé, M.R., 2000. Phase Behavior. In: Monograph, vol. 20. Henry L. Doherty Series, Society of Petroleum Engineers Inc., USA.

Wieland, D.R., Kennedy, H.T., 1957. Measurements of Bubble Frequency in Cores. Trans. AIME 210, 122–125.

Wilhite, G.P., 1986. Waterflooding. SPE, Richardson, TX.

Worthington, P.F., Cosentino, L., 2005. The role of cutoffs in integrated reservoir studies. SPE Reservoir Eval. Eng. 8 (4), 276–290.

Wyllie, M.R.J., Gregory, A.R., Gardner, L.W., 1956. Elastic wave velocities in heterogeneous and porous media. Geophysics 21 (1), 41–70.

Yale, D.P., Nabor, G.W., Russell, J.A., Pham, H.D., Yousef, M., October 3–6, 1993. Application of variable formation compressibility for improved reservoir analysis. SPE Paper 26647. In: 68th Annual Technical Conference, Houston.

Yang, W., Wattenbarger, R.A., October 6–9, 1991. Water coning calculations for vertical and horizontal wells. SPE Paper 22931. In: 66th SPE Annual Technical Conference and Exhibition, Dallas.

Yarranton, H.W., van Dorp, J.J., Verlaan, M.L., Lastovka, V., May 2013. Wanted dead or live – crude cocktail viscosity: a pseudo-component method to predict the viscosity of dead oils, live oils, and mixtures. J. Can. Pet. Technol. 178–191.

Zhang, Y., Chakma, A., Maini, B.B., June 14–18, 1999. Effects of temperature on foamy oil flow in solution gas drive in cold lake field. CIM Paper 99–41. In: CIM Annual Technical Meeting, Calgary.

Index

Note: Page numbers followed by "f", "t" and "b" indicate figures, tables and boxes respectively.

C

R

S

T

Printed in the United States
By Bookmasters